On Advances in Robot Kinematics

On Advances in Robot Kinematics

Edited by

J. LENARČIČ
Jožef Stefan Institute,
Ljubljana, Slovenia

and

C. GALLETTI
University of Genoa,
Genoa, Italy

KLUWER ACADEMIC PUBLISHERS
DORDRECHT / BOSTON / LONDON

A C.I.P. Catalogue record for this book is available from the Library of Congress.

ISBN 1-4020-2248-4 (HB)
ISBN 1-4020-2249-2 (e-book)

Published by Kluwer Academic Publishers,
P.O. Box 17, 3300 AA Dordrecht, The Netherlands.

Sold and distributed in North, Central and South America
by Kluwer Academic Publishers,
101 Philip Drive, Norwell, MA 02061, U.S.A.

In all other countries, sold and distributed
by Kluwer Academic Publishers,
P.O. Box 322, 3300 AH Dordrecht, The Netherlands.

Printed on acid-free paper

Printed in the Netherlands.

Contents

Methods in Kinematics

Properties of Mechanisms

Analysis of Mechanisms

Design of Mechanisms

Motion Synthesis and Mobility

Preface

In the last decade, we have seen an extraordinary progress in the theory and applications of robot kinematics. This has been motivated especially by the development of complex parallel and humanoid robots. The present book reports the most recent research advances in the theory, design, control and application of robotic systems, which are intended for a variety of purposes such as manipulation, manufacturing, automation, surgery, locomotion and biomechanics. The issues addressed are fundamentally kinematic in nature, including synthesis, calibration, redundancy, force control, dexterity, inverse and forward kinematics, kinematic singularities, as well as over-constrained systems. Methods used include line geometry, quaternion algebra, screw algebra, and linear algebra. These methods are applied to both parallel and serial multi-degree-of-freedom systems. The results should interest researchers, teachers and students, in fields of engineering and mathematics related to robot theory, design, control and application. This is the sixth book of the series Advances in Robot Kinematics published by Kluwer.

The contributions in this book had been rigorously reviewed by independent reviewers and fifty one articles had been recommended for publication. They were introduced in seven chapters. These articles were also reported and discussed at the ninth international symposium on Advances in Robot Kinematics which was held in June 2004 in Sestri Levante in Italy. The symposium was organized by the University of Genova and co-organized by the Jozef Stefan Institute, Ljubljana, under the patronage of IFToMM - International Federation for the Promotion of Mechanism and Machine Science.

We are grateful to the authors of the articles for their valuable contributions and for their efficiency in preparing their manuscripts in time, and to the reviewers for their timely reviews. We are also indebted to the personnel at Kluwer for their excellent technical and editorial support.

Jadran Lenarčič and Carlo Galleti

Methods in Kinematics

R. Vertechy and V. Parenti-Castelli
A fast and accurate method for the singular value decomposition of 3×3 matrices

P.M. Larochelle, A.P. Murray and J. Angeles
SVD and PD based projection metrics on SE(n)

P. Fanghella and C. Galletti
Choice of independent coordinates in modular kinematics

F. Thomas, J.M. Porta and L. Ros
Distance constraints solved geometrically

M.J.D. Hayes, T. Luu and X.-W. Chang
Kinematic mapping application to approximate type and dimension synthesis of planar mechanisms

I.A. Parkin and C. Freeman
On the ruled surface generated by a linear combination screw

A. Kecskeméthy
Lie-group first-order operations in rigid-body kinematics

A. Kecskeméthy
First-order intrinsic properties of the rotation parameters SO(3), quaternion and rotation vector

A FAST AND ACCURATE METHOD FOR THE SINGULAR VALUE DECOMPOSITION OF 3X3 MATRICES

Rocco Vertechy
DIEM, University of Bologna
Viale Risorgimento, 2, 40136 BOLOGNA, ITALY
rocco.vertechy@mail.ing.unibo.it

Vincenzo Parenti-Castelli
DIEM, University of Bologna
Viale Risorgimento, 2, 40136 BOLOGNA, ITALY
vincenzo.parenticastelli@mail.ing.unibo.it

Abstract This paper deals with the Singular Value Decomposition (SVD) of 3x3 matrices. A customized algorithm is devised which proves to be fast and accurate. Slight modifications of the method also provide very efficient schemes that may be used when fewer unknowns than the ones normally found by the standard SVD are of interest or to seek for the null space of 3x3 singular matrices.

Keywords: 3x3 Matrices, SVD, Null Space.

1. Introduction

In the last decade, the Singular Value Decomposition (SVD) has become a very popular tool for analysing both kinematic and dynamic properties of robotic manipulators (Yoshikawa and Nagai, 1990). The reason for this is that the SVD gives quantitative and qualitative information about linear maps. Practically, the technique allows solving systems of linear equations with equality constraints, solving homogeneous systems of linear equations, computing the Moore-Penrose pseudoinverse and other generalized inverses, and determining dependencies or near-dependencies among the columns or rows of a matrix. It is worth noting that, while the SVD plays a useful role in analyzing square, invertible matrices, its full power is realized in both the analysis of non-square matrices which arise, for example, in linear least squares as well as in over-determined problems, and in the examination of possibly rank-deficient matrices which, in robotics, occur

J. Lenarčič and C. Galletti (eds.), On Advances in Robot Kinematics, 3-12.

in special (singular) configurations of the manipulator. Indeed, unlike methods such as Gaussian elimination and LU decomposition which fail to give satisfactory results even close to singularities, the SVD always works correctly no matter how degenerate the matrix is.

Thus, the SVD always makes it possible not only to diagnose problems, but also to solve them in the sense of giving useful numerical answers. Indeed, besides the condition number, i.e. the ratio between the maximum and minimum singular values of a matrix which monitors the accurateness of the related linear map, whenever a matrix is singular, the SVD, as the inherent property, explicitly constructs orthonormal bases for both the null-space and the range of the associated mapping.

Three degrees of freedom (dof) manipulators form a very important class of robots. Indeed, many daily tasks may be performed by them, which, after remaining for a long time only at a research level, seem now to be the closest ones to be effectively applied in the industrial world. Due to the number of dof, three is also the dimension of the manipulator square Jacobian matrix whose analysis by means of an SVD algorithm is very important for predicting and controlling the robot's behaviour.

Although the SVD has been very elegantly solved for general $n \times m$ matrices by Golub and Reinsch (Golub and Van Loan, 1983), their algorithm is too expensive when applied to matrices of reduced dimension, e.g. 3×3 matrices, and thus it is inappropriate for real-time applications such as manipulator control. Then, the need to have a new customized procedure is an important issue. In this context, the main objective of this paper is that of proposing a fast algorithm for the SVD of 3×3 matrices.

The algorithm is based on the one-dimensional Newton-Raphson (NR) iterative scheme and will be named as NRSVD. In Section 2.1 and Section 2.2 we will show how, if $n = 3$, singular values and singular vectors can be very easily obtained, while in Section 2.4 we compare the computational cost of such procedure with the computational burden required by other available methods, in particular the one implemented by Matlab (Redfern and Campbell, 1998) and the one described in Golub and Van Loan (1983). Section 2.3 describes how to find the null space of 3×3 singular matrices.

Meanwhile, we will discuss accuracy issues when we have to deal with matrices affected by errors, i.e. when only an estimate $\tilde{\mathbf{A}}$ of a matrix $\mathbf{A}$ is known. This matter is very important in robotics, where matrix $\mathbf{A}$ may be the Jacobian matrix of the manipulator which is often computed from measurement data and therefore it is consistently affected by noise.

Since in many robotic problems the application of the SVD algorithm gives more information than needed and, in practice, only the computation of fewer singular values and/or vectors is required, then it is usually better to replace the SVD with the so called "Special Methods" (SM). Indeed, by exploiting the rich mathematical structure of the symmetric eigenvalue problem, SM require a lower computational burden for evaluating the required reduced number of unknowns. Thus, in Section 3 we present two SM that allow the evaluation of the smallest singular values of 3×3 matrices. This may be useful, in robotics, in order to check the manipulator vicinity to a singular configuration. In particular, in Section 3.1 we propose a special method, called Newton-Raphson Special Method (NRSM), derived by the NRSVD, while in Section 3.2 we review the well known Power Method (PM) (Golub and Van Loan, 1983) which has been exploited, in robotics, by McAree and Daniel (1996). Comparison between NRSM and PM is provided in Section 3.3.

2. The SVD and the Proposed NRSVD Algorithm

Given a $n\times n$ matrix $\mathbf{A}$ with elements a_{ij}, $i,j=1,\ldots,n$, i.e. $\mathbf{A}=\left[a_{ij}\right]$, the SVD provides

$$\mathbf{A}=\mathbf{U}\boldsymbol{\Sigma}\mathbf{V}^T \tag{1}$$

$\mathbf{U}$ and $\mathbf{V}$ being unitary matrices, while $\boldsymbol{\Sigma}$ is a diagonal matrix, i.e. $\Sigma=\mathrm{diag}(\sigma_1,\ldots,\sigma_n)$ with $\sigma_1\geq\ldots\geq\sigma_n\geq0$. The σ_i's stand for the singular values of $\mathbf{A}$, while the columns of $\mathbf{U}$ and $\mathbf{V}$ regard, respectively, the corresponding left and right singular vectors.

2.1 NRSVD: Determination of Singular Values

Given a 3×3 matrix $\mathbf{A}$, we can evaluate the symmetric matrix $\mathbf{B}$

$$\mathbf{B}=\left[b_{ij}\right]=\mathbf{A}^T\mathbf{A}, \tag{2}$$

such that, according to Eq. (1),

$$\mathbf{B}=\mathbf{V}\boldsymbol{\Sigma}^2\mathbf{V}^T, \tag{3}$$

where $\boldsymbol{\Sigma}^2=\boldsymbol{\Sigma}\boldsymbol{\Sigma}$. Eq. (3) corresponds to the Eigen Value Decomposition (EVD) of $\mathbf{B}$, whose eigenvalues λ_i satisfy

$$\lambda_i=\sigma_i^2,\ i=1,2,3. \tag{4}$$

By definition, the λ_i's are found as the roots of the polynomial

$$p(x)=(x-\lambda_1)(x-\lambda_2)(x-\lambda_3)=x^3-ax^2+bx-c, \tag{5}$$

where, since $\lambda_i\geq0$, $i=1,2,3$, then the coefficients a, b and c are functions of the λ_i's and are all greater than or equal to zero. In practice, these coefficients are determined by the well known relation

$$p(x) = \det(\mathbf{B} - \mathbf{I}x), \tag{6}$$

where $\mathbf{I}$ stands for the 3×3 identity matrix.

A typical shape of the function $p(x)$ versus x is plotted in Figure 1. It can be seen that the roots of such a polynomial can be found easily by means of the one dimensional Newton-Raphson method. Indeed, due to the nature of its coefficients, $p(x)$ admits three positive solutions for x.

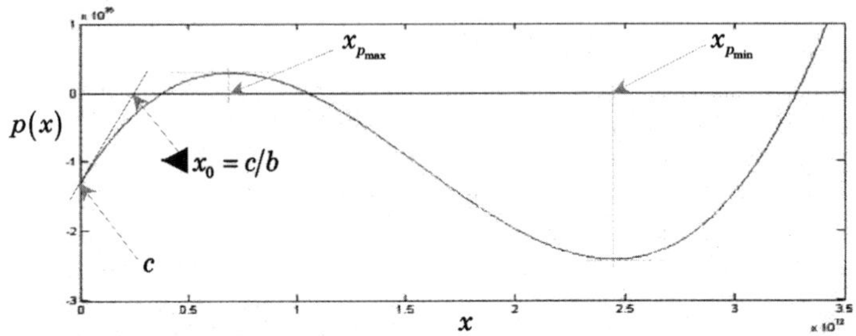

Figure 1. Characteristic Polynomial $p(x)$ versus x

Further, in order to obtain at least three solutions (that clearly exist, i.e. λ_i, $i = 1,2,3$), the smallest eigenvalue λ_3, is subject to $\lambda_3 \le x_{p_{max}}$, where $x_{p_{max}}$ is such that $p(x_{p_{max}}) = \max_x[p(x)]$; while, on the other side,

$$\lambda_3 \ge c/b, \tag{7}$$

$x_0 = c/b$ being the abscissa at which the tangent to $p(x)$ at $x = 0$ crosses the x axis. Therefore, choosing x_0 as an initial guess, the NR scheme surely converges on the effective solution λ_3. In fact, according to the iteration

$$x_{k+1} = x_k + \Delta x_k \tag{8.1}$$

$$\Delta x_k = p(x_k)/p'(x_k) = \left[\prod_{i=1}^{3}(x_k - \lambda_i)\right] \Bigg/ \left\{\sum_{i=1}^{3}\left[(x_k - \lambda_i)(x_k - \lambda_j)\right]\right\}, \tag{8.2}$$

where Δx_k is the correction term, $p'(x_k)$ is the first x-derivative of $p(x)$ evaluated at $x = x_k$ and $j = 1 + \mathrm{mod}(i,3)$, $\mathrm{mod}(n,m)$ being the operator that gives the reminder of the division n/m. From Eq. (8.2) it can be seen that if $x_k < \lambda_3$, then $\Delta x_k > 0$ and $x_{k+1} \le \lambda_3$; besides, $\Delta x_k = 0$ only when $x_k = \lambda_3$. Furthermore, the error ε_{k+1} related to the $(k+1)$-th iteration, i.e. $\varepsilon_{k+1} = x_{k+1} - \lambda_3$, can be expressed as a function of the error ε_k at the k-th iteration as

$$\varepsilon_{k+1} = \left\{\left[3p''(x_k)\varepsilon_k^2 - p'''(x_k)\varepsilon_k^3\right]\right\} \Big/ \left[6p'(x_k)\right], \tag{9}$$

$p''(x_k)$ and $p'''(x_k)$ being the second and third x-derivatives of $p(x)$ evaluated at $x = x_k$. Eq. (9) shows a good quadratic convergence.

If matrix $\mathbf{A}$ is affected by errors, i.e. we have its estimate $\tilde{\mathbf{A}} = \mathbf{A} + \Delta\mathbf{A}$, $\Delta\mathbf{A}$ being an error matrix, the evaluation of the effective singular values of $\mathbf{A}$ can be done as reported in the following. From Eq. (2), if $\tilde{\mathbf{A}}$ is used instead of $\mathbf{A}$, matrix $\mathbf{B}$ will be replaced by $\tilde{\mathbf{B}} = \mathbf{B} + \Delta\mathbf{B}$, and the above reported procedure will give the estimates $\tilde{\sigma}_i$. Due to the properties of the SVD, it can be shown (Golub and Van Loan, 1983) that the estimated and effective singular values satisfy the relation

$$\left|\tilde{\sigma}_i\left(\tilde{\mathbf{A}}\right) - \sigma_i\left(\mathbf{A}\right)\right| \le \left\|\Delta\mathbf{A}\right\|_2, \; i = 1,2,3. \tag{10}$$

Upon manipulation of Eq. (10) and recalling Eq. (4), as for the corresponding eigenvalues of $\mathbf{B}$ and $\tilde{\mathbf{B}}$, we can write

$$\left|\tilde{\lambda}_i\left(\tilde{\mathbf{B}}\right) - \lambda_i\left(\mathbf{B}\right)\right| \le \left\|\Delta\mathbf{A}\right\|_2^2 + 2\left\|\Delta\mathbf{A}\right\|_2 \min\left[\tilde{\sigma}_i\left(\tilde{\mathbf{A}}\right), \sigma_i\left(\mathbf{A}\right)\right] \tag{11}$$

and thus it is meaningless to iterate Eqs. (8) until exact convergence, i.e. until $\Delta x_k = 0$. In particular, it is not convenient to proceed beyond

$$\tilde{\lambda}_3 - x_{k+1} \le \left\|\Delta\mathbf{A}\right\|_2^2 \tag{12}$$

and therefore, as usual, the method is assumed to converge when

$$\Delta x_k = x_{k+1} - x_k \le \xi_{NRSVD} \left\|\Delta\mathbf{A}\right\|_2^2, \tag{13}$$

where $0 \le \xi_{NRSVD} \le 1$, ξ_{NRSVD} being chosen in order to optimize both accuracy and computation requirements. In both Eq. (12) and Eq. (13) we have omitted the absolute value operator on the left hand side because, due to the initial guess $x_0 = c/b$, those quantities are always positive. However, although in general NR schemes it is not true that Eq. (13) implies Eq. (12), this is the case of our problem. Indeed, Eq. (12) holds if

$$\tilde{\lambda}_3 - x_{k+1} \le x_{k+1} - x_k \tag{14}$$

or, equivalently, if

$$\varepsilon_k \le 2\Delta x_k. \tag{15}$$

It is worth noting that Eq. (13) is verified when x_k is near to $\tilde{\lambda}_3$. Indeed, in such instances, both the leftmost and rightmost expressions zero for $x_k = \tilde{\lambda}_3$, they have the same sign, while ε_k has a rate of change lower than $2\Delta x_k$. In fact, as for their derivatives, we have

$$\left|\varepsilon_k'\right| \le \left|2\Delta x_k'\right| \tag{16}$$

which, for $x_k = \tilde{\lambda}_3$, leads to

$$\left|-1\right| = \left|\varepsilon_k'\left(\tilde{\lambda}_3\right)\right| \le \left|2\Delta x_k'\left(\tilde{\lambda}_3\right)\right| = \left|-2\right|. \tag{17}$$

Once the method has converged on x_{k+1}, we get the estimate $\hat{\lambda}_3$ of λ_3, i.e. $\hat{\lambda}_3 = x_{k+1}$. Note that $\tilde{\lambda}_3$ differs from λ_3 not only due to the error $\left\|\Delta\mathbf{A}\right\|_2$ (i.e. $\tilde{\lambda}_3 \ne \lambda_3$), but also for the convergence tolerance of the NR scheme (i.e.

$\hat{\lambda}_3 \neq \tilde{\lambda}_3$). Proceeding further, the estimation of the other eigenvalues of $\mathbf{B}$, i.e. $\hat{\lambda}_1$ and $\hat{\lambda}_2$, can be performed using Eq. (5). Indeed, once $\hat{\lambda}_3$ is known, from those equations we get

$$\hat{\lambda}_{1,2} = \hat{\lambda} \pm \Delta\hat{\lambda}\,, \tag{18}$$

with

$$\hat{\lambda} = \left(\tilde{a} - \hat{\lambda}_3\right)/2 \text{ and } \Delta\hat{\lambda} = \sqrt{\hat{\lambda}\left(\hat{\lambda} + 2\hat{\lambda}_3\right) - \tilde{b}}\,, \tag{19}$$

where $\tilde{a}$ and $\tilde{b}$ are the coefficients of the characteristic polynomial of $\tilde{\mathbf{B}}$.

Finally, according to Eq. (4), we can obtain the sought estimates of the singular values of $\mathbf{A}$ as

$$\hat{\sigma}_i = \sqrt{\hat{\lambda}_i}\,,\ i = 1,2,3\,. \tag{20}$$

It is worth noting that, upon evaluating the smallest singular value $\hat{\sigma}_3$, it is then possible to investigate for rank deficiencies of the linear systems of equations with coefficient matrix $\mathbf{A}$. In particular, note that, when such matrix coincides with the manipulator Jacobian, it makes it possible to identify manipulator singularities. Indeed, in the ideal case, these configurations are obtained when $\sigma_3(\mathbf{A}) = 0$. Likewise, dealing with errors and according to Eq. (10) and Eq. (11), we may assume that a singular configuration occurs when

$$\hat{\lambda}_3 \leq \xi_{NRSVD} \left\|\mathbf{A}\right\|_2^2, \tag{21}$$

where, as reported before, $0 < \xi_{NRSVD} \leq 1$.

2.2 Determination of Singular Vectors

As for the determination of matrix $\mathbf{V} = [\mathbf{v}_1 \quad \mathbf{v}_2 \quad \mathbf{v}_3]$, we have to recall that by definition of the EVD, the singular vectors $\mathbf{v}_i$ satisfy

$$(\mathbf{B} - \mathbf{I}\lambda_i)\mathbf{v}_i = 0\,,\ i = 1,2,3\,. \tag{22}$$

It is worth noting that $\mathbf{C} = (\mathbf{B} - \mathbf{I}\lambda_i) = [c_{kl}]$ is rank deficient; this means that the $\mathbf{v}_i$'s define vector sub-spaces whose dimensions are m_i, this latter being the multiplicity of the eigenvalue λ_i. Thus, depending on m_i we have the following cases:

1) Case $m_i = 1$. If we define d_{kl}, for $k,l = 1,2,3$, as the determinants of the nine 2×2 minors of matrix $\mathbf{C}$, i.e.

$$d_{kl} = \det\left(\begin{bmatrix} c_{kl} & c_{kn} \\ c_{ml} & c_{mn} \end{bmatrix}\right), \tag{23}$$

where $m = 1 + \text{mod}(k,3)$ and $n = 1 + \text{mod}(l,3)$, then, upon determining the indices K and L such that

$$d_{KL} = \max_{k,l=1,2,3} \left(\left| d_{kl} \right| \right), \tag{24}$$

for every λ, the corresponding one-dimensional singular vector space can be defined by the vector $\mathbf{v}_i = \left[v_{1i} \quad v_{2i} \quad v_{3i} \right]^T$ such that, after posing $N = 1 + \text{mod}(L,3)$ and $P = 1 + \text{mod}(L+1,3)$, its elements are defined as

$$v_L = d_K \, / d_{KL} \, , \; v \quad = d_{KP} / d_{KL} \, , \; v_{Pi} = 1 \, . \tag{25}$$

Note that d_{KN} and d_K are determinants obtained by means of Eq. (23) where $k \equiv K$ and $l \equiv N,P$, respectively.

2) Case $m \; = 2$. If the multiplicity of λ is two, certainly a q exists such that $\lambda_i \equiv \lambda_q$, where q is either $q = 1 + \text{mod}(i,3)$ or $q = 1 + \text{mod}(i+1,3)$. Then, if we define the indices K and L so as

$$c_K \; = \max_{,=1,2,3} \left(\left| c \quad \right| \right), \tag{26}$$

the two-dimensional null-space is given by vectors $\mathbf{v}_i = \left[v_{1i} \quad v_{2i} \quad v_{3i} \right]$ and $\mathbf{v}_q = \left[v_{1q} \quad v_{2q} \quad v_{3q} \right]^T$ such that, after posing $N = 1 + \text{mod}(L,3)$ and $P = 1 + \text{mod}(L+1,3)$,

$$v \quad = -a \quad / a \quad , \; v_{Ni} = 1 \, , \; v_{Pi} = 0 \, , \tag{27.1}$$

$$v_{Lq} = -a_{KP} / a_{KL} \, , \; v_{Nq} = 1 \, , \; v_{Pq} = 0 \, , \tag{27.2}$$

where $a \quad$, $a_{\ P}$ and a_{KN} are the elements of $\mathbf{A} = \left[a_{ij} \right]$ for $i \equiv K$ and $j \equiv L,P,N$, respectively.

3) Case $m_i = 3$. Now the dimension of the singular vector space is three and, of course, it can be easily described by the following three orthogonal vectors $\mathbf{v}_1 = [1 \quad 0 \quad 0]^T$, $\mathbf{v}_2 = [0 \quad 1 \quad 0]^T$ and $\mathbf{v}_3 = [0 \quad 0 \quad 1]^T$.

Thus, whatever the multiplicity of each eigenvalue λ_i, we get the sought right singular vector matrix $\mathbf{V} = [\mathbf{v}_1 \quad \mathbf{v}_2 \quad \mathbf{v}_3]$. Besides, after manipulating Eq. (3), the left singular vector matrix $\mathbf{U}$ is

$$\mathbf{U} = \mathbf{A} \mathbf{V} \boldsymbol{\Sigma}^{-1} \, . \tag{28}$$

The above described procedure works well also when the estimated matrix $\tilde{\mathbf{A}}$ is available instead of $\mathbf{A}$. The only issue regards the determination of the multiplicity of each eigenvalue. In this context, according to Eq. (11), a satisfactory criterion is to check for each $\hat{\lambda}_i$ if

$$\left| \hat{\lambda}_i - \hat{\lambda}_{1+\text{mod}(i,3)} \right| \leq 2 \xi_{NRSVD} \left\| \Delta \mathbf{A} \right\|_2^2 \text{ and } \left| \hat{\lambda}_i - \hat{\lambda}_{1+\text{mod}(i+1,3)} \right| \leq 2 \xi_{NRSVD} \left\| \Delta \mathbf{A} \right\|_2^2 \tag{29}$$

are satisfied. When none of Eqs. (29) is verified, then $m_i = 1$; if just one of them holds, then $m_i = 2$; otherwise $m_i = 3$.

2.3 Null Space Decomposition of a Singular Matrix

Besides the evaluation of all the singular elements and the rank of whatever conditioned matrix, another important property of the SVD when applied to singular matrices is that of allowing the determination of both the range and the null-space of these matrices.

Indeed, according to the decomposition described by Eq. (3), it is known that the columns of $\mathbf{V}$, whose same-numbered elements σ_i are zero, are an orthonormal basis for the null-space of $\mathbf{A}$; while the columns of $\mathbf{U}$, whose same-numbered elements are nonzero, are an orthonormal set of basis vectors that span the range of $\mathbf{A}$.

Thus, if one or both of these subspaces are required, the method proposed in Sections 2.1 and 2.2 can be used or adapted in order to solve even these issues. In particular, when, due to errors, we have the estimated matrix $\tilde{\mathbf{A}}$, while, at the same time, $\tilde{\mathbf{A}}$ is singular, the dimension of the null-space, i.e. the multiplicity of the singular values which are equal to zero, may be defined as the number z such that

$$\lambda_{z-1} \le 2\xi_{NRSVD} \left\|\Delta\mathbf{A}\right\|_2^2 \text{ and } \lambda_z \ge 2\xi_{NRSVD} \left\|\Delta\mathbf{A}\right\|_2^2 . \tag{30}$$

2.4 Computational Complexity

The proposed NRSVD requires a total of $221 + 14 \cdot n_{NRSVD}$ *flops* and $4 + n_{NRSVD}$ transcendental operations, where n_{NRSVD} is the number of iterations required by the scheme to converge. Compared to other procedures found in the literature, it is worth noting that, since $n_{NRSVD} \le 1$, as obtained by numerical simulations, our NRSVD requires less than both the 600 *flops* required by the procedure implemented within the Matlab code (Redfern and Campbell, 1998) and the 282 *flops** required by the Golub-Reinsch algorithm (Golub and Van Loan, 1983). Note that, according to the definition reported in Golub and Van Loan, (1983), $1\,flop^* \ge 1\,flop$. Indeed, $1\,flop^*$ stands for the effort of doing together a floating point addition and a floating point multiplication; while, according to us and the Matlab characterization, $1\,flop$ is the effort spent for each of those operations.

3. Special Methods for Singularity Detection

This section describes and compares two methods, i.e. the NRSM and the PM for evaluating the smallest singular value of 3×3 matrices.

3.1 A Proposed Special Method: the NRSM

The NRSM is the special method we propose. Practically, it is a reduced version of the previous NRSVD scheme. Indeed, it can be devised just by implementing the procedure explained in Section 2.1 up to Eq. (13).

Regarding its computational requirements, this algorithm needs a total of $56+14\cdot n_{NRSM}$ *flops* and n_{NRSM} transcendental operations.

3.2 Power Method Review

By definition (Golub and Van Loan, 1983) the Power Method (PM) finds the biggest eigenvalue of a matrix. Thus, if we have the estimated matrix $\tilde{\mathbf{A}}$ and we want to determine its smallest eigenvalue, i.e. $\hat{\sigma}_3$, we have to apply the PM to a matrix $\tilde{\mathbf{D}}$ defined as

$$\tilde{\mathbf{D}} = \left(\tilde{\mathbf{A}}^{-T}\right)\left(\tilde{\mathbf{A}}^{-1}\right) = \tilde{\mathbf{U}}\tilde{\mathbf{\Sigma}}^{-2}\tilde{\mathbf{U}}^{T}. \tag{31}$$

The PM has an iterative scheme. Indeed, it starts with an initial guess vector $\mathbf{z}_0$ and, after having repeated the loop

$$\mathbf{z}_{k+1} = \tilde{\mathbf{D}}\mathbf{z}_k\,,\; y_{k+1} = \left\|\mathbf{z}_{k+1}\right\|_{\infty}, \tag{32.1}$$

$$\mathbf{z}_{k+1} = \mathbf{z}_{k+1}/y_{k+1} \text{ and } x_{k+1} = 1/y_{k+1}\,, \tag{32.2}$$

until

$$\left|x_{k+1} - x_k\right| \le \xi_{PM}\left\|\Delta\mathbf{A}\right\|_2^2, \tag{33}$$

it makes the variable x_{k+1} converge towards $\hat{\lambda}_3$, and the vector $\mathbf{z}_{k+1}$ converge towards the left eigenvector $\mathbf{u}_3$. Similarly to ξ_{NRSVD}, for ξ_{PM} is $0<\xi_{PM}\le 1$. The sought smallest singular value $\hat{\sigma}_3$ of $\tilde{\mathbf{A}}$ can then be obtained as

$$\hat{\sigma}_3 = \sqrt{\hat{\lambda}_3}\,. \tag{34}$$

As for the computational requirement the PM needs a total of $56+20\cdot n_{PM}$ *flops* and n_{PM} transcendental operations.

3.3 Comparison: Computation and Accuracy Issues

Comparing the burden of the two methods presented, respectively, in Section 3.1 and 3.2, it is easy to prove that just considering $n_{PM}=n_{NRSM}>0$, our scheme requires less calculations than the PM.

Furthermore, as obtained in numerical simulations $n_{PM} > n_{NRSM}$. In fact, since the NRSM is based on the one dimensional NR scheme applied to the third order function $p(x)$, it shows very good convergence characteristic. On the contrary, the PM has a worse behavior. Indeed, indicating with $\tilde{\mathbf{u}}_1$, $\tilde{\mathbf{u}}_2$ and $\tilde{\mathbf{u}}_3$ the left singular vectors which correspond to the exact singular values of $\tilde{\mathbf{A}}$, i.e. $\tilde{\sigma}_1 = \tilde{\eta}_3^{-2}$, $\tilde{\sigma}_2 = \tilde{\eta}_2^{-2}$ and $\tilde{\sigma}_3 = \tilde{\eta}_1^{-2}$, and writing the initial guess as $\mathbf{z}_0 = a_1\tilde{\mathbf{u}}_1 + a_2\tilde{\mathbf{u}}_2 + a_3\tilde{\mathbf{u}}_3$, after some manipulations we get

$$\mathbf{z}_k = a_3\tilde{\eta}_1^k\left[\tilde{\mathbf{u}}_3 + \frac{a_2}{a_3}\left(\frac{\tilde{\eta}_2}{\tilde{\eta}_1}\right)^k\tilde{\mathbf{u}}_2 + \frac{a_1}{a_3}\left(\frac{\tilde{\eta}_3}{\tilde{\eta}_1}\right)^k\tilde{\mathbf{u}}_1\right] \text{ and } |y_k - \tilde{\eta}_1| = O\left[\left(\frac{|\tilde{\eta}_2|}{|\tilde{\eta}_1|}\right)^k\right]. \tag{35}$$

This highlights that the convergence of the PM mainly depends on the ratio between the singular values of $\tilde{\mathbf{A}}$, i.e. $\tilde{\sigma}_1/\tilde{\sigma}_3 = \sqrt{\tilde{\eta}_3/\tilde{\eta}_1}$ and $\tilde{\sigma}_3/\tilde{\sigma}_2 = \sqrt{\tilde{\eta}_2/\tilde{\eta}_1}$, and, only marginally, on the distance between $\mathbf{z}_0$ and $\tilde{\mathbf{u}}_3$, i.e. on the ratios a_1/a_3 and a_2/a_3.

Unfortunately, in many robotics problems the quantities $\tilde{\eta}_3/\tilde{\eta}_1$ and $\tilde{\eta}_2/\tilde{\eta}_1$ may be far from being small; rather, it could even be advisable to use $\tilde{\eta}_1 \approx \tilde{\eta}_2 \approx \tilde{\eta}_3$, i.e. near isotropic configurations where the real eigenvalues satisfy $\eta_1 = \eta_2 = \eta_3$. Thus, the computation burden of the PM may be very consistent. In addition, unlike n_{NRSM} which may become zero, n_{PM} is always greater than one. All the above statements have been proven by simulations which we have omitted for the sake of brevity.

4. Conclusions

This paper presented an efficient method for solving the SVD of 3×3 matrices. A special method which finds the smallest singular value of 3×3 matrices is also derived. Both the algorithms have been compared with other procedures proposed in the literature and considerations are provided which show our method to be superior in terms of both efficiency and accuracy.

Acknowledgments

The funds of the Italian MIUR and CNR are gratefully acknowledged.

References

Golub, G.H. and Van Loan, C.F. (1983), *Matrix Computation*, The John Hopkins University Press.

McAree, P.R. and Daniel, R.W. (1996), A Fast, Robust Solution to the Stewart Platform Forward Kinematics, *Journal of Robotics Systems*, n. 7, vol. 13, pp. 407-427.

Redfern, D. and Campbell, C. (1998), New York, *The MATLAB 5 Handbook*, Springer-Verlag.

Yoshikawa, T. (1992), *Foundations of Robotics*, The MIT Press, Cambridge.

SVD AND PD BASED PROJECTION METRICS ON SE(N)

Pierre M. Larochelle
Robotics & Spatial Systems Lab
Department of Mechanical and Aerospace Engineering
Florida Institute of Technology
pierrel@fit.edu

Andrew P. Murray
Department of Mechanical and Aerospace Engineering
University of Dayton
andrew.murray@notes.udayton.edu

Jorge Angeles
Center for Intelligent Machines
Department of Mechanical Engineering
McGill University
angeles@cim.mcgill.ca

Abstract An open research question is how to define a metric on SE(n) that is as invariant as possible with respect to (1) the choice of coordinate frames and (2) the units used to measure linear and angular distances. We present two techniques for approximating elements of the special Euclidean group SE(n) with elements of the special orthogonal group SO(n+1). These techniques are based on the singular value and polar decompositions (denoted as SVD and PD respectively) of the homogeneous transform representation of the elements of SE(n). The projection of the elements of SE(n) onto SO(n+1) yields hyperdimensional rotations that approximate the rigid-body displacements. Any of the infinite bi-invariant metrics on SO(n+1) may then be used to measure the *distance* between any two spatial displacements. The results are PD and SVD based projection techniques that yield two approximately bi-invariant metrics on SE(n). These metrics have applications in motion synthesis, robot calibration, motion interpolation, and hybrid robot control.

Keywords: Displacement metrics, metrics on the special Euclidean group, rigid-body displacements

J. Lenarčič and C. Galletti (eds.), On Advances in Robot Kinematics, 13-22.

1. Introduction

Simply stated a metric measures the distance between two points in a set. There exist numerous useful metrics for defining the distance between two points in Euclidean space, however, defining similar metrics for determining the distance between two locations of a finite rigid body is still an area of ongoing research, see Kazerounian and Rastegar, 1992, Bobrow and Park, 1995, Park, 1995, Martinez and Duffy, 1995, Larochelle and McCarthy, 1995, Etzel and McCarthy, 1996, Gupta, 1997, Tse and Larochelle, 2000, Chirikjian, 1998, and Belta and Kumar, 2002. In the cases of two locations of a finite rigid body in either SE(3) (spatial locations) or SE(2) (planar locations) any metric used to measure the distance between the locations yields a result which depends upon the chosen reference frames, see Bobrow and Park, 1995 and Martinez and Duffy, 1995. However, a metric that is independent of these choices, referred to as being bi-invariant, is desirable. Interestingly, for the specific case of orienting a finite rigid body in SO(n) bi-invariant metrics do exist. For example, Ravani and Roth, 1983 defined the distance between two orientations of a rigid body in space as the magnitude of the difference between the associated quaternions and a proof that this metric is bi-invariant may be found in Larochelle and McCarthy, 1995.

Larochelle and McCarthy, 1995 and Larochelle, 1994 presented an algorithm for approximating displacements in SE(2) with spherical orientations in SO(3). By utilizing the bi-invariant metric of Ravani and Roth, 1983 they arrived at an approximate bi-invariant metric for planar locations in which the error induced by the spherical approximation is of the order $\frac{1}{\mathsf{R}^2}$, where R is the radius of the approximating sphere (hence the term *projective displacement metric*). Their algorithm for an approximately bi-invariant metric is based upon an algebraic formulation which utilizes Taylor series expansions of *sine*() and *cosine*() terms in homogeneous transforms, see McCarthy, 1983. Etzel and McCarthy, 1996 extended this work to spatial displacements by using orientations in $SO(4)$ to approximate locations in $SE(3)$. Their algorithm is also based upon Taylor series expansions of *sine*() and *cosine*() terms, see Ge, 1994, and here too the error is of the order $\frac{1}{\mathsf{R}^2}$.

This paper presents an efficient alternative approach for defining approximately bi-invariant projection metrics on SE(n) to those presented by Larochelle and McCarthy, 1995 and Etzel and McCarthy, 1996. Here, the underlying geometrical motivations are the same- to approximate displacements with hyperspherical rotations. However, an alternative approach for reaching the same goal is presented. We utilize the singular value and polar decompositions to yield projections of planar and spatial finite displacements onto hyperspherical orientations.

2. Projecting SE(n) onto SO(n+1)

First, we review how spherical displacements may be used to approximate planar displacements with some finite error associated with the radius R of the sphere, see Larochelle, 1999 and Larochelle and McCarthy, 1995. This approach is based upon the work of McCarthy, 1983 in which he examined spherical and 3-spherical motions with instantaneous invariants approaching zero and showed that these motions may be identified with planar and spatial motions, respectively.

Recall that a general planar displacement (a, b, α) in the $z = \mathsf{R}$ plane (an element of SE(2)) may be expressed as a homogeneous coordinate transformation (an element of H(2)),

$$\begin{pmatrix} X \\ Y \\ Z \end{pmatrix} = [A_p] \begin{pmatrix} x \\ y \\ 1 \end{pmatrix} = \begin{bmatrix} \cos\alpha & -\sin\alpha & a \\ \sin\alpha & \cos\alpha & b \\ 0 & 0 & \mathsf{R} \end{bmatrix} \begin{pmatrix} x \\ y \\ 1 \end{pmatrix}. \tag{1}$$

Now consider a general spherical displacement in which the parameters used to describe the displacement are the three angles longitude(θ), latitude(ϕ), and roll(ψ), see Fig. 1. Using these parameters a general spherical displacement may be written as,

$$\begin{pmatrix} X \\ Y \\ Z \end{pmatrix} = [A_s] \begin{pmatrix} x \\ y \\ z \end{pmatrix} = Rot(y, \theta) Rot(x, -\phi) Rot(z, \psi) \begin{pmatrix} x \\ y \\ z \end{pmatrix}. \tag{2}$$

We now define $\hat{a} = \mathsf{R}\theta$ as the longitudinal arc length and $\hat{b} = \mathsf{R}\phi$ as the latitudinal arc length. If we consider displacements in the $z = \mathsf{R}$ plane and expand the trigonometric functions *sine*() and *cosine*() using a Taylor series about 0 and substitute the angles θ and ϕ from above into the expansions then we may rewrite Eq. 2 as,

$$\begin{aligned} \begin{pmatrix} X \\ Y \\ Z \end{pmatrix} &= \begin{bmatrix} \cos\psi & -\sin\psi & \hat{a} \\ \sin\psi & \cos\psi & \hat{b} \\ 0 & 0 & \mathsf{R} \end{bmatrix} \begin{pmatrix} x \\ y \\ 1 \end{pmatrix} \\ &+ \frac{1}{\mathsf{R}} \begin{bmatrix} 0 & 0 & 0 \\ 0 & 0 & 0 \\ -\hat{a}\cos\psi - \hat{b}\sin\psi & \hat{a}\sin\psi - \hat{b}\cos\psi & -\frac{1}{2}\left(\hat{a}^2 + \hat{b}^2\right) \end{bmatrix} \begin{pmatrix} x \\ y \\ 1 \end{pmatrix} \\ &+ O\left(\frac{1}{\mathsf{R}^2}\right). \end{aligned} \tag{3}$$

Note that the first term of Eq. 3 is identical to Eq. 1 and we may approximate planar displacements (a, b, ψ) with some finite error that is

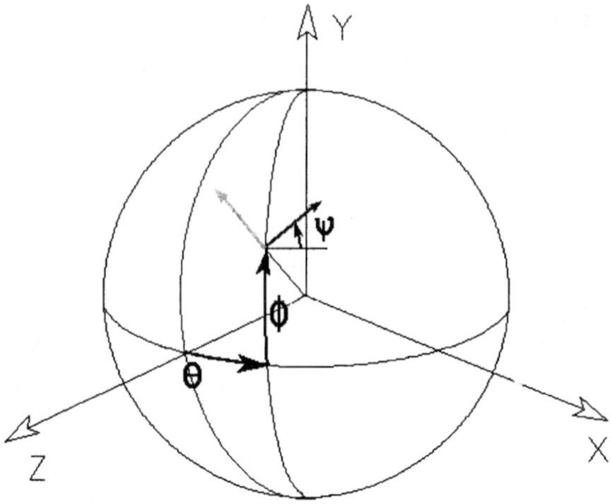

Figure 1. Planar Case: SE(2) ⇒ SO(3)

associated with the radius of the sphere. From Eq. 3 we make the following identifications: $\hat{a} \Rightarrow a$, $\hat{b} \Rightarrow b$, and, $\psi \Rightarrow \alpha$. Using the definition of the arc lengths and the radius of the sphere we obtain the three angles; θ, ϕ, and ψ, which describe the spherical displacement on the sphere of radius R that approximates the prescribed planar displacement: $\theta = \frac{a}{\mathsf{R}}$, $\phi = \frac{b}{\mathsf{R}}$, and, $\psi = \alpha$.

Etzel and McCarthy, 1996 extended the above methodology to spatial displacements by using orientations in SO(4) to approximate locations in SE(3). They showed that a 4x4 homogeneous transform representation of SE(3) can be approximated by a pure rotation $[D]$ in SO(4),

$$[D] = [J(\alpha, \beta, \gamma)][K(\theta, \phi, \psi)] \tag{4}$$

where,

$$J(\alpha, \beta, \gamma)] = \begin{bmatrix} \cos\alpha & 0 & 0 & \sin\alpha \\ -\sin\beta\sin\alpha & \cos\beta & 0 & \sin\beta\cos\alpha \\ -\sin\gamma\cos\beta\sin\alpha & -\sin\gamma\sin\beta & \cos\gamma & \sin\gamma\cos\beta\cos\alpha \\ -\cos\gamma\cos\beta\sin\alpha & -\sin\beta\cos\gamma & -\sin\gamma & \cos\gamma\cos\beta\cos\alpha \end{bmatrix}$$

and,

$$K(\theta, \phi, \psi)] = \begin{bmatrix} & & & 0 \\ & [A_s] & & 0 \\ & & & 0 \\ 0 & 0 & 0 & 1 \end{bmatrix}.$$

The angles α, β and γ are defined as follows: $\tan(\alpha) = \frac{d_x}{\mathsf{R}}$, $\tan(\beta) = \frac{d_y}{\mathsf{R}}$, and $\tan(\gamma) = \frac{d_z}{\mathsf{R}}$ where d_x, d_y, and d_z are the components of the translation vector **d** of the displacement and R is the radius of the hypersphere. A conceptual representation, analgous to Fig. 1, can be seen in Fig. 2.

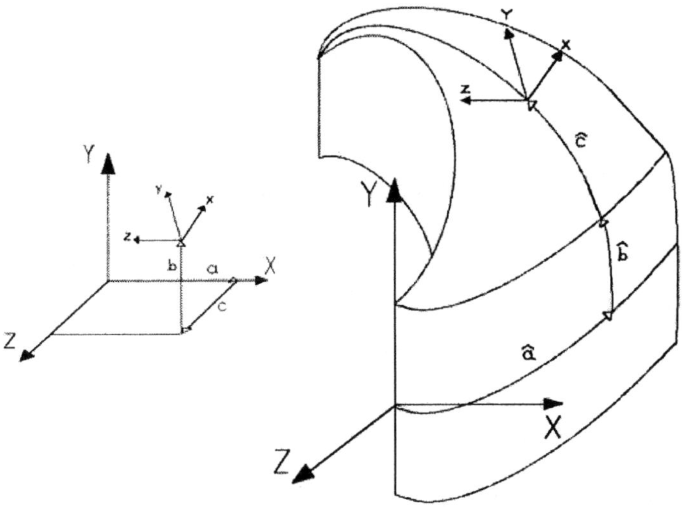

Figure 2. Spatial Case: SE(3) $\Rightarrow$ SO(4) (figure from McCarthy, 1983)

3. The SVD Based Projection

This approach, analagous to the works reviewed above, also uses hyperdimensional rotations to approximate displacements. However, this new technique uses products derived from the singular value decomposition (SVD) of the homogeneous transform to realize the projection of SE(n-1) onto SO(n). The general approach here is based upon preliminary works reported in Larochelle and Dees, 2002 and Dees, 2001.

Consider the space of $(n \times n)$ matrices as shown in Fig. 3. Let $[T]$ be a $(n \times n)$ homogeneous transform that represents an element of SE(n-1). Note that $[T]$ defines a point in R^{n^2}. $[A]$ is the desired element of SO(n) nearest $[T]$ when it lies in a direction orthogonal to the tangent plane of SO(n) at $[A]$.

The following theorem, based upon related works by Hanson and Norris, 1981 provides the foundation for the projection,

Theorem 1 *Given any* $(n \times n)$ *matrix* $[T]$ *the closest element of SO(n) is given by:* $[A] = [U][V]^T$ *where* $[T] = [U][diag(s_1, s_2, \ldots, s_n)][V]^T$ *is the SVD of* $[T]$.

Shoemake and Duff, 1992 prove that matrix $[A]$ satisfies the following optimization problem: *Minimize:* $||[A]-[T]||_F^2$ subject to: $[A]^T[A]-[I] =$

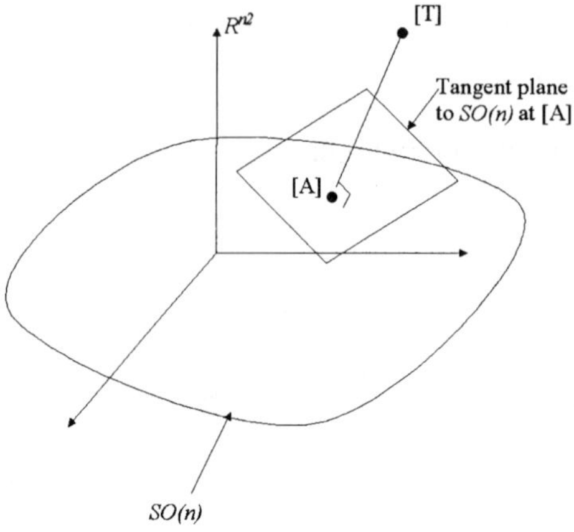

Figure 3. General Case: SE(n-1) ⇒ SO(n)

[0], where $||[A]-[T]||_F^2 = \sum_{i,j}(a_{ij}-t_{ij})^2$ is used to denote the Frobenius norm. Since $[A]$ minimizes the Frobenius norm in R^{n^2} it is the element of SO(n) that lies in a direction orthogonal to the tangent plane of SO(n) at $[R]$. Hence, $[A]$ is the closest element of SO(n) to $[T]$. Moreover, for full rank matrices the SVD is well defined and unique. We now restate Th. 1 with respect to the desired SVD based projection of SE(n-1) onto SO(n),

Theorem 2 *For* $[T] \in$ *SE(n-1) and* $[T] = [U][diag(s_1, s_2, \ldots, s_{n-1})][V]^T$ *if* $[A] = [U][V]^T$ *then* $[A]$ *is the unique element of SO(n) nearest* $[T]$.

Recall that $[T]$, the homogenous representation of SE(n), is full rank (McCarthy, 1990) and therefore $[A]$ exists, is well defined, and unique.

4. The PD Based Projection

The polar decomposition, though perhaps less known than the SVD, is quite powerful and actually provides the foundation for the SVD. The polar decomposition theorem of Cauchy states that "a non-singular matrix equals an orthogonal matrix either pre or post multiplied by a positive definite symmetric matrix", see Halmos, 1958. With respect to our application, for $[T] \in$ SE(n-1) its PD is $[T] = [P][Q]$, where $[P]$ and $[Q]$ are $(n \times n)$ matrices such that $[P]$ is orthogonal and $[Q]$ is positive definite and symmetric. Recalling the properties of the SVD, the decomposition of $[T][U][\text{diag}(s_1, s_2, \ldots, s_{n-1})][V]^T$, yields matrices $[U]$ and $[V]$ that are orthogonal and matrix $[\text{diag}(s_1, s_2, \ldots, s_{n-1})]$ which is positive definite and symmetric. Hence, for $[A] = [U][V]^T$ we have $[A]$

= $[P]$ and conclude that the polar decomposition yields the same element of SO(n).

5. Computational Issues

Often, the evaluation of the singular value decomposition is implemented in code by computing the eigenvalues and eigenvectors of the matrix since the singular values are the positive square roots of the eigenvalues of $[T][T]^T$ and the columns of $[U]$ and $[V]$ are the normed eigenvectors of $[T][T]^T$ and $[T]^T[T]$ respectively. However, we are computing the SVD of a homogeneous transform representing SE(n-1). The eigenvalue and eigenvectors of SE(2) and SE(3) are well known and should be exploited to facilitate the computations, see McCarthy, 1990.

With regard to the PD, a simple and efficient iterative algorithm exists for its evaluation. Dubrulle, 1999 provides an algorithm that produces monotonic convergence in the Frobenius norm that "... generally delivers an IEEE double-precision solution in $\sim$ 10 or fewer steps". A MatLab implementation of Dubrulle's algorithm is shown in Fig. 4.

```
function P=polar(T)
%
% initialization
%
P=T;
limit = (1 + eps) * sqrt(size(T,2));
T = inv(P');
g = sqrt(norm(T,'fro')/norm(P,'fro'));
P=0.5*(g*P+(1/g)*T);
f = norm(P,'fro');
pf = inf;
%
% iteration
%
while (f>limit) & (f<pf)
    pf = f;
    T = inv(P');
    g=sqrt(norm(T,'fro')/f);
    P=0.5*(g*P+(1/g)*T);
    f=norm(P,'fro');
end
return
```

Figure 4. Dubrulle's PD Algorithm: MatLab Implmentation

Finally, it is important to recall that both the SVD and PD based projections of SE(n-1) onto SO(n) are coordinate frame and unit dependent. This is true for all metrics on spatial and planar displacements as no bi-invariant metric exists, see Bobrow and Park, 1995 and Martinez and Duffy, 1995. Note however that these mappings project SE(n-1) onto SO(n) and bi-invariant metrics do exist on SO(n).

6. One metric on SO(n)

One useful and easily computed metric d on SO(n) follows. Given two elements $[A_1]$ and $[A_2]$ of SO(n) we can define a metric using the Frobenius norm as,

$$d = \|[I] - [A_2][A_1]^T\|_F. \tag{5}$$

It is straightforward to verify that this is a valid metric on SO(n), see Schilling and Lee, 1988.

7. Case Study-1

Consider a planar displacement $(a, b, \alpha) = (1, 1, 45)$. Its corresponding element of SE(2) is $[T]$ and we compute its projection $[A]$ onto SO(3) using either technique presented here and yield:

$$[T] = \begin{bmatrix} 0.7071 & 0.7071 & 1 \\ 0.7071 & 0.7071 & 1 \\ 0 & 0 & 1 \end{bmatrix} \tag{6}$$

and

$$[A] = \begin{bmatrix} 0.5774 & 0.7071 & 0.4082 \\ 0.5774 & 0.70714 & 0.4082 \\ -0.5774 & 0.0 & 0.8165 \end{bmatrix}. \tag{7}$$

It is illustrative compute the angle and axis of rotation $\eta = 56.60$(deg) and $\mathbf{s} = [-0.2445\ 0.5903\ 0.7693]^T$, see Fig. 1. Moreover, the longitude, latitude, and roll angles associated with $[A]$ are: $\theta = 25.56$, $\phi = 24.09$, and $\psi = 39.23$(deg). Finally, using the definitions of the longitudinal and latitudinal arc lengths $\mathsf{R} = 2.2674$ and from Eq. 5, we have $\|[T]\| = 1.3409$.

8. Case Study-2

Consider a spatial displacement $(d_x, d_y, d_z, \theta, \phi, \psi) = (1, 2, 3, 10, 30, 75)$. We proceed as above and yield the following:

$$[T] = \begin{bmatrix} 0.1710 & -0.9737 & 0.1540 & 1.0000 \\ 0.8365 & 0.2241 & 0.5000 & 2.0000 \\ -0.5206 & 0.0403 & 0.8529 & 3.0000 \\ 0 & 0 & 0 & 1 \end{bmatrix} \tag{8}$$

and

$$[A] = \begin{bmatrix} 0.1604 & -0.9584 & 0.0103 & 0.2357 \\ 0.8152 & 0.2547 & 0.2199 & 0.4714 \\ -0.5526 & 0.0861 & 0.4327 & 0.7071 \\ 0 & 0 & 0 & 1 \end{bmatrix} \tag{9}$$

and from Eq. 5, we have $||[T]|| = 2.3155$.

9. Conclusions

We have presented two new methods for approximate bi-invariant metrics on SE(n). These methods are based on projections of SE(n) onto SO(n+1) that utilize the singular value and polar decompositions of the homogeneous transform representations of SE(n). It was shown that both methods yield the same projection that determines the element of SO(n+1) nearest the given element of SE(n). Any of the infinite bi-invariant metrics on SO(n+1) may then be used to measure the *distance* between any two spatial displacements SE(n). The results are PD and SVD based projection techniques that yield two approximately bi-invariant metrics on SE(n). These metrics have applications in motion synthesis, robot calibration, motion interpolation, and hybrid robot control.

10. Acknowledgements

This work was made possible by NSF Grant No. #9816611. Moreover, the generous welcome and support extended to the 1st author during his sabbatical stay at the Center for Intelligent Machines at McGill University during Fall of 2002 (when the work reported here was in progress) are sincerely acknowledged. The collaborations and discussions with Profs. Angeles and Zsombor-Murray made the entire sabbatical experience most enjoyable and productive.

References

Belta, C. and Kumar, V. (2002), An svd-based projection method for interpolation on SE(3), *IEEE Transactions on Robotics and Automation*, vol 18, no 3, pp. 334-345.

Bobrow, J.E., and Park, F.C. (1995), On computing exact gradients for rigid body guidance using screw parameters, *Proc. of the ASME Design Engineering Technical Conferences*, Boston, MA, USA.

Bodduluri, R.M.C., (1990), *Design and planned movement of multi-degree of freedom spatial mechanisms*, PhD Dissertation, University of California, Irvine.

Chirikjian, G.S. (1998), Convolution metrics for rigid body motion, *Proc. of the ASME Design Engineering Technical Conferences*, Atlanta, USA.

Dees, S.L. (2001), *Spatial mechanism design using an svd-based distance metric*, Master's Thesis, Florida Institute of Technology.

Dubrulle, A.A. (2001), An optimum iteration for the matrix polar decomposition, *Electronic Transaction on Numerical Analysis*, vol. 8, pp. 21-25.

Etzel, K., and McCarthy, J.M. (1996), A metric for spatial displacements using biquaternions on SO(4), *Proc. of the IEEE International Conference on Robotics and Automation*, Minneapolis, USA.

Ge, Q.J. (1994), On the matrix algebra realization of of the theory of biquaternions, *Proc. of the ASME Design Engineering Technical Conferences*, Minneapolis, USA.

Gupta, K.C. (1997), Measures of positional error for a rigid body, *ASME Journal of Mechanical Design*, vol. 119, pp. 346-349.

Halmos, P.R., (1990), *Finite Dimensional Vector Spaces*, Van Nostrand.

Hanson and Norris, (1981), Analysis of measurements based upon the singular value decomposition, *SIAM Journal of Scientific and Computations*, vol. 2, no. 3, pp. 308-313.

Kazerounian, K., and Rastegar, J., (1992), Object norms: A class of coordinate and metric independent norms for displacements, *Proc. of the ASME Design Engineering Technical Conferences*, Scotsdale, USA.

Larochelle, P. (1999), On the geometry of approximate bi-invariant projective displacement metrics, *Proc. of the World Congress on the Theory of Machines and Mechanisms*, Oulu, Finland.

Larochelle, P. (2000), Approximate motion synthesis via parametric constraint manifold fitting, *Proc. of Advances in Robot Kinematics*, Piran, Slovenia.

Larochelle, P. (1998), Spades: software for synthesizing spatial 4c mechanisms, *Proc. of the ASME Design Engineering Technical Conferences*, Atlanta, USA.

Larochelle, P., (1994), *Design of cooperating robots and spatial mechanisms*, PhD Dissertation, University of California, Irvine.

Larochelle, P., and Dees, S., (2002), Approximate motion synthesis using an svd based distance metric, *Proc. of Advances in Robot Kinematics*, Caldes de Malavella, Spain.

Larochelle, P., and McCarthy, J.M. (1995), Planar motion synthesis using an approximate bi-invariant metric, *ASME Journal of Mechanical Design*, vol. 117, no. 4, pp. 646-651.

Martinez, J.M.R., and Duffy, J. (1955), On the metrics of rigid body displacements for infinite and finite bodies, *ASME Journal of Mechanical Design*, vol. 117, pp. 41-47.

McCarthy, J.M., (1983), Planar and spatial rigid body motion as special cases of spherical and 3-spherical motion, *ASME Journal of Mechanisms, Transmissions, and Automation in Design*, vol. 105, pp. 569-575.

McCarthy, J.M., (1990), *An Introduction to Theoretical Kinematics*, MIT Press.

Park, F.C., (1995), Distance metrics on the rigid-body motions with applications to mechanism design, *ASME Journal of Mechanical Design*, vol. 117, no. 1, pp. 48-54.

Ravani, B., and Roth, B., (1983), Motion synthesis using kinematic mappings, *ASME Journal of Mechanisms, Transmissions, and Automation in Design*, vol. 105, pp. 460-467.

Schilling, R.J., and Lee, H., (1988), *Engineering Analysis- a Vector Space Approach*, Wiley & Sons.

Shoemake, K., and Duff, T. (1992), Matrix animation and polar decomposition, *Proc. of Graphics Interface '92*, pp. 258-264.

Tse, D.M., Larochelle, P.M. (2000), Approximating spatial locations with spherical orientations for spherical mechanism design, *ASME Journal of Mechanical Design*, vol. 122, pp. 457-463.

CHOICE OF INDEPENDENT COORDINATES IN MODULAR KINEMATICS

Pietro Fanghella
Department of Mechanics and Design of Machines (DIMEC)
University of Genoa, Genova, Italy
pietro.fanghella@unige.it

Carlo Galletti
Department of Mechanics and Design of Machines (DIMEC)
University of Genoa, Genova, Italy
carlo.galletti@unige.it

Abstract Modular kinematics is considered by several authors a valuable tool in robot simulation. The paper describes how to verify if a given set of kinematic variables that allow a modular solution of the constraint equations of a robot can form a set of independent coordinates to be used in dynamic analysis. Then, some strategies to select a topologically valid set are provided. The proposed method is applied to a planar robot example.

Keywords: independent coordinates, modular kinematics, closure equations.

1. Introduction

Dynamic models of mechanisms and robots can be developed in many different ways, one of which, followed by many researchers (see, for example, Sudarsan and Sathiya Keerthi, 1998, Yen and Petzold, 1998, Von Schwerin, 1999, for a general discussion and numerous references, and Hiller and Kecskeméthy, 1994, Fanghella et al., 2003 for dynamic modeling in conjunction with modular kinematics) is to transform the native DAE model into an ODE form, sometimes denominated "state space," "minimal coordinates" or "independent variables." A key step to obtain a minimal model is the choice of the set $\mathbf{v}$ of independent variables that can be used to compute, by constraint equations $\mathbf{\Phi}$, all other (dependent) kinematic variables $\mathbf{u}$. For general mechanisms, like robots, the choice of a valid $\mathbf{v}$ depends on the actual position of the system, and therefore the ODE set can be constructed only locally; when the construction is no longer valid, new independent coordinates must be selected.

Most commonly, the choice of $\mathbf{v}$ is based on the local, tangent properties of the constraint manifold, represented by its Jacobian [$\mathbf{\Phi}_{,\mathrm{u}}$

J. Lenarčič and C. Galletti (eds.), On Advances in Robot Kinematics, 23-30.

$\mathbf{\Phi}_{,\mathbf{v}}$]; using this scheme, starting from the well-known paper of Wehage and Haug, 1982, a large number of approaches well suited to numerical implementations have been proposed in the last 20 years (see cited refs.). In this paper, we follow a different scheme derived from the fact that, by adopting a modular approach, (see Section 2) very often it is possible to obtain, systematically, the symbolic, closed form solution of the constraint equations, expressed in triangular form with respect to the unknown variables. It is well-known that the possibility of achieving such a significant result depends on the choice of the independent variables. Therefore, the selection of the set $\mathbf{v}$ within the framework of modular kinematics must meet two distinct requirements: a) selection of a set of truly independent variables (Section 3), and b) selection of a set that allows a modular solution of the constraint equations (Section 4).

2. An outline of modular kinematics

The purpose of modular kinematics is twofold: i) to transform a complex mechanism into a hierarchy of single-loop chains, whose kinematics (position, velocity and acceleration analyses) can be performed sequentially; ii) then, for each chain, to obtain a set of kinematic equations that can be solved sequentially, often with only one unknown in each equation.

The first problem has been studied by several authors for many years (references can be found in Kecskeméthy, 1993, Kovaks and Hommel, 1999, Ceresole et al. 1996) and various useful strategies have been devised to solve it. The resulting hierarchy depends on the choice of the independent variables in the chains. It has been shown that, for many mechanisms of technical interest, the hierarchy can be made up of a number n_s of *single-loop determined chains* (SLDCs), so that the single loops can be analyzed sequentially and simpler procedures of kinematic analysis can be adopted (Fanghella and Galletti, 1992).

Concerning the second problem, systematic results on spatial mechanism analysis have been obtained (Hiller and Kecskeméthy, 1994, Fanghella and Galletti, 1995). Let us consider an SLDC. An approach based on metric relations is very useful in deriving closure equations for the chain. This approach, which has been developed in a general way by Fanghella and Galletti, 1995, requires that two *kinematic constraints* in the chain be considered (for instance, kinematic pairs), say KCa and KCb. A kinematic constraint is characterized by its *displacement group* GR and by a set of *invariant properties* IP of the group. *Metric relations* MRs can be stated between the IPs of the constraints.

The kinematic chain is divided into two subchains, constituted by the paths of bodies connecting the constraints KCa and KCb. Metric relations can be expressed for KCa and KCb on both subchains. Each expression for MRs, say {MR}' for one subchain and {MR}" for the other, depends on the shapes of the bodies of the subchain, on the unknown variables in the same subchains, except those in KCa and KCb, on the independent variables **v** in the SLDC, and on other known time-varying transformations **T** that are provided by the solutions of the previous SLDCs in the hierarchy. As MRs are scalar reference-independent quantities, by equating the two sets we obtain the non-linear closure equations for the closed chain

$$\mathbf{\Phi}(\mathbf{u},\mathbf{v}) \equiv \{\mathrm{MR}'(\mathbf{u}', \mathbf{v}', \mathbf{T}')\} - \{\mathrm{MR}''(\mathbf{u}'', \mathbf{v}'', \mathbf{T}'')\} = 0 \qquad (1)$$

which contain the unknown variables **u** of the chain, except those in KCa and KCb.

It has also been shown by Fanghella, 1995, that, for many mechanisms of technical interest one or more couples of constraints KCa and KCb can be discovered in the chain so that a metric relations with only one unknown u are found, while all other kinematic unknowns of the chain can be found sequentially by direct expressions.

3. Properties of the closure equations

Once a modular model of the kinematics of a robot mechanism has been obtained, for a certain choice of independent coordinates **v**, its closure equations $\mathbf{\Phi}(\mathbf{u},\mathbf{v})=0$ are expressed in triangular form with respect to the unknown dependent variables **u**. Then, the complete system of n_Φ closure equations takes on the form:

$$\mathbf{\Phi}(\mathbf{u},\mathbf{v}) = \begin{cases} \begin{cases} \Phi_{11}(u_{11};\mathbf{v}_1) = 0 \\ \Phi_{12}(u_{12};u_{11},\mathbf{v}_1) = 0 \\ \ldots\ldots\ldots\ldots \end{cases} & \text{! SLDC1} \\ \begin{cases} \Phi_{21}(u_{21};\mathbf{u}_1,\mathbf{v}_1,\mathbf{v}_2) = 0 \\ \ldots\ldots\ldots\ldots \end{cases} & \text{! SLDC2} \\ \ldots\ldots\ldots\ldots \\ \begin{cases} \Phi_{n_s 1}(u_{n_s 1};\mathbf{u}_1,..,\mathbf{u}_{n_s-1},\mathbf{v}_1,...,\mathbf{v}_{n_s}) = 0 \\ \ldots\ldots\ldots\ldots \\ \Phi_{n_s n_{MR}}(u_{n_s n_{MR}};u_{n_s 1},...,u_{n_s n_{MR}-1},\mathbf{u}_1,...,\mathbf{u}_{n_s-1},\mathbf{v}_1,...,\mathbf{v}_{n_s}) = 0 \end{cases} & \text{! SLDCn}_s \end{cases} \qquad (2)$$

Equations in (2) can be grouped according to the SLDC hierarchy into subsets $\mathbf{\Phi}_i$ ($\mathbf{\Phi}\equiv\{\mathbf{\Phi}_i,\ i=1,n_s\}$), each containing the sequence Φ_{ij} ($\mathbf{\Phi}_i\equiv\{\Phi_{ij},\ j=1,n_{MR}\}$) of MRs used to solve SLDCi; the number of relations n_{MR} in each subset $\mathbf{\Phi}_i$ depends on the SLDC type. The vector of independent coordinates $\mathbf{v}$ can be divided into subsets $\mathbf{v}_i$, i=1,n_s according to the level of the SLDC hierarchy where the independent coordinates are located. In many practical mechanisms, some subsets $\mathbf{v}_i$ can be empty. One new dependent variable u_{ij} is introduced as an unknown into each closure equation Φ_{ij}; dependent variables can be grouped into subsets $\mathbf{u}\equiv\{\mathbf{u}_i, i=1,n_s\}\equiv\{u_{ij}\}$.

The method adopted, i.e., derivation of closure equations Φ_{ij} from MRs, allows us to state a priori the final form of each equation Φ_{ij} with respect to its unknown variable u_{ij} (Su and Radcliffe, 1978, Hiller and Woernle, 1988, Fanghella, 1995). Three main cases must be considered:

1) if u_{ij} is an angle (typically, the variable of an R, C or H pair), the general form is (with few exceptions):

$$\Phi_{ij} \equiv A_{ij}\cos(u_{ij}) + B_{ij}\sin(u_{ij}) + C_{ij} = 0 \tag{3}$$

2) if u_{ij} is a length (typically, the variable of a P or C pair, or the distance between two points), depending on the types of SLDC and MR, the form can be:

$$2.1)\ \Phi_{ij} \equiv A_{ij}u_{ij}^2 + B_{ij}u_{ij} + C_{ij} = 0 \quad \text{or} \quad 2.2)\ \Phi_{ij} \equiv B_{ij}u_{ij} + C_{ij} = 0 \tag{4}$$

In order to establish if, given the variables $\mathbf{v}$ and their time derivatives (t.d.), system (2) and its t.d. can be used to determine the variables $\mathbf{u}$ and their t.d., we can rely on the implicit function theorem; therefore, we analyze the properties of the Jacobian of Eq. (2)

$$\mathbf{\Phi}_{,\mathbf{u}} = \begin{bmatrix} \Phi_{11,11} & 0 & 0 & 0 & 0 \\ \Phi_{12,11} & \Phi_{12,12} & 0 & 0 & 0 \\ \cdots & \cdots & \cdots & 0 & 0 \\ \cdots & \cdots & \cdots & \cdots & 0 \\ \cdots & \cdots & \cdots & \cdots & \Phi_{n_s n_{MR}, n_s n_{MR}} \end{bmatrix} \quad \text{with } \Phi_{ij,hk} = \frac{\partial \Phi_{ij}}{\partial u_{hk}} \tag{5}$$

The variables $\mathbf{u}, \dot{\mathbf{u}}$ and $\ddot{\mathbf{u}}$ can be determined if the determinant of $\mathbf{\Phi}_{,\mathbf{u}}$ is not zero, i.e., if all n_Φ coefficients on the main diagonal $\Phi_{ij,ij}$ are different from zero. If a coefficient $\Phi_{ij,ij}$ is zero, then the corresponding SLDCi (the i-th one of the hierarchy) is in a singular configuration.

Since the form of each closure equation is known, explicit expressions for such coefficients can be obtained by differentiating relations (3) and (4) with respect to their unknowns. For example, if we consider eq. (3), we obtain:

$$\Phi_{ij,ij} = -A_{ij}\sin(u_{ij}) + B_{ij}\cos(u_{ij}) \tag{6}$$

In general, the coefficients A_{ij} and B_{ij} can be complicated functions of all the independent $\mathbf{v}_k$ and dependent $\mathbf{u}_k$ coordinates associated with SLDCs in higher positions in the hierarchy ($k \le i$). A favorable situation arises when the subchain used to generate the closure equation from an MR and containing the unknown variable u_{ij} is made-up of two rigid bodies. In this case, the coefficients A_{ij} and B_{ij} are constant, and the singular configurations of the considered SLDC can be found as functions of u_{ij}. In a general case, the value of each $\Phi_{ij,ij}$ must be checked while performing each step of the dynamic analysis.

In both cases, if a near-singular configuration of the i-th SLDC is detected, within a given tolerance, i.e., a near-singular value of a variable u_{ij} or a small value of a $\Phi_{ij,ij}$, then: a) for the present choice of independent coordinates **v**, the subset of constraint equations from the first to the ij-th and the corresponding part of the SLDC hierarchy become overconstrained; b) the remaining lower part of the hierarchy becomes (locally) movable; c) in order to re-establish a correct constraint representation, a new set of independent coordinates must be adopted.

4. Selection of a set of independent coordinates

As outlined in the introduction, the most common approaches to the selection of independent coordinates rely on forms of numerical partitioning of dependent and independent coordinates based on the properties of the tangent space of the constraint manifold, but, in general, do not alter the system of constraint equations. Our goal is to have a set of independent coordinates **v** that allow the representation of constraints in the modular form described in Section 2.

Let us suppose that the i-th SLDC (i.e., SLDCi) has reached a singular configuration; then two main situations may occur:

a) The subset $\mathbf{v}_i$ of independent coordinates directly related to SLDCi is not empty, so it may be feasible to exchange variables between $\mathbf{v}_i$ and $\mathbf{u}_i$. Two subcases are possible: a.1) the exchange of variables can eliminate the singularity without changing the set of constraint equations: this is the easiest solution but is applicable only to simple chains (e.g., planar dyads) when the variable switch does not alter the type of SLDC constraints; a.2) a more complex solution, which may find wider use, requires that the SLDCi constraints be reformulated contextually with the interchange of various elements between $\mathbf{u}_i$ and $\mathbf{v}_i$, and that a new set of MRs be used to solve them. In both cases, the SLDC hierarchy is not affected by the change of independent coordinates.

b) If $\mathbf{v}_i$ is empty or the methods considered in a.1) and a.2) fail, then one or more independent coordinates in the subsets $\mathbf{v}_1 ... \mathbf{v}_{i-1}$ must be converted into dependent variables; the same number of dependent variables in $\mathbf{u}_1 ... \mathbf{u}_i$ become independent coordinates. The SLDC hierarchy is restructured, at least partially. More than one solution is usually feasible in order to obtain a new hierarchical scheme.

In some cases, no choice of independent coordinates compatible with a hierarchical SLDC model exist.

5. Example

A planar parallel manipulator with 3 dof is shown in Fig. 1. Several choices for a set of independent coordinates are possible. In Fig. 1-a the manipulator is sketched in a generic position. A "natural" choice for the independent coordinates is illustrated in Fig. 1-b, where the variables of the pairs A, B and C are used. An analogous possibility results from choosing the pairs G, H and I, or the pairs D, E and F. In these cases, no SLDC hierarchy can be devised and the closure equations of the whole multiloop chain must be solved. On the contrary, if one chooses the generalized coordinates of the ternary link as independent coordinates $\mathbf{v}$ (Fig. 1-c), three SLDCs (of the "dyad" type) can be recognized, and their closure equations independently stated and solved. These SLDCs are: SLDC1 ≡ Frame, A, G, D, Frame; SLDC2 ≡ Frame, C, I, F, Frame; SLDC3 ≡ Frame, B, H, E, Frame).

In Fig. 1-d the same manipulator is shown in a particular position, with the pairs A, G, D lying on the same straight line. The choice of independent coordinates $\mathbf{v}$ in Fig. 1-e (i.e., the generalized coordinates of the ternary link) that allowed us to obtain the SLDC hierarchy for the generic manipulator position cannot be adopted for the present position. SLDC1 has $\mathbf{v}_1 \equiv \mathbf{v}$ and is in a singular position. This can be verified automatically, as outlined in Section 4, by considering for this SLDC a closure equation based on metric relations. Three different possibilities exist, each providing one of the unknown variables of the pair A or G or D. A first possibility is to use the subchain AGD to form MR' in (1), and the subchain A-Frame-D to form MR". The expression of MR' depends on the variable $u_{11} \equiv \varphi_G$ of the pair G, and that of MR" depends on the chosen independent coordinates $\mathbf{v}_1$. In such a way, an explicit equation in the form (3) is obtained. Its derivative in the form (6) establishes the value of u_{11} that made it impossible to adopt the set $\mathbf{v}_1$ as independent coordinates. Due to the fact that only MR" depends on $\mathbf{v}_1$, this value of u_{11} is independent of the actual manipulator position, so it can be computed only once. The other two possibilities of obtaining the closure equation use the subchain GA-Frame-D (or A-Frame-DG) for MR' and the link GD (or AG) for MR" in equation (1). The unknown variable $u_{11} \equiv \varphi_A$ of the pair A (or $u_{11} \equiv \varphi_D$ of D) can now be obtained from that equation. Also in these cases, the condition required by (6) can be established, but it now depends on the actual values of $\mathbf{v}_1$, and must be evaluated for any manipulator position.

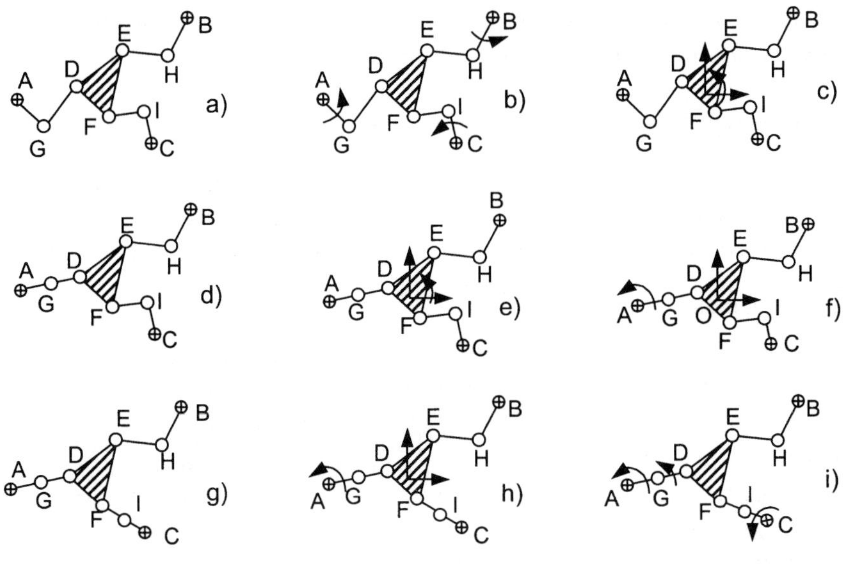

Figure 1. Choices of independent variables for a planar manipulator

In order to verify the existence of a different set **v** that furnishes a modular solution without singularities, we observe that $\mathbf{v}_1$ is not empty, and so we can apply the scheme a) of Section 4. According to this scheme, a possible solution is shown in Fig. 1-f, for which the scheme a.1) or a.2) can be adopted, depending on the previous choice made for u_{11} and for the metric relation to solve SLDC1. At any rate, the choice of independent coordinates in Fig. 1-e is wrong, but the choice made in Fig. 1-f is right. SLDC2 and SLDC3 can be solved without problems.

In the position shown in Fig. 1-g, two dyads (AGD and FIC) lie on two straight lines. It is easy to ascertain that the choice of $\mathbf{v}_1 \equiv \mathbf{v}$ valid for the previous situation (Fig. 1-f), is now wrong (Fig. 1-h), due to the singular position of SLDC2, for which similar conclusions to those drawn from the previous discussion for SLDC1 hold. In this case, $\mathbf{v}_2$ is empty and the scheme b) of Section 4 must be adopted. A possible solution is shown in Fig 1-i, with **v** consisting of the three variables of the pairs A, G, C. The hierarchy of SLDCs is rebuilt and provides the following path: SLDC1 ≡ Frame, A, G, D, F, I, C, Frame; SLDC2 ≡ Frame, C, I, F, E, H, B, Frame.

Finally, if all the dyads AGD, EHB, and FIC lie on straight lines, no hierarchy of SLDCs without singularities can be obtained.

6. Conclusions

The method presented in Section 3 allows to verify whether a set of driving variables that allow a modular solution of constraint equations of a robot mechanism can be adopted as a set of independent coordinates for its dynamic analysis. The proposed method can be applied to any robot allowing modular solutions. The check for coordinate independence must be performed during dynamic simulation, but, in many favorable cases, the range of valid values for variables can be state in advance.

Section 4 introduces some criteria to adopt a new set of independent coordinates in case the selected one fails.

References

Ceresole, E., Fanghella, P., and Galletti, C. (1996), Assur's Groups, AKCS, Basic Trusses, SOCS, etc.: Modular Kinematics of Planar Linkages, *Proc. of the 1996 ASME DETC and CIE Conference*, Irvine, CA, CD:96-DETC/MECH-1027.

Fanghella P., and Galletti C. (1992), Hierarchical Generation of Independent Loops for Symbolic Kinematics of Robot Mechanisms, *Proc. of the Third International Workshop on Advances in Robot Kinematics* (V. Parenti-Castelli and J. Lenarcic eds.) Felloni ed., pp. 96-103.

Fanghella, P., Galletti, C., and Torre G., (2003), An Explicit Independent Coordinate Formulation for the Equations of Motion of Flexible Multibody Systems, *Mechanism and Machine Theory*, vol. 38, pp. 417–437.

Fanghella, P. (1995), Kinematics of Single-Loop Mechanisms and Serial Robot Arms: A Systematic Approach, *Meccanica*, vol. 30, 685-705.

Hiller, M., and Kecskeméthy, A. (1994), Dynamics of Multibody Systems with Minimal Coordinates, in *Computer-Aided Analysis of Rigid and Flexible Mechanical Systems*, (M. Seabra Pereira and J. Ambrosio, Eds.), NATO ASI Series, vol. 268, pp. 61-100.

Hiller, M., and Woernle, C. (1988), The Characteristic Pair of Joints - an Effective Approach for the Solution of Inverse Kinematics Problems for Robots, *Proc. of IEEE International Conference on Robotics and Automation.*

Kecskeméthy, A. (1993), On Closed Form Solution of Multiple-Loop Mechanisms, *Computational Kinematics*, Kluwer, pp. 263-274.

Kovacs, P., and Hommel, G. (1999), An algorithm for the kinematic analysis and optimal input joint assignment of recursive mechanisms, *Mechanism and Machine Theory*, vol. 34, pp. 615-635.

Su, C. H., and Radcliffe, C. W. (1978), *Kinematics and Mechanisms Design*, John Wiley & Sons.

Sudarsan, R., and Sathiya Keerthi, S. (1998), Numerical Approaches for Solution of Differential Equations on Manifolds, *Applied Mathematics and Computation*, vol. 92, pp. 153-193.

Von Schwerin, R., (1999), *Multibody System Simulation*, Springer

Wehage, R.A., and Haug, E.J. (1982), Generalized coordinate partitioning for dimension reduction in analysis of constrained dynamic systems, *Trans. ASME J. Mech. Design*, vol. 104, pp. 247-255.

Yen, J., and Petzold, L. (1998), An Efficient Newton-Type Iteration for the Numerical Solution of Highly Oscillatory Constrained Multibody Dynamic Systems, *SIAM J. Sci. Comput.*, vol. 19, pp. 1513-1534.

DISTANCE CONSTRAINTS SOLVED GEOMETRICALLY

F. Thomas, J.M. Porta, and L. Ros

Institut de Robòtica i Informàtica Industrial (CSIC-UPC)

Llorens i Artigas 4-6, 08028-Barcelona, Spain

{fthomas,jporta,llros}@iri.upc.es

Abstract Most geometric constraint problems can be reduced to give coordinates to a set of points from a subset of their pairwise distances. By exploiting this fact, this paper presents an algorithm that solves distance constraint systems by iteratively reducing and expanding the dimension of the problem. In general, these projection/backprojection iterations permit tightening the ranges for the possible solutions but, if at a given point no progress is made, the algorithm bisects the search space and proceeds recursively for both subproblems. This branch-and-prune strategy is shown to converge to all solutions.

Keywords: Distance constraints, distance geometry, geometric constraint systems.

1. Introduction

The resolution of systems of geometric constraints has aroused interest in many areas of Robotics (contact formation between polyhedra, assembly planning, forward/inverse kinematics of parallel/serial manipulators, path planning of closed-loop kinematic chains, etc.) and CAD/CAM (constraint-based sketching and design, interactive placement of objects, etc.). The solution of such problems entails finding object positions and orientations that satisfy all established constraints simultaneously. A review of the methods proposed to solve this problem in the context of Robotics can be found in (Nielsen and Roth, 1997).

Most of the proposed methods consist in translating the original geometric problem into a system of algebraic equations. In this paper, we depart from this usual formulation in that our algorithm does not rely on an algebrization of the problem. Contrarily, all operations have a direct interpretation in terms of geometric transformations in the embedding space of the problem and variables are *distances* instead of degrees of freedom linked to artificial reference frames.

Most direct and inverse kinematics problems can be expressed in terms of systems of distance constraints between points. Consider, for example, the problem of finding all valid configurations of a closed 6R linkage, a cycle of of six binary links pairwise articulated with revolute joints (Fig.

J. Lenarčič and C. Galletti (eds.), On Advances in Robot Kinematics, 31-40.

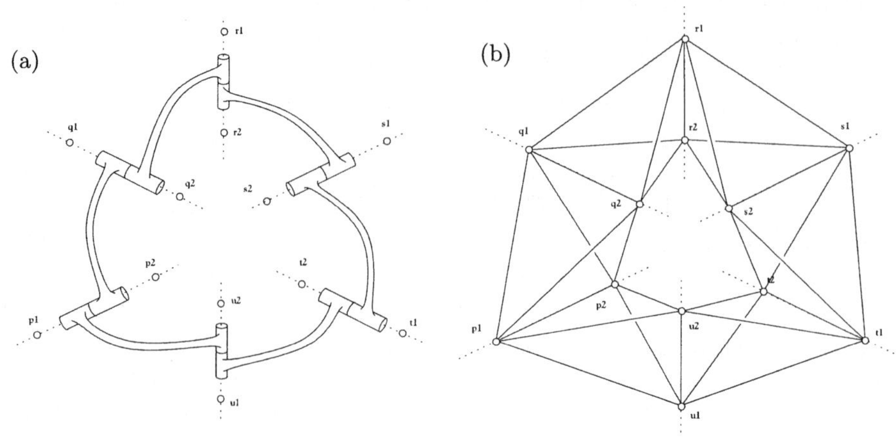

Figure 1. A general 6R linkage (a) and its translation into a tetrahedral ring (b).

1a). A binary link can be modelled by taking two points on each of its two revolute axes and connecting them all with rigid bars to form a tetrahedron. By doing so, a 6R linkage is easily translated into a ring of six tetrahedra, pairwise articulated through a common edge (Fig. 1b) which can be simply regarded as a set of points that keep some pairwise distances.

The total number of pairwise distances between n points are $\frac{n(n-1)}{2}$. The above example involves 12 points and 66 distances from which 30 are known. Likewise, it can be checked that the translation of the forward kinematics of the general Gough-Stewart platform into distance constraints between points also involves 12 points and 66 distances, 36 of which are known. Therefore, if a computer program, able to obtain all sets of values for the unknown distances compatible with the known ones, would be available, obtaining all possible solutions of the inverse kinematics of a 6R robot or the forwards kinematics of a Gough-Stewart platform would become trivial.

Our problem can be more formally stated as follows: Given a graph $G = (V_n, E)$, with node set $V_n = \{1, \dots, n\}$ and edge set E, and a real partial symmetric matrix $\mathbf{A} = (a_{ij})$ with pattern G and with zero diagonal entries, determine whether $\mathbf{A}$ can be completed to a *Euclidean distance matrix* (Laurent, 1998); that is, whether there exist vectors $\mathbf{x}_1, \dots, \mathbf{x}_n \in \Re^d$ for a given $d \geq 1$ such that $a_{ij} = \|\mathbf{x}_i - \mathbf{x}_j\|^2$ for all $ij \in E$, where $\|\mathbf{x}\|$ denotes the Euclidean norm of $\mathbf{x} \in \Re^d$. The vectors $\mathbf{x}_1, \dots, \mathbf{x}_n$ are then said to form a *realization* of $\mathbf{A}$. Unfortunately, this problem has been shown to be NP-complete if $d = 1$ and NP-hard if

$d \geq 2$ (Saxe, 1979). Thus, the best we can expect is to come up with an algorithm that, despite its unavoidable exponential complexity, works well for problems of reasonable size.

Given an arbitrary symmetric matrix with 0 entries in its diagonal and strictly positive entries in the rest, two different criteria have been found to decide whether it corresponds to a Euclidean distance matrix. One derives from the theory of Cayley-Menger determinants and the other from the theory of positive semidefinite matrices. These two criteria lie at the innermost of the two main families of algorithms that have been proposed for solving the problem at hand. It is worth saying here that all these algorithms lead to a rapid algebrization of the problem while the one proposed in this paper is based on elementary geometric operations, thus keeping the geometric flavor of the problem. Despite its geometric nature, though, it is closely related to the algorithmic tools developed for positive semidefinitive matrices. This connection permits proving the correctness of the proposed algorithm.

This paper is structured as follows. Section 2 prepares the ground for the sections that follow. To keep it short we give the needed material in intuitive terms without going into mathematical details. Section 3 describes the two basic geometric operations in which our algorithm is based; namely, projection and backprojection. Section 4 presents the algorithm and examples to clarify the main points. Finally, section 5 contains the conclusions.

2. Preliminaries

The pairwise distances between n points, say $\mathbf{x}_1, \mathbf{x}_2, \ldots, \mathbf{x}_n$, will be arranged in a symmetric matrix of the form $\mathbf{D}^0 = (d_{ij}^2)$, $i, j = 1, \ldots, n$, where d_{ij} is the Euclidean distance between $\mathbf{x}_i$ and $\mathbf{x}_j$. In our case, not all elements of $\mathbf{D}^0$ are known but lower and upper bounds on them can readily be obtained. For example, all unknown distances are necessarily smaller than the sum of all known distances, say σ. Then, all unknown distances can be bound to lie in the interval $[0, \sigma^2]$. But this trivial bound can be further tightened. In the literature this process is referred to as *bound-smoothing*: given the upper and lower bounds on a subset of pairwise distances, triangle and/or tetrangle inequalities are used to further tighten these bounds (Crippen and Havel, 1988). As a consequence, in what follows, it is convenient to assume that all entries in vectors and matrices are real compact intervals, and that all ordinary arithmetic operations on them are carried out according to the standard interval arithmetics. That is, intervals are added, subtracted, multiplied, etc. in such a way that each computed interval is generated to contain the unknown value of the quantity it represents.

Although it introduces a slight abuse of language, vectors and matrices will be treated as sets which, under certain circumstances, could be operated as such. For example, two matrices of the same size can be intersected provided that the result is also a matrix with real compact intervals.

The elements of $\mathbf{D}^0$ have to satisfy certain algebraic conditions, derived from the theory of Cayley-Menger determinants, to be the set of pairwise squared Euclidean distances between n points in $\Re^d$ (Porta et al., 2003). Unfortunately, this characterization of realizability leads to a rapid algebrization of the problem and we are here interested on obtaining an algorithm based on purely geometric constructions. To find an alternative characterization of realizability, let us organize, without loss of generality, the coordinates of $\mathbf{x}_1, \ldots, \mathbf{x}_{n-1}$ with reference to $\mathbf{x}_n$ in the following $(n-1) \times d$ matrix:

$$\mathbf{X} = \begin{pmatrix} \mathbf{x}_1 - \mathbf{x}_n \\ \mathbf{x}_2 - \mathbf{x}_n \\ \vdots \\ \mathbf{x}_{n-1} - \mathbf{x}_n \end{pmatrix},$$

where row i contains the coordinates of the vector pointing from $\mathbf{x}_n$ to $\mathbf{x}_i$. Then, the element (i, j) of the $(n-1) \times (n-1)$ matrix $\mathbf{X}\mathbf{X}^t$ contains the inner product $\langle \mathbf{x}_i - \mathbf{x}_n, \mathbf{x}_j - \mathbf{x}_n \rangle$. Actually, $\mathbf{G} = \mathbf{X}\mathbf{X}^t$ is known as a *Gram matrix*, a positive semidefinite matrix whose rank is equal to the dimension of the space in which the n points are embedded (in our case d). Then, using the cosine theorem, we can directly obtain the elements of $\mathbf{G}$ from those of $\mathbf{D}^0$ as follows:

$$\mathbf{G}[i, j] = \langle \mathbf{x}_i - \mathbf{x}_n, \mathbf{x}_j - \mathbf{x}_n \rangle = \frac{1}{2}(d_{in}^2 + d_{jn}^2 - d_{ij}^2). \tag{1}$$

Schoenberg showed that $\mathbf{D}^0$ is a proper Euclidean matrix if, and only if, $\mathbf{G}$, obtained using (1), is positive semidefinite and its rank gives the dimension of the space in which the points can be embedded (Schoenberg, 1935). This characterization of realizability has led to several algorithms for solving the matrix completion problem by semidefinite programming and variations (Alfakih et al., 1999), (Laurent, 1998), (Nikitopoulos and Emiris, 2002). The problem with these algorithms is that they are only able to find *one* realization within the provided ranges for the distances, while we are actually interested in *all* possible realizations.

What is important for us is that Schoenberg's characterization of realizability permits concluding that the Gram matrix $\mathbf{G}$ can be uniquely factorized into the product $\mathbf{L}\mathbf{L}^t$, where $\mathbf{L}$ is an $(n-1)\times d$ lower triangular matrix, because it is positive semidefinite of rank d. This factorization can be obtained by the application of d steps of the Cholesky factorization algorithm (Steward, 1998, p. 188). Then, the rows of $\mathbf{L}$ can be directly seen as the coordinates of $\mathbf{x}_1, \ldots, \mathbf{x}_{n-1}$ and therefore it is equivalent to $\mathbf{X}$ up to rotations. This algebraic fact has a nice geometric interpretation on which the algorithm given below is based but, before we explain it, let us introduce the basic operations in which it relies.

3. Basic operations: projection and backprojection

Let us take as a reference the coordinate axis defined by $\mathbf{x}_n\mathbf{x}_1$, with origin at $\mathbf{x}_n$ and pointing to $\mathbf{x}_1$ in the positive direction. Then, the projection $\overline{d}_{in}$ of d_{in} on this axis is, using the cosine theorem,

$$\overline{d}_{in} = d_{in}\cos\beta_i = \frac{1}{2d_{1n}}(d_{in}^2 + d_{1n}^2 - d_{i1}^2). \tag{2}$$

Hence,

$$\overline{d}_{ij} = \overline{d}_{in} - \overline{d}_{jn} = \frac{1}{2d_{1n}}(d_{in}^2 - d_{jn}^2 + d_{j1}^2 - d_{i1}^2), \tag{3}$$

and the orthogonal component $d_{ij}^{\perp}$ of d_{ij} is

$$(d_{ij}^{\perp})^2 = d_{ij}^2 - \overline{d}_{ij}^2 = d_{ij}^2 - \frac{1}{4d_{1n}^2}(d_{in}^2 - d_{jn}^2 + d_{j1}^2 - d_{i1}^2)^2. \tag{4}$$

Then, we can construct the vector

$$\mathbf{v}^1[i] = \overline{d}_{in}, \quad i = 1, \ldots, n, \tag{5}$$

that contains the projections of d_{in} on the axis defined by $\mathbf{x}_n\mathbf{x}_1$, and the matrix

$$\mathbf{D}^1[i,j] = (d_{i,j}^{\perp})^2, \quad i,j = 1, \cdots, n, \tag{6}$$

which is a distance matrix containing all squared distances involved in $\mathbf{D}^0$ projected onto the hyperplane orthogonal to the axis defined by $\mathbf{x}_n\mathbf{x}_1$. Note that $\mathbf{D}^1[1,n] = 0$, and, as a consequence, $\mathbf{D}^1[i,1] = \mathbf{D}^1[i,n]$, $i = 1, \ldots, n$, so that we can define $\widetilde{\mathbf{D}}^1[i,j] = \mathbf{D}^1[i,j]$, $i,j = 1, \cdots, n-1$, from which $\mathbf{D}^1$ can be straightforwardly recovered. Then, deciding whether n points can be embedded in $\Re^d$ from their interpoint distances boils down to decide whether $n-1$ points can be embedded in $\Re^{d-1}$ from a new set of interpoint distances obtained by projection. The above projection process can be iteratively repeated so that $\widetilde{\mathbf{D}}^1$ yields $\mathbf{D}^2$ and $\mathbf{v}^2$ and so on. After $d-1$ iterations, if our points can be embedded in $\Re^d$, $\widetilde{\mathbf{D}}^{d-1}$ can be embedded on a line. After d iterations, $\widetilde{\mathbf{D}}^d$ should be identically null. Space limitations prevent us from showing that, if $\mathbf{D}^0$ is a point matrix, this is a necessary and sufficient condition of realizability because $\mathbf{v}^1, \ldots, \mathbf{v}^d$ provide a Cholesky factorization of the Gram matrix associated with $\mathbf{D}^0$ (remind that all distances should be treated as intervals and, as a consequence, all expressions above should be evaluated using interval arithmetics).

Now, it can be checked that $\widetilde{\mathbf{D}}^{k-1}[i,j] = \mathbf{D}^k[i,j] + (\mathbf{v}^k[i] - \mathbf{v}^k[j])^2$, for $i, j = 1, \ldots, n-1$. Therefore, it is clear that $\mathbf{D}^{d-1}$ can be recovered from $\mathbf{D}^d$ and $\mathbf{v}^d$. This backprojection operation can take as input $\mathbf{D}^d$ and may be repeated till $\mathbf{D}^0$ is recovered. Since $\mathbf{D}^d$ must be identically null for solutions embedded in $\Re^d$, one concludes that $\mathbf{v}^1, \ldots, \mathbf{v}^d$ encodes all the information required to recover $\mathbf{D}^0$.

4. The algorithm and examples

To find all realizations in $\Re^d$ contained in a given distance matrix $\mathbf{D}^0$, the proposed algorithm projects this matrix $d-1$ times yielding $\mathbf{D}^{d-1}$. As a byproduct of these projections, we get a set of coordinate vectors that permit recovering $\mathbf{D}^0$ from $\mathbf{D}^{d-1}$. Then, all elements in $\mathbf{D}^{d-1}$ that cannot be embedded on a line can be ruled out because they correspond to realizations that cannot be embedded in $\Re^d$. Thus, $\mathbf{D}^{d-1}$ is pruned accordingly and backprojected, using the obtained coordinate vectors, yielding $\hat{\mathbf{D}}^0$. Then, if $\hat{\mathbf{D}}^0 \cap \mathbf{D}^0$ provides a reduction in the original bounds, this process can be iterated for all possible projections by permuting the indices of $\mathbf{D}^0$ until either (1) one of the entries in the obtained distance matrix gets empty, in which case we can conclude that our distance matrix contain no realization, or (2) the matrix is "sufficiently" small, in which case it is considered a solution, or (3) the matrix cannot be "significantly" reduced, in which case it is split into two matrices by bisection. If the latter occurs, the whole process is repeated onto the newly created matrices, and to the matrices recursively created thereafter, until we end up with a collection of "small" matrices containing all solutions.

The presented algorithm has been implemented in MATLAB using the INTLAB toolbox (Hargreaves, 2002) that implements the standard interval arithmetics. As an example, let us consider four points in $\Re^2$ so that all but one distances between them are known. In general, this leads to two possible realizations, as shown in Fig. 2, up to congruences.

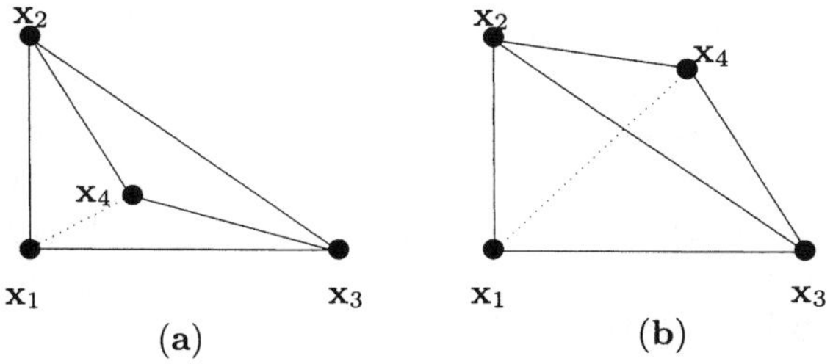

Figure 2. If there is only an unknown distance between four points in $\Re^2$ – represented here by a dotted segment–, two possible realizations arise, up to congruences.

If the corresponding distance matrix is

$$\mathbf{D}^0 = \begin{pmatrix} 0 & 16 & 36 & d_{14}^2 \\ 16 & 0 & 52 & 13 \\ 36 & 52 & 0 & 17 \\ d_{14}^2 & 13 & 17 & 0 \end{pmatrix},$$

the two realizations correspond to the values for d_{14}^2 of 5 and 23.461.

Let us suppose that $d_{14}^2 = [4.9, 5.1]$. This interval contains the realization in Fig.2a. The application of two consecutive projections leads to:

$$\widetilde{\mathbf{D}}^1 = \begin{pmatrix} [0,0] & [12.77, 12.82] & [6.86, 7.53] \\ [12.77, 12.82] & [0,0] & [38.94, 39.45] \\ [6.86, 7.53] & [38.94, 39.45] & [0,0] \end{pmatrix},$$

and

$$\widetilde{\mathbf{D}}^2 = \begin{pmatrix} [0,0] & [0, 1.35] \\ [0, 1.35] & [0,0] \end{pmatrix},$$

respectively. The last projection contains the null matrix. In other words, $\mathbf{D}^0$ might contain a realization in $\Re^2$ within the analyzed interval.

Now, let us suppose that $d_{14}^2 = [5.1, 5.2]$ where no realization exists. This is readily detected because, after the same two projections, one gets

$$\widetilde{\mathbf{D}}^2 = \begin{pmatrix} [0,0] & [0.26, 1.50] \\ [0.26, 1.5] & [0,0] \end{pmatrix}$$

which does not contain the null matrix. Nevertheless, note that after a single projection, the result can be efficiently pruned by embedding it on a line. Then, let us assume that $d_{14}^2 = [0, 100]$. By iterating for all possible projections, and after a single bisection and 64 projections, our algorithm is able to find the two solutions with an error bounded to 0.001. The number of projections is highly dependant on the chosen ordering for all possible projections. In this example, this number drops to 12 by properly choosing this ordering.

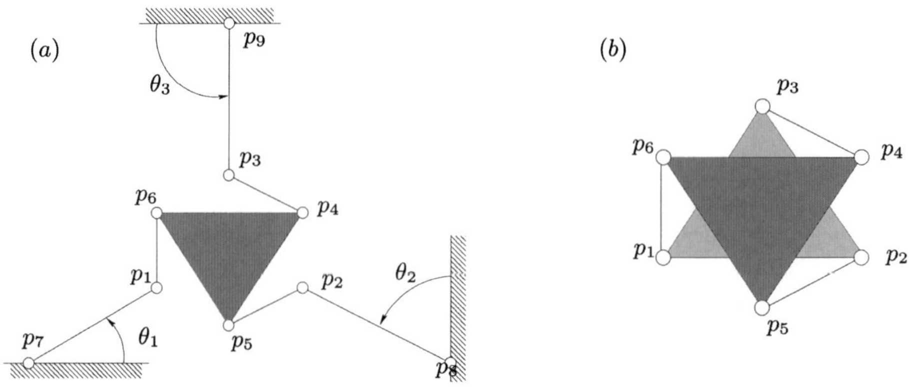

Figure 3. The forward analysis of the 3RRR planar manipulator (a) reduces to that of a 3RPR device (b).

As another example, let us consider the forward kinematics of the 3RRR planar manipulator in Fig. 3a which can be straightforwardly reduced to that of the 3RPR planar device shown in Fig. 3b. Using our formulation, this problem can be translated into 9 pairwise distance constraints between 6 points, and the solution entails finding the 6 unknown distances. For generic dimensions, it can have up to 8 different discrete solutions. An eight-degree polynomial to compute them all is derived in (Tsai, 1999, Section 3.4), and a numerical method of solution can be found in (Gosselin and Sefrioui, 1991). For particular dimensions, a one-dimensional solution set can arise for which standard numerical approaches may fail to provide a proper description. For example, if the distances between the three points forming the two triangles in Fig. 3b are 1 and the three leg lengths are set to $\sqrt{3/2}$, one gets the solution space shown in Fig. 4 using the presented algorithm. It has 6 discrete solutions, contained in 6 isolated boxes, and a one-dimensional solution set, obtained here as a discretization of 538 boxes. Since the shown solution space is actually a projection of a solution space embedded in $\Re^6$, these boxes overlap at some points. This space has been obtained after 50054 projections and 1108 bisections for a maximum accepted error of 0.05.

5. Conclusions

An algorithm for solving systems of pairwise distance constraints between points, based on two simple geometric operations, has been presented. The cluster effect near singularities has not been observed, contrarily to what happened in (Porta et al., 2003), because this algorithm is

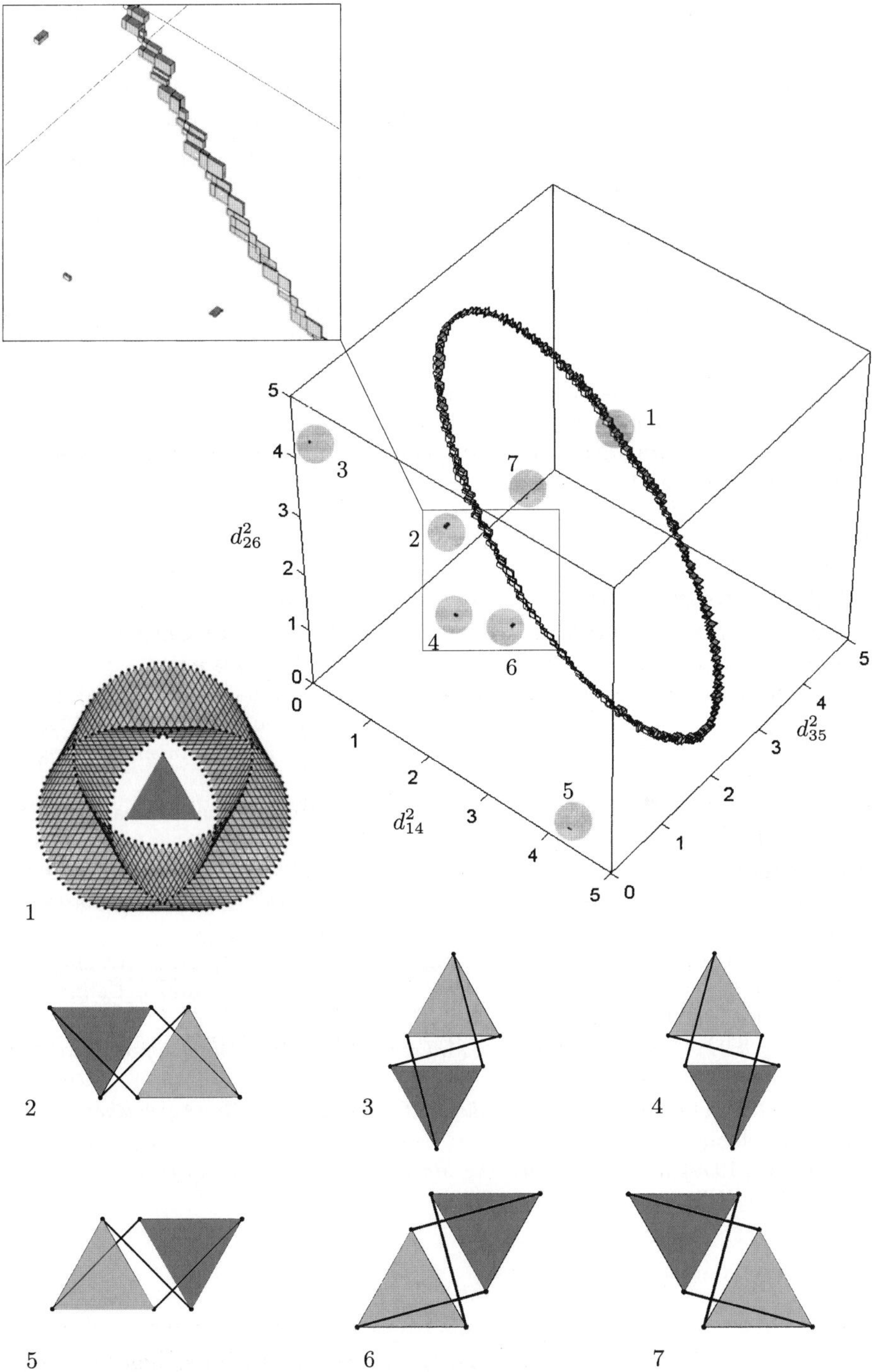

Figure 4. Feasible configurations of the 3RPR planar manipulator. Each configuration corresponds to the same-numbered box in the plot above. See the text for details.

based on the application of a sequence of projection/backprojection operations, each of them derived from a necessary and sufficient condition of realizability. The speed of convergence to the solutions heavily depends on the chosen sequence of projections that are iteratively repeated. The development of heuristics for properly choosing this sequence is a point that concentrates our current efforts.

References

Alfakih, A.Y., Khandani, A., and Wolkowicz, H. (1999),"Solving Euclidean Distance matrix completion problems via Semidefinite Programming," *Computational Optimization and Applications*, Vol. 12, No. 1-3, pp. 13-30.

Crippen, G.M., and Havel, T. (1988), *Distance Geometry and Molecular Conformation*, John Wiley and Sons Inc., New York.

Gosselin, C., and Sefrioui, J. (1991), "Polynomial Solutions for the Direct Kinematic Problem of Planar Three-Degree-of-Freedom Parallel Manipulators," in *Proc. International Conf. on Advanced Robotics*, Pisa, Italy, pp. 1124-1129, 1991.

Hargreaves, G.I. (2002), "Interval Analysis in MATLAB," Numerical Analysis Report No. 416, Manchester Center for Computational Mathematics, 2002.

Laurent, M. (1998), "A Connection Between Positive Semidefinite and Euclidean Distance Matrix Completion Problems," *Linear Algebra and its Applications*, Vol. 273, No. 1-3, pp. 9-22.

Nielsen, J., and Roth, B. (1997), "Formulation and Solution for the Direct and Inverse Kinematics Problems for Mechanisms and Mechatronic Systems," *Proc. of the NATO Advanced Institute on Computational Methods in Mechanisms,* J. Angeles and E. Zakhariev, Eds., Vol. I, pp. 233-252.

Nikitopoulos, T.G., and Emiris, I.Z. (2003), "Molecular Conformation Search by Matrix Perturbations," *preprint.*

Porta, J.M., Ros, L., Thomas, F., and Torras, C. (2004), "A Branch-and-Prune Algorithm for Solving Systems of Distance Constraints," *IEEE Trans. on Robotics and Automation*, to appear.

Saxe, J.B. (1979), "Embedadbility of Weighted Graphs in k-Space is Strongly NP-Hard," *Proc. of the 17th Allerton Conference in Communications, Control and Computing*, pp. 480-489.

Schoenberg, I.J. (1935), "Remarks to a M. Fréchet's article," *Annals of Mathematics*, Vol. 36, pp. 724-732.

Steward, G.W. (1998), *Matrix Algorithms. Volume I: Basic Decompositions*, SIAM, Philadelphia.

Tsai, L.-W. (1999) *Robot Analysis. The Machanics of Serial and Parallel Manipulators*, John Wiley and Sons Inc., 1999.

KINEMATIC MAPPING APPLICATION TO APPROXIMATE TYPE AND DIMENSION SYNTHESIS OF PLANAR MECHANISMS

M. John D. Hayes, Tim Luu

Deptartment of Mechanical and Aerospace Engineering, Carleton University
1125 Colonel By Drive, Ottawa, ON, K1S 5B6, Canada
jhayes@mae.carleton.ca, tim_luu@hotmail.com

Xiao-Wen Chang

School of Computer Science, McGill University
3480 University Street, Montréal, QC, H3A 2A7, Canada
chang@cs.mcgill.ca

Abstract Kinematic mapping is used for preliminary development of an algorithm for the approximate synthesis of planar four-bar mechanisms for rigid body guidance. Both dyad type and dimensions are determined. Planar mechanism coupler motions are represented as the curves of intersection of a pair of quadric constraint surfaces, one for each of two dyads. The problem reduces to identifying the two best constraint surfaces in the pencil of quadrics containing the curve. The overdetermined synthesis equations are linear in the unknown surface shape coefficients, and their products. Non-trivial solutions exist only in the nullspace of the coefficient matrix. While the algorithm remains incomplete, results presented herein are encouraging.

Keywords: Approximate type and dimensional synthesis, rigid body guidance, kinematic mapping, singular value decomposition.

1. Introduction

The kinematic synthesis of robot mechanical systems has been the focus of a significant volume of research; in particular, robots whose mechanical systems are parallel kinematic chains. For example Shirkhodaie and Soni, 1987 use a reasonably straightforward Cartesian approach for kinematic synthesis of planar parallel robots with three degrees-of-freedom (DOF). Murray and Pierrot, 1998 present an algebraic algorithm, based on quaternions, for synthesis of planar three-legged platforms given n-positions. We intend to develop an n-pose synthesis algorithm for both dyad type and dimensions based on the geometry of the kinematic image of the desired coupler positions and orientations. But,

J. Lenarčič and C. Galletti (eds.), On Advances in Robot Kinematics, 41-48.

as with many things kinematic, while the proposed concept is elegantly simple, the devil is in the details.

2. Kinematic Mapping

Kinematic mapping was introduced independently by Blaschke, 1911, and Grünwald, 1911. One can consider the relative displacement of two rigid-bodies in the plane as the displacement of a Cartesian reference coordinate frame E attached to one of the bodies with respect to a Cartesian reference coordinate frame Σ attached to the other. Without loss of generality, Σ may be considered fixed with E free to move, as is the case with the four-bar mechanism illustrated by Figure 1.

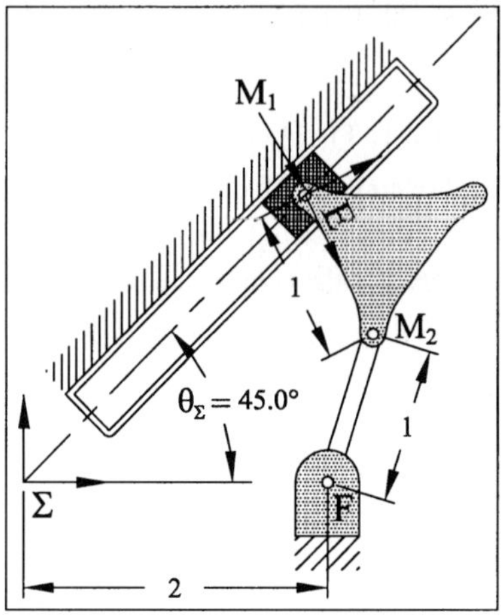

Figure 1. Planar *RRRP* linkage.

The homogeneous coordinates of points represented in E are given by the ratios $(x : y : z)$. Those of the same points represented in Σ are given by the ratios $(X : Y : Z)$. The position of a point $(X : Y : Z)$ in E in terms of the basis of Σ can be expressed compactly as

$$\begin{bmatrix} X \\ Y \\ Z \end{bmatrix} = \begin{bmatrix} \cos\varphi & -\sin\varphi & a \\ \sin\varphi & \cos\varphi & b \\ 0 & 0 & 1 \end{bmatrix} \begin{bmatrix} x \\ y \\ z \end{bmatrix}, \qquad (1)$$

where the pair $(a,\ b)$ are the $(X/Z, Y/Z)$ Cartesian coordinates of the origin of E expressed in Σ, and φ is the orientation of E relative to Σ, respectively.

The essential idea of kinematic mapping is to map the three homogeneous coordinates of the pole of a planar displacement, in terms of (a, b, φ), to the points of a three dimensional projective image space. The image space coordinates are defined as:

$$\begin{aligned} X_1 &= a\sin(\varphi/2) - b\cos(\varphi/2); \qquad & X_3 &= 2\sin(\varphi/2) \\ X_2 &= a\cos(\varphi/2) + b\sin(\varphi/2); \qquad & X_4 &= 2\cos(\varphi/2). \end{aligned} \qquad (2)$$

The mapping is injective, not bijective: *there is at most one pre-image for each image point.* Any image point on the real line l, defined by the intersection of the coordinate planes $X_3 = X_4 = 0$, has no pre-image and therefore does not correspond to a real displacement of E. See Bottema and Roth, 1979, for a detailed analysis of the geometry of the image space.

To be practical, we can remove the one parameter family of image points for coupler orientations of $\varphi = \pi$, and normalize the image space coordinates by setting $X_4 = 1$. Conceptually, this implies dividing the X_i by $X_4 = 2\cos(\varphi/2)$ giving

$$
\begin{aligned}
X_1 &= \frac{1}{2}\left(a\tan(\varphi/2) - b\right); \qquad & X_3 &= \tan(\varphi/2) \\
X_2 &= \frac{1}{2}\left(a + b\tan(\varphi/2)\right); \qquad & X_4 &= 1.
\end{aligned} \tag{3}
$$

Since each distinct displacement described by (a, b, φ) has a corresponding unique image point, the inverse mapping can be obtained from Eqs. (3): for a given point of the image space, the displacement parameters are

$$
\begin{aligned}
\tan(\varphi/2) &= X_3, \\
a &= 2(X_1X_3 + X_2)/(X_3^2 + 1), \\
b &= 2(X_2X_3 - X_1)/(X_3^2 + 1).
\end{aligned} \tag{4}
$$

By virtue of the relationships expressed in Eqs. (3), the transformation matrix from Eq. (1) may be expressed in terms of the homogeneous coordinates of the image space. After setting $z = 1$, which is done because no practical coupler will have a point at infinity, one obtains a linear transformation to express a displacement of E with respect to Σ in terms of the coordinates of the image point:

$$
\begin{bmatrix} X \\ Y \\ Z \end{bmatrix} = \begin{bmatrix} 1 - X_3^2 & -2X_3 & 2(X_1X_3 + X_2) \\ 2X_3 & 1 - X_3^2 & 2(X_2X_3 - X_1) \\ 0 & 0 & X_3^2 + 1 \end{bmatrix} \begin{bmatrix} x \\ y \\ 1 \end{bmatrix}. \tag{5}
$$

2.1 Planar Constraint Equations

Corresponding to the kinematic constraints imposed by *RR*- and *PR*-dyads are quadric constraint surfaces in the image space. A general equation is obtained when $(X : Y : Z)$ from Eqs. (5) are substituted into the general equation of a circle, the form of the most general constraint, Hayes, et al., 2004:

$$
K_0(X^2 + Y^2) + 2K_1XZ + 2K_2YZ + K_3Z^2 = 0. \tag{6}
$$

The result is that the constraint surfaces corresponding to *RR*, and *PR*-dyads can be represented by *one* equation (see Hayes, et al., 2004, for how to include *RP*- and *PP*-dyads as well). After re-arranging in terms of the constraint surface *shape* parameters K_0, K_1, K_2, K_3, x, and y, treating the image space coordinates X_1, X_2, and X_3 as constants yields:

$$
\begin{aligned}
&\left[\tfrac{1}{4}(X_3^2+1)x^2 + (X_2 - X_1X_3)x + \tfrac{1}{4}(X_3^2+1)y^2 - (X_1 + X_2X_3)y + X_2^2 + X_1^2\right]K_0 + \\
&\left[\tfrac{1}{2}(1 - X_3^2)x - X_3y + X_1X_3 + X_2\right]K_1 + \left[X_3x + \tfrac{1}{2}(1 - X_3^2)y - X_1 + X_2X_3\right]K_2 + \\
&\tfrac{1}{4}(X_3^2 + 1)K_3.
\end{aligned} \tag{7}
$$

For a particular dyad the associated $[K_0 : K_1 : K_2 : K_3]$, along with the design values of the coordinates of the coupler attachment point (x, y), expressed in reference frame E, are substituted into Eq. (7) revealing the image space constraint surface for the given dyad. The K_i in Eqs. (6) and (7) depend on the constraints imposed by the dyad.

For RR-dyads $K_0 = 1$ and the surface is a hyperboloid of one sheet that intersects planes parallel to $X_3 = 0$ in circles, Hayes and Husty, 2003. The K_i are termed *circular coefficients* and are defined as:

$$[K_0 : K_1 : K_2 : K_3] = [1 : -X_c : -Y_c : (K_1^2 + K_2^2 - r^2)], \tag{8}$$

where the ungrounded R-pair in an RR-dyad is constrained to move on a circle of constant radius, r, and fixed centre coordinates in Σ, (X_c, Y_c).

Linear constraints result when PR-dyads are employed. In this case $K_0 = 0$ and the constraint surface is an hyperbolic paraboloid with one regulus ruled by skew lines that are all parallel to the plane $X_3 = 0$, Hayes and Husty, 2003. The *linear coefficients* are defined as

$$[K_0 : K_1 : K_2 : K_3] = [0 : \tfrac{1}{2}L_1 : \tfrac{1}{2}L_2 : L_3], \tag{9}$$

where the L_i are line coordinates obtained by Grassmann expansion of the determinant of any two distinct points on the line, Klein, 1939. We obtain

$$[K_0 : K_1 : K_2 : K_3] = [0 : -\frac{1}{2}\sin\vartheta_\Sigma : \frac{1}{2}\cos\vartheta_\Sigma : F_{X/\Sigma}\sin\vartheta_\Sigma - F_{Y/\Sigma}\cos\vartheta_\Sigma], \tag{10}$$

where ϑ_Σ is the angle the direction of translation makes with respect to the X-axis, expressed in Σ (see Figure 1), $F_{X/\Sigma}$, $F_{Y/\Sigma}$, represent the homogeneous coordinates $(X : Y : 1)$, expressed in reference frame Σ, of a point on the line that is fixed relative to Σ.

3. Singular Value Decomposition

Singular value decomposition (SVD) decomposes any given $m \times n$ matrix $\mathbf{C}$ into the product of three matrix *factors* (see Goulb and Van Loan, 1996, for example) such that $\mathbf{C}_{m\times n}$=$\mathbf{U}_{m\times m}\mathbf{S}_{m\times n}\mathbf{V}^T_{n\times n}$, where $\mathbf{U}$ and $\mathbf{V}$ are orthogonal ($\mathbf{U}^T\mathbf{U} = \mathbf{I}_m$ and $\mathbf{V}^T\mathbf{V} = \mathbf{I}_n$), $\mathbf{S}$ is a diagonal matrix whose diagonal elements are the singular values of $\mathbf{C}$ arranged in descending order (singular values are lower bounded by 0).

SVD explicitly constructs orthonormal bases spanning the nullspace and the range of a matrix. This has very appealing applications to any set of linear equations (or equations that may be treated as linear in the *unknowns*) of the form $\mathbf{Ck} = \mathbf{0}$. In fact, if $\mathbf{C}_{m\times n}$ is not of full column rank, then the last n − rank($\mathbf{C}$) columns of $\mathbf{V}$ span the nullspace of $\mathbf{C}$ and any of these columns is a non-trivial solution of $\mathbf{Ck} = \mathbf{0}$. For example, if $\mathbf{C}_{m\times n}$ has rank $n-1$, then the nth column of $\mathbf{V}$ is a solution.

4. Combining Type and Dimension Synthesis

In Hayes and Zsombor-Murray, 2002, it was shown that kinematic mapping can be used for exact dimension synthesis for rigid body guidance by solving the resulting five equations generated by Eq. (7) assuming a dyad type, i.e., assigning a value to K_0. By solving the system of equations for the shape parameters K_1, K_2, K_3, x, and y all in terms of K_0, one can simultaneously perform type and dimension synthesis, see Hayes and Zsombor-Murray, 2004. If K_0 is unspecified, Eq. (7) can be used to generate the homogeneous equations $\mathbf{Ck} = \mathbf{0}$. The product $\mathbf{Ck}$ is linear in the products of K_i, x, and y. Examining Eq. (7), the corresponding vector $\mathbf{k}$ is:

$$\begin{bmatrix} K_0 & K_1 & K_2 & K_3 & K_0x & K_0y & K_0x^2 & K_0y^2 & K_1x & K_1y & K_2x & K_2y \end{bmatrix}^{\mathrm{T}}.$$

It is well known that, in general, nine non-coplanar points in space define a quadric. However, if we allow K_0 to be variable (not set a priori to either 1 or 0), then it would appear that 6 points will determine one, possibly more, image space constraint quadrics. But, the coefficient matrix has 12 columns, suggesting that the model does not fit the physics. Hence, an exploratory experiment was devised.

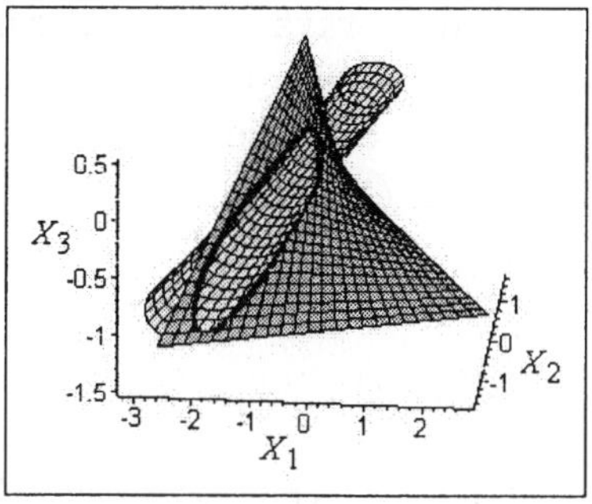

Figure 2. Image space constraint curve.

Table 1. Generating mechanism shape parameters.

Parameter	*PR-dyad*	*RR-dyad*
K_0	0	1
K_1	-1	-2
K_2	1	0
K_3	0	3
x	0	1
y	0	0

The *PRRR* mechanism illustrated in Figure 1 was used to generate 20 poses. The *PR*- and *RR*-dyad parameters for this linkage are listed in Table 1, while the resulting 20 poses are listed, but only up to four decimal places, in Table 2. The curve of intersection of the constraint hyperbolic paraboloid and hyperboloid is easily parameterized and used to generate the poses, see Hayes and Husty, 2003. Figure 2 shows the intersection of the two generating surfaces. The properties of quartic image space curves of this kind are well known, and discussed in Bottema and Roth, 1979. They are the images of fourth order coupler curves.

Table 2. The 20 generated poses.

a	b	φ (deg.)
1.0956	1.0956	-5.7248
1.1005	1.1005	-6.0256
1.1058	1.1058	-6.3597
1.1117	1.1117	-6.7329
1.1184	1.1184	-7.1527
1.1259	1.1259	-7.6281
1.1344	1.1344	-8.1712
1.1441	1.1441	-8.7974
1.1554	1.1554	-9.5273
1.1687	1.1687	-10.3889
1.1844	1.1844	-11.4212
1.2034	1.2034	-12.6804
1.2268	1.2268	-14.2500
1.2563	1.2563	-16.2602
1.2949	1.2949	-18.9246
1.3474	1.3474	-22.6199
1.4229	1.4229	-28.0725
1.5403	1.5403	-36.8699
1.7403	1.7403	-53.1301
2.0000	2.0000	-90.0000

The coefficient matrix $\mathbf{C}$, whose elements are all functions of the image space coordinates corresponding to this set of poses is 20×12. Careful examination of Eq. (7) reveals several of the X_i terms are identical. For example the products K_0x^2, K_0y^2, and K_3 all share $1/4(X_3^2+1)$ as a coefficient. An important outcome is that $\text{rank}(\mathbf{C}) \leq 8$, here $\text{rank}(\mathbf{C}) = 6$. The resulting nullspace basis vectors include linear combinations that correspond to the generating dyads. But, there is no obvious way to determine the optimal linear combinations in a way that presents a computational advantage over Cartesian-based numerical methods.

We are looking for the two dyads that best generate the desired poses. As long as we can identify all surface shape parameters, we can add certain columns in $\mathbf{C}$ to reduce its nullity, but retain its rank. Thus we reduce the nullity of $\mathbf{C}$ to 1, but in two different ways. Two different column combinations lead to SVD exactly identifying the generating PR- and RR-dyads. We start with a new 20×8 coefficient matrix, $\mathcal{C}$, obtained by collecting all like coefficients in Eq. (7):

$$\begin{array}{llll} \mathcal{C}_1 = X_1^2+X_2^2 & \mathcal{C}_3 = X_2X_3-X_1 & \mathcal{C}_5 = -(X_1+X_2X_3) & \mathcal{C}_7 = \frac{1}{2}(1-X_3^2) \\ \mathcal{C}_2 = X_2+(X_1X_3) & \mathcal{C}_4 = X_2-(X_1X_3) & \mathcal{C}_6 = \frac{1}{4}(1+X_3^2) & \mathcal{C}_8 = X_3 \end{array} \quad (11)$$

while the corresponding vector $\mathbf{k}$ becomes a new 8×1 vector $\boldsymbol{\kappa}$:

$$\left[K_0 \;\; K_1 \;\; K_2 \;\; K_0x \;\; K_0y \;\; (K_0(x^2+y^2)+K_3) \;\; (K_1x+K_2y) \;\; (K_2x-K_1y)\right]^{\mathrm{T}}. \quad (12)$$

The 20×8 coeffieient matrix $\mathcal{C}$ can be altered by strategically combining columns yielding a 20×7 coefficient matrix which still has rank 6. When different columns are added, the resulting nullspace vector either represents the generating PR-, or RR-dyad. In turn, the vector elements in $\boldsymbol{\kappa}$ corresponding to the added terms in Eq. (7) are transformed into a 7×1 vector, $\boldsymbol{\kappa}'$. But, the combined κ_i' loses its utility for parameter identification.

To extract the PR-dyad columns 4 and 5 of $\mathcal{C}$ are added. This can be done without affecting the system because $K_0 = 0$. The $\mathbf{V}$ matrix factor from the SVD of the resulting coefficient matrix yields the solution. The corresponding elements of $\boldsymbol{\kappa}'$ together with the 7th column in $\mathbf{V}$ are listed in the first two columns of Table 3.

The RR-dyad is extracted by adding columns 2 and 3 of $\mathcal{C}$. This can be done when $(X_1 - X_2X_3)/(X_1X_3 + X_2)$ has the same scalar value for every X_1, X_2, and X_3 in the data set. This happens only when the PR-dyad design parameters contain $K_3 = x = y = 0$. This condition makes the constraint surface curve of intersection symmetric in X_1 and X_2. An image space curve with rank($\mathcal{C}$)=6, but with PR-dyad design parameters $K_3 \neq x \neq y \neq 0$, can be transformed to one with symmetry in X_1 and X_2. The corresponding elements of $\boldsymbol{\kappa}'$ together with the 7th, and the normalized 7th column in $\mathbf{V}$ are listed in the last three columns of Table 3. The new elements of $\boldsymbol{\kappa}'$ resulting from adding columns are indicated by K_{45} and K_{23}, respectively.

Table 3. Nullspace vectors obtained by adding different columns of $\mathcal{C}$.

Column 4+5	*Value*	*Column 2+3*	*Value*	*Value/K_0*
K_0	0	K_0	-0.2085	1
K_1	0.7071	K_{23}	0.2085	-1
K_2	-0.7071	K_0x	-0.2085	1
K_{45}	0	K_0y	0	0
$K_0(x^2+y^2)+K_3$	0	$K_0(x^2+y^2)+K_3$	-0.8340	4
K_1x+K_2y	0	K_1x+K_2y	0.4170	-2
K_2x-K_1y	0	K_2x-K_1y	0	0

Recall that scalar multiples of a triple of line coordinates represent one and the same line. In particular, the line coordinates for the PR-dyad line constraint that passes through the origin are $\lambda[-1 : 1 : 0]$. Note that SVD has determined that $K_0 = 0$, the remaining K_i therefore determine a PR-dyad. In this case, $\lambda = -.7071$.

For the RR-dyad note that $K_0 \neq 0$, indicating other than a PR-dyad. The elements of $\boldsymbol{\kappa}'$ are normalized leading to the last column in Table 3. From these values we see that the Cartesian coordinates, expressed in the moving frame E, of the point constrained to move on a fixed circle are $(x, y) = (1, 0)$. Knowing these two coordinates, and substituting them into the last two elements of $\boldsymbol{\kappa}'$, it is seen that $K_1 = -2$ and $K_2 = 0$, which concurs with the generating RR-dyad.

5. Conclusions

In this paper we have presented preliminary results that will be used in the development of an algorithm combining type and dimensional synthesis of planar mechanisms for n-pose rigid body guidance. While the results are preliminary, they nonetheless suggest the proposed approach of using the nullspace of the synthesis equations can form the backbone for an algorithm that combines both type and dimensional synthesis for the n-pose rigid body guidance problem. We can now identify generating mechanisms when rank($\mathcal{C}$)=5 ($PRRP$'s) and certain rank($\mathcal{C}$)=6 ($PRRR$'s). For general $PRRR$'s we must transform the given image space curve into one symmetric in X_1 and X_2. We have yet to establish the conditions on the columns of $\mathcal{C}$ that will enable synthesis when rank($\mathcal{C}$)=7 ($RRRR$'s) and rank($\mathcal{C}$)=8. The last case is the most interesting as it requires some form of error minimization because no planar four-bar linkage can yield rank($\mathcal{C}$)=8.

References

Blaschke, W., (1911), "Euklidische Kinematik und nichteuklidische Geometrie", *Zeit. Math. Phys.*, Vol. 60, pp. 61-91 and 203-204.

Bottema, O. and Roth, B. (1979), *Theoretical Kinematics*, Dover Publications, Inc. (1990), New York, N.Y., U.S.A..

Goulb, G.H. and Van Loan, C. (1996), *Matrix Computations*, The Johns Hopkins University Press, Baltimore, U.S.A..

Grünwald, J., (1911), "Ein Abbildungsprinzip, welches die ebene Geometrie und Kinematik mit der räumlichen Geometrie verknüpft", *Sitzber. Ak. Wiss. Wien*, Vol. 120, pp. 677-741.

Hayes, M.J.D. and Husty, M.L., (2003), "On the Kinematic Constraint Surfaces of General Three-Legged Planar Robot Platforms", *Mechanism and Machine Theory*, vol. 38, no. 5, pp. 379-394.

Hayes, M.J.D., Zsombor-Murray, P.J., and Chen, C., (2004), "Kinematic Analyais of General Planar Parallel Manipulators", *ASME, Journal of Mechanical Design*, in press.

Hayes, M.J.D., Zsombor-Murray, P.J., (2002), "Solving the Burmester Problem Using Kinematic Mapping", *Proc. of the ASME Design Engineering Technical Conferences: Mechanisms Conference*, Montréal, QC, Canada, on CD.

Hayes, M.J.D., Zsombor-Murray, P.J., (2004), "Kinematic Mapping Synthesis of Planar Linkages", *Proceedings of the CSME Forum 2004*, University of Western Ontario, London, ON., Canada, on CD.

Klein, F., (1939), *Elementary Mathematics from an Advanced Standpoint: Geometry*, Dover Publications, Inc., New York, N.Y., U.S.A..

Murray, A.P. and Pierrot, F., (1998), N-Position Synthesis of Parallel Planar RPR Platforms, *Advances in Robot Kinematics: Analysis and Control*, eds. Lenarčič, J., Husty, M.L., Kluwer Academic Publishers, Dordrecht, The Netherlands, pp. 69-78.

Ravani, B. and Roth, B. (1983), Motion Synthesis Using Kinematic Mappings, *ASME, J. of Mechanisms, Transmissions, & Automation in Design*, Vol. 105, pp. 460-467.

Shirkhodaie, A.H. and Soni, A.H., (1987), Forward and Inverse Synthesis for a Robot with Three Degrees of Freedom, *Summer Computation Simulation Conference*, Montréal, Qué, Canada, pp. 851-856.

ON THE RULED SURFACE GENERATED BY A LINEAR COMBINATION SCREW

I.A. Parkin
School of Information Technologies
The University of Sydney, N.S.W. 2006, Australia
ian@it.usyd.edu.au

C. Freeman
School of Information Technologies
The University of Sydney, N.S.W. 2006, Australia
cfreeman@it.usyd.edu.au

Abstract A variable screw $\hat{\mathbf{S}}$ defined as a linear combination of three fixed orthogonal screws $\hat{\mathbf{X}}$, $\hat{\mathbf{Y}}$, $\hat{\mathbf{Z}}$ with respective real coefficient functions $\mathcal{X}(t)$, $\mathcal{Y}(t)$, $\mathcal{Z}(t)$ typically generates a ruled surface when the parameter t is varied. Treating those functions as unspecified, the paper provides: (a). equations from which the parameter t may be eliminated to provide the polynomial equation of the ruled surface in coordinates x, y, z defined on the lines of the screws as axes, (b). an expression in t for the typical point of the *curve of striction* of the surface, and (c). an expression for the variation of the surface normal direction along a ruling.

Keywords: screw theory, ruled surface, computer algebra, computer graphics

1. Introduction

Situations frequently occur in spatial kinematic analysis where some variable screw of interest finds expression as a linear combination

$$\hat{\mathbf{S}} = \mathcal{X}(t)\,(1+\varepsilon p_X)\,\hat{\mathbf{x}} \;+\; \mathcal{Y}(t)\,(1+\varepsilon p_Y)\,\hat{\mathbf{y}} \;+\; \mathcal{Z}(t)\,(1+\varepsilon p_Z)\,\hat{\mathbf{z}} \qquad (1)$$

of three orthogonal basis screws – of pitches p_X, p_Y, p_Z on lines $\hat{\mathbf{x}}$, $\hat{\mathbf{y}}$, $\hat{\mathbf{z}}$ – in which the real weighting coefficients $\mathcal{X}(t)$, $\mathcal{Y}(t)$, $\mathcal{Z}(t)$ are functions of a single parameter t. As this parameter varies, the (line of the) screw $\hat{\mathbf{S}}$ moves in space, so to speak, and generates a ruled surface which – for the purpose of better visualising the kinematic process – can be studied as a mathematical structure or depicted using computer-graphic methods.

This paper outlines the geometric basis of computer-based aids to these ends. In the first place, in Section 3, a starting point is established for discovering the multi-variate polynomial which is satisfied by the

J. Lenarčič and C. Galletti (eds.), On Advances in Robot Kinematics, 49-56.

coordinates x, y, z of each point of the ruled surface as measured on the orthogonal axial lines $\hat{\mathbf{x}}$, $\hat{\mathbf{y}}$, $\hat{\mathbf{z}}$ of the basis screws.

An effective aid to visualisation is to depict the 3-dimensional surface on a computer screen as if it were illuminated by a number of light sources. The way in which the brightness varies as the surface turns toward or away from the light provides useful visual clues to its structure.

In order to achieve this methodically, it is necessary to identify certain structural features of the surface. The *curve of striction* of the rulings, defined explicitly in Section 4, is a centralising feature which, beside providing a datum for other measurement, identifies a region of the surface toward which the "gaze" may be directed when initiating exploration. The determination of the *surface-normal* direction at a general point, discussed in Section 5, is crucial to calculation of the surface brightness under illumination.

An example of applying the formulations in a particularly simple case is given in Section 6.

2. Notation

Typically, we write a *screw* as a 3-vector of dual numbers

$$\hat{\mathbf{S}} = \mathbf{S} + \varepsilon\,\mathbf{S_p} = |\hat{\mathbf{S}}|\,(1+\varepsilon\,p)\,\hat{\mathbf{s}}$$
$$\hat{\mathbf{s}} = \mathbf{l} + \varepsilon\,\mathbf{M}, \quad \hat{\mathbf{s}}^2 = \mathbf{l}^2 + \varepsilon\,2\,\mathbf{l}\cdot\mathbf{M} = 1 + \varepsilon\,0, \quad \mathbf{l}\times\mathbf{M} = \mathbf{R}, \tag{2}$$

in which ε is a quasi-sacalar satisfying $\varepsilon^2 = 0$ and such that for all real a, b, c, d, $(a+\varepsilon\,b = c+\varepsilon\,d) \Leftrightarrow (a = c)$ *and* $(b = d)$ We use emboldened letters to represent 3-vectors, lower case letters indicating *unit* vectors and the overwritten 'hat' symbol indicating dual quantities. $|\hat{\mathbf{S}}|$ is the real *magnitude* and p is the real *pitch* of the screw $\hat{\mathbf{S}}$, and $\hat{\mathbf{s}}$ is its *normalised line* which, regarded as a screw in its own right, has unit magnitude and zero pitch. The screw $\hat{\mathbf{S}}$ is spatially located by its *direction vector* $\mathbf{S}$ and *moment vector* $\mathbf{S_p}$, and these are more fully resolved by stating that the line $\hat{\mathbf{s}}$ of the screw has the direction of the vector of direction cosines $\mathbf{l} = (l,\ m,\ n)$ and is located by the *orthogonal* moment vector $\mathbf{M} = (P,\ Q,\ R)$ which determines its *origin-radius* vector $\mathbf{R}$, the real values l, m, n, P, Q, R being *Plucker* coordinates of the line.

For any two screws

$$\hat{\mathbf{S}}_1 = \mathbf{S}_1 + \varepsilon\,\mathbf{S}_{\mathbf{p}1} = |\mathbf{S}_1|\,(1+\varepsilon\,p_1)\,\hat{\mathbf{s}}_1, \quad \hat{\mathbf{S}}_2 = \mathbf{S}_2 + \varepsilon\,\mathbf{S}_{\mathbf{p}2} = |\mathbf{S}_2|\,(1+\varepsilon\,p_2)\,\hat{\mathbf{s}}_2,$$

their *scalar product* is given by

$$\hat{\mathbf{S}}_1\cdot\hat{\mathbf{S}}_2 = \mathbf{S}_1\cdot\mathbf{S}_2 \;+\; \varepsilon\,\hat{\mathbf{S}}_1\otimes\hat{\mathbf{S}}_2,$$

in which

$$\hat{\mathbf{S}}_1\otimes\hat{\mathbf{S}}_2 = \mathbf{S}_1\cdot\mathbf{S}_{\mathbf{p}2} \;+\; \mathbf{S}_2\cdot\mathbf{S}_{\mathbf{p}1}$$

is the *mutual moment* of the screws $\hat{\mathbf{S}}_1$ and $\hat{\mathbf{S}}_2$. Two screws $\hat{\mathbf{S}}_1$, $\hat{\mathbf{S}}_2$ are said to be *perpendicular* if $\mathbf{S}_1 \cdot \mathbf{S}_2 = 0$, to be *reciprocal* if $\hat{\mathbf{S}}_1 \otimes \hat{\mathbf{S}}_2 = 0$, and to be *orthogonal* if both of these are true, *i.e.* if $\hat{\mathbf{S}}_1 \cdot \hat{\mathbf{S}}_2 = 0$, which implies that each intersects the other at right angles. Their *cross product* screw $\hat{\mathbf{S}}_1 \times \hat{\mathbf{S}}_2$ is sited in the *common perpendicular line* of $\hat{\mathbf{S}}_1$ and $\hat{\mathbf{S}}_2$, and has pitch $p_1 + p_2 + d \cot \theta$ where d is the distance and θ is the angle from $\hat{\mathbf{S}}_1$ to $\hat{\mathbf{S}}_2$ as measured along and about that line.

The typical right-handed axial frame consists of orthogonal unit lines $\hat{\mathbf{x}}$, $\hat{\mathbf{y}}$, $\hat{\mathbf{z}}$ for which

$$\hat{\mathbf{x}} \cdot \hat{\mathbf{y}} = \hat{\mathbf{y}} \cdot \hat{\mathbf{z}} = \hat{\mathbf{z}} \cdot \hat{\mathbf{x}} = 0, \qquad \hat{\mathbf{x}}^2 = \hat{\mathbf{y}}^2 = \hat{\mathbf{z}}^2 = 1,$$
$$\hat{\mathbf{x}} \times \hat{\mathbf{y}} = \hat{\mathbf{z}}, \quad \hat{\mathbf{y}} \times \hat{\mathbf{z}} = \hat{\mathbf{x}}, \quad \hat{\mathbf{z}} \times \hat{\mathbf{x}} = \hat{\mathbf{y}}. \tag{3}$$

Since each of these lines contains the origin point, $\mathbf{V}$ say, we may write

$$\hat{\mathbf{x}} = \mathbf{x} + \varepsilon \, \mathbf{x_0} = \mathbf{x} + \varepsilon \, \mathbf{V} \times \mathbf{x} \qquad \textit{i.e. } \mathbf{x_0} = \mathbf{V} \times \mathbf{x}, \quad \textit{etc.},$$

from which we derive

$$\mathbf{x_0} \cdot \mathbf{y} = -\mathbf{y_0} \cdot \mathbf{x} = \mathbf{V} \cdot \mathbf{x} \times \mathbf{y} = \mathbf{V} \cdot \mathbf{z}, \quad \textit{etc.}$$

So whenever three such orthonormal axial lines $\hat{\mathbf{x}}$, $\hat{\mathbf{y}}$, $\hat{\mathbf{z}}$ are provided, we may always derive the origin point as

$$\mathbf{V} = \mathbf{y_0} \cdot \mathbf{z}\mathbf{x} \; + \; \mathbf{z_0} \cdot \mathbf{x}\mathbf{y} \; + \; \mathbf{x_0} \cdot \mathbf{y}\mathbf{z}$$

or, if given mutually orthogonal screws $\hat{\mathbf{X}} = \mathbf{X} + \varepsilon \, \mathbf{X_p}$, $\hat{\mathbf{Y}} = \mathbf{Y} + \varepsilon \, \mathbf{Y_p}$, $\hat{\mathbf{Z}} = \mathbf{Z} + \varepsilon \, \mathbf{Z_p}$ of arbitrary magnitudes and pitches, we find

$$\mathbf{V} = \frac{1}{\mathbf{X} \cdot \mathbf{Y} \times \mathbf{Z}} (\mathbf{Y_p} \cdot \mathbf{Z}\mathbf{X} \; + \; \mathbf{Z_p} \cdot \mathbf{X}\mathbf{Y} \; + \; \mathbf{X_p} \cdot \mathbf{Y}\mathbf{Z}) \tag{4}$$

as their point of common intersection.

3. The Ruled Surface

We assume given three mutually orthogonal basis screws $\hat{\mathbf{X}}$, $\hat{\mathbf{Y}}$, $\hat{\mathbf{Z}}$ each meeting the others at right angles in a single point which we adopt as the origin of measurement. These screws have the forms of eqns. (2):

$$\hat{\mathbf{X}} = |\mathbf{X}|(1 + \varepsilon p_X)\,\hat{\mathbf{x}}, \quad \hat{\mathbf{Y}} = |\mathbf{Y}|(1 + \varepsilon p_Y)\,\hat{\mathbf{y}}, \quad \hat{\mathbf{Z}} = |\mathbf{Z}|(1 + \varepsilon p_Z)\,\hat{\mathbf{z}},$$

in which $|\mathbf{X}|$, $|\mathbf{Y}|$, $|\mathbf{Z}|$ are their real magnitudes, p_X, p_Y, p_Z are their pitches, and $\hat{\mathbf{x}} = \mathbf{x} + \varepsilon \mathbf{0}$, $\hat{\mathbf{y}} = \mathbf{y} + \varepsilon \mathbf{0}$, $\hat{\mathbf{z}} = \mathbf{z} + \varepsilon \mathbf{0}$ are their unit axial lines which satisfy the orthonormality conditions of eqns. (3). The unit direction 3-vectors $\mathbf{x}$, $\mathbf{y}$, $\mathbf{z}$ of these lines indicate the axial directions and their null moment vectors (**0**) show that the lines, of zero pitch, pass through the origin.

Our interest lies in the variable screw $\hat{\mathbf{S}}$ formed by combining those basis screws linearly with coefficients of combination provided by the respective real-valued functions $\mathcal{X}(t)$, $\mathcal{Y}(t)$, $\mathcal{Z}(t)$ in a variable parameter t. Nothing is lost, and considerable simplicity is gained, if we consider the real magnitudes $|\mathbf{X}|$, $|\mathbf{Y}|$, $|\mathbf{Z}|$ to be respectively incorporated into those functions so that the variable screw may be written as in eqn. (1), that is, as

$$\hat{\mathbf{S}} = \mathbf{L}+\varepsilon\mathbf{M} \tag{5}$$

whose direction vector is

$$\mathbf{L} = \mathcal{X}(t)\,\mathbf{x}+\mathcal{Y}(t)\,\mathbf{y}+\mathcal{Z}(t)\,\mathbf{z}, \quad \mathbf{L^2} = \mathcal{X}^2(t)+\mathcal{Y}^2(t)+\mathcal{Z}^2(t), \tag{6}$$

and whose moment vector is

$$\mathbf{M} = \mathcal{X}(t)\,p_X\mathbf{x}+\mathcal{Y}(t)\,p_Y\mathbf{y}+\mathcal{Z}(t)\,p_Z\mathbf{z}. \tag{7}$$

If, on the line of the variable screw $\hat{\mathbf{S}}$, $\mathbf{V} = (\mathbf{x}, \mathbf{y}, \mathbf{z})$ is a general point and $\mathbf{Q} = (\mathbf{x_Q}, \mathbf{y_Q}, \mathbf{z_Q})$ is a particular point, we may equate representations of the unit direction vector of the screw to obtain

$$\frac{(x-x_Q)\mathbf{x}+(y-y_Q)\mathbf{y}+(z-z_Q)\mathbf{z}}{\sqrt{(x-x_Q)^2+(y-y_Q)^2+(z-z_Q)^2}} = \pm\frac{\mathbf{L}}{\sqrt{\mathbf{L^2}}}$$

in which the +ve. or -ve. sign is to be taken according to whether or not the point **V** is beyond the point **Q** in the direction of the vector **L**. So, on isolating individual components for consideration, we find

$$\frac{x-x_Q}{\mathcal{X}(t)} = \frac{y-y_Q}{\mathcal{Y}(t)} = \frac{z-z_Q}{\mathcal{Z}(t)} = \pm\frac{\sqrt{(x-x_Q)^2+(y-y_Q)^2+(z-z_Q)^2}}{\sqrt{\mathbf{L^2}}}.$$

in which the value of right-hand side is independent of component choice. The origin radius point of the screw $\hat{\mathbf{S}}$, given by $\mathbf{R} = \mathbf{L}\times\mathbf{M}/\mathbf{L^2}$, has components $\mathbf{R} = (x_R, y_R, z_R)$ which evaluate respectively to

$$\frac{\mathcal{Y}(t)\,\mathcal{Z}(t)\,(p_Z-p_Y)}{\mathbf{L^2}}, \quad \frac{\mathcal{Z}(t)\,\mathcal{X}(t)\,(p_X-p_Z)}{\mathbf{L^2}}, \quad \frac{\mathcal{X}(t)\,\mathcal{Y}(t)\,(p_Y-p_X)}{\mathbf{L^2}}.$$

So, on adopting $\mathbf{R} = (\mathbf{x_R}, \mathbf{y_R}, \mathbf{z_R})$ in the role of $\mathbf{Q} = (\mathbf{x_Q}, \mathbf{y_Q}, \mathbf{z_Q})$ above, we derive three polynomial equations of which only two are independent:

$$\begin{aligned}
&\mathbf{L^2}\,[\mathcal{Y}(t)\,x-\mathcal{X}(t)\,y]+\mathcal{Z}(t)\,[\mathcal{X}^2(t)\,(p_X-p_Z)-\mathcal{Y}^2(t)\,(p_Z-p_Y)] = 0,\\
&\mathbf{L^2}[\mathcal{Z}(t)\,y-\mathcal{Y}(t)\,z]+\mathcal{X}(t)\,[\mathcal{Y}^2(t)\,(p_Y-p_X)-\mathcal{Z}^2(t)\,(p_X-p_Z)] = 0,\\
&\mathbf{L^2}\,[\mathcal{X}(t)\,z-\mathcal{Z}(t)\,x]+\mathcal{Y}(t)\,[\mathcal{Z}^2(t)\,(p_Z-p_Y)-\mathcal{X}^2(t)\,(p_Y-p_X)] = 0,
\end{aligned} \tag{8}$$

the first being a multiple of $\mathcal{X}(t)$ times the second plus $\mathcal{Y}(t)$ times the third.

To discover the surface polynomial in x, y, z which is generated by the variable screw, we may suppose that the coefficient functions $\mathcal{X}(t)$, $\mathcal{Y}(t)$, $\mathcal{Z}(t)$ are each provided as a polynomial in t and that the parameter t is then eliminated from a pair of these equations. A suitable elimination process is provided by the Buchberger Algorithm (Cox *et al*, 1991) which identifies the Grobner basis associated with the pair.

We may observe in eqn. (1) that the ruled surface generated by $\hat{\mathbf{S}}$ is unaffected if the coefficient functions $\mathcal{X}(t)$, $\mathcal{Y}(t)$, $\mathcal{Z}(t)$ are all multiplied by a dual function of the form $\mathcal{A}(t)\,[1+\varepsilon\mathcal{D}(t)]$, where $\mathcal{A}(t)$, $\mathcal{D}(t)$ are real. For this merely multiplies the real magnitude of the screw $\hat{\mathbf{S}}$ by $\mathcal{A}(t)$ and increases its pitch by $\mathcal{D}(t)$ but has no effect on the location of $\hat{\mathbf{S}}$. Correspondingly, the equations just derived show that any factor common to all three coefficient functions $\mathcal{X}(t)$, $\mathcal{Y}(t)$, $\mathcal{Z}(t)$ – whether or not it is a function of t – has no effect on the structure of the surface; and that the polynomial for the surface depends only upon differences of the axial pitches p_X, p_Y, p_Z and not upon their absolute values.

4. The Curve of Striction

The *curve of striction* of the ruled surface passes, at every ruling, through the foot of the common perpendicular of that ruling and its infinitesimally separated neighbour rulings. We identify that point, parameterised by t, by first constructing a screw $\hat{\mathbf{T}}$ that lies in the common perpendicular of the typical screw $\hat{\mathbf{S}}$ and a screw $\hat{\mathbf{S}}'$ which is infinitesimally separated from it. We then identify the screw $\hat{\mathbf{N}} = \hat{\mathbf{S}} \times \hat{\mathbf{T}}$ which is normal to the surface and, making use of the results at eqn. (4), find the desired point as the point of common intersection of $\hat{\mathbf{S}}$, $\hat{\mathbf{T}}$ and $\hat{\mathbf{N}}$.

Just as the specification of $\hat{\mathbf{S}}$ at eqn. (1) employs the coefficient functions $\mathcal{X}(t)$, $\mathcal{Y}(t)$, $\mathcal{Z}(t)$, it is convenient to define another variable screw in which the coefficients of combination are their derivatives, thus

$$\dot{\hat{\mathbf{S}}} = \frac{d\mathcal{X}(t)}{dt}\,(1+\varepsilon p_X)\,\hat{\mathbf{x}} + \frac{d\mathcal{Y}(t)}{dt}\,(1+\varepsilon p_Y)\,\hat{\mathbf{y}} + \frac{d\mathcal{Z}(t)}{dt}\,(1+\varepsilon p_Z)\,\hat{\mathbf{z}}. \qquad (9)$$

Then a screw infinitesimally separated from $\hat{\mathbf{S}}$ is approximated by

$$\begin{aligned}\hat{\mathbf{S}}' &= \left\{\mathcal{X}(t)+\frac{d\mathcal{X}(t)}{dt}\,\delta t\right\}(1+\varepsilon p_X)\,\hat{\mathbf{x}}+\left\{\mathcal{Y}(t)+\frac{d\mathcal{Y}(t)}{dt}\,\delta t\right\}(1+\varepsilon p_Y)\,\hat{\mathbf{y}}\\ &\qquad +\left\{\mathcal{Z}(t)+\frac{d\mathcal{Z}(t)}{dt}\,\delta t\right\}(1+\varepsilon p_Z)\,\hat{\mathbf{z}}\\ &= \hat{\mathbf{S}}+\dot{\hat{\mathbf{S}}}\delta t \qquad (10)\end{aligned}$$

where δt is an infinitesimal change from the value of t which applies for the screw $\hat{\mathbf{S}}$. The common perpendicular of $\hat{\mathbf{S}}$ and $\hat{\mathbf{S}}'$ is sited in the screw

$$\hat{\mathbf{S}} \times \hat{\mathbf{S}}' = \hat{\mathbf{S}} \times \dot{\hat{\mathbf{S}}}\delta t = \hat{\mathbf{T}}\delta t, \tag{11}$$

in which $\hat{\mathbf{T}}$ is a screw, tangent to the surface, given by

$$\begin{aligned} \hat{\mathbf{T}} \;=\; & \left\{\mathcal{Y}(t)\,\frac{d\mathcal{Z}(t)}{dt} - \mathcal{Z}(t)\,\frac{d\mathcal{Y}(t)}{dt}\right\}[1+\varepsilon(p_Y+p_Z)]\,\hat{\mathbf{x}} \\ & + \left\{\mathcal{Z}(t)\,\frac{d\mathcal{X}(t)}{dt} - \mathcal{X}(t)\,\frac{d\mathcal{Z}(t)}{dt}\right\}[1+\varepsilon(p_Z+p_X)]\,\hat{\mathbf{y}} \\ & + \left\{\mathcal{X}(t)\,\frac{d\mathcal{Y}(t)}{dt} - \mathcal{Y}(t)\,\frac{d\mathcal{X}(t)}{dt}\right\}[1+\varepsilon(p_X+p_Y)]\,\hat{\mathbf{z}}. \end{aligned}$$

Symbolic evaluation reveals

$$\hat{\mathbf{S}} \cdot \hat{\mathbf{T}} = 0+\varepsilon 0,$$

showing that $\hat{\mathbf{S}}$ intersects $\hat{\mathbf{T}}$ at right angles. We construct a screw which is normal to the surface at the point of intersection of $\hat{\mathbf{S}}$ and $\hat{\mathbf{T}}$ by evaluating

$$\hat{\mathbf{N}} = \hat{\mathbf{S}} \times \hat{\mathbf{T}}, \tag{12}$$

If we distinguish direction and moment parts of these screws by writing

$$\hat{\mathbf{S}} = \mathbf{S}+\varepsilon\mathbf{S_p}, \quad \hat{\mathbf{T}} = \mathbf{T}+\varepsilon\mathbf{T_p}, \quad \hat{\mathbf{N}} = \mathbf{N}+\varepsilon\mathbf{N_p},$$

we may exploit eqn. (4) to write the point of intersection of $\hat{\mathbf{S}}$, $\hat{\mathbf{T}}$, $\hat{\mathbf{N}}$ as

$$\mathbf{V} = \frac{1}{\mathbf{S}\cdot\mathbf{T}\times\mathbf{N}}\left[\mathbf{T_p}\cdot\mathbf{N}\mathbf{S} \;+\; \mathbf{N_p}\cdot\mathbf{S}\mathbf{T} \;+\; \mathbf{S_p}\cdot\mathbf{T}\mathbf{N}\right] \tag{13}$$

which is the point of the curve of striction at the ruling $\hat{\mathbf{S}}$.

5. Rotation of the Normal along a Ruling

For computer-graphic display, a curved surface is typically modelled by a covering of planar triangular facets whose vertex points lie in the true surface and whose extents are chosen small enough to simulate smooth curvature. The surface brightness is evaluated accurately at each vertex, the surface geometry entering the calculation by way of the *normal direction* of the true surface at that point; and then – in the *Gouraud shading* method – the brightness at each screen point is determined by linear interpolation of those vertex values (Hill, 2001).

To determine the normal direction at a chosen point on a ruling $\hat{\mathbf{S}}$, we proceed as follows. Consider coordinates σ, τ, η to be defined on the respective lines of the screws $\hat{\mathbf{S}}$, $\hat{\mathbf{T}}$, $\hat{\mathbf{N}}$ as axes. Then the surface local to the ruling $\hat{\mathbf{S}}$ may be modelled as a doubly-ruled *parabolic hyperboloid*

$$\eta = \frac{\tau}{P}\sigma = \frac{\sigma}{P}\tau, \tag{14}$$

which, for some constant value P, shows – in its first form – rulings which, for each given τ, intersect $\hat{\mathbf{T}}$ at right angles and – in its second – shows rulings which, for given σ, intersect $\hat{\mathbf{S}}$ at right angles.

The surface normal direction at distance σ along $\hat{\mathbf{S}}$ from the curve of striction is found as the common perpendicular of the screw $\hat{\mathbf{S}}$ itslf and the ruling for value σ chosen from the latter set. Observably, in the σ, τ, η frame, that normal has direction numbers

$$(\ 0,\ \ -\sigma,\ P\ ).$$

We determine the value of P by requiring that a screw $\hat{\mathbf{S}}'$ of the surface which, in the limit, is infinitesimally separated from $\hat{\mathbf{S}}$ (eqn. (10)) should lie within the former set of rulings. Restating eqn. (14), we find

$$P = \tau\,\frac{\sigma}{\eta} = \tau\cot\theta$$

where τ is the distance and θ is the angle through which such an $\hat{\mathbf{S}}'$ is displaced from $\hat{\mathbf{S}}$ along and about the axis $\hat{\mathbf{T}}$. Discussion of the screw cross product in Section 2 shows that P can be measured, therefore, as the amount by which the pitch of the screw $\hat{\mathbf{T}}$ exceeds the sum of pitches of the screws $\hat{\mathbf{S}}$, $\dot{\hat{\mathbf{S}}}$ from which $\hat{\mathbf{T}}$ is constituted at eqn. (11).

6. Example Application for $\mathcal{Z}(t) = 0$

As example, we examine a well-known situation which can be solved by simple observation. We consider the case in which $\mathcal{Z}(t) = 0$, which implies that the variable screw $\hat{\mathbf{S}}$ is a linear combination of the axial screws $\hat{\mathbf{X}}$, $\hat{\mathbf{Y}}$ only, and thereby generates, at least potentially, a *2-system* of screws, each of which intersects the z-axis at right angles. Under the constraint $\mathcal{Z}(t) = 0$, the polynomial equations (8) simplify to

$$\mathcal{Y}(t)\,x - \mathcal{X}(t)\,y = 0, \qquad \mathbf{L^2}z\ -\ (p_Y - p_X)\,\mathcal{X}(t)\,\mathcal{Y}(t) = 0, \tag{15}$$

the latter deriving from both the second and third of those equations. From the first of these we obtain progressively

$$\frac{y}{x} = \frac{\mathcal{Y}(t)}{\mathcal{X}(t)} \qquad \textit{i.e.}\ \frac{x}{\sqrt{x^2+y^2}} = \frac{\mathcal{X}(t)}{\sqrt{\mathbf{L^2}}} \quad \textit{and} \quad \frac{y}{\sqrt{x^2+y^2}} = \frac{\mathcal{Y}(t)}{\sqrt{\mathbf{L^2}}}.$$

So, after manipulation, the second polynomial at eqn. (15) reads

$$(x^2+y^2)z-(p_Y-p_X)xy = 0 \tag{16}$$

which is the cubic equation of the cylindroidal surface in which the screws of this 2-system lie (Hunt, 1990). Under the constraint $\mathcal{Z}(t) = \frac{d\mathcal{Z}(t)}{dt} = 0$, the factors entering into eqn. (13) take the values

$$\mathbf{T_p} \cdot \mathbf{N} = 0, \qquad \mathbf{S_p} \cdot \mathbf{T} = 0, \tag{17}$$

$$\mathbf{N_p} \cdot \mathbf{S} = \mathcal{X}(t)\,\mathcal{Y}(t)\left[\mathcal{X}(t)\,\frac{d\mathcal{Y}(t)}{dt} - \mathcal{Y}(t)\,\frac{d\mathcal{X}(t)}{dt}\right](p_Y-p_X), \tag{18}$$

which implies, through eqn. (13), that the typical point of the curve of striction lies always in the direction of the tangent screw

$$\hat{\mathbf{T}} = \left[\mathcal{X}(t)\,\frac{d\mathcal{Y}(t)}{dt} - \mathcal{Y}(t)\,\frac{d\mathcal{X}(t)}{dt}\right][1+\varepsilon(p_X+p_Y)]\,\hat{\mathbf{z}} \tag{19}$$

which, itself, lies always upon the z-axis.

We observe that these deductions have been made without knowledge of the functions $\mathcal{X}(t)$, $\mathcal{Y}(t)$ beyond the assumption that, under variation of the parameter t, these functions cause full-cycle variation of the angle θ so that the cubic surface is generated in its entirety.

7. Closing Remarks

It is anticipated that the various expressions derived here will aid the construction of algorithmic interfaces between an input process – by which the coefficient functions $\mathcal{X}(t)$, $\mathcal{Y}(t)$, $\mathcal{Z}(t)$ would be read, presumably in symbolic form – and standard computer packages for either algebraic or computer-graphic manipulation.

8. Acknowledgment

The authors acknowledge the assistance of the School of Information Technologies, The University of Sydney, in providing computing facilities used in preparation of this paper.

References

D. Cox, J. Little and D. O'Shea. *Ideals, Varieties, and Algorithms*, Springer-Verlag, New York 1991.

F.S. Hill, Jr. *Computer Graphics Using Open GL*, Prentice Hall, 2001.

K.H. Hunt. *Kinematic Geometry of Mechanisms*, Dover, 1990.

LIE-GROUP FIRST-ORDER OPERATIONS IN RIGID-BODY KINEMATICS

Andrés Kecskeméthy
Chair for Mechanics, Faculty of Engineering
University Duisburg-Essen, D-47048 Duisburg, Germany
a.kecskemethy@uni-duisburg.de

Abstract Rigid-body kinematics is a well-understood and highly developed field of mechanics. Nevertheless, due to the non-commutative nature of rigid-body motions, different parametrizations exist whose derivation is not always intuitive and for which commonalities are not immediately evident. The present paper is aimed at describing a unified description of first-order kinematics, i.e., position and first-order differentials, for rigid-body motion based on generic Lie group concepts. The derivations are performed in a coordinate-free manner in order to be later applied to arbitrary parametrizations.

Keywords: Kinematics, Lie groups

1. Introduction

Kinematics of rigid-body motion is a well-studied and highly developed field of mechanics. In this setting, a rich set of algorithms and theorems exists for analyzing and computing rigid-body motion to great detail, see, e.g. Angeles, 1988, Bottema and Roth, 1970. On the other hand, due to the non-commutative nature of rigid-body transformations, the ensuing operations are not always intuitive, and there exist several parametrizations with specialized formulas that are difficult to relate. For this reason, there has been intensive research in defining the generic properties of rigid-body transformations from the perspective of general Lie group theory (Chevallier, 1991, Karger and Novák, 1985 and Park, 1995). While one can not expect Lie-group theory to provide new procedures that can compete with the specialized formulas developed at detail over centuries for rigid-body kinematics, the Lie-group theoretic approach may help to structure computations and to discern physical from numerical properties when applying nonlinear parameters to rigid-body motion. In the existing literature, Lie groups are either described on an abstract level (see, e.g., Choquet Bruhat, DeWitt Morette, 1989, Varadarajan, 1984), or they are related to rigid-body motions for the case of linear or matrix representations only (see, e.g., McCarthy, 1990). This has made it difficult to apply them to other types

J. Lenarčič and C. Galletti (eds.), On Advances in Robot Kinematics, 57-66.

of parametrizations. The present paper revises the first-order Lie-group theoretic properties related to rigid-body motion using kinematically intuitive notions. While no new algorithms are derived in this context, a framework is provided that allows one to discern physical from numerical properties of parametrizations. Such concepts can be of use for example for object-oriented, implementation-independent kinematic libraries that are based on generic properties at the superclass level and individual parametrization computations at the subclass level. These generic properties comprise: (1) the existence of tangent elements at the identity element, which correspond to "physical" notions such as angular velocity and velocity twists, and (2) the description of the basic differential mappings between parameter space and point space as a two-step procedure involving only derivatives at the identity element, leading to the notions of left and right translations or body-fixed and inertially-fixed representations, respectively. The corresponding intrinsic operations are developed independently of parametrization. An application of the generic concepts to concrete parametrizations is presented in Kecskemethy, 2004 for the case of rigid-body rotations.

2. Generic Lie-Group Operations

A *Lie group* G is a set of elements embodying simultaneously the properties of a group and a smooth manifold, Hamermesh, 1964 (Fig. 1). By

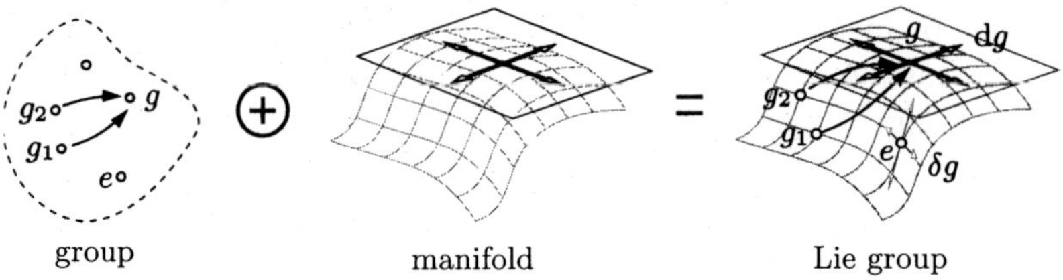

Figure 1. Basic structure of a Lie group

the group property, (1) *composition* of any two elements $g_1, g_2 \in G$ is defined, yielding a composite element g that belongs also to G; (2) there exists a neutral element e in G such that the composition of e with any other element g yields the element g itself, and (3) for any element g in G there exists an inverse g^{-1} such that the composition of g and g^{-1} yields the neutral element. By the manifold property, any two points in G can be connected by a smooth trajectory, and at any point g one can define a differential $\mathrm{d}g$ that is tangent to G. Of these, differentials δg at the neutral element play a special role: they represent infinitesimal deviations from the identity that act as incremental perturbations from an actual position, such as angular velocities and twist vectors. Besides acting on themselves, Lie groups can also have an effect on other objects, such as it is the case for rigid-body motions. They are then termed *Lie groups of transformations* (Lie, 1896). The space on which the group elements act is termed the point space X. When group elements along a trajectory $C_g(t) \in G$ are applied to X, points $x \in X$ move smoothly along trajectories C'_x under the action of the elements $g \in C_g(t)$ (Fig. 2). The relationships between points and tangent vectors along these trajectories is the topic of first-order Lie group analysis.

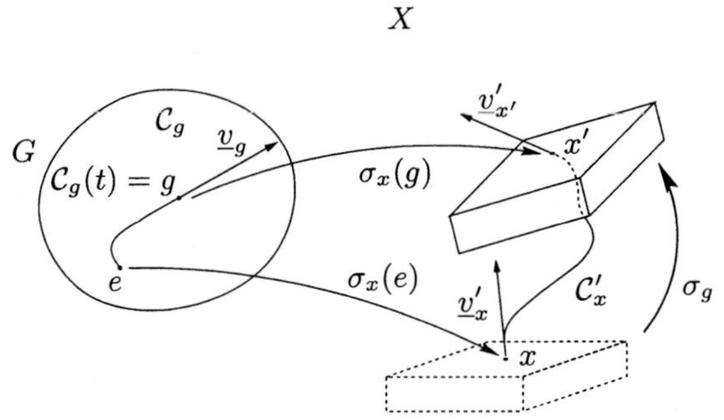

Figure 2. Action of a Lie group G of transformations on the point space X

Composition and Transformation Functions. Lie groups of transformation involve two characteristic functions: (a) a *group composition function*

$$\phi : \; G \times G \rightarrow G : \; (g_1 , g_2) \mapsto g = \phi(g_1 , g_2) \in G \; ; \qquad (1)$$

with which the resulting transformation of a composition of two individual transformations g_1, g_2 is computed, and (b) a *transformation* or *point-mapping function*

$$\sigma : \; G \times X \rightarrow X \; ; \; (g , x) \mapsto x' = \sigma(g , x) \; ; \qquad (2)$$

where the image x' of a point $x \in X$ under the action of a transformation g is computed. From this two-argument function, two further monadic functions can be extracted by fixing one of the arguments (Fig. 2)

$$\begin{array}{llll} \text{the } \textit{space mapping} & \sigma_g : X \rightarrow X & ; & x \mapsto \sigma_g(x) \equiv \sigma(g , x) \\ \text{the } \textit{point mapping} & \sigma_x : G \rightarrow X & ; & g \mapsto \sigma_x(g) \equiv \sigma(g , x) \end{array} \qquad (3)$$

The space mapping σ_g maps all points of X to their new position; hence σ_g represents the *pose* of the point space under the action of group element g. On the other hand, the point mapping σ_x describes *trajectories* of points $x \in X$ as images of corresponding trajectories $g(t)$ in G. *Linear parametrizations* realize the functions (1) and (2) by purely linear mappings both in g and x. They give rise to the matrix representations of rigid-body motions ($SO(3)$ or orthogonal matrices for pure rotations, homogeneous 4×4 matrices for general rigid-body motions). It can be shown (Choquet Bruhat, DeWitt Morette, 1989) that *any* Lie group of transformations allows for a linear parametrization.

3. Differential Relationships

Differential relationships relate small parameter variations within the group G to tangent vectors along a trajectory in point space X. Regard the rigid-body motion as being generated by a curve $\mathcal{C}_g : \mathbb{R} \to G : \tau \mapsto g(\tau)$ in G which passes through the point g for $\tau = t$ (Fig. 2). Denote by $\underline{v}_g$ a differential variation along this curve, i.e., an increment in direction of the tangent vector of the trajectory $\mathcal{C}_g(t)$ at g

$$\underline{v}_g = \left.\frac{\mathrm{d}\,\mathcal{C}_g(\tau)}{\mathrm{d}\tau}\right|_{\tau=t} \mathrm{d}\tau \ . \tag{4}$$

As the trajectory $\mathcal{C}_g(\tau)$ is mapped through the point mapping $\sigma_x : G \to X$ to a corresponding trajectory $\mathcal{C}'_x(\tau)$ in X, the image $\underline{v}'_{x'}$ in X of the tangent vector $\underline{v}_g$ under the action of σ_x yields the tangent vector in X

$$\underline{v}'_{x'} = \left.\frac{\mathrm{d}\sigma_x(\mathcal{C}_g(\tau))}{\mathrm{d}\tau}\right|_{\tau=t} \mathrm{d}\tau = \left.\frac{\partial\sigma_x(\mu)}{\partial\mu}\right|_{\mu=g} \left.\frac{\mathrm{d}\,\mathcal{C}_g(\tau)}{\mathrm{d}\tau}\right|_{\tau=t} \mathrm{d}\tau = \mathrm{D}\sigma_x(g)\,\underline{v}_g \ . \tag{5}$$

This gives rise to the Jacobian of the point mapping function

$$\mathrm{D}\sigma_x(g) = \left.\frac{\partial\sigma_x(\mu)}{\partial\mu}\right|_{\mu=g} , \tag{6}$$

which maps tangent vectors $\underline{v}_g$ in parameter space to tangent vectors $\underline{v}'_{x'}$ in point space evaluated at the image x' of a point x. This Jacobian may be difficult to compute at a general group element g, while it is trivial at the neutral element. By the Lie group approach, it can be shown that it is always to decompose this Jacobian into two intermediate Jacobians evaluated at the identity transformation, simplifying computations.

One-Parameter Subgroups. A one-parameter subgroup is a special trajectory $\mathcal{C}_g(t)$ within the group G which has the property that the composition of any two elements within the trajectory is again an element of the trajectory. A key property of one-parameter subgroups is that they are commutative (Choquet Bruhat, DeWitt Morette, 1989), i.e, $\mathcal{C}_g(t)\,\mathcal{C}_g(s) = \mathcal{C}_g(s)\,\mathcal{C}_g(s) = \mathcal{C}_g(s+t)$ for any t and s. Choosing parameter τ such that $\mathcal{C}_g(\tau) = e$ for $\tau = 0$, one can factor differentials in the neighborhood of t as

$$\left.\frac{\mathrm{d}\mathcal{C}_g(\tau)}{\mathrm{d}\tau}\right|_{\tau=t} = \left.\frac{\mathrm{d}\mathcal{C}_g(t+\tau)}{\mathrm{d}\tau}\right|_{\tau=0} = \mathcal{C}_g(t)\left.\frac{\mathrm{d}\mathcal{C}_g(\tau)}{\mathrm{d}\tau}\right|_{\tau=0} = \left.\frac{\mathrm{d}\mathcal{C}_g(\tau)}{\mathrm{d}\tau}\right|_{\tau=0} \mathcal{C}_g(t) \ .$$

Introducing the tangent vector $\underline{\gamma}$ at the identity element e renders

$$\underline{\gamma} = \left.\frac{\mathrm{d}\mathcal{C}_g(\tau)}{\mathrm{d}\tau}\right|_{\tau=0} \tag{7}$$

$$\underline{v}_g = g\,\underline{\gamma}\,\mathrm{d}\tau = \underline{\gamma}\,g\,\mathrm{d}\tau \iff \underline{\gamma}\,\mathrm{d}\tau = g^{-1}\,\underline{v}_g = \underline{v}_g\,g^{-1} \ . \tag{8}$$

Thus the tangent vector at the finite element g results from the commutative product of the group element with the tangent vector at the neutral element. These properties are independent of parametrization.

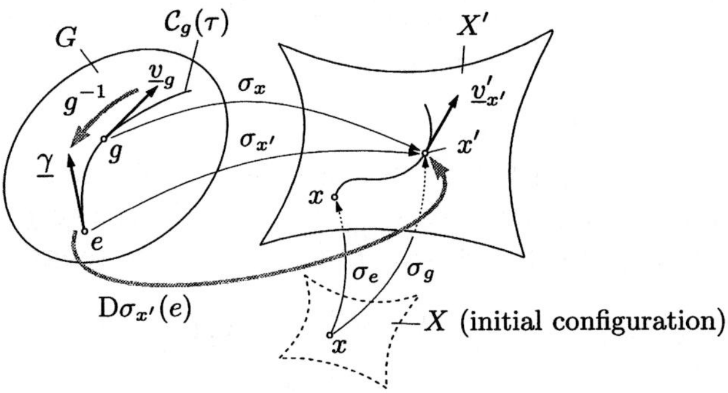

Figure 3. Basic differential mappings for a one-parameter Lie subgroup

In point space, the differential $\underline{v}_g$ of a group element gives rise to a differential $\underline{v}'_{x'}$ (Fig. 3) that is tangent to the trajectory of point x at a current location $x' = \sigma(\mathcal{C}_g(t), x)$. By the compatibility condition for composition and transformation functions and the commutativity of one-parameter motion, one has (Hamermesh, 1964)

$$\sigma(\mathcal{C}_g(t+\tau), x) = \sigma(\mathcal{C}_g(\tau)\,\mathcal{C}_g(t), x) = \sigma(\mathcal{C}_g(\tau), \sigma(g, x)) = \sigma(\mathcal{C}_g(\tau), x') \ ,$$

$$\left.\frac{\mathrm{d}\sigma(\mathcal{C}_g(\tau), x)}{\mathrm{d}\,\tau}\right|_{\tau=t} = \left.\frac{\mathrm{d}\sigma(\mathcal{C}_g(\tau), x')}{\mathrm{d}\,\tau}\right|_{\tau=0} = \left.\frac{\partial\,\sigma(\mu, x')}{\partial\mu}\right|_{\mu=e} \left.\frac{\mathrm{d}\mathcal{C}_g(\tau)}{\mathrm{d}\tau}\right|_{\tau=0} \ .$$

Thus it follows

$$\underline{v}'_{x'} = \left.\frac{\mathrm{d}\sigma(\mathcal{C}_g(\tau), x)}{\mathrm{d}\,\tau}\right|_{\tau=t} \mathrm{d}\tau = \mathrm{D}\sigma_{x'}(e)\,\underline{\gamma}\,\mathrm{d}\tau \ , \tag{9}$$

where $\mathrm{D}\sigma_{x'}(e)$ is the point-mapping Jacobian at the neutral element

$$\mathrm{D}\sigma_{x'}(e) = \left.\frac{\partial\sigma(\mu, x')}{\partial\mu}\right|_{\mu=e} \ . \tag{10}$$

Eqs.(8) and (9) render the mapping of group parameter differentials $\underline{v}_g$ to tangent vectors in point space by a two-step procedure via the tangent vector $\underline{\gamma}$ at the identity element

$$\underline{v}_g \overset{g^{-1}}{\longrightarrow} \underline{\gamma}\,\mathrm{d}\tau \overset{\mathrm{D}\sigma_{x'}(e)}{\longrightarrow} \underline{v}'_{x'} \tag{11}$$

This not only simplifies calculations, but also introduces the notion of *infinitesimal transformation increments* $\underline{\gamma}$, which are uniquely determined infinitesimal perturbations of the identity transformation related to differentials at the current position, such as angular velocities or twists.

Exponential Representation. For *linear representations* of the Lie group, one-parameter subgroups give rise to a simple analytic expression. Denote by $[\,g\,]$ the matrix representation of a group element. Then, the trajectory $\mathcal{C}_g(t)$ can be regarded as a solution of the initial value problem

$$\frac{\mathrm{d}[\,\mathcal{C}_g(t)\,]}{\mathrm{d}t} = [\,\underline{v}_g(t)\,] \;\;,\;\; \mathcal{C}_g(0) = e \;\;, \tag{12}$$

which with Eq. (8) can be rewritten as

$$\frac{\mathrm{d}[\,\mathcal{C}_g(\tau)\,]}{\mathrm{d}t} = [\,\underline{\gamma}\,][\,\mathcal{C}_g(t)\,] \;\;,\;\; [\,\mathcal{C}_g(0)\,] = [\,e\,] = \mathbf{I} \;\;, \tag{13}$$

where $[\mathcal{C}_g(\tau)]$ and $[\underline{\gamma}]$ are matrices. This is a linear differential equation in $[\mathcal{C}_g(\tau)]$, whose solution can be established via the matrix-exponential function known also in control theory (see, e.g., Belanger, 1995)

$$[\,\mathcal{C}_g(t)\,] = \mathrm{e}^{[\,\underline{\gamma}\,]t}\,[\,\mathcal{C}_g(0)\,] \tag{14}$$

$$\mathrm{e}^{[\,\underline{\gamma}\,]t} = \mathbf{I} + [\,\underline{\gamma}\,]t + \frac{1}{2}[\,\underline{\gamma}\,]^2 t^2 + \cdots + \frac{1}{n!}[\,\underline{\gamma}\,]^n t^n + \cdots \tag{15}$$

where e denotes here the natural number. Thus matrix representations of a one-parameter Lie group of transformations (but *only* these) can be described as an exponential series based on the tangent vector at the unit element. This tangent vector is hence denoted as the *generator* of the one-parameter subgroup. The matrix representations of rigid-body motions are SO(3) (Special Orthogonal Group) for the case of pure rotations and SE(3) (Special Euclidean Group) for the case of general rigid-body motions. For these cases, the generators are the unit rotation vector and the screw axis, and the exponential series representation renders Euler's and Chasles' theorem, respectively. The commutativity relationship Eq. (8) shows that the infinitesimal generators are always eigenvectors of the matrix representing the transformation.

General Motions. For general trajectories $\mathcal{C}_g(t)$, the assumptions of one-parameter subgroup motions do not hold anymore, as the composition of two transformations along a general trajectory will in general not belong to the trajectory. However, one can regard the neighborhood of a point $g \in \mathcal{C}_g(t)$ as being assembled from a finite transformation g and a small variable transformation $h(t)$ along a one-parameter subgroup. This assembly is termed a "*translation*" of $h(t)$, which can be accomplished in one of two ways (Fig. 4):

$$\begin{array}{lllll} \text{(a)} & \underline{\textit{left translation:}} & L_g(h) & = & g\,h = \phi(\,g\,,\,h\,)\;, \\ \text{(b)} & \underline{\textit{right translation:}} & R_g(h) & = & h\,g = \phi(\,h\,,\,g\,)\;. \end{array} \tag{16}$$

In the *left translation*, the variable transformation $h(t)$ is *appended* to the finite (actual) position B. This corresponds to the *body-fixed* representation, as variations are measured with respect to the actual position B. In the *right translation*, the variable transformation $h(t)$ is *pre-inserted* to the finite one. This corresponds to the *spatially-fixed* representation, as variations are measured with respect to the inertial frame A.

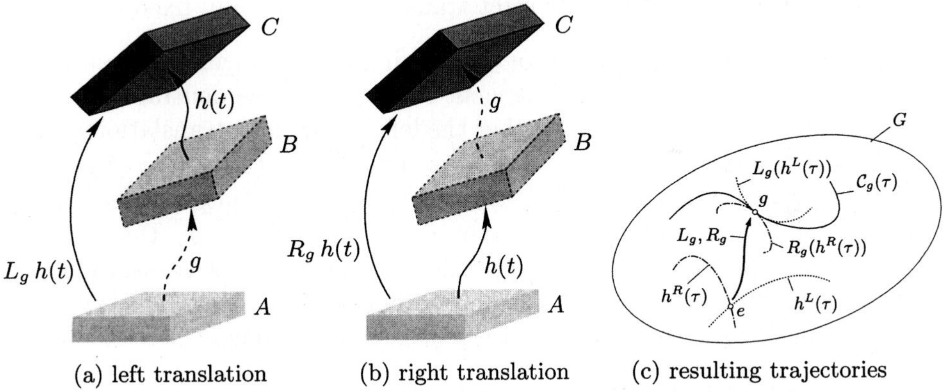

Figure 4. Assembly of a finite element g and a one-parameter subgroup $h(t)$

Let the path parameter τ be defined such that $\mathcal{C}_g(\tau)$ passes through g at the origin, i.e., $\mathcal{C}_g(0) = g$. One can match $\mathcal{C}_g(\tau)$ in the neighborhood of $\mathcal{C}_g(0)$ to first order by composing a fixed finite g with a small piece of the one-parameter subgroup $h(\tau)$. Depending on whether left or right translation is used, one obtains two possible one-parameter subgroups $h^L(\tau)$ and $h^R(\tau)$ (Fig. 4):

$$\mathcal{C}_g(\tau) \approx \begin{cases} L_g(h^L(\tau)) & = \quad \phi(g, h^L(\tau)) \quad \text{for left translation} \\ R_g(h^R(\tau)) & = \quad \phi(h^R(\tau), g) \quad \text{for right translation} \end{cases} \tag{17}$$

The tangent vector in the group space at a general point g can be obtained by evaluating the differential of the trajectory at $\tau = 0$ with the assumption that for the one-parameter subgroups $h^L(0) = h^R(0) = e$:

$$\underline{v}_g = \left.\frac{\mathrm{d}\,\mathcal{C}_g(\tau)}{\mathrm{d}\tau}\right|_{\tau=0} \mathrm{d}\tau = \begin{cases} \left.\dfrac{\partial L_g(h)}{\partial h}\right|_{h=e} \left.\dfrac{\mathrm{d}h^L(\tau)}{\mathrm{d}\tau}\right|_{\tau=0} \mathrm{d}\tau & \text{for left translation} \\[2ex] \left.\dfrac{\partial R_g(h)}{\partial h}\right|_{h=e} \left.\dfrac{\mathrm{d}h^R(\tau)}{\mathrm{d}\tau}\right|_{\tau=0} \mathrm{d}\tau & \text{for right translation} \end{cases}$$

After setting

$$\mathrm{D}L_g(e) = \left.\frac{\partial L_g(h)}{\partial h}\right|_{h=e} , \quad \underline{\gamma}^L = \left.\frac{\mathrm{d}h^L(\tau)}{\mathrm{d}\tau}\right|_{\tau=0} \quad \text{for left translation}$$

$$\mathrm{D}R_g(e) = \left.\frac{\partial R_g(h)}{\partial h}\right|_{h=e} , \quad \underline{\gamma}^R = \left.\frac{\mathrm{d}h^R(\tau)}{\mathrm{d}\tau}\right|_{\tau=0} \quad \text{for right translation}$$

this can be written in compact form as

$$\underline{v}_g = \begin{cases} \mathrm{D}L_g(e) \cdot \underline{\gamma}^L \, \mathrm{d}\tau & \text{for left translation (body fixed)} \\ \mathrm{D}R_g(e) \cdot \underline{\gamma}^R \, \mathrm{d}\tau & \text{for right translation (inertially fixed)} \end{cases} \tag{18}$$

This describes the tangent vector $\underline{v}_g$ at a finite group element as a linear mapping of a uniquely defined tangent element $\underline{\gamma}$ at the neutral element, whose direction depends on whether the left or the right translation (i.e., body-fixed or inertial representation) is chosen. For the action of group parameter variations on point space, one needs to evaluate the Jacobian $\mathrm{D}\sigma_x(g)$ of Eq. (10) at a general group element $\mathcal{C}_g(t) = g$. Depending on whether the expansion of $\mathcal{C}_g(t)$ is done by left or right translation of the one-parameter subgroup $h(t)$, two different Jacobians $\mathrm{D}\sigma^L_{x'}(g)$ and $\mathrm{D}\sigma^R_{x'}(g)$ arise. For the case of right translation, it holds:

$$\left.\frac{\partial \sigma(\phi(h,g)\,,\,x)}{\partial h}\right|_{h=e} = \left.\frac{\partial \sigma(\mu, x)}{\partial \mu}\right|_{\mu=g} \left.\frac{\partial \phi(h,g)}{\partial h}\right|_{h=e} .$$

As $\sigma(\phi(h,g),x) \equiv \sigma(h,\sigma(g,x)) \equiv \sigma(h,x')$, the left side simplifies to

$$\left.\frac{\partial \sigma(\phi(h,g)\,,\,x)}{\partial h}\right|_{h=e} = \left.\frac{\partial \sigma(\mu\,,\,x')}{\partial \mu}\right|_{h=e} \equiv \mathrm{D}\sigma_{x'}(e) \ . \tag{19}$$

The first factor on the right side of Eq. (4) is equal to the seeked Jacobian $\mathrm{D}\sigma^R_{x'}(g)$, while the second factor corresponds to the Jacobian $\mathrm{D}R_g(e)$ derived above. Hence, one obtains for right translation or inertial representation

$$\mathrm{D}\sigma^R_{x'}(g) = \mathrm{D}\sigma_{x'}(e)\,[\mathrm{D}R_g(e)]^{-1} \ . \tag{20}$$

A similar derivation yields for the *left translation*

$$\mathrm{D}\sigma_x^L(g) = \mathrm{D}\sigma_x(e) \circ [\mathrm{D}L_g(e)]^{-1} \quad , \tag{21}$$

which describes the result in the *body-fixed* frame. The transformation sequence shows that the overall mapping from differentials of group parameters to differentials of point trajectories can be decomposed in two steps (Fig. 5): first, differentials of group parameters are mapped to infinitesimal perturbations of the identity transformation, which, depending on whether the left (body-fixed) or right (inertially-fixed) translation is used, yields different tangent vectors $\underline{\gamma}^L$ and $\underline{\gamma}^R$. Then, from these, differentials of point trajectories are established. Hereby, only derivatives at the neutral element are required. The tangent vectors at the neutral element correspond to "physical" quantities such as angular velocities or velocity twists, which can be evaluated in any parametrization.

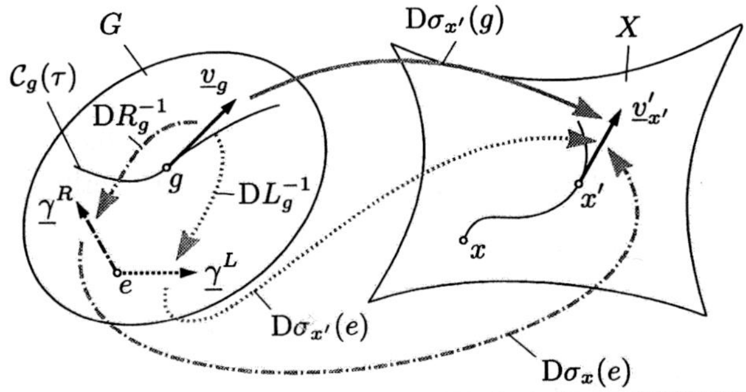

$\underline{v}_g$	$\xrightarrow{[\mathrm{D}L_g(e)]^{-1}}$	$\underline{\gamma}^L\,\mathrm{d}\tau$	$\xrightarrow{\mathrm{D}\sigma_x(e)}$	$\underline{v}'_{x'}$	(body-fixed representation)
$\underline{v}_g$	$\xrightarrow{[\mathrm{D}R_g(e)]^{-1}}$	$\underline{\gamma}^R\,\mathrm{d}\tau$	$\xrightarrow{\mathrm{D}\sigma_{x'}(e)}$	$\underline{v}'_{x'}$	(inertially-fixed representation)

Figure 5. Mappings of tangential vectors from G to X

4. Conclusions

The paper regards a Lie-group theoretic interpretation of first-order rigid-body kinematics that can be used to derive the main operations of first-order rigid-body kinematics independently of parametrization. As shown, the mapping from differentials in parameter space to differentials in point space can always be accomplished through a two-step procedure involving only derivatives at the neutral element. These derivatives are not only more simple to evaluate, but they also characterize the relationship between differentials of transformation parameters at a finite element and those at the identity element. As differentials at the identity element represent physical notions like angular velocity or twist, in this way the relationship between numerical and physical properties of a parametrization can be established. Such generic concepts could be used to realize object-oriented kinematics libraries. Application of these formulas to rotational parameters is shown in Kecskemethy, 2004.

5. Acknowledgments

Support of this work by the German Research Foundation under grant Ke 526/3-1 is gratefully acknowledged. The paper is part of a dedication to Professor Bernard Roth on occasion of the Special Celebratory Symposium in honor of his 70th birthday held in Stanford, 2003, in appreciation for his invaluable contributions in the field of kinematics.

References

Angeles, J. (1988), *Rational Kinematics.* Springer Tracts in Natural Philosophy 34. Springer-Verlag, New York.

Bélanger, P.R. (1995), *Control Engineering: A Modern Approach.* Saunders College Publishing, Orlando.

Bottema, O., and Roth, B. (1970), *Theoretical Kinematics.* Applied Mathematics and Mechanics 24. North-Holland Publishing Company, Amsterdam, Oxford, New York.

Chevallier, D.P. (1991), Lie algebras, modules, dual quaternions and algebraic methods in kinematics. *Mechanism and Machine Theory,* 26(6):613–627.

Choquet-Bruhat, Y., and DeWitt-Morette, C. (1989), *Analysis, Manifolds and Physics. Part I: Basics.* North-Holland, Amsterdam, New York.

Hamermesh, M. (1964), *Group Theory and its Applications to Physical Problems.* Addison-Wesley Series in Physics. Addison-Wesley Publishing Company.

Isidori, A. (1989), *Nonlinear Control Theory.* Springer-Verlag, Berlin, New York.

Karger, A., and Novák, J. (1985), *Space Kinematics and Lie Groups.* Gordon and Breach Science Publishers, Montreux, Schweiz.

Kecskeméthy, A. (2004), *First-order intrinsic properties of the rotation parameters SO(3), Quaternion and rotation vector.* 9th International Symposium on Advances in Robot Kinematics June 28 - July 1, 2004, Sestri Levante.

Lie, S. (1896). *Theorie der Berührungstransformationen.* Leipzig.

McCarthy, J.M. (1990), *Introduction to Theoretical Kinematics.* Robotics Science. The MIT Press, Cambridge, Massachusetts.

Park, F.C., Bobrow, S.R., and Ploen, A. (1995), A Lie group formulation of robot dynamics. *International Journal of Robotics Research,* 14(6): 609-618 (1995)

Varadarajan, V.S. (1984), *Lie Groups, Lie Algebras, and Their Representations.* Graduate Texts in Mathematics, 102. Springer-Verlag, New York.

FIRST-ORDER INTRINSIC PROPERTIES OF THE ROTATION PARAMETERS SO(3), QUATERNION AND ROTATION VECTOR

Andrés Kecskeméthy
Chair for Mechanics, Faculty of Engineering
University Duisburg-Essen, D-47048 Duisburg, Germany
a.kecskemethy@uni-duisburg.de

Abstract The present paper compares the intrinsic properties of the first-order relationships for the three parametrizations of rotations SO(3) or matrix representation, the Euler-parameter or Quaternion parametrization, and the rotational-vector description. It is shown that the condition numbers, i.e., space distortions for the first-order transformation matrices, are optimal for SO(3) and Euler parameters, but poor for the rotational-vector approach.

Keywords: Rotations, Lie groups

1. Introduction

Kinematics of rigid-body rotation involves some well-known intuitive terms such as rotation matrices and angular velocities, which can be found in classical kinematics textbooks (see, e.g., Angeles, 1988, Bottema and Roth, 1970). On the other hand, quantities such as time-derivatives of the rotation parameters are still considered to be of "algorithmic" nature, making their kinematic interpretation difficult. Application of Lie group theory (see, e.g., Varadarajan, 1984) to kinematics has shown that by an intrinsic view of the ensuing operations, insight can be found in some design and analysis issues (Chevallier, 1991, Karger and Novák, 1985, Park, 1995). However, the intrinsic relationships between differentials of motion parameters and velocities in point space has not been addressed up to date, making is difficult to discern between physical and numerical properties. In this paper, the first-order differential properties of the three rotation parameters SO(3) (3×3 orthogonal matrices), Euler-Rodrigues and rotation vectors are analyzed using Lie group operations. In is shown that the condition numbers of the intrinsic Jacobians may differ from one parametrization to the other, hence leading to intrinsic numerical problems that are independent of physics. The derivations can be easily extended to other types of parametrizations.

J. Lenarčič and C. Galletti (eds.), On Advances in Robot Kinematics, 67-76.

2. Intrinsic Mappings of First-Order Kinematics

The operations arising in first-order kinematics of rigid-body motion can be summarized by regarding the transmission of tangent vectors $\underline{v}_g$ in motion parameter space G (e.g., Euler angles, Quaternions, rotation vector) to tangent vectors $\underline{v}'_{x'}$ in point space X. By considering the intrinsic property of the set of rigid-body motions to form a Lie group of transformations (Kecskeméthy, 2004), one obtains for this transmission

$$\underline{v}'_{x'} = \mathrm{D}\sigma_x(g)\,\underline{v}_g \quad , \tag{1}$$

where $\underline{v}_g$ is an infinitesimal change of rotation parameters, g is the actual value for the rotation parameters, $\underline{v}'_{x'}$ is the corresponding tangent vector for the point trajectory at the current point location x' , and $\mathrm{D}\sigma_x(g) = \partial\sigma_x(\mu)\,/\,\partial\mu|_{\mu=g}$ is the Jacobian of the function $\sigma_x(g)$ mapping rotation parameter variations to point location variations for one point $x \in X$. Depending on whether a body-fixed or inertially-fixed representations is seeked, two possible decompositions result for the aforementioned Jacobian:

$$\mathrm{D}\sigma^R_{x'}(g) = \mathrm{D}\sigma_{x'}(e)\,[\mathrm{D}R_g(e)]^{-1} \quad . \tag{2}$$

$$\mathrm{D}\sigma^L_{x}(g) = \mathrm{D}\sigma_{x}(e)\circ[\mathrm{D}L_g(e)]^{-1} \quad , \tag{3}$$

where $\mathrm{D}L_g(e), \mathrm{D}R_g(e)$ are the intermediate Jacobians mapping infinitesimal transformation increments $[\,\underline{\gamma}\,]$ at the identity, such as angular velocity vectors, to corresponding differentials $\underline{v}_g$ of the parameters at a finite position via left and right translation (Kecskeméthy, 2004).

In order to assess relative precision of the ensuing kinematic transmission matrices, one can apply the notion of the spectral condition number of a matrix A (Golub and Van Loan, 1983)

$$\mathrm{cond}(A) = |\lambda_{\mathrm{max}}(A)|\;/\;|\lambda_{\mathrm{min}}(A)| \tag{4}$$

where $\lambda_{\mathrm{max}}(A)$ and $\lambda_{\mathrm{min}}(A)$ denote the maximum and minimum eigenvalues of A, respectively. This is carried out below for three specific rotation parametrizations.

3. SO(3) Matrix Representation

The SO(3) parametrization of rotations is *the* linear representation of rotations, named after their description by 3×3 orthogonal transformation matrices $\mathbf{R}$ with

$$\mathbf{R}\,\mathbf{R}^{\mathrm{T}} = \mathbf{R}^{\mathrm{T}}\,\mathbf{R} = \mathbf{I}_3 \tag{5}$$

The *tangent vector* at a *finite* position is a matrix of differentials

$$\underline{v}_g \overset{SO(3)}{\equiv} \mathsf{d}\mathbf{R} = \begin{bmatrix} \mathrm{d}\rho_{11} & \mathrm{d}\rho_{12} & \mathrm{d}\rho_{13} \\ \mathrm{d}\rho_{21} & \mathrm{d}\rho_{22} & \mathrm{d}\rho_{23} \\ \mathrm{d}\rho_{31} & \mathrm{d}\rho_{32} & \mathrm{d}\rho_{33} \end{bmatrix} . \tag{6}$$

This matrix does not render an orthogonal matrix when added to the identity matrix, hence clearly $\mathrm{d}\mathbf{R}$ is not the perturbation part of a "small" rotation. Such a small rotation matrix can be assembled for example by inserting small Cardan angles $\delta\phi, \delta\theta, \delta\psi$ in the general yaw-pitch-roll (x-y-z) form $\mathbf{R}^{(RPY)}$ of the rotation matrix and expanding up to the linear terms. It follows

$$\mathbf{R}^{(RPY)}(\delta\phi, \delta\theta, \delta\psi) = \mathbf{I}_3 + \begin{bmatrix} 0 & -\delta\psi & \delta\theta \\ \delta\psi & 0 & -\delta\phi \\ -\delta\theta & \delta\phi & 0 \end{bmatrix} + O([\delta\phi + \delta\theta + \delta\psi]^2) . \tag{7}$$

from where one obtains for the seeked infinitesimal rotation increment

$$\delta\mathbf{R} = \begin{bmatrix} 0 & -\delta\psi & \delta\theta \\ \delta\psi & 0 & -\delta\phi \\ -\delta\theta & \delta\phi & 0 \end{bmatrix} . \tag{8}$$

Hence, tangent vectors at the identity element of the SO(3) parametrization are skew-symmetric matrices. From $\delta\mathbf{R}$ one obtains, after dividing by $\delta\varphi = \sqrt{\delta\phi^2 + \delta\theta^2 + \delta\psi^2}$, the unit tangent vector at the identity

$$[\,\underline{\gamma}\,] = \begin{bmatrix} 0 & -u_z & u_y \\ u_z & 0 & -u_x \\ -u_y & u_x & 0 \end{bmatrix} \quad , \quad u_x^2 + u_y^2 + u_z^2 \equiv 1 . \tag{9}$$

The three non-vanishing components of the skew-symmetric matrices can be assembled in vectors using the operator "vect$(\cdot)$" (Angeles, 1988)

$$\delta\underline{\varphi} = \begin{bmatrix} \delta\phi \\ \delta\theta \\ \delta\psi \end{bmatrix} \doteq \mathrm{vect}(\delta\mathbf{R}) \quad , \quad \underline{u} = \begin{bmatrix} u_x \\ u_y \\ u_z \end{bmatrix} \doteq \mathrm{vect}([\,\underline{\gamma}\,]) \quad , \tag{10}$$

where $\delta\underline{\varphi}$ is the vector of an infinitesimal rotation and $\underline{u}$ is the unit vector in the corresponding direction. Dividing the infinitesimal rotation vector by the corresponding time increment $\mathrm{d}t$ renders the angular velocity $\underline{\omega} = \dot{\varphi}\,\underline{u}$.

As composition of rotation matrices is equal to their products, the relationship between the differential $\mathrm{d}\mathbf{R}$ of the rotation matrix and the right and left translations $L_g(h), R_g(h)$ of an infinitesimal rotation is simply the linear matrix multiplication (see Table 1)

$$\mathrm{d}\mathbf{R}_g = \delta\mathbf{R}^R\,\mathbf{R}_g \;\; ; \;\; \mathrm{d}\mathbf{R}_g = \mathbf{R}_g\delta\,\mathbf{R}^L \;\; . \tag{11}$$

With the orthogonality of $\mathbf{R}$ and the linearity of composition it follows

$$\mathrm{cond}\left(\mathrm{D}R_g^{-1}(\cdot)\right) = \mathrm{cond}\left(\mathrm{D}L_g^{-1}(\cdot)\right) = \mathrm{cond}\,(\mathbf{R}) = 1 \;\; . \tag{12}$$

Table 1. Differential mappings for SO(3)

	left translation (body fixed)	right translation (inertially fixed)
differential at general element	$\underline{v}_g\,\mathrm{d}\tau \equiv \mathrm{d}\mathbf{R}_g$	$\underline{v}_g\,\mathrm{d}\tau \equiv \mathrm{d}\mathbf{R}_g$
differential at neutral element	$\underline{\gamma}^L\,\mathrm{d}\tau \equiv \delta\mathbf{R}^L$	$\underline{\gamma}^R\,\mathrm{d}\tau \equiv \delta\mathbf{R}^R$
translation	$L_g(h) \equiv \mathbf{R}_g\,\mathbf{R}_h$	$R_g(h) \equiv \mathbf{R}_h\,\mathbf{R}_g$
translation differential	$\mathrm{D}L_g(\delta\mathbf{R}^L) \equiv \mathbf{R}_g\,\delta\mathbf{R}^L$	$\mathrm{D}R_g(\delta\mathbf{R}^R) \equiv \delta\mathbf{R}^R\,\mathbf{R}_g$
inverse translation differential	$\mathrm{D}L_g^{-1}(\mathrm{d}\mathbf{R}_g) \equiv \mathbf{R}_g^{\mathrm{T}}\,\mathrm{d}\mathbf{R}_g$	$\mathrm{D}R_g^{-1}(\mathrm{d}\mathbf{R}_g) \equiv \mathrm{d}\mathbf{R}_g\,\mathbf{R}_g^{\mathrm{T}}$

Due to their skew-symmetry, application of tangent vectors of SO(3) at the identity element to points $\underline{x}'$ corresponds to the cross product of the infinitesimal rotation vector with the radius vector $\underline{p}$ from the rotation center to the actual position of point $\underline{x}'$. Comparing this to the generic formula (Kecskeméthy, 2004) yields

$$\mathrm{d}\underline{x} = \mathrm{D}\sigma_{x'}(e)\,\underline{\gamma}\,\delta\varphi \overset{\mathrm{SO(3)}}{\equiv} \underline{u}\times\underline{p}\,\delta\varphi \tag{13}$$

Hence $\mathrm{D}\sigma_{x'}(e)$ is the negative cross-product operator of radius vector $\underline{p}$

$$\mathrm{D}\sigma_{x'}(e) \overset{\mathrm{SO(3)}}{\equiv} -\underline{\widetilde{p}} = -\begin{bmatrix} 0 & -p_z & p_y \\ p_z & 0 & -p_x \\ -p_y & p_x & 0 \end{bmatrix} \;\; . \tag{14}$$

Calculating the condition number of a skew-symmetric matrix $\underline{\widetilde{p}}$ is not meaningful as such a matrix has an eigenvalue zero. This is consistent with the property that resolving Eq. (13) with respect to $\underline{u}$ is only unique for components $\underline{u}^\perp$ perpendicular to $\underline{p}$. Restricting the resolution to such perpendicular components $\underline{u}^\perp$, i.e., to the range of $\underline{\widetilde{p}}$, gives a regular mapping. This is equivalent to formulating a constraint $\underline{p}^{\mathrm{T}}\underline{u}^\perp = 0$ as additional equation. Let $\underline{e} = \underline{p}/p$, where p is the length of radius vector $\underline{p}$. Then, Eq. (13) is equivalent to

$$-p\,\widehat{\mathbf{P}}\,\underline{\widehat{u}}^\perp = -p\begin{bmatrix} 0 & \underline{e}^{\mathrm{T}} \\ \underline{e} & \underline{\widetilde{e}} \end{bmatrix}\begin{bmatrix} 0 \\ \underline{u}^\perp \end{bmatrix} = \begin{bmatrix} 0 \\ \mathrm{d}\underline{x} \end{bmatrix} \tag{15}$$

where $\widehat{\underline{u}}^{\perp}$ is vector $\underline{u}^{\perp}$ with a 0 as additional first element, and matrix $\widehat{\mathbf{P}}$ is obviously orthogonal. Hence, in the given range, the condition number of $\mathrm{D}\sigma_{x'}(e)$ is equal to 1. This restricted-range condition number can be regarded to be the characteristic condition number of the mapping $\mathrm{D}\sigma_{x'}(e)$, as there is no space distortion involved by $\mathrm{D}\sigma_{x'}(e)$, and the singularity of the resolution can be circumvented by taking the differentials $d\underline{x}$ at two different points. By combining point-space and group-composition Jacobian one obtains for the final transmission from differentials of group parameters to differentials in point space

$$\underline{v}'_{x'} = \begin{cases} \operatorname{vect}(\delta\mathbf{R}^L) \times \underline{p} & = \quad \operatorname{vect}(\mathbf{R}_g^{\mathrm{T}}\,\mathrm{d}\mathbf{R}) \times \underline{p} \quad \text{(for left translation)} \\ \operatorname{vect}(\delta\mathbf{R}^R) \times \underline{p}' & = \quad \operatorname{vect}(\mathrm{d}\mathbf{R}\,\mathbf{R}_g^{\mathrm{T}}) \times \underline{p}' \quad \text{(for right translation)} \end{cases}$$

As both arising Jacobians have unit condition number, SO(3) renders a well-behaved parametrization with stable numeric properties.

4. Exponential Map/Rotation Vector

By the linearity of the SO(3) parametrization, it is possible to compute the exponential mapping. With $t \equiv \varphi$ and $[\,\underline{\gamma}\,] \equiv \widetilde{\underline{u}}$ one obtains

$$\mathbf{R} = \mathtt{Rot}\,[\,\underline{u}\,,\,\varphi\,] = \mathbf{I}_3 + \widetilde{\underline{u}}\,\varphi + \frac{1}{2}\,\widetilde{\underline{u}}^2\,\varphi^2 + \cdots + \frac{1}{n!}\,\widetilde{\underline{u}}^n\,\varphi^n + \cdots \quad , \qquad (16)$$

from which it follows

$$\mathbf{R} = \mathtt{Rot}\,[\,\underline{u}\,,\,\varphi\,] = \mathbf{I}_3 + \sin\varphi\,\widetilde{\underline{u}} + (1-\cos\varphi)\,\widetilde{\underline{u}}^2 \quad . \qquad (17)$$

After multiplying the axis vector $\underline{u}$ with the rotation angle φ one obtains a three-dimensional vector $\underline{\beta} \doteq \varphi\underline{u}$ called the rotation vector. The corresponding exponential mapping then becomes in terms of this rotation vector

$$\mathbf{R} = \mathbf{I}_3 + \frac{\sin\|\,\underline{\beta}\,\|}{\|\,\underline{\beta}\,\|}\,\widetilde{\underline{\beta}} + \frac{1}{2}\,\frac{\sin^2(\|\,\underline{\beta}\,\|/2)}{(\|\,\underline{\beta}\,\|/2)^2}\,\widetilde{\underline{\beta}}^2 \quad , \qquad (18)$$

where $\|\cdot\|$ denotes the Euclidean vector norm. The derivation of the differential relationships for the rotation vector is cumbersome and can not be described here due to lack of space. In summary, one obtains, for right translation

$$\delta\mathbf{R}^R = \mathrm{d}\mathbf{R}\,\mathbf{R}^{\mathrm{T}} = \int_{\lambda=0}^{1} \mathtt{Rot}\,[\,\lambda\,\underline{\beta}\,] \cdot \mathrm{d}\widetilde{\underline{\beta}} \cdot \mathtt{Rot}\,[\,\lambda\,\underline{\beta}\,]^{-1}\,\mathrm{d}\lambda \quad , \qquad (19)$$

$$\delta\underline{\varphi}^R = \mathrm{vect}(\delta \mathbf{R}^R) = \left[\int_{\lambda=0}^{1} \mathtt{Rot}\,[\lambda\,\underline{\beta}]\,\mathrm{d}\lambda \right] \mathrm{d}\underline{\beta} = \mathbf{T}(\underline{\beta})\,\mathrm{d}\underline{\beta} \quad . \tag{20}$$

For the transmission matrix $\mathbf{T}(\underline{\beta})$ one has (Géradin and Cardona, 1989)

$$\mathbf{T}(\underline{\beta}) = \frac{\sin\|\underline{\beta}\|}{\|\underline{\beta}\|}\,\mathbf{I}_3 + \frac{1}{2}\,\frac{\sin^2(\|\underline{\beta}\|/2)}{(\|\underline{\beta}\|/2)^2}\,\underline{\tilde{\beta}} + \left(1 - \frac{\sin\|\underline{\beta}\|}{\|\underline{\beta}\|}\right)\frac{\underline{\beta}\otimes\underline{\beta}}{\|\underline{\beta}\|^2} \quad . \tag{21}$$

The corresponding Jacobian for left translation results by multiplying Eq. (20) from the left by $\mathbf{R}^{\mathrm{T}}$. Hence one obtains

$$\begin{array}{lcll} \mathrm{D}L_g^{-1}(e) & = & \mathbf{R}^{\mathrm{T}}\,\mathbf{T}(\underline{\beta}) & \text{for left translation} \\ \mathrm{D}R_g^{-1}(e) & = & \mathbf{T}(\underline{\beta}) & \text{for right translation} \end{array} \tag{22}$$

As $\mathbf{R}$ is orthogonal, the condition numbers of both Jacobians equal the condition number of $\mathbf{T}(\underline{\beta})$. In order to evaluate this, note first from the expansion (20) that $\mathbf{T}(\underline{\beta})$ commutes with the rotation matrix, i.e., $\mathbf{T}(\underline{\beta})\,\mathbf{R} = \mathbf{R}\,\mathbf{T}(\underline{\beta})$. Thus it holds $\mathrm{cond}(\mathbf{T}(\underline{\beta})) = \mathrm{cond}(\mathbf{R}^{\mathrm{T}}\,\mathbf{T}(\underline{\beta})\,\mathbf{R})$. On the other hand, one can easily verify that the expanded term on the right hand side represents the evaluation of $\mathbf{T}(\underline{\beta})$ in a rotated coordinate frame. Hence, the condition number of $\mathbf{T}(\underline{\beta})$ is invariant to coordinate transformations. Selecting $\underline{u}$ parallel to axis x renders

$$\mathbf{T}(\underline{\beta})\Big|_{\underline{u}\|x} = \begin{bmatrix} \varphi & 0 & 0 \\ 0 & \sin\varphi & -(1-\cos\varphi) \\ 0 & (1-\cos\varphi) & \sin\varphi \end{bmatrix} \tag{23}$$

This matrix has the eigenvalues $\lambda_1 = \varphi$ and $\lambda_{2,3} = \sin\varphi \pm i\,(1-\cos\varphi)$, where i is here the imaginary unit. Hence the condition number is

$$\mathrm{cond}(\mathbf{T}(\underline{\beta})) = \frac{|\lambda_1|}{|\lambda_{2,3}|} = \frac{\varphi}{\sqrt{2(1-\cos\varphi)}} \quad . \tag{24}$$

This condition number increases from 1 for $\varphi = 0$ over $\pi/2$ for $\varphi = \pi$ to infinity for a full turn. Hence, the rotation vectors have a disadvantageous numeric behavior, critical conditions arising in the neighborhood of a full rotation. Regarding the Jacobian of the point mapping function, it holds as in Eq. (13)

$$\mathrm{d}\underline{x} = \delta\varphi\,\underline{u}\times\underline{p} \quad , \tag{25}$$

hence one obtains as in the case of SO(3) $\mathrm{D}\sigma_x(e) = -\underline{\tilde{p}}$ with unit condition number within the range of $\mathrm{D}\sigma_x(e)$.

5. Euler-Rodrigues Parameters/Quaternions

By expressing Eq. (17) in terms of the half angle $\varphi/2$, one obtains for the rotation matrix (Nikravesh, 1988)

$$\mathbf{R} = \mathbf{I}_3 + 2\left(q_0\,\tilde{\mathbf{q}} + \tilde{\mathbf{q}}^2\right) \quad , \tag{26}$$

with

$$q_0 = \cos\frac{\varphi}{2} \quad , \quad \mathbf{q} = \sin\frac{\varphi}{2}\,\underline{u} \quad . \tag{27}$$

Here, q_0 and $\mathbf{q} = [q_1, q_2, q_3]^{\mathrm{T}}$ are termed the Euler-Rodrigues parameters of rotation, where q_0 is the real part and $\mathbf{q}$ is the vector part. The Euler-Rodrigues parameters obey the normalizing condition

$$q_0^2 + ||\mathbf{q}||^2 = 1 \quad . \tag{28}$$

Hamilton, 1844 recognized that these parameters fulfill special composition relationships, which were later related to rotations by Cayley, 1845. These relationships lead to a special algebra, which is termed the quaternion algebra. Let there be given two rotations I and II by corresponding Euler-Rodrigues parameters

$$Q^I = q_0^I + \mathbf{q}^I \quad , \quad Q^{II} = q_0^{II} + \mathbf{q}^{II}$$

In the quaternion algebra, the Euler-Rodrigues parameters of the resulting rotation $Q = Q^I\,Q^{II}$ are computed by the following relationships, Altmann, 1986

$$\begin{aligned} q_0 &= q_0^I\,q_0^{II} - \mathbf{q}^I \cdot \mathbf{q}^{II} \quad , \\ \mathbf{q} &= q_0^I\,\mathbf{q}^{II} + q_0^{II}\,\mathbf{q}^I + \mathbf{q}^I \times \mathbf{q}^{II} \quad . \end{aligned} \tag{29}$$

Here, '$\cdot$' denotes the dot product and '$\times$' the vector product. Define a unit quaternion E as one with pure real part $q_0^E = 1, \mathbf{q}^E = 0$. Define for a quaternion $Q = q_0 + \mathbf{q}$ the conjugated quaternion $\overline{Q} = q_0 - \mathbf{q}$. Then,

$$Q\,\overline{Q} = \overline{Q}\,Q = E \tag{30}$$

so $Q^{-1} \equiv \overline{Q}$. Hence, the set of quaternions forms a group. As the set of quaternions is open, this group is also a manifold. Hence, quaternions represent a parametrization of the Lie group of rigid-body motions. The action of a group element of quaternions on the point space can be represented by the operation

$$\underline{p}' = Q\,\underline{p}\,\overline{Q} \tag{31}$$

where $\underline{p}' \in \mathbb{R}^3$ is a point vector in Euclidean space. Here, $\underline{p}$ is regarded as a quaternion with vanishing real part. Factoring out $\underline{p}$, one obtains the representation (Nikravesh, 1988)

$$\underline{p}' = \left[\mathbf{I}_3 + 2\left(q_0\,\widetilde{\mathbf{q}} + \widetilde{\mathbf{q}}^2\right)\right]\underline{p} \ , \tag{32}$$

which again yields the expression Eq. (26) for the rotation matrix $\mathbf{R}$. In order to obtain the differential relationships for quaternions, one regards first the relationship between the differential of a quaternion and the infinitesimal rotation increment for a left and right translation, respectively. It holds $\underline{v}_g = \mathrm{d}Q$ and $\delta Q = 0 + \delta\mathbf{q}$, where due to (28)

$$q_0\,\mathrm{d}q_0 + \mathbf{q}\cdot\mathrm{d}\mathbf{q} = 0 \ . \tag{33}$$

For the relationship between differentials and infinitesimal rotations it follows

$$\mathrm{d}Q = \begin{cases} Q(1+\delta\mathbf{q}^L) - Q &= -\mathbf{q}\cdot\delta\mathbf{q}^L + q_0\,\delta\mathbf{q}^L + \mathbf{q}\times\delta\mathbf{q}^L \quad \text{(left translation)} \\ (1+\delta\mathbf{q}^R)Q - Q &= -\mathbf{q}\cdot\delta\mathbf{q}^R + q_0\,\delta\mathbf{q}^R - \mathbf{q}\times\delta\mathbf{q}^R \quad \text{(right translation)} \end{cases}$$

These expressions can be put in the form

$$\begin{bmatrix} \mathrm{d}q_0 \\ \mathrm{d}\mathbf{q} \end{bmatrix} = \begin{cases} \begin{bmatrix} q_0 & -\mathbf{q}^{\mathrm{T}} \\ \mathbf{q} & q_0\,\mathbf{I}_3 + \widetilde{\mathbf{q}} \end{bmatrix} \begin{bmatrix} 0 \\ \delta\mathbf{q}^L \end{bmatrix} & \text{for left translation} \\[2ex] \begin{bmatrix} q_0 & -\mathbf{q}^{\mathrm{T}} \\ \mathbf{q} & q_0\,\mathbf{I}_3 - \widetilde{\mathbf{q}} \end{bmatrix} \begin{bmatrix} 0 \\ \delta\mathbf{q}^R \end{bmatrix} & \text{for right translation} \end{cases} \tag{34}$$

where the first column in the coefficient matrix has been introduced artificially in order to assemble a full 4×4 matrix. As it can be easily verified, this resulting matrix is orthogonal. Hence, one obtains for the Jacobian of the composition function the corresponding inverses

$$\mathrm{D}L_g(e) = \begin{bmatrix} q_0 & \mathbf{q}^{\mathrm{T}} \\ -\mathbf{q} & q_0\,\mathbf{I}_3 - \widetilde{\mathbf{q}} \end{bmatrix} \quad \text{for left translation} \tag{35}$$

$$\mathrm{D}R_g(e) = \begin{bmatrix} q_0 & \mathbf{q}^{\mathrm{T}} \\ -\mathbf{q} & q_0\,\mathbf{I}_3 + \widetilde{\mathbf{q}} \end{bmatrix} \quad \text{for right translation} \tag{36}$$

As these Jacobians are orthogonal, they feature unit condition number.

Considering the point mapping function, note that at the neutral element E it holds $q_0^E = 1, \mathbf{q}^E = 0$ and $\delta q_0^E = 0$. The infinitesimal increment δQ at the neutral element is thus a pure vector $\delta\mathbf{q}$. After substituting $q_0 = 1$, $\mathbf{q} = \delta\mathbf{q}$, $\underline{p}' = \underline{p} + \mathrm{d}\underline{p}$ in Eq. (32), one obtains

$$\mathrm{d}\underline{p} = -2\,\underline{p}\times\delta\mathbf{q} + O\left(\|\,\delta\mathbf{q}\,\|^2\right) \tag{37}$$

Comparing coefficients with the general expansion

$$\sigma_x(E + \delta Q) - \sigma_x(E) = \mathrm{D}\sigma_x(E)\,\delta Q + O\,(\,\|\,\delta Q\,\|^2\,) \tag{38}$$

yields for the Jacobian of the point mapping function at the neutral element

$$\mathrm{D}\sigma_x(E) = -2\,\widetilde{\underline{p}}\ , \tag{39}$$

which again can be extended to a matrix with unit condition number within the range of $\mathrm{D}\sigma_x(E)$.

By combining Eqs. (34), (35), (36), (37), (39) one finally obtains

$$\mathrm{d}\underline{p} = \begin{cases} 2\left[\ \underline{p}\times\mathbf{q}\quad -q_0\,\widetilde{\underline{p}} + \widetilde{\underline{p}}\,\widetilde{\mathbf{q}}\ \right]\begin{bmatrix}\mathrm{d}q_0\\ \mathrm{d}\mathbf{q}\end{bmatrix} & \text{for left translation}\\[2ex] 2\left[\ \underline{p}\times\mathbf{q}\quad -q_0\,\widetilde{\underline{p}} - \widetilde{\underline{p}}\,\widetilde{\mathbf{q}}\ \right]\begin{bmatrix}\mathrm{d}q_0\\ \mathrm{d}\mathbf{q}\end{bmatrix} & \text{for right translation}\end{cases} \tag{40}$$

As both involved Jacobians in the differential mappings feature unit condition number, quaternions embody, as in the case of the SO(3) parametrization, a numerically well-behaved parametrization.

6. Conclusions

By the Lie-group theoretic interpretation of first-order rigid-body kinematics, the mapping from differentials in parameter space to differentials in point space can be accomplished through a two step procedure involving only derivatives at the neutral element. These derivatives are not only more simple to evaluate, but they also are useful to unveil the sources of numerical distortions. As can be seen in Table 2, the parametrizations using orthogonal matrices SO(3) and Quaternions are completely distortion-free. On the other hand, the rotation-vector

parameter type	condition numbers $\mathrm{D}R_g(e)$	$\mathrm{D}L_g(e)$	$\mathrm{D}\sigma_x(e)$
SO(3)	1	1	1
Quaternions	1	1	1
rotational vector	$\dfrac{\varphi}{\sqrt{2(1-\cos\varphi)}}$	$\dfrac{\varphi}{\sqrt{2(1-\cos\varphi)}}$	1

Table 2. Condition numbers for the three basic Jacobians of rigid-body rotations

parametrization induces for rotations with angles greater than zero condition numbers greater than one for the left and right translation Jacobian, showing that the parametrization distorts space in the transmission from angular velocities to time derivatives of the rotation parameters. Hence, such parameterizations may lead to inconvenient numerical properties. The analysis performed here shows that these properties stem from the minimal Lie group theoretic mappings between differentials and thus are inherent to the parametrization and independent of special coordinate systems.

7. Acknowledgments

Support of this work by the German Research Foundation under grant Ke 526/3-1 is gratefully acknowledged. The paper is part of a dedication to Professor Bernard Roth on occasion of the Special Celebratory Symposium in honor of his 70th birthday, held in Stanford, 2003, in appreciation for his invaluable contributions in the field of kinematics.

References

Altmann, S.L. (1986), *Rotations, Quaternions and Double Groups.* Clarendon Press, Oxford.

Angeles, J. (1988), *Rational Kinematics.* Springer Tracts in Natural Philosophy 34. Springer-Verlag, New York.

Bottema, O., and Roth, B. (1970), *Theoretical Kinematics.* Applied Mathematics and Mechanics 24. North-Holland Publishing Company, Amsterdam, Oxford, New York, 1970.

Cayley, A. (1945), On certain results relating to quaternions. *Philosophical Magazine and Journal of Science*, 25:141–145.

Chevallier, D.P. (1991), Lie algebras, modules, dual quaternions and algebraic methods in kinematics. *Mechanism and Machine Theory*, 26(6):613–627.

Géradin, M., and Cardona, A. (1989), Kinematics and dynamics of rigid and flexible mechanisms using finite elements and quaternion algebra. *Computational Mechanics*, 4:115–135.

Golub, G.H., and Van Loan, C.F. (1983), *Matrix computations.* The John Hopkins University Press, Baltimore.

Hamilton, W.R. (1844), On quaternions; or on a new system of imaginaries in algebra. *Philosophical Magazine and Journal of Science*, 25(169):10–13, 489–492.

Karger, A., and Novák, J. (1985), *Space Kinematics and Lie Groups.* Gordon and Breach Science Publishers, Montreux, Schweiz.

Kecskeméthy, A. (2004), Lie-group first-order operations in rigid-body kinematics. In *9th International Symposium on Advances in Robot Kinematics*, Sestri Levante, Italy, June 28 - July 1 2004. Kluwer Academic Publishers, Dordrecht, Boston, London.

Nikravesh, P.E. (1988), *Computer-Aided Analysis of Mechanical Systems.* Prentice-Hall.

Park, F.C., Bobrow, S.R., and Ploen, A. (1995), A Lie group formulation of robot dynamics. *International Journal of Robotics Research*, 14(6): 609-618 (1995)

Varadarajan, V.S. (1984), *Lie Groups, Lie Algebras, and Their Representations.* Graduate Texts in Mathematics, 102. Springer-Verlag, New York.

Properties of Mechanisms

TIME-INVARIANT PROPERTIES OF PLANAR MOTIONS

Bernard Roth
Stanford University
Stanford, CA, USA
broth@stanford.edu

Abstract This paper illustrates the use of canonical coordinates and instantaneous invariants to derive geometric properties from time-based motions. It is shown how, for a given mechanism, the same geometric trajectories can be generated by different actuator motions. Relationships are derived between those parameters that define a time-dependent motion and the parameters that define its equivalent time-independent motion. These relationships are used to determine all time-dependent motions that generate trajectories with identical geometric properties. The numerical example determines the actuator motions for open-loop, series, R-R-R planar kinematic chains.

Keywords: Geometric kinematics, planar motions, canonical coordinates, instantaneous invariants, actuator motions

1. Introduction

This paper deals with the planar motion of a rigid body relative to a reference system. As a body moves, all of its points trace trajectories in the reference system. These trajectories are studied in terms of their instantaneous differential properties. When time is used as the motion parameter, the study is referred to as *time-dependent* kinematics. Although it is most common to think of motions as being functions of time, the concept of time is not necessary for defining many important properties that characterize the trajectories produced by a motion. Eliminating time from the development allows for much simpler derivations and clearer understanding of the geometric properties of the trajectory. The term *geometric kinematics* is used to describe the study of *time-independent* kinematic properties.

When studying the instantaneous differential properties of a motion, it is possible to take advantage of the intrinsic properties of the motion in order to choose a coordinate system that yields the simplest analytical description. This system is called the *canonical coordinate system.* The parameters that define a specific motion in such a system are called the

J. Lenarčič and C. Galletti (eds.), On Advances in Robot Kinematics, 79-88.

instantaneous invariants of the motion. This paper shows how to use the instantaneous invariants to determine all actuator motions that produce exactly the same geometric trajectories[1]. In principle, the same methods can be used for spherical and spatial motions. However, the details are so different, that their treatment requires a separate presentation.

2. Canonical Coordinate Systems and Instantaneous Invariants

The basic concepts for planar canonical coordinates and instantaneous invariants are derived from Veldkamp, 1963. The canonical coordinate systems, instantaneous invariants and notations used in this paper are described in detail in the paper by Roth and Yang, 1977, and the book by Bottema and Roth, 1990. It is well known that if a moving body

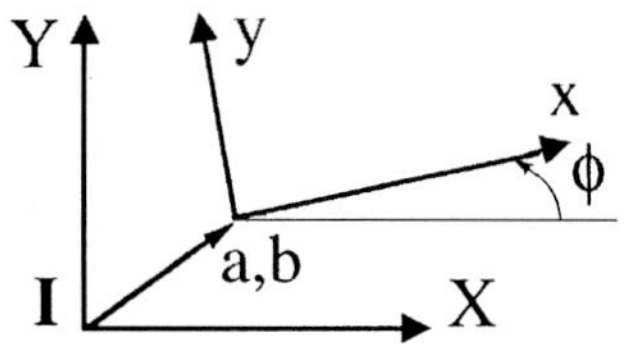

Figure 1. Fixed and moving canonical coordinate axes

is undergoing a general planar motion there exists, at each position, a unique point which is instantaneously at rest. This point is known as the instantancous center of the motion (also, the instant center, the centro, the velocity pole or the first order pole.)

In the instantaneous position under study, two initially coinciding coordinate systems are chosen: one is fixed to the moving body and one is fixed to the reference body. Initially both systems have their common origin at the instantaneous center of the motion, and their Y-axes are along the instantaneous center's acceleration vector. The positive sense of the X-axis then follows from the rule that the angle from positive X-axis to positive Y-axis is measured in the counterclockwise sense, Fig. 1. As the motion proceeds away from the instantaneous position under study, the fixed axes (X, Y) remain stationary while the moving axes (x, y) move so that their origin's coordinates, measured in the (X, Y) system, are at (a, b), and the x-axis makes and angle of ϕ with the X-axis. If a point's coordinates in the fixed system are (X, Y), and its coordinates in the moving system are (x, y), then the coordinates are related by:

$$X = x\cos\phi - y\sin\phi + a, \qquad Y = x\sin\phi + y\cos\phi + b \tag{1}$$

Taking the derivatives with respect to time, t, of Eq. 1, and evaluating them in the position where both $t = 0$ and $\phi = 0$, the results are (Roth and Yang, 1977 [2]):

$$\begin{aligned} X &= x & Y &= y \\ \dot{X} &= -\omega y & \dot{Y} &= \omega x \\ \ddot{X} &= -\omega^2 x - \alpha y & \ddot{Y} &= \alpha x - \omega^2 y + \omega^2 b_2 \\ \dddot{X} &= -3\omega\alpha x + (\omega^3 - \dot{\alpha})y & \dddot{Y} &= -(\omega^3 - \dot{\alpha})x - 3\omega\alpha y \\ &\quad + \omega^3 a_3 & &\quad + 3\omega\alpha b_2 + \omega^3 b_3 \end{aligned} \tag{2}$$

In Eq. 2 we have used the following definitions:

$$a_n = \left(\frac{d^n a}{d\phi^n}\right)_{\phi=o}, b_n = \left(\frac{d^n b}{d\phi^n}\right)_{\phi=o}, \tag{3}$$

$$\omega = \left(\frac{d\phi}{dt}\right)_{t=o}, \alpha = \left(\frac{d^2\phi}{dt^2}\right)_{t=o}, \dot{\alpha} = \left(\frac{d^3\phi}{dt^3}\right)_{t=o} \tag{4}$$

The constants a_n and b_n are called the nth order instantaneous invariants of the motion when the canonical axes are chosen, as above, so that $a_0 = b_0 = a_1 = b_1 = a_2 = 0$. Eq. 2 describes *time-based* kinematics. To get the geometric-kinematics description, derivatives of Eq. 1 are taken with respect to the rotation angle ϕ. This yields, in the position $\phi = 0$:

$$\begin{aligned} X_0 &= x & Y_0 &= y \\ X_1 &= -y & Y_1 &= x \\ X_2 &= -x & Y_2 &= -y + b_2 \\ X_3 &= y + a_3 & Y_3 &= -x + b_3 \end{aligned} \tag{5}$$

In Eq. 5 we have used the following notation:

$$X_n = \left(\frac{d^n X}{d\phi^n}\right)_{\phi=o}, Y_n = \left(\frac{d^n Y}{d\phi^n}\right)_{\phi=o} \tag{6}$$

Any calculation or derivation based on Eq. 5 will be much simpler than corresponding calculations made using Eq. 2. This is the essential advantage of geometric kinematics over time-based kinematics.

3. Time-Based Inputs that Preserve Geometric Properties

By setting $x = 0, y = 0$ in Eq. 2, the time-based derivatives of the instant center's motion are immediately obtained as functions of the geometric instantaneous invariants:

$$\ddot{a} = 0 \qquad \ddot{b} = \omega^2 b_2$$
$$\dddot{a} = \omega^3 a_3 \qquad \dddot{b} = 3\omega\alpha b_2 + \omega^3 b_3 \tag{7}$$

Inverting Eq. 7 gives the geometric invariants as functions of the instant center's time-based motion derivatives:

$$a_2 = 0 \qquad b_2 = \frac{\ddot{b}}{\omega^2}$$
$$a_3 = \frac{\dddot{a}}{\omega^3} \qquad b_3 = \frac{\dddot{b}}{\omega^3} - 3\alpha\frac{\ddot{b}}{\omega^4} \tag{8}$$

With the foregoing, it is possible to determine all the inputs that allow a given multi-degree-of-freedom mechanism to have a motion such that one its links generates desired geometric trajectories to any given order. The basic idea of how to accomplish this is explained in the next paragraph, using properties up to the third derivative.

Assume there exists an n-degree-of-freedom planar mechanism, and that every link's motion is known to the mth order. This means that the canonical coordinate system for any link can be obtained, and the corresponding mth order time-based derivatives of the instant center's motion can be evaluated. With this information, Eq. 8 can be used to determine the corresponding geometric instantaneous invariants. With the geometric invariants known, it is possible to use Eq. 7 to determine $\ddot{b}, \dddot{b}, \dddot{a}$ corresponding to any values of ω, α. Each choice of a set ω, α represents a different time-based motion, and yields a different set $\ddot{b}, \dddot{b}$, $\dddot{a}$. The important thing to realize is that all such different time-based motion, have exactly the same geometric properties. The exact details of how to choose the different sets of ω, α depend upon the specific type of linkage. In the next sections the details are given for a series, open-loop R-R-R chain.

4. R-R-R chains

Fig. 2 gives the nomenclature used for a three-link R-R-R chain. The angles θ_1, θ_2 and θ_3 describe the linkage's position. The derivatives of these angles, with respect to time, define the time-based motion of all three moving links. Here we are interested in the motion of the third link. The theory is developed in conjunction with a numerical example, where the lengths are $a_1 = 5, a_2 = 3$. The position under study is taken to be $\theta_1 = 30\,\text{deg}, \theta_2 = 15\,\text{deg}$. Take $\dot{\theta}_1 = 1.5\text{rad}/\sec, \dot{\theta}_2 = 0.3\text{rad}/\sec, \dot{\theta}_3 = 0.5\text{rad}/\sec, \ddot{\theta}_1 = -0.5\text{rad}/\sec^2, \ddot{\theta}_2 = 0.1\text{rad}/\sec^2, \ddot{\theta}_3 = -0.2\text{rad}/\sec^2$.

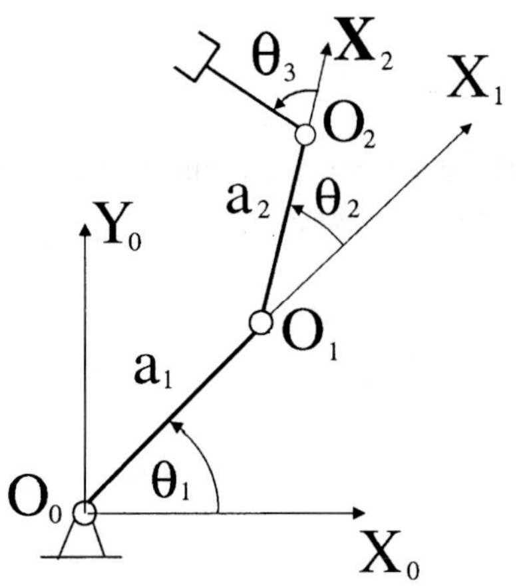

Figure 2. A three-degree-of-freedom R-R-R chain

The canonical coordinate system is determined as follows: Let $\mathbf{Z}$ be the unit axis pointing out of the paper, and define the vectors $\mathbf{A}_1 = a_1\mathbf{X}_1, \mathbf{A}_2 = a_2\mathbf{X}_2$. The origin of the canonical system is the point in link 3 with zero velocity. If $\mathbf{L}$ is the vector from the third joint (O_2) to the zero velocity point, then:

$$\dot{\theta}_1\mathbf{Z}\times\mathbf{A}_1 + (\dot{\theta}_1+\dot{\theta}_2)\mathbf{Z}\times\mathbf{A}_2 + (\dot{\theta}_1+\dot{\theta}_2+\dot{\theta}_3)\mathbf{Z}\times\mathbf{L} = 0, \tag{9}$$

Solving for $\mathbf{L}$, from Eq. 9, yields

$$\mathbf{L} = -\frac{\mathbf{A}_1 + (1+r_1)\mathbf{A}_2}{1+r_1+q_1} \tag{10}$$

Where, $r_1 = \dot{\theta}_2/\dot{\theta}_1, q_1 = \dot{\theta}_3/\dot{\theta}_1$. In this example, the numerical values for the velocity ratios are $r_1 = 0.2, q_1 = 0.3333$. The result is $\mathbf{L} = (-4.484, -3.291, 0)^T$. Since the vector to the third joint is known, $\mathbf{O}_2 = \mathbf{A}_1 + \mathbf{A}_2$, it is easy to compute that the instantaneous center, $\mathbf{I}$, is at $\mathbf{I} = \mathbf{O}_2 + \mathbf{L} = (1.967, 1.331, 0)^T$.

Next, the direction of canonical axes can be determined by using the fact that the fixed canonical Y-axis coincides with the acceleration vector of the instant center. It is straightforward to find the acceleration of the velocity pole, $\ddot{\mathbf{I}}$, since it is a point fixed in the moving system:

$$\begin{aligned}\ddot{\mathbf{I}} &= \dot{\theta}_1\mathbf{Z}\times(\dot{\theta}_1\mathbf{Z}\times\mathbf{A}_1) + \ddot{\theta}_1\mathbf{Z}\times\mathbf{A}_1\\ &\quad + (\dot{\theta}_1+\dot{\theta}_2)\mathbf{Z}\times((\dot{\theta}_1+\dot{\theta}_2)\mathbf{Z}\times\mathbf{A}_2) + (\ddot{\theta}_1+\ddot{\theta}_2)\mathbf{Z}\times\mathbf{A}_2\\ &\quad + \omega\mathbf{Z}\times(\omega\mathbf{Z}\times\mathbf{L}) + \alpha\mathbf{Z}\times\mathbf{L}\end{aligned} \tag{11}$$

Where, $\omega = \dot{\theta}_1+\dot{\theta}_2+\dot{\theta}_3, \alpha = \ddot{\theta}_1+\ddot{\theta}_2+\ddot{\theta}_3$.

For this example, the results is $\ddot{\mathbf{I}} = (7.229, 4.586, 0)^T$. Since, $\ddot{b} = |\ddot{\mathbf{I}}|$, it follows that $\ddot{b} = 8.561$. Since $b_2 = \ddot{b}/\omega^2$ and $\omega = 2.3, b_2 = 1.618$. Furthermore, the direction cosines of the canonical Y-axis are given by the unit vector along $\ddot{\mathbf{I}}$, which in this case is $\mathbf{Y}_c = (0.8444, 0.5357, 0)^T$, and since the canonical X-axis is given by $\mathbf{Y}_c \times \mathbf{Z}$, the canonical X-axis is $\mathbf{X}_c = (0.5357, -0.8444, 0)^T$.

To obtain $\dddot{a}$ and $\dddot{b}$, the derivative, with respect to time, of Eq. 11 needs to be computed. The formula for the change of acceleration of the velocity pole, $\dddot{\mathbf{I}}$, obtained by direct differentiation, is:

$$\begin{aligned}
\dddot{\mathbf{I}} &= 2\ddot{\theta}_1\mathbf{Z} \times (\dot{\theta}_1\mathbf{Z} \times \mathbf{A}_1) + \dot{\theta}_1\mathbf{Z} \times (\ddot{\theta}_1\mathbf{Z} \times \mathbf{A}_1) \\
&\quad + \dot{\theta}_1\mathbf{Z} \times (\dot{\theta}_1\mathbf{Z} \times (\dot{\theta}_1\mathbf{Z} \times \mathbf{A}_1)) + \dddot{\theta}_1\,\mathbf{Z} \times \mathbf{A}_1 \\
&\quad + 2(\ddot{\theta}_1 + \ddot{\theta}_2)\mathbf{Z} \times ((\dot{\theta}_1 + \dot{\theta}_2)\mathbf{Z} \times \mathbf{A}_2 \\
&\quad + (\dot{\theta}_1 + \dot{\theta}_2)\mathbf{Z} \times ((\dot{\theta}_1 + \dot{\theta}_2)\mathbf{Z} \times (\dot{\theta}_1 + \dot{\theta}_2)\mathbf{Z} \times \mathbf{A}_2)) \\
&\quad + (\dot{\theta}_1 + \dot{\theta}_2)\mathbf{Z} \times ((\ddot{\theta}_1 + \ddot{\theta}_2)\mathbf{Z} \times \mathbf{A}_2) + (\dddot{\theta}_1 + \dddot{\theta}_2)\mathbf{Z} \times \mathbf{A}_2 \\
&\quad + 2(\alpha\mathbf{Z} \times (\omega\mathbf{Z} \times \mathbf{L}) \\
&\quad + \omega\mathbf{Z} \times (\omega\mathbf{Z} \times \omega\mathbf{Z} \times \mathbf{L})) \\
&\quad + \omega\mathbf{Z} \times (\alpha\mathbf{Z} \times \mathbf{L}) + \dot{\alpha}\mathbf{Z} \times \mathbf{L})
\end{aligned} \tag{12}$$

Where, $\dot{\alpha} = \dddot{\theta}_1 + \dddot{\theta}_2 + \dddot{\theta}_3$.

Choosing the numerical values $\dddot{\theta}_1 = -0.7\text{rad}/\sec^3, \dddot{\theta}_2 = 0.15\text{rad}/\sec^3$ and $\dddot{\theta}_3 = -0.1\text{rad}/\sec^3$, Eq. 12 yields $\dddot{\mathbf{I}} = (-22.69, 22.87, 0)^T$ for the present example. Projecting this vector onto the canonical directions gives $\dddot{a} = -31.47$ and $\dddot{b} = -6.907$. With these values, it follows from Eq. 8 that, with $\omega = 2.3$ and $\alpha = -0.6$, the third order invariants are $a_3 = -2.586$ and $b_3 = -0.0170$.

4.1 Equivalent motions

With the canonical system and the invariants b_2, a_3, b_3 known, the geometric properties of the motion of link 3 (in Fig. 2) are completely determined up to the third order. Now, it is possible to use these same invariants to determine which other values of the inputs $\dot{\theta}_1, \dot{\theta}_2, \dot{\theta}_3, \ddot{\theta}_1, \ddot{\theta}_2, \ddot{\theta}_3, \dddot{\theta}_1, \dddot{\theta}_2, \dddot{\theta}_3$ will result in exactly the same geometric properties for the motion up to the third order. The requirement is to determine which values of ω, α and $\dot{\alpha}$ yield exactly the same canonical coordinate system and the same instantaneous invariants as the original system.

First it is necessary to deal with the location of I. Obviously, the vector $\mathbf{L}$ must be the same for all geometrically equivalent motions. Thus, it follows from Eq. 10 that r_1 and q_1 are both required to be constants for all geometrically equivalent motions. This implies that the motion's angular velocity must be such that:

$$\omega = \dot{\theta}_1(1 + r_1 + q_1) = \dot{\theta}_1(\text{constant}) \tag{13}$$

Thus, for all equivalent motions, the ratios of $\dot{\theta}_1, \dot{\theta}_2, \dot{\theta}_3$ are fixed to be the ratios of the original motion. Next, it is necessary to assure that the canonical axes are in the same directions as for the original motion. The direction of $\mathbf{Y}_c$ is given by the direction of the vector $\ddot{\mathbf{I}}$. Also, any geometrically equivalent motion has the same value for the invariant b_2 as the original motion. Hence, from Eq. 2,

$$b_2\mathbf{Y}_c = \ddot{\mathbf{I}}/\omega^2 =^e \ddot{\mathbf{I}}/^e\omega^2 \tag{14}$$

Here, the leading superscript e indicates the values are for an equivalent new motion; the unsuperscripted values are for the original motion. Assuming that an arbitrary choice has been made for the value of ${}^e\dot{\theta}_1$ for the new motion, ${}^e\omega$ is known from Eq. 13, and of course then ${}^e\dot{\theta}_2$ and ${}^e\dot{\theta}_3$ are also known for the equivalent new motion. So, Eq. 14 can be used to determine the restrictions on the new motion's joint accelerations: First, Eq. 11 is rewritten as:

$${}^e\ddot{\mathbf{I}} =^e \ddot{\theta}_1\mathbf{Z} \times \mathbf{A}_1 + ({}^e\ddot{\theta}_1 +^e \ddot{\theta}_2)\mathbf{Z} \times \mathbf{A}_2 +^e \alpha\mathbf{Z} \times \mathbf{L} +^e \mathbf{A}_f \tag{15}$$

Where, all the terms which contain the known first derivative of the input angles have been group into the vector ${}^e\mathbf{A}_f$. Then, substituting for ${}^e\ddot{\mathbf{I}}$ from Eq. 15 into Eq. 14 yields:

$${}^e\ddot{\theta}_1\mathbf{Z}\times(\mathbf{A}_1+\mathbf{A}_2+\mathbf{L})+{}^e\ddot{\theta}_2\mathbf{Z}\times(\mathbf{A}_2+\mathbf{L})+{}^e\ddot{\theta}_3\mathbf{Z}\times\mathbf{L} = -{}^e\mathbf{A}_f+{}^e\omega^2 b_2\mathbf{Y}_c \tag{16}$$

This can be rewritten as:

$${}^e r_2\mathbf{Z} \times (\mathbf{A}_2 + \mathbf{L}) +^e q_2\mathbf{Z} \times \mathbf{L} = \frac{-{}^e\mathbf{A}_f +^e \omega^2 b_2\mathbf{Y}_c}{{}^e\ddot{\theta}_1} - \mathbf{Z} \times \mathbf{I} \tag{17}$$

Where ${}^e r_2 =^e \ddot{\theta}_2/{}^e\ddot{\theta}_1$ and ${}^e q_2 =^e \ddot{\theta}_3/{}^e\ddot{\theta}_1$. It then follows, using either Eq. 16 or Eq. 17, that if one of the input angular accelerations is chosen arbitrarily, the other two are uniquely determined. All such motions will have b_2 and their canonical directions equal to those of the original motion.

For the third-derivative properties, the original motion's values for the geometric invariants b_2, a_3, b_3 and the new motion's values for ω and α are used in Eq. 7 to compute the new motion's values for $\ddot{a}$ and $\dddot{b}$. These are components of a vector measured in the canonical coordinate system. Since it is easier to do the next calculation in the mechanism's base coordinate system, the following transformation of coordinates is performed:

$$\begin{bmatrix} \dddot{a} \\ \dddot{b} \\ 0 \end{bmatrix}_0 = \begin{bmatrix} \mathbf{X}_c & \mathbf{Y}_c & 0 \\ & & 0 \\ & & 0 \end{bmatrix} \begin{bmatrix} \dddot{a} \\ \dddot{b} \\ 0 \end{bmatrix}_c$$

The subscripts c and 0 have been added to make it clear whether the vector's components are measured along the canonical or the base coordinate direction, respectively.

Using the new values for the first and second derivatives of θ_1, θ_2 and θ_3, Eq. 12 can be written as:

$$[\dddot{a}, \dddot{b}, 0]_0^T = \mathbf{K} + \dddot{\theta}_1 \mathbf{Z} \times \mathbf{A}_1 + (\dddot{\theta}_1 + \dddot{\theta}_2)\mathbf{Z} \times \mathbf{A}_2 + \dot{\alpha}\mathbf{Z} \times \mathbf{L} \tag{18}$$

K represents all the terms on the right-hand-side of Eq. 12 that do not contain the third derivatives of the joint angles.The structure of Eq. 18 is directly analogous to the structure of Eq. 15, and so it follows that, in analogy to Eq. 16 and 17, it can be written as:

$$^e\dddot{\theta}_1 \mathbf{Z} \times (\mathbf{A}_1 + \mathbf{A}_2 + \mathbf{L}) + ^e\dddot{\theta}_2 \mathbf{Z} \times (\mathbf{A}_2 + \mathbf{L}) + ^e\dddot{\theta}_3 \mathbf{Z} \times \mathbf{L} = -^e\mathbf{K} + ^e[\dddot{a}, \dddot{b}, 0]_0^T \tag{19}$$

or alternatively

$$^er_3\mathbf{Z} \times (\mathbf{A}_2 + \mathbf{L}) + ^eq_3\mathbf{Z} \times \mathbf{L} = \frac{-^e\mathbf{K} + ^e[\dddot{a}, \dddot{b}, 0]_0^T}{^e\dddot{\theta}_1} - \mathbf{Z} \times \mathbf{I} \tag{20}$$

Where $^er_3 = ^e\dddot{\theta}_2 \,/\, ^e\dddot{\theta}_1$ and $^eq_3 = ^e\dddot{\theta}_3 \,/\, ^e\dddot{\theta}_1$. If an arbitrary value is chosen for one of the third derivatives, say, $^e\dddot{\theta}_2$, Eq. 19 (or 20) contains two linear unknowns and two (non trivial) scalar equations. So, these equations yield unique values for $^e\dddot{\theta}_2$ and $^e\dddot{\theta}_3$.

The third derivative establishes the pattern that holds for the fourth and higher orders. Namely, for the nth order matching of geometric properties, each order involves using the known geometric invariants for that order and all lower orders, and the angular velocity and its derivatives up to the $(n-1)$th order, which are also known[3]. These values are then substituted into Eq. 7 to compute the nth time derivatives of a and b. Which are then transformed into the base coordinate directions, yielding the vector in the base coordinate system that is the nth derivative of I. This vector is the left-hand side for the $(n-3)$th derivative of Eq. 12. The new Eq. 18 yields two non-trivial linear scalar equations in the unknown nth derivatives of θ_1, θ_2 and θ_3. One of these may be chosen arbitrarily, and the other two are then uniquely determined.

4.2 Geometric motions

For a geometric motion, it is required that ω be constant and equal to unity. Hence, the conditions $\omega = 1, \alpha = 0, \dot{\alpha} = 0, \ldots$ need to be to added to the foregoing analysis: Eq. 13 becomes $1 = \dot{\theta}_1(1 + r_1 + q_1)$, hence $\dot{\theta}_1$ is unique. It always must be $\dot{\theta}_1 = 1/(1 + r_1 + q_1)$. So $\dot{\theta}_1, \dot{\theta}_2, \dot{\theta}_3$ are all unique for the given motion. The condition $\alpha = \ddot{\theta}_1 + \ddot{\theta}_2 + \ddot{\theta}_3 = 0$ means that $\ddot{\theta}_1$ must always be chosen as equal to $-\ddot{\theta}_2 - \ddot{\theta}_3$ in Eq. 16 (or Eq. 17). So, again there is no free choice, and $\ddot{\theta}_1, \ddot{\theta}_2, \ddot{\theta}_3$ are all unique for the given motion. This pattern exist at every order. Hence, corresponding to any given nth order time-based motion of an R-R-R chain, there is a unique nth order geometric motion.

4.3 Numerical example

As an example, for the above set of numbers, a new motion with the same geometric motion properties is obtained:
1) The first order, the ratios $r_1 = 0.3/1.5 = 0.2$ and $q_1 = 0.5/1.5 = 0.3333$ must be maintained. For the arbitrary choice of $\dot{\theta}_1 = 0.625$, it follows that $\dot{\theta}_2 = 0.125, \dot{\theta}_3 = 0.2083$.
2) The second order, both the direction $\mathbf{Y}_c = [0.8444, 0.5357, 0]^T$ and $b_2 = 1.6184$ must be maintained. For an arbitrary choice of $\ddot{\theta}_1 = 0.6$, Eq. 16 yields, $\ddot{\theta}_2 = 0.1546, \ddot{\theta}_3 = 0.1942$.
3) To the third order, with $a_3 = -2.586, b_3 = -0.0170$, and the new values of $\omega = 0.625{+}0.125{+}0.2083 = 0.9583, \alpha = 0.6{+}0.1546{+}0.1942 = 0.9489$, Eq. 7 yields $\ddot{a} = -2.276, \dddot{b} = 4.400$. From which it follows from Eq. 19 that, for the arbitrary choice of $\dddot{\theta}_1 = 0.2$, the values of the other joint variables are $\dddot{\theta}_2 = 0.1754, \dddot{\theta}_3 = 0.0572$.
Geometric motion: For this example, with $\omega = 1, \alpha = 0, \dot{\alpha} = 0$, $\dot{\theta}_1 = 1/(1 + r_1 + q_1)$ gives $\dot{\theta}_1 = 0.6522$, and therefore $\dot{\theta}_2 = 0.1304, \dot{\theta}_3 = 0.2174$. For the second order, Eq. 16 yields, $\ddot{\theta}_2 = 0.0336, \ddot{\theta}_3 = -0.0131$, and therefore $\ddot{\theta}_1 = -0.0205$. Finally, since $\ddot{a} = a_3 = -2.586, \dddot{b} = b_3 = -0.0170$, from Eq. 19, $\dddot{\theta}_2 = 0.0310, \dddot{\theta}_3 = -0.0012$. From which, by virtue of $\dot{\alpha} = 0, \dddot{\theta}_1 = -0.0298$.

5. Conclusions

The path tangent's slope, the curvature, and the first change of curvature, of the trajectory of an arbitrary point, $P(1, 2, 0)^T$, were computed for the original motion, the new motion and the geometric motion of the numerical example. As expected, all three sets of values were exactly the same (respectively, $-2.364, 1.306, 3.238$).

The pattern established up to the third order holds for the fourth and higher orders. Namely, for the R-R-R chain of Fig. 2, at each order, an infinite number of motions exist with matching geometric trajectory properties. For each order, one of the three input angular motions can be chosen arbitrarily, and then the other two are determined uniquely by the geometric properties of the motion. There is a unique value that corresponds to the geometric motion.

The ideas in this paper can be applied to a sequence of closely spaced positions along a finite portion or even an entire motion trajectory. In this way, it is possible to produce set-points for controllers that control both the position and derivatives of the joint actuators over the entire motion cycle. Moreover, this method has the advantage that the joint angles of the new motion can be computed "exactly" to any order.

The basic method of this paper is to first determine the geometric instantaneous invariants of a given time-based motion. Then to determine all the other time-based motions that have exactly the same geometric properties as the original motion. Clearly, this method is applicable to any planar mechanism with any topology that has three or more independently controllable inputs.

Notes

1. In private correspondence, Dr. C. W. Wampler has pointed out that solving this problem using the mathematics of re-parameterization is much simpler than using canonical coordinates, since it does not require the detailed analysis of the forward kinematics. Re-parameterization is universally applicable to any order and for any type of linkage be it planar, spherical or spatial. Clearly it is a superior method for the problem considered in this paper. However, it does not provide the desired insights into the use of canonical coordinates. It is these insights that are the main motivation for the writing of this paper.

2. A small typographical error in Eq. (14) of Roth and Yang, 1977, is corrected in Eq. 2: in the $\dddot{X}$ term, α_3 has been replaced by a_3. Also in Eq. (14), in the $\overset{IV}{Y}$ term, ω multiplies the $4\dot{\alpha}$ term in the coefficient of b_2. Elsewhere in that paper, the right-hand side of Eq. (16c) should read $-H(x^2+y^2)+5xb_2G$, and, in the definition of H, the term with F should read $3b_2^2F$. Also, obviously, ω is the derivative of ϕ not θ.

3. For the fourth derivative, in Eq. 8 $a_4 = \overset{IV}{a}/\omega^4 - 6\alpha\dddot{a}/\omega^5, b_4 = \overset{IV}{b}/\omega^4 - 6\alpha\dddot{b}/\omega^5 - (4\dot{\alpha} - 15\alpha^2/\omega)\ddot{b}/\omega^5$; in Eq. 7 $\overset{IV}{a} = 6\omega^2\alpha a_3 + \omega^4 a_4$, $\overset{IV}{b} = (4\dot{\alpha}\omega + 3\alpha^2)b_2 + 6\omega^2\alpha b_3 + \omega^4 b_4$.

References

Bottema, O. and Roth, B. (1990), *Theoretical Kinematics*, New York, Dover Publications.

Roth, B. and Yang, A.T. (1977), Application of Instantaneous Invariants to the Analysis and Synthesis of Mechanisms, *ASME Journal of Engineering for Industry*, vol. 99, no. 1, pp. 97–103.

Veldkamp, G.R. (1963), *Curvature Theory in Plane Kinematics*, Doctoral dissertation, T.H. Delft, Groningen.

PROJECTIVE PROPERTIES OF PARALLEL MANIPULATORS

Adolf Karger
Charles University, Faculty of Mathematics and Physics
Praha, Czech Republic
karger@karlin.mff.cuni.cz

Abstract Geometry of a parallel manipulator is determined by concepts of Euclidean geometry - distances and angles. In spite of this, parallel manipulators have some properties which are projectively invariant. This contribution is devoted to one of them, to the projective invariance of singular positions. We explain at first the projective invariance of singular positions. Using this property we can use projective coordinate systems to reduce the number of parameters determining the parallel manipulator. This enables to simplify the equation for singular positions of a parallel manipulator and using computer algebra we can give purely geometric characterization of singular positions of some special parallel manipulators. We obtain complete characterization of singular positions for 3-3 manipulators and for planar manipulators with projective correspondence between platform and base.

Keywords: parallel manipulators, singular positions, projective transformatios

1. Introduction

Some notions of Euclidean geometry preserve their meaning even if a larger group than the group of Euclidean congruences is used. For instance a sphere is defined as the set of points which have given distance from a fixed point, but this notion is well defined even in the conformal geometry, which does not preserve distances. Similarly, ratio of three colinear points is defined as a quotient of their distances but it is preserved by affine transformations, which do not preserve distances of points.

In this paper we shall discuss concepts which are defined in Euclidean geometry but preserve their meaning even if projective transformations are used. A simple example is given by the cross ratio of four colinear points which is defined by distances and which is projectively invariant.

From the very beginning we see that in comparison to the affine case we are facing some difficulties. Projective transformations which are not affinities do not map the Euclidean space onto itself. We have a problem with the plane in space, which has no image ("it goes to infinity"). We overcome this complication by considering only some projective trans-

J. Lenarčič and C. Galletti (eds.), On Advances in Robot Kinematics, 89-96.

formations and not on the whole space, but only at its (sufficiently large) part. We shall not develop a detailed theory of these concepts, as this is not the aim of the paper. Nevertheless let us present the following

Definition. Let E_3 be the Euclidean space considered as the set of triples of real numbers $A = [x, y, z]$ with the usual distance

$d(A_1, A_2) = \sqrt{(x_1 - x_2)^2 + (y_1 - y_2)^2 + (z_1 - z_2)^2}$

for $A_i = [x_i, y_i, z_i]$, $i = 1, 2 \in E_3$.

Local group of projective transformations in E_3 is a set of transformations of the type

$$\tilde{x} = \frac{a_{11}x + a_{12}y + a_{13}z + a_{14}}{a_{41}x + a_{42}y + a_{43}z + a_{44}}$$

$$\tilde{y} = \frac{a_{21}x + a_{22}y + a_{23}z + a_{24}}{a_{41}x + a_{42}y + a_{43}z + a_{44}}$$

$$\tilde{z} = \frac{a_{31}x + a_{32}y + a_{33}z + a_{34}}{a_{41}x + a_{42}y + a_{43}z + a_{44}},$$

where $\det(a_{ij}) = 1, i, j = 1, ..., 4$, defined on some open neighbourhood of identity transformation such that it maps some open set of E_3 again into E_3. (Let us remark that local group is not a group in the sense of group theory, but it is a standard concept in local geometry.)

This takes care of the problem with points at infinity, to a given (bounded) object we choose transformations which are near to identity in such a way that none of its points "goes to infinity". Similar considerations are true also for the Euclidean line or Euclidean plane.

We see that on the line we have a locally projective notion of a line segment with end points A and B as the set of points X such that

$$d(A, X) + d(X, B) = d(A, B).$$

This follows immediately that the notion of a convex set in plane or of a convex body in space is also a locally projective notion.

The following example is less trivial. Let us define ellipse as the set k of points X in the Euclidean plane with the property that there exist two different points F_1 a F_2 (foci) such that

$$d(X, F_1) + d(X, F_2) = const.$$

Because projective transformations map conic sections into conic sections, ellipses are locally maped into ellipses and therefore ellipse is a locally projective notion. On the other side a circle or a parabola are not locally projective notions.

2. Infinitesimal transformations

To prove local invariance of some concept it is enough to show its invariance with respect to infinitesimal transformations. This follows from the theory of local Lie groups or the theory of dynamical systems (partial differential equations). Let us explain in short what are infinitesimal transformations.

Let a Lie group G with unit e act as a transformation group on a manifold M.

A function $f : M \to R$ is called invariant with respect to G iff $f(m) = f(g(m))$ for all $g \in G,\ m \in M$.

For such a function and for any curve $g(t)$ on G, $g(0) = e$, it is $F(t) = f(g(t))$ constant and therefore,

$$F'(0) = f(g(t)(m))'_{t=0} = \frac{\partial f}{\partial \tilde{x}_i}.\frac{d\tilde{x}_i}{dt} = \frac{\partial f}{\partial x_i}.\frac{\partial \phi_i}{\partial a_\alpha}.\frac{da_\alpha}{dt} = 0,$$

where $\tilde{x}_i = \phi_i(a_\alpha, x_j)$ is coordinate expression of the action of G with parameters a_α and local coordinates x_i on M.

Operators

$$X = \frac{\partial \phi_i}{\partial a_\alpha}.\frac{da_\alpha}{dt}.\frac{\partial}{\partial x_i}$$

are called infinitesimal transformations (fundamental vector fields). If $Xf = 0$ for all infinitesimal transformations, then f is invariant with respect to a local group of G.

For instance for $M = E_3$ and for the local projective group for curve $\tilde{x} = x+ty,\ \tilde{y} = y,\ \tilde{z} = z$ we obtain $f(x+ty, y, z)'_{t=0} = \frac{\partial f}{\partial x}.y$, and therefore the corresponding infinitesimal transformation is $X = y.\frac{\partial}{\partial x}$. For the projective group we obtain 15 basic infinitesimal transformations,

$$\frac{\partial}{\partial x}, x.\frac{\partial}{\partial x}, y.\frac{\partial}{\partial x}, x^2.\frac{\partial}{\partial x} + xy.\frac{\partial}{\partial y} + xz.\frac{\partial}{\partial z}, ...,$$

others are their linear combinations.

3. Projective invariance of singular positions

Let us consider a screw motion in E_3 given by $\tilde{x} = x\cos\phi + y\sin\phi, \tilde{y} = -x\sin\phi + y\cos\phi, \tilde{z} = z + v_o\phi$. The tangent line of the trajectory of the point $A(x, y, z)$ is determined by this point and by the tangent vector

$A'(0) = (-x\sin\phi + y\cos\phi, -x\cos\phi - y\sin\phi, v_o) = (y, -x, v_o)$.

The normal line of the trajectory is determined by point A and vector $\vec{u} = (u_1, u_2, u_3)$, for which $u_1 y - u_2 x + u_3 v_0 = 0$. Plücker coordinates of this normal line are

$(\vec{u}; A \times \vec{u}) = (u_1, u_2, u_3; yu_3 - zu_2, -xu_3 + zu_1, xu_2 - yu_1)$.

If we denote the Plücker coordinates of the normal line of the trajectory by $(u_1, u_2, u_3; w_1, w_2, w_3)$, we obtain $w_3 = xu_2 - yu_1 = u_3 v_o$.

This shows that normal lines of trajectories of our screw motion belong to the linear complex described in Plücker coordinates by the equation (Bottema, Roth, 1990).

$$v_o u_3 - w_3 = 0.$$

Euclidean congruences of the space map linear line complexes into themselves and every linear line complex contains a set of normals of trajectories of some Euclidean screw motion. Moreover, the situation here is similar to the situation with ellipses in the plane. The set of normals of trajectories of a screw motion is by projective transformations mapped again into the set of normals of some other screw motion (axes and parameters do not correspond to each other). We consider only the local group, in general some normals can "go to infinity". This means that the set of normals of a screw motion is a locally projective concept. The use of a local group will avoid the undesired phenomena of some leg of the parallel manipulator dissappering at infinity or having legs of infinite length.

The projective invariance has applications in the geometry of parallel manipulators. A parallel manipulator (a Stewart-Gough type platform) is in a singular position if its legs are normals of some screw motion. This follows that if we perform a projective transformation in the space (near to identity), the parallel manipulator will transform to another parallel manipulator with totally different geometry, but it will be again in a singular position.

As an example how to use the concept of a local group, we can prove directly the projective invariance of singular positions (for linear complexes it is known and this is almost the same). Let us consider a parallel manipulator with six points $m_1, ..., m_6$ in the platform and six points $M_1, ..., M_6$ in the base, $m_i = (a_i, b_i c_i), M_i = (A_i, B_i, C_i)$. Plücker (unnormed) coordinates of the line $m_i M_i$ are given by

$$(A_i - a_i, B_i - b_i, C_i - c_i, A_i b_i - B_i a_i, A_i c_i - C_i a, B_i c_i - C_i b_i).$$

We combine them into a 6×6 matrix J. Then $\det(J) = 0$ is the equation for singular positions. It is a function from $E_3 \times ... \times E_3$, 12-times, into real numbers. The action of the group of projective transformations is extended in a natural way on Cartesian products. Infinitesimal operators are for instance

$X_{12} = \sum_{i=1}^{6} (b_i \frac{\partial}{\partial a_i} + B_i \frac{\partial}{\partial A_i})$, $X_{14} = \sum_{i=1}^{6} (\frac{\partial}{\partial a_i} + \frac{\partial}{\partial A_i})$,
$X_{11} = \sum_{i=1}^{6} (a_i \frac{\partial}{\partial a_i} + A_i \frac{\partial}{\partial A_i})$,
$X_{42} = \sum_{i=1}^{6} (a_i b_i \frac{\partial}{\partial a_i} + A_i B_i \frac{\partial}{\partial A_i} + b_i c_i \frac{\partial}{\partial c_i} + B_i C_i \frac{\partial}{\partial C_i} + b_i^2 \frac{\partial}{\partial b_i} + B_i^2 \frac{\partial}{\partial B_i})$.

Computation shows that

$X_{12} \det(J) = 0,\ X_{14} \det(J) = 0,\ X_{11} \det(J) = 3 \det(J)$,
$X_{42} \det(J) = \sum_{i=1}^{6} (b_i + B_i) \det(J)$

For remaining operators we obtain similar results. This proves the locally projective invariance of the set $\det(J) = 0$.

4. Applications to S.-G type of platforms

Let us consider the projective space Π^3 with homogeneous coordinates $[x, y, z, t]$. We are interested in singular configurations of parallel manipulators and therefore we can use projective transformations to choose special projective coordinate systems. We take two different planes α and β. We can choose the coordinate system in such a way that they are planes $x = 0$ and $y = 0$. A point $m \in \alpha$ ($M \in \beta$) is given by $[0, b, c, d]$ ($[A, 0, C, D]$), respectively. A projective correspondence P from α to β is given by
$A = p_{11}b + p_{12}c + p_{13}d,\ C = p_{21}b + p_{22}c + p_{23}d, D = p_{31}b + p_{32}c + p_{33}d.$

Let us choose points $m_i \in \alpha$ and $M_i \in \beta, i = 1, ...6$. Lines $m_iM_i, i = 1, ..., 6$ belong to a linear complex iff the determinant of the matrix J of their Plücker coordinates is equal to zero (Ma, Angeles, 1992, Merlet, 1992). The expression for $\det(J)$ factorizes into 3 factors, $\det(J) = Q_1.Q_2.Q_3$.

$Q_1 = 0$ iff points m_i lie on a conic section, Q_2 is the determinant of the matrix $p_{rs}, r, s = 1, 2, 3$ and therefore it is different from zero and $Q_3 = 0$ iff P has an invariant point. (see Karger, 2002 for details of the proof, see also Karger, 2001). This proves the following

Theorem 1. Let us have two different planes in Π^3 with a projective correspondence between them. Choose six points in one of the planes and join them with corresponding points in the other one. Then the connecting lines belong to a linear complex iff either choosen points lie on a conic or the correspondence has an invariant point (and the property is independent of the choice of points).

Remark. Condition $Q_1 = 0$ is very well known from the beginning of the 20-th century (Borel,1908, Bricard, 1906),$Q_3 = 0$ is new.

In the following we shall present a similar theorem, but in this case we shall use a more detailed approach. We proceed as above, no correspondence between planes is required but we choose a special configuration (see Fig 1). We ask again the same question as above. We shall try to bring coordinates of some points to a simpler form by using projective transformations of the space which preserve chosen planes. Such a projective transformation is given by equations

$x' = p_{11}x,\ y' = p_{22}y,$

$z' = p_{31}x + p_{32}y + p_{33}z + p_{34}t,\ t' = p_{41}x + p_{42}y + p_{43}z + p_{44}t,$
where $p_{11}p_{22}(p_{33}p_{44} - p_{34}p_{43}) \neq 0$.

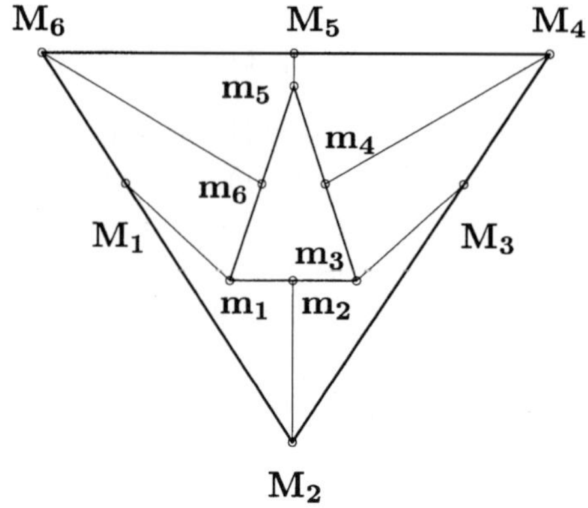

Figure 1. Configuration of points for Theorem 2.

We take point m in α with coordinates $(0, b, c, d)$ and point M in β with coordinates $(A, 0, C, D)$. We choose $p_{33} = p_{44} = 1, p_{34} = p_{43} = 0$. Then the choice $p_{32} = -c/b, p_{42} = -d/b, p_{31} = -C/A, p_{41} = -D/A$ yields a projective map, which maps points m and M into points with $c = d = C = D = 0$. (We have to select points with $b \neq 0, A \neq 0$.) Now we are able to prove the following (see Husty, Karger, 2000).

Theorem 2. Let us have planes α and β in Π^3, points $m_1, ..., m_6$ in α and points $M_1, ..., M_6$ in β such that m_1, m_2, m_3 and m_3, m_4, m_5 and m_5, m_6, m_1 are colinear, similarly let M_2, M_3, M_4 and M_4, M_5, M_6 and M_6, M_1, M_2 be colinear (see Figure 1). **Then lines m_iM_i belong to a linear complex iff**
1) $rR = 1$ for the characteristic cross ratios r of $m_i, i = 1, ..., 6$ and R of $M_i, i = 1, ..., 6$ or
2) the projective transformation induced on the common line of α and β is an involution.

Remark. The Jacobian of the parallel manipulator in Theorems 1 and 2 factorizes into two factors - the first one is purely architectural, it depends only on the mutual position of points of the platform and of the base and therefore it yields the condition for architectural singularity of the platform (Ma, Angeles, 1992, Karger, 2001). The other one describes singular positions for all non architectural singularities. The reason for this behavior lies in the fact that the chosen manipulator is very special, it has to satisfy only one condition to be architecturally singular.

Remark. The characteristic cross ratio (CR) of the configuration of points m_i is seen at the Figure 2, where we have

$$CR(m_1, m_3, m_2, Y) = CR(m_5, m_3, X, m_4) = CR(m_3, m_5, m_4, X).$$

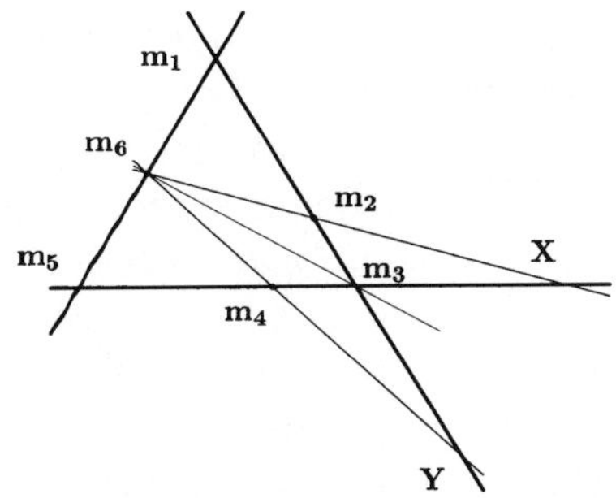

Figure 2. The characteristic cross ratio from Theorem 2.

Teorem 3. Let us consider a parallel manipulator of type 3-3, where

$m_1 = m_2,\ m_3 = m_4,\ m_5 = m_6$ **and** $M_2 = M_3,\ M_4 = M_5,\ M_6 = M_1$. **Denote** p_{12} **the line connecting points** $m_3 = m_4$ **and** $m_5 = m_6$, **similarly** $p_{34}(p_{56})$ **be the line connecting points** $m_1 = m_2, m_5 = m_6$, $(m_1 = m_2,\ m_3 = m_4)$, **respectively. In a similar way we denote** $P_{12}(P_{34}, P_{56})$ **the line connecting points** $M_2 = M_3, M_6 = M_1$ $(M_2 = M_3, M_4 = M_5 \quad M_4 = M_5, M_6 = M_1)$, **respectively. Let the common line of planes of the platform and base be** q. **Denote** $q_{ij}(Q_{ij})$, **the intersection point of the line** q **with** $p_{ij}(P_{ij})$, **respectively. Then the manipulator is in a singular position iff the projective correspondence** $\alpha : q_{ij} \to Q_{ij}, ij = 12, 34, 56$ **is an involution.**

Remarks. The situation is to be considered in the projective extension of the Euclidean space, the denotation of the lines in the base is defined in such a way that the corresponding line is the only line which contains only two common indices with the upper one, an involution is a projectivity different from identity such that its square is the identity. The matrix of an involution on a straight line is characterized by the property that its trace is zero.

All the proofs of the mentioned theorems were done on the computer using MapleV and cannot be presented because of the complexity of all formulas. In fact proofs are quite elementary, the complicated part is to find out the statement.

5. Conclusions

It is intuitively recognized that singular positions of parallel manipulators are projectively invariant. It follows from the general fact that the notion of a linear complex is a concept from projective line geometry. Nevertheless there is a difference between a singular positions of manipulators and linear complexes - joints or legs of a manipulator should not go to infinity. We have used the projective invariance to characterize singular positions of some Stewart-Gough manipulators with planar platform and base. The projective invariance enables to use a projective system of coordinates (in the Euclidean space!) which simplifies all computations - the group of Euclidean congruences depends of 6 parameters, the projective group has 15 parameters and so the number of unknowns is smaller by 9. Using it, we have shown that singular positions can be characterized by very simple geometric criteria - by existence of invariant points or by a involution on the common line. It explains for instance why a parallel manipulator has a large nonsingular region around the home position - it is because involutions are very far from the identity. Because of the linear nature of the singular positions there should exist some similar properties as in Theorems 1,2,3, the projective invariance couls help with the proof. The local projective invariance could be used also for some other purposes - for instance any statement about rigidity of convex bodies is also locally projectively invariant.

6. Acknowledgements

The research of this contribution was supported by the Grant nr. 201/02/0616 of the Czech Grant Agency and partially by grant MSM113200007 of the Ministery of Education of the Czech Republic.

References

Borel E. (1908), Mémoire sur les déplacements à trajectoires sphérigues. *Mém.présentés par divers savants*, Paris (2), 33, 1-128.

Bottema, O., Roth, B (1990), Theoretical Kinematics, *Dover Publishing.*

Bricard R. (1906), Mémoire sur les déplacements à trajectoires sphérigues. *Journ. École Polyt. (2)* 11, 1-96.

Husty, M. L., Karger A.(2000), Self-motions of Griffis-Duffy type parallel manipulators. *Proceedings of the 2000 IEEE international Conference on Robotics & Automation* San Francisco, CA, pp. 7-12.

Karger A. (2001), Singularities and self-motions of equiform platforms, *MMT* **36,** 801-815.

Karger A. (2002), Singularities ans Self-Motions of a special type of platforms. *J. Lernarčič and F. Thomas, Advances in Robot Kinematics* Kluwer Acad. Publ., ISBN 1-4020-0696-9, pp.155-164.

Ma O., Angeles J. (1992), Architecture Singularities of Parallel Manipulators. *Int. Journal of Robotics and Automation* 7, 23-29.

Merlet J.P. (1992), Singular Configurations of Parallel Manipulators and Grasssmann Geometry. *Int. Journal of Robotics Research,* 8, 45-56.

SCREW SPACES AND CONNECTIVITIES IN MULTILOOP LINKAGES

K. Wohlhart
Graz University of Technology
A-8010 Graz, Austria
wohlhart@mechanik.tu-graz.ac.at

Abstract In this paper a procedure is decribed which allows determining the relative screw spaces between any two links in a multilooped linkage most efficiently. The leading role is played by the null space of the general constraint matrix of the linkage, which is generated quasi automatically by a planar graph with edges oriented according to the joint screws. The null space of the constraint matrix provides a computer program numerically. With the joint screws on an arbitrary path between any two vertices in the graph a path matrix can be composed which, multiplied by the null matrix gives the screws which constitute the screw space between the two links the path connects in the graph. This space is independent of the chosen path. The rank of the null space matrix equals the overall degrees of freedoms of the linkage and the ranks of the relative screw spaces between any two links yield their relative degrees of freedom, i.e. their connectivity. All connectivities are finally compiled in a symmetric connectivity matrix. As the procedure takes into account the dimensions of the linkages right from the beginning, paradoxical linkages are not excluded.

Keywords: Multi-body linkages, screw spaces, connectivities.

1. Introduction

In many applications the relative mobility between two specific links in a mechanism (e.g. the link on the ground and the end effector link) is often of greater interest than the overall mobility of the mechanism itself, which in the case of non-paradoxical mechanisms can be found by the topological structure formula of Grübler-Kutzbach.

Waldron showed that the "connectivity" of any two links (i.e. the relative degrees of freedom) in a mechanism can be determined by considering the linkage as a simple joint between the two links and by finding its screw system (Waldron, K.J.,1966). Davies and Primrose showed that two laws, the law of the addition of spaces for kinematical chains in series and the law of the intersection of spaces for parallel chains, are not sufficient for the determination of the connectivities between links located in different loops. So they proposed a special solution for this specific problem (T.H. Davies and J.F. Primrose 1971). Other solutions have

J. Lenarčič and C. Galletti (eds.), On Advances in Robot Kinematics, 97-104.

been presented subsequently by Baker (Baker, J.E., 1981) and by Davies (Davies, T.H., 1981). Although the methods proposed are general, these authors limit themselves to examples of linkages with not more than two loops. To overcome the difficulties in connection with many loops in a mechanism Fayet proposed in three detailed papers in French another general method which he calls "méthode projective triangulaire" (Fayet, M.,1995). This method proceeds progressively and allows carrying out all calculations within the space of the six-dimensional screw vectors.

In the present paper higher dimensional vector spaces are not avoided and it will be shown that the determination of the relative screw vector spaces or the connectivity between two bodies with the null-space matrix of the general constraint matrix becomes more transparent than in any other procedure.

2. Demonstration: First Example

To ease the comparison especially of Fayet's method with the method followed in the present paper, we choose an example similar to the mechanism M_3 in Fayet's first paper (Fig.1).

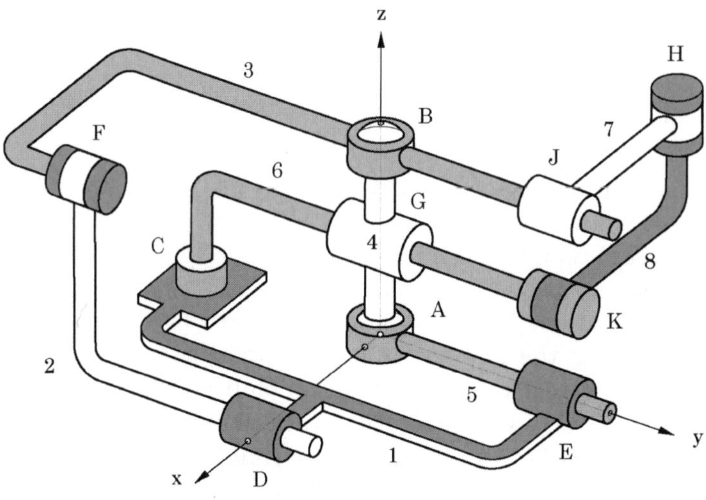

Figure 1. Mechanism with three Fundamental Loops

The mechanism consists of 8 links which are held together by 9 joints forming three fundamental loops. Its oriented graph is shown in Fig. 2. For each loop a planar polygon can be drawn in which the vertices represent the rigid bodies of the mechanism and the edges (polygon sides) the joints which connect them. The polygons of the loops can be united to a planar graph by making equal vertices and edges coincide. By choosing an equal sense of circulation for its loop polygons I, II and III an orientation of the graph is created in which the inner edges are doubly oriented (Fig.2).

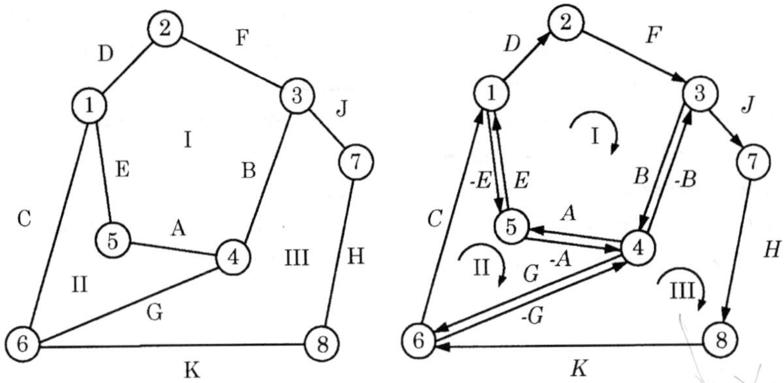

Figure 2. The Planar Graph of the Mechanism and its Orientation

The most general infinitesimal motion of a rigid body is a screwing motion, i.e. an infinitesimal translatorial motion along an axis and simultaneously an infinitesimal rotation around this axis. If the screw axis is given with a position vector $\bar{x}$ of one of its points together with the unit vector $\bar{n}$ in the direction of it and if furtheron the angular velocity ω and the translatorial velocity v in the direction of $\bar{n}$ or the pitch $p = v/\omega$ are known, then the infinitesimal motion of the body is mathematically described by the velocity motor V. If related to the origin O of the coordinate system this motor (then written: V_O) can be represented by the pair of vectors $\bar{\omega} = \omega\bar{n}$ and $\bar{v}_O$, the velocity of the body point which coincides at the moment with the origin of the coordinate system. Therefore we have

$$V_O = \begin{Bmatrix} \bar{\omega} \\ \bar{v}_O \end{Bmatrix} = \begin{Bmatrix} \bar{\omega} \\ \bar{x} \times \bar{\omega} + \bar{v} \end{Bmatrix} = \omega \begin{Bmatrix} \bar{n} \\ \bar{x} \times \bar{n} + p\bar{n} \end{Bmatrix} = \omega S_O. \tag{1}$$

With the angular velocity ω going to zero and simultaneously the pitch progressing to infinity we get with $p\omega = v$ and $\bar{0}^{\mathrm{T}} = \{0,0,0\}$ for the translation velocity motor V_O or with $\omega \to 0$ and $v \to 0$ for the zero velocity motor:

$$V_O = v\begin{Bmatrix} \bar{0} \\ \bar{n} \end{Bmatrix} = vT_O, \quad \theta = \begin{Bmatrix} \bar{0} \\ \bar{0} \end{Bmatrix}. \tag{2}$$

For geometric motors (joint screws) as S_O or T_O in formulas (1), (2) we shall simply write $A,B,C...$ according to the joints A,B,C.... they are attributed to (Fig.2).

As ω in (1) and v in (2) are time derivatives of internal position parameters (angles or lengths). Being both speed coordinates, we shall write for them $s_A, s_B, s_C...$ according to their joint screws $A,B,C...$

A multiple joint can always be interpreted as a compound of elementary joints, e.g. a spherical joint is mathematically equivalent to three rotor joints with intersecting axes or a cylindrical joint equals a rotor joint and a prismatic joint on an identical axis.

The positions of the joints in the mechanism are given in the Cartesian coordinate system $\{O = A; e_1, e_2, e_3\}$ with the vectors:

$$x_A^T = \{0,0,0\},\quad x_B^T = \{0,0,1\},\quad x_C^T = \{0,-1,0\},\quad x_D^T = \{1,0,0\},\quad x_E^T = \{0,1,0\},$$
$$x_F^T = \{1,-1,1\},\quad x_G^T = \{0,0,1/2\},\quad x_H^T = \{-1,1,1\},\quad x_J^T = \{0,1,1\},\quad x_K^T = \{0,1,1/2\}.$$

The basic geometric motors in the mechanism shown in Fig.1 are given with

$$A_1^T = \{1,0,0,0,0,0\},\quad A_2^T = \{0,1,0,0,0,0\},\quad A_3^T = \{0,0,1,0,0,0\},$$
$$B_1^T = \{1,0,0,0,1,0\},\quad B_2^T = \{0,1,0,-1,0,0\},\quad B_3^T = \{0,0,1,0,0,0\},$$
$$C_1^T = \{0,0,0,1,0,0\},\quad C_2^T = \{0,0,0,0,1,0\},\quad C_3^T = \{0,0,1,-1,0,0\},$$
$$D_1^T = \{0,0,0,0,1,0\},\quad D_2^T = \{0,1,0,0,0,1\},\quad E_1^T = \{0,0,0,0,1,0\},$$
$$E_2^T = \{0,1,0,0,0,0\},\quad F^T = \{0,1,0,-1,0,1\},\quad G_1^T = \{0,0,0,0,1,0\},$$
$$G_2^T = \{0,1,0,-.5,0,0\},\quad H^T = \{0,0,1,1,1,0\},\quad J_1^T = \{0,0,0,0,1,0\},$$
$$J_2^T = \{0,1,0,-1,0,0\},\quad K^T = \{0,1,0,-.5,1,0\}.$$

Finally all the speed coordinates are collected in a column matrix s :

$$s^T = \{s_A, s_B, s_C, s_D, s_E, s_F, s_G, s_H, s_J, s_K\}, \tag{3}$$

where

$$s_A = \{s_{A_1}, s_{A_2}, s_{A_2}\},\quad s_B = \{s_{B_1}, s_{B_2}, s_{B_2}\},\quad s_C = \{s_{C_1}, s_{C_2}, s_{C_2}\},$$
$$s_D = \{s_{D_1}, s_{D_2}\},\quad s_E = \{s_{E_1}, s_{E_2}\},\quad s_G = \{s_{G_1}, s_{G_2}\},\quad s_J = \{s_{J_1}, s_{J_2}\}.$$

With the matrices:

$$A = \left\{A_1, A_2, A_3\right\},\quad B = \left\{B_1, B_2, B_3\right\},\quad C = \left\{C_1, C_2, C_3\right\},$$
$$D = \left\{D_1, D_2\right\},\quad E = \left\{E_1, E_2\right\},\quad G = \left\{G_1, G_2\right\},\quad J = \left\{J_1, J_2\right\},$$

the three motor closure equations can be formulated according to the three fundamental loops in the mechanism:

$$\begin{aligned} &A\, s_A + B\, s_B + D\, s_D + E\, s_S + F\, s_F = \theta, \\ -&A\, s_A + C\, s_C - E\, s_E + G\, s_G = \theta, \\ -&B\, s_B - G\, s_G + H\, s_H + J\, s_J + K\, s_K = \theta, \end{aligned} \tag{4}$$

where

$$\begin{aligned} &A\, s_A = A_1\, s_{A_1} + A_2\, s_{A_2} + A_3\, s_{A_3}, \quad B\, s_B = B_1\, s_{B_1} + B_2\, s_{B_2} + B_3\, s_{B_3}, \\ &C\, s_C = C_1\, s_{C_1} + C_2\, s_{C_2} + C_3\, s_{C_3}, \quad D\, s_D = D_1\, s_{D_1} + D_2\, s_{D_2}, \\ &E\, s_E = E_1 s_{E_1} + E_2 s_{E_2}, \ G\, s_G = G_1 s_{G_1} + G_2 s_{G_2}, \ J s_E = J_1 s_{J_1} + J_2 s_{J_2}. \end{aligned}$$

These three motor equations can be written as one matrix equation:

$$\begin{Bmatrix} A & B & \theta_3 & D & E & F & \theta_2 & \theta & \theta_2 & \theta \\ -A & \theta_3 & C & \theta_2 & -E & \theta & G & \theta & \theta_2 & \theta \\ \theta_3 & -B & \theta_3 & \theta_2 & \theta_2 & \theta & -G & H & J & K \end{Bmatrix} s = \begin{Bmatrix} \theta \\ \theta \\ \theta \end{Bmatrix}, \tag{5}$$

where θ_2, θ_3 stands for the following two zero matrices

$$\theta_2 = \{\theta, \theta\}, \quad \theta_3 = \{\theta, \theta, \theta\}.$$

With the system parameters the global constraint matrix M (of dimensions 18×20) is numerically given:

$$M = \begin{Bmatrix} A & B & \theta_3 & D & E & F & \theta_2 & \theta & \theta_2 & \theta \\ -A & \theta_3 & C & \theta_2 & -E & \theta & G & \theta & \theta_2 & \theta \\ \theta_3 & -B & \theta_3 & \theta_2 & \theta_2 & \theta & -G & H & J & K \end{Bmatrix}, \tag{6}$$

and its null space matrix nM can be determined numerically or found directly with the computer software system "Mathematica" by the command: Transpose[NullSpace[M]]. The dimensions of the null space matrix are 20×5 and its transposed matrix $(nM)^{\mathsf{T}}$ reads:

$$\begin{Bmatrix} 0 & 1 & 0 & 0 & -1 & 0 & 1/2 & -1 & 0 & 0 & -1 & 0 & 0 & 1 & 1 & 1 & 0 & 0 & -1 & 1 \\ 0 & 0 & 0 & 0 & 0 & 0 & 0 & -1 & 0 & 0 & 0 & 0 & 0 & 0 & 1 & 0 & 0 & 1 & 0 & 0 \\ 0 & -1 & 0 & 0 & 0 & 0 & 0 & 0 & 0 & 0 & 0 & 0 & 1 & 0 & 0 & 0 & 0 & 0 & 0 & 0 \\ 0 & 0 & 0 & 0 & 0 & 0 & 0 & 1 & 0 & -1 & 0 & 0 & 0 & 0 & 0 & 0 & 0 & 0 & 0 & 0 \\ 0 & -1 & 1 & 0 & 0 & 0 & 0 & 0 & 0 & 0 & 0 & 0 & 0 & 0 & 0 & 0 & 0 & 0 & 0 & 0 \end{Bmatrix}.$$

3. Relative Screw Spaces and Connectivities

The Grüber-Kutzbach formula yields for the mechanism in Fig.1

$$f \geq f_0 = \Sigma f_i - 6l = 20 - 3 \times 6 = 2 .$$

According to the number of independent columns in the null space matrix nM the global degrees of freedom are in the given position: $f = 4$.

To find the velocity motor spaces and therewith the connectivities between two links in the mechanism an arbitrary path is chosen in its oriented graph which connects the vertices corresponding to these two links. The geometric motors on this path are then put into a matrix of the dimensions $6 \times \Sigma f_i$ in an ordered way. This matrix premultiplied with the null space matrix nM gives the velocity motors, a linear combination of which gives the velocity motor of one link if the other one is held fixed. The connectivity equals the number of independent motors.

The shortest path leading from vertex 5 to vertex 8 in Fig.2 goes over the edges $-A,G,-K$. Ordered in a $6 \times \Sigma f_i$ path-matrices we get

$$P_{58} = \left\{ -A, \quad \theta_3, \quad \theta_3, \quad \theta_2, \quad \theta_2, \quad \theta, \quad G, \quad \theta, \quad \theta_2 \quad -K \right\}. \tag{7}$$

Multiplication with the null space matrix nM gives for the relative motor space S_{58} and for the connectivity f_{58} of link 5 and the link 8:

$$P_{58} \,.\, nM = \left\{ \begin{matrix} 0 & 0 & 0 & 0 & 0 \\ -1 & 0 & 1 & 0 & 1 \\ 0 & 0 & 0 & 0 & -1 \\ 0 & 0 & 0 & 0 & 0 \\ 0 & 1 & 0 & 0 & 0 \\ 0 & 0 & 0 & 0 & 0 \end{matrix} \right\} \Rightarrow S_{58} = \left\{ \begin{matrix} 0 & 0 & 0 \\ 0 & 1 & 1 \\ 0 & 0 & -1 \\ 0 & 0 & 0 \\ 1 & 0 & 0 \\ 0 & 0 & 0 \end{matrix} \right\} \Rightarrow f_{58} = 3. \tag{8}$$

For any other chosen path leading from vertex 5 to vertex 8 e.g. for $-A.-B,J,H$ or $E,D,F,B,G,-K$ with the path matrices

$$P'_{58} = \left\{ -A, \quad -B, \quad \theta_3, \quad \theta_2, \quad \theta_2, \quad \theta, \quad \theta_2, \quad H, \quad J \quad \theta \right\}$$
$$P''_{58} = \left\{ \theta_3, \quad B, \quad \theta_3, \quad D, \quad E, \quad F, \quad G, \quad \theta, \quad \theta_2 \quad -K \right\}$$

the same results are obtained as $P'_{58} . nM = P''_{58} . nM = P_{58} . nM$. For the path from the vertex 8 to 5 over the edges $A, -G, K$ with the path matrix

$$P_{85} = -P_{58} = \left\{ A, \quad \theta_3, \quad \theta_3, \quad \theta_2, \quad \theta_2, \quad \theta, \quad -G, \quad \theta, \quad \theta_2 \quad K \right\}$$

$S_{85} = -S_{58}$ and $f_{85} = f_{58} = 3$ are obtained.

If for all link pairs the motorspaces $S_{ij} = -S_{ji} \Rightarrow S_{ii} = 0$ and the connectivities are determined in this way, the latter can be collected in a symmetric connectivity matrix with zeros in its diagonal:

$$F = \| f_{ij} \|, \quad (f_{ij} = f_{ji}) \wedge (f_{ii} = 0).$$

For the mechanism in Fig 1 the connectivity matrix is found to be:

$$F = \begin{Bmatrix} 0 & 2 & 2 & 2 & 3 & 2 & 2 & 2 \\ 2 & 0 & 1 & 1 & 3 & 2 & 2 & 2 \\ 2 & 1 & 0 & 1 & 3 & 2 & 2 & 2 \\ 2 & 1 & 1 & 0 & 2 & 2 & 1 & 1 \\ 3 & 3 & 3 & 2 & 0 & 4 & 3 & 3 \\ 2 & 2 & 2 & 2 & 4 & 0 & 1 & 1 \\ 2 & 2 & 2 & 1 & 3 & 1 & 0 & 0 \\ 2 & 2 & 2 & 1 & 3 & 1 & 0 & 0 \end{Bmatrix}. \quad (8)$$

4. Second Example

If the joints C and J are reduced to $C = \{C_1\}$ and $J = \{J_2\}$ the mechanism becomes overconstrained (Fig.3) as the Grübler-Kutzbach formula yields $f \geq f_0 = \Sigma f_i - 6l = 17 - 3 \times 6 = -1$.

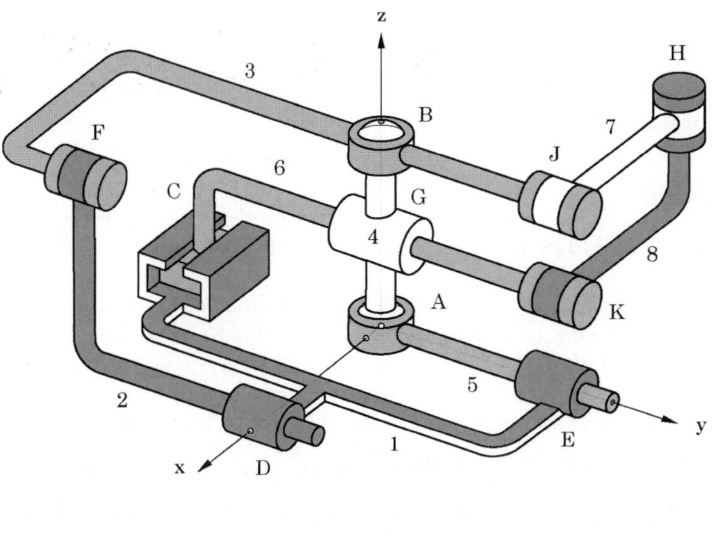

Figure 3. Overconstrained Mechanism with three Fundamental Loops

For this mechanism the oriented loop matrix reads:

$$M = \left\{ \begin{array}{cccccccccc} A & B & \theta_3 & D & E & F & \theta_2 & \theta & \theta & \theta \\ -A & \theta_3 & C & \theta_2 & -E & \theta & G & \theta & \theta & \theta \\ \theta_3 & -B & \theta_3 & \theta_2 & \theta_2 & \theta & -G & H & J & K \end{array} \right\}$$

and its null space matrix nM has two independent columns

$$nM^{\mathsf{T}} = \left\{ \begin{array}{ccccccccccccccccc} 0 & 1 & 0 & 0 & -1 & 0 & 1/2 & 0 & -1 & 0 & 0 & 1 & 0 & 1 & 0 & -1 & 1 \\ 0 & -1 & 0 & 0 & 0 & 0 & 0 & 0 & 0 & 0 & 1 & 0 & 0 & 0 & 0 & 0 & 0 \end{array} \right\}.$$

Therefore the mechanism in the given position is mobile with two degrees of freedom: $f = 2$. With the null space matrix the connectivities can be determined. The connectivity matrix is found to be:

$$F = \left\{ \begin{array}{cccccccc} 0 & 1 & 1 & 1 & 1 & 1 & 1 & 1 \\ 1 & 0 & 1 & 1 & 2 & 1 & 1 & 1 \\ 1 & 1 & 0 & 1 & 2 & 1 & 1 & 1 \\ 1 & 1 & 1 & 0 & 1 & 1 & 2 & 0 \\ 1 & 2 & 2 & 1 & 0 & 2 & 1 & 1 \\ 1 & 1 & 1 & 1 & 2 & 0 & 1 & 1 \\ 1 & 1 & 1 & 2 & 1 & 1 & 0 & 0 \\ 1 & 1 & 1 & 0 & 1 & 1 & 0 & 0 \end{array} \right\}.$$

5. Conclusions

It is shown that the null space matrix of the oriented loop matrix immediately leads to the motor spaces or the connectivities between any two links in a multilooped linkage.

References

Waldron,K.J., (1966), The Constrained Analysis of Mechanisms, *J. Mechanism*, Vol. 1, pp. 101-114.

Davies, T.H., and Primerose, E.J.F., (1971), *Proceedings of the Third World Congress for the Theory of Machines and Mechanisms*, Kupary, Yougoslavia. Vol.D, paper D-14, pp.199-212.

Baker, J.E, (1981), On Mobility and Relative Freedoms in Multiloop Linkages and Structures, *Mechanism and Machine Theory*, Vol.16, pp.583-597.

Davies, T.H, (1981), Kirchhoff's Circulation Law Applied to Multi-Loop Kinematic Chains, *Mechanism and Machine Theory*, Vol.16, pp.171-183.

Fayet, M, (1995), Mécanismes multi-boucles I, II, III, *Mechanism and Machine Theory*, Vol. 30, No. 2, pp. 201-217, pp. 219-232, pp. 233-252.

KINEMATICS EQUATIONS OF A CLASS OF 4DOF PARALLEL WRISTS

Matteo Zoppi[a], Dimiter Zlatanov[b], Clément M. Gosselin[c]
[a] *Università di Genova, via Opera Pia 15A, 16145, Genova, Italia*
[b] *Universität Innsbruck, Technikerstr. 13, A-6020, Innsbruck, Austria*
[c] *Département de Génie Mécanique, Université Laval, Québec, QC, Canada*
zoppi@dimec.unige.it, [zlatanov, gosselin]@gmc.ulaval.ca

Abstract The paper discusses forward and inverse kinematics of a class of 4-dof 4-legged parallel mechanisms providing three rotational and one translational freedoms. A fully parametric analytical form solution to the inverse position problem is provided. All working modes of the mechanism are shown and discussed. The equations of the forward position problem are obtained under different leg arrangement and a numerical example is provided. New special geometries in the class are proposed.

Keywords: forward kinematics, inverse kinematics, 4dof mechanisms

1. Introduction

The paper studies inverse and forward kinematics of the class of 4-dof, parallel mechanisms (PMs) shown in Fig. 1, and shows some new special geometries in the class. When they were introduced (Zlatanov and Gosselin, 2001), these were the first known 4-dof PMs. Figure 1(a) is the geometry presented at that time. Later, other 4-dof mechanisms have been presented, Huang and Li, 2002.

The end-effector is connected to the base by four 5R serial chains (legs), labelled $A, \ldots, D$. The joints of each leg are numbered from the base. In leg L ($L \in \{A, B, C, D\}$), the joint axes are $\boldsymbol{\xi}_1^L, \ldots, \boldsymbol{\xi}_5^L$ and the corresponding direction vectors $\mathbf{k}_1^L, \ldots, \mathbf{k}_5^L$. The first three axes intersect in the fixed *rotation center*, O, common to all four legs. The fourth and

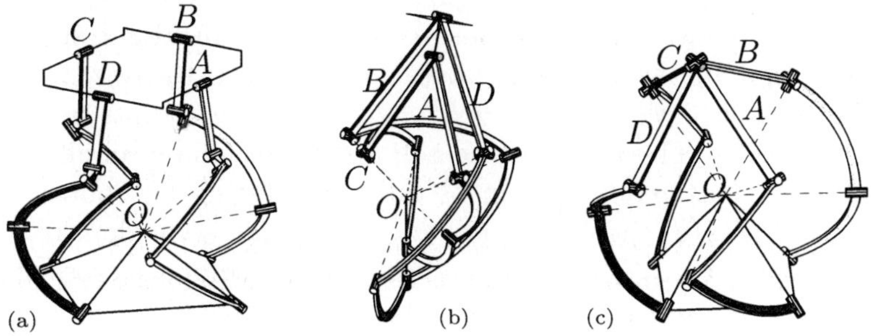

Figure 1. Mechanisms in the class.

J. Lenarčič and C. Galletti (eds.), On Advances in Robot Kinematics, 105-112.

fifth joints are parallel. The platform joint axes, $\boldsymbol{\xi}_5^A, \ldots, \boldsymbol{\xi}_5^D$, are parallel to the *platform plane*, π_h, which can be chosen arbitrarily among all planes fixed in the end-effector with the same normal, $\mathbf{k}$. Assuming that the $\boldsymbol{\xi}_5^L$ are not all parallel, the platform has a full rotational mobility around O and translates orthogonally to π_h (Zlatanov and Gosselin, 2001). The choice of four legs allows to control the PM by actuating base joints only. The mechanism works with any number of legs greater than one and part of the results can be extended to the general, m-legs layout. The extrusion (or heave), h, of the platform (representing the amount of translation) is the distance between π_h and the parallel 0-*extrusion plane*, π_0, through O.

2. Geometry

We use: a platform reference frame $P\mathbf{ijk}$ ($\mathbf{k} \perp \pi_h$), at the projection, P, of O on π_h; a rotating frame $O\mathbf{ijk}$; a base frame, $O\mathbf{i}_b\mathbf{j}_b\mathbf{k}_b$. The leg L *heave plane*, π_e^L, is through O and orthogonal to $\boldsymbol{\xi}_5^L$. Points P_4^L and P_5^L are the intersections of π_e^L with $\boldsymbol{\xi}_4^L$ and $\boldsymbol{\xi}_5^L$, respectively. The distance from P_4^L to P_5^L is l_{45}^L; r_4^L is between P_4^L and O. The constant angle α_{34}^L, spanned by the third leg link, is from $\boldsymbol{\xi}_3^L$ to $\overrightarrow{OP_4^L}$ (Fig. 2(b)). The angles of the first two links, $\alpha_{i\,i+1}^L$, $i = 1, 2$, are between $\mathbf{k}_i^L$ and $\mathbf{k}_{i+1}^L$. The set $\{l_{45}^L, r_4^L, \alpha_{34}^L, \alpha_{12}^L, \alpha_{23}^L\}$ gives the geometry of leg L standing alone.

How leg L is placed in the mechanism is described by: the distances r_5^L from P_5^L to $\overrightarrow{OP}$, and h^L from π_h to $\boldsymbol{\xi}_5^L$; the angles θ^L between π_e^L and $O\mathbf{ik}$; and the tilt, α_1^L, and azimuth, β_1^L, angles that place $\boldsymbol{\xi}_1^L$ in $O\mathbf{i}_b\mathbf{j}_b\mathbf{k}_b$.

The leg can be divided into *heave section* and *spherical section*, formed by the third and fourth, and first two links, respectively. The *PM's heave* and *spherical sections* are the end-effector with all leg heave sections, and the base and the leg spherical sections.

A PM is *rhombic*, if its leg heave sections are two by two equal ($l_{45}^M = l_{45}^N$, $r_4^M = r_4^N$, $\alpha_{34}^M = \alpha_{34}^N$, $\{M, N\} = \{A, C\}, \{B, D\}$) and symmetrical, with π_e^L coincident ($r_5^M = r_5^N$, $\theta^M - \theta^N = \pi$). Also, $\pi_e^M \equiv \pi_e^N$, $\psi_4^M = \psi_4^N$ ($\psi_4^L = \angle P_4^L O P^L$) and $(\mathbf{k}_4^M - \mathbf{k}_4^N) \perp \mathbf{k}$. A *rectangular* PM is rhombic with $\pi_e^A \perp \pi_e^B$; a *square* PM is a rectangular with four identical legs.

Fig. 1 shows two new mechanisms with two nontrivial special geometries. In 1(b) the platform joints coincide, two by two, along two skew axes ($r_5^L = 0\ \forall L$, $h^A = h^C \neq h^B = h^D$); the base joints are coaxial ($\alpha_1^L = \pi/2\ \forall L$). The end-effector has a straight shape along the axis through $O\mathbf{k}$. This geometry can be quite interesting for robotic applications. In Fig. 1(c) the fifth joints are two by two coincident and form a cross($r_5^L = h^L = 0\ \forall L$). The end-effector has a compact cross shape.

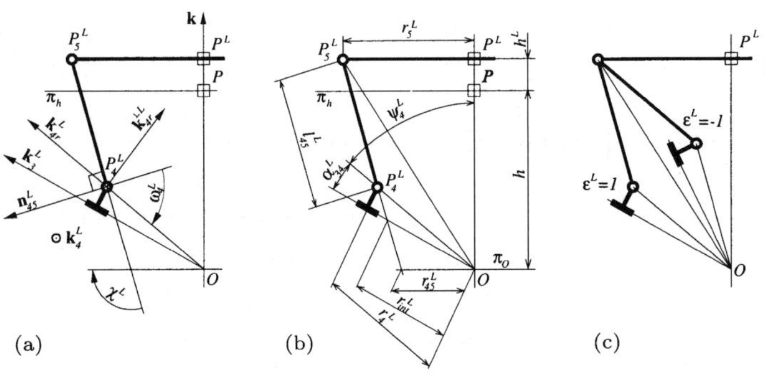

Figure 2. Heave section of leg L: vectors (a); parameters (b); working modes (c).

3. Inverse Kinematics

The end-effector pose is given by the extrusion, h, and the platform orientation matrix, $\mathbf{R}$. The aim is to compute the PM's entire configuration. First, we solve the heave section, then the spherical section. Unless noted otherwise, vectors are given in the end-effector frame.

Heave-section. Consider the leg L heave section. Point P^L is the projection of O on the plane through $\boldsymbol{\xi}_5^L$ parallel to π_h (Fig. 2); the quadrilateral $OP^LP_5^LP_4^L$ is a virtual 1-dof mechanism with parameter h. We obtain $\mathbf{k}_3^L$ in the reference frame $O(\mathbf{k}_4^L{\times}\mathbf{k})\,\mathbf{k}_4^L\,\mathbf{k}$, and then in $O\mathbf{ijk}$.

First, we get $c_4^L = \cos\psi_4^L$, $s_4^L = \sin\psi_4^L$. We have

$$2r_4^LH^Lc_4^L + 2r_4^Lr_5^Ls_4^L - {H^L}^2 - N^L = 0 \tag{1}$$

where $H^L = h + h^L$, $N^L = {r_5^L}^2 + {r_4^L}^2 - {l_{45}^L}^2$. The solutions of (1) are:

$$\begin{bmatrix} c_4^L & s_4^L \end{bmatrix}^T = (2r_4^L({H^L}^2 + {r_5^L}^2))^{-1} \begin{bmatrix} H^L & -r_5^L \\ r_5^L & H^L \end{bmatrix} \begin{bmatrix} {H^L}^2 + N^L & \varepsilon^L C_\varepsilon^L \end{bmatrix}^T \tag{2}$$

where $\varepsilon^L = \pm 1$ distinguishes between the two feasible working modes of the quadrilateral $OP^LP_5^LP_4^L$ (Fig. 2), and:

$$C_\varepsilon^L = \sqrt{-({H^L}^2 - m_{45}^L)\,({H^L}^2 - M_{45}^L)} \tag{3}$$

with $M_{45}^L = (r_4^L + l_{45}^L)^2 - {r_5^L}^2$ and $m_{45}^L = (r_4^L - l_{45}^L)^2 - {r_5^L}^2$.

The quadrangle takes symmetric configurations for positive and negative values of h. Consider $h \geq 0$. At maximal extrusion, P_5^L, P_4^L and O are collinear and $H^L = (M_{45}^L)^{\frac{1}{2}}$. At minimal extrusion (if allowed by the geometry of the whole mechanism), P_5^L, P_4^L and O are collinear, with P_5^L between P_4^L and O; $H^L = (m_{45}^L)^{\frac{1}{2}}$. (These are transition configurations

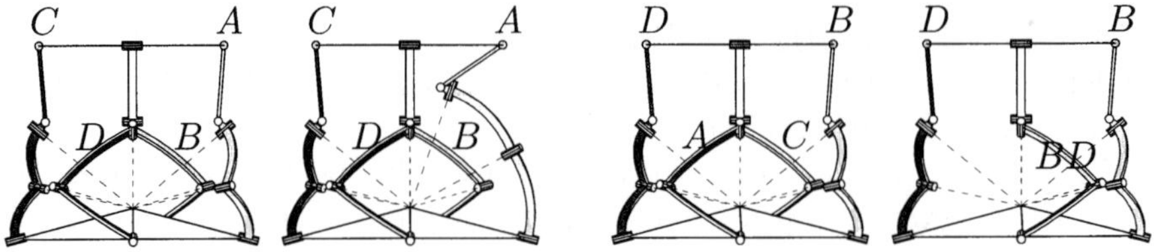

(a) $\varepsilon^A = +1$ (b) $\varepsilon^A = -1$ (c) $\delta^A = +1$ (d) $\delta^A = -1$

Figure 3. Working modes of the heave section ((a) and (b)) and of the spherical section ((c) and (d)) of leg A.

between the two working modes of the quadrilateral.) It is clear from (3) that for C_ε^L to be real, it is necessary that $m_{45}^L \leq {H^L}^2 \leq M_{45}^L$.

To obtain $\mathbf{k}_4^L$ in $P\mathbf{ijk}$, we rotate $\mathbf{j}$ about $\mathbf{k}$ at θ^L, $\mathbf{k}_4^L = \mathbf{R}_z(\theta^L)\,\mathbf{j}$. Vector $\mathbf{k}_{4r}^L$ is computed analogously:

$$\mathbf{k}_{4r}^L = \mathbf{R}_z(\theta^L)\left[\, s_4^L \;0\; c_4^L \,\right]^T \tag{4}$$

Now an α_{34}^L rotation about $\mathbf{k}_4^L \perp \pi_e^L$ gets $\mathbf{k}_3^L = \mathbf{R}_e(\alpha_{34}^L)\,\mathbf{R}_z(\theta^L)\left[\, s_4^L \;0\; c_4^L \,\right]^T$.

The four ε^L, one for each leg, distinguish between four different working modes of the heave section of the mechanism. Fig. 3 shows the two working modes of the heave section of one leg.

We compute the components of $\mathbf{n}_{45}^L$ along $\mathbf{k}_{4r}^L$ and $\mathbf{k}_{4r}^{\perp L}$ and add them:

$$\mathbf{n}_{45}^L = \frac{\mathbf{R}_z(\theta^L)}{2\, l_{45}^L\, r_4^L}\left(\varepsilon^L C_\varepsilon^L\, \mathbf{k}_{4r}^L + ({l_{45}^L}^2 + {r_4^L}^2 - {r_5^L}^2 - {H^L}^2)\, \mathbf{k}_{4r}^{\perp L}\right) \tag{5}$$

where $\mathbf{k}_{4r}^L$ is provided by (4), and $\mathbf{k}_{4r}^{\perp L} = \mathbf{k}_4^L \times \mathbf{k}_{4r}^L$.

Spherical section. The poses of the first and second links of leg L depend both on $\mathbf{R}$ and h through the direction vectors $\mathbf{k}_1^L$ and $\mathbf{k}_3^L$. Vector $\mathbf{k}_3^L$ is a result of the inverse kinematics of the heave section, while vector $\mathbf{k}_1^L$ is easily obtained by a change of reference frame: $\mathbf{k}_1^L = \mathbf{R}\,\mathbf{k}_{1\,\mathrm{b}}^L$.

It remains to compute vector $\mathbf{k}_2^L$. The axis $\boldsymbol{\xi}_2^L$ of the second joint lies at the intersection of two right circular cones with vertex, axes and half-angle $O, \boldsymbol{\xi}_1^L, \alpha_{12}^L$ and $O, \boldsymbol{\xi}_3^L, \alpha_{23}^L$, respectively. For $\boldsymbol{\xi}_2^L$ to lie on both cones, $\mathbf{k}_2^L$ must satisfy $\mathbf{k}_2^L \cdot \mathbf{k}_1^L = c_{12}^L$ and $\mathbf{k}_2^L \cdot \mathbf{k}_3^L = c_{23}^L$, where $c_{ij}^L = \cos\alpha_{ij}^L$. Then, $\mathbf{k}_2^L$ is obtained as a linear combination of $\mathbf{k}_1^L$, $\mathbf{k}_3^L$ and $\mathbf{k}_1^L \times \mathbf{k}_3^L$,

$$\mathbf{k}_2^L = (1 - (\mathbf{k}_1^L \cdot \mathbf{k}_3^L)^2)^{-1}\left[\, \mathbf{k}_1^L \;\mathbf{k}_3^L \;\mathbf{k}_1^L \times \mathbf{k}_3^L \,\right]\left[\, t_1^L \; t_2^L \; t_3^L \,\right]^T \tag{6}$$

The coefficients are: $t_1^L = c_{12}^L - c_{23}^L\, \mathbf{k}_1^L \cdot \mathbf{k}_3^L$, $t_2^L = c_{23}^L - c_{12}^L\, \mathbf{k}_1^L \cdot \mathbf{k}_3^L$, $t_3^L = \delta^L\, C_\delta^L$,

$$C_\delta^L = \sqrt{-(\mathbf{k}_1^L \cdot \mathbf{k}_3^L - m_{AB}^L)\,(\mathbf{k}_1^L \cdot \mathbf{k}_3^L - M_{AB}^L)} \tag{7}$$

with $m^L_{AB}=c^L_{12}c^L_{23}-|s^L_{12}s^L_{23}|$, $M^L_{AB}=c^L_{12}c^L_{23}+|s^L_{12}s^L_{23}|$, $s^L_{ij}=\sin\alpha^L_{ij}$. The boolean parameter $\delta^L=\pm 1$ distinguishes between the two working modes of the spherical section of the leg (Fig. 3).

The rotation angles of the actuated (base) joints, the solution of the inverse kinematics problem, are easily computed from the four $\mathbf{k}^L_2$.

A leg configuration is feasible (C^L_δ in (7) is real) iff $|\alpha^L_{12}-\alpha^L_{23}| \leq \alpha^L_{13} \leq |\alpha^L_{12}+\alpha^L_{23}|$, where $\alpha^L_{13}=\angle(\boldsymbol{\xi}^L_1,\boldsymbol{\xi}^L_3)$, i.e. $m^L_{AB} \leq \mathbf{k}^L_1{\cdot}\mathbf{k}^L_3 \leq M^L_{AB}$.

An expressions for the unit-length vector $\mathbf{n}^L_{23}$, orthogonal to the plane π^L_{23} defined by $\boldsymbol{\xi}^L_2$ and $\boldsymbol{\xi}^L_3$ is:

$$\mathbf{n}^L_{23}=|s^L_{23}|^{-1}\left(-t^L_3\,\mathbf{k}^L_1+t^L_3\,\mathbf{k}^L_1{\cdot}\mathbf{k}^L_3\,\mathbf{k}^L_3+t^L_1\,\mathbf{k}^L_1{\times}\mathbf{k}^L_3\right) \tag{8}$$

Things simplify if $c^L_{12}=c^L_{23}=0$: (7) becomes $C^L_\delta=\sqrt{1-(\mathbf{k}^L_1{\cdot}\mathbf{k}^L_3)^2}$, $\mathbf{k}^L_2 \parallel \mathbf{k}^L_1{\times}\mathbf{k}^L_3$ ($t^L_1=t^L_2=0$) and $T^L_3=0$.

4. Forward Kinematics

The four $\mathbf{k}^L_2$ are given and the pose, $(h,\mathbf{R})$, is unknown.

In the general case, to write a system of equations for the forward kinematics is quite complicated; a general discussion can be found in Zlatanov and Gosselin, 2001. Here, we consider rhombic PMs with $\alpha^L_{34}=0$, $h^L=0$, $\forall L$. Even for this case, writing a reasonable system of equations is not obvious. Using, as is usual, h and some rotation parameters as variables, results in very complex formulations. We have succeeded in deriving relatively simple equations, however, an analytical form solution is not known. Such systems can be solved numerically (see Saaty, 1981). We provide an even simpler system of equations for the square mechanisms. An analytical form solution in this case may be achievable for particular geometries (see Salmon, 1859, Gohberg et al., 1982), however, this remains an open problem.

To get four simpler equations, we use as intermediate unknowns the angles θ^L_2 ($s^L_2=\sin\theta^L_2$, $c^L_2=\cos\theta^L_2$) of the second joints of the legs, rather than h and three rotation parameters for $\mathbf{R}$. In leg L, θ^L_2 is the angle about $\mathbf{k}^L_2$ from π^L_{12} to π^L_{23}. Since $\boldsymbol{\xi}^L_3$ lies on the cone with axis $\boldsymbol{\xi}^L_2$ and half angle α^L_{23}, an expression of $\mathbf{k}^L_3$ as a function of θ^L_2 is:

$$\mathbf{k}^L_3(\theta^L_2)=\mathbf{a}^L_c c^L_2+\mathbf{b}^L_c s^L_2+\mathbf{g}^L_c \tag{9}$$

where $\mathbf{a}^L_c=s^L_{23}\,\mathbf{k}^L_1{\times}\mathbf{k}^L_2$, $\mathbf{b}^L_c=s^L_{23}\,(c^L_{12}\mathbf{k}^L_2-\mathbf{k}^L_1)$, $\mathbf{g}^L_c=c^L_{23}\,\mathbf{k}^L_2$.

Rhombic mechanisms. Two equations state that the right, rectangular pyramid with base edges along $\boldsymbol{\xi}^L_3$ has equal opposite faces, i.e.,

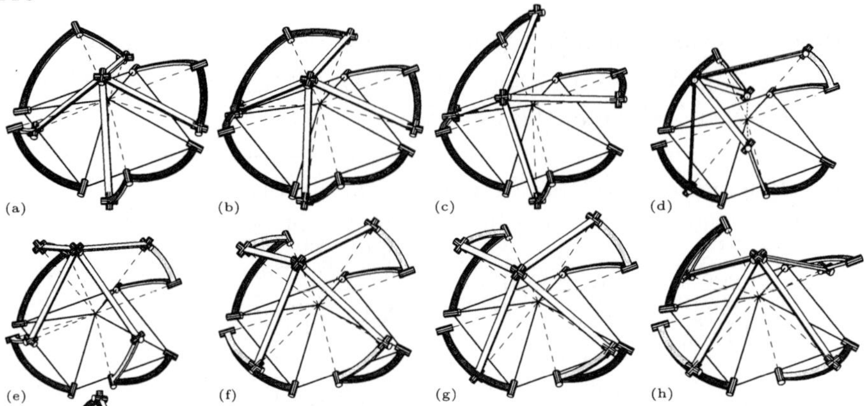

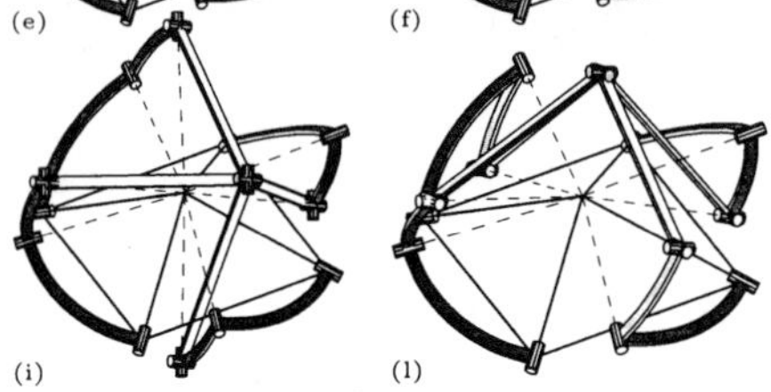

fig.	θ_2^A	θ_2^B	θ_2^C	θ_2^D
a	.0221	.3947	.3535	6.1686
b	.5082	1.0353	.8155	.1905
c	1.3246	2.0550	.7475	6.0834
d	2.7094	.0067	5.5956	2.4968
e	2.9029	3.4042	3.7732	3.3844
f	3.1853	3.8419	4.4627	3.9232
g	3.1513	3.8820	4.7720	4.1761
h	4.8807	4.8715	4.6196	4.6091
i	5.5429	2.2388	3.1101	6.2231
l	5.6334	5.2362	3.2778	3.4593

$\mathbf{R}=\begin{bmatrix} .7846 & -.552 & .2823 \\ .552 & .8292 & .0873 \\ -.2823 & .0873 & .9553 \end{bmatrix}$, $h/l_{45}=1.04$, $\varepsilon^L=\delta^L=1\ \forall L$; $r_4/l_{45}=.8$, $\beta_1=\alpha_{43}=0$, $\alpha_{12}=1.0472$, $\alpha_{23}=.6283$

Figure 4. Example of solutions for the forward kinematics of a square mechanism. To improve visibility, base pyramids are represented wire-frame.

$\angle\boldsymbol{\xi}_3^A\boldsymbol{\xi}_3^B = \angle\boldsymbol{\xi}_3^C\boldsymbol{\xi}_3^D$ and $\angle\boldsymbol{\xi}_3^A\boldsymbol{\xi}_3^D = \angle\boldsymbol{\xi}_3^C\boldsymbol{\xi}_3^B$:

$$\mathbf{k}_3^A \cdot \mathbf{k}_3^B - \mathbf{k}_3^C \cdot \mathbf{k}_3^D = 0 \tag{10}$$

$$\mathbf{k}_3^A \cdot \mathbf{k}_3^D - \mathbf{k}_3^C \cdot \mathbf{k}_3^B = 0 \tag{11}$$

The third equation describes the angle, $\theta^B - \theta^A$, between π_e^A and π_e^B:

$$\left(\mathbf{k}_3^A - \mathbf{k}_3^C\right) \cdot \left(\mathbf{k}_3^B - \mathbf{k}_3^D\right) = \cos(\theta^B - \theta^A) \tag{12}$$

By developing Eq. (12), simplifying and substituting in Eqs. (10) and (11), we obtain the following second degree equations in s_2^L and c_2^L:

$$\mathbf{k}_3^A(\theta_2^A) \cdot \mathbf{k}_3^B(\theta_2^B) - \mathbf{k}_3^B(\theta_2^B) \cdot \mathbf{k}_3^C(\theta_2^C) = \cos(\theta^B - \theta^A)/2 \tag{13}$$

$$\mathbf{k}_3^C(\theta_2^C) \cdot \mathbf{k}_3^D(\theta_2^D) - \mathbf{k}_3^B(\theta_2^B) \cdot \mathbf{k}_3^C(\theta_2^C) = \cos(\theta^B - \theta^A)/2 \tag{14}$$

$$\mathbf{k}_3^A(\theta_2^A) \cdot \mathbf{k}_3^B(\theta_2^B) - \mathbf{k}_3^A(\theta_2^A) \cdot \mathbf{k}_3^D(\theta_2^D) = \cos(\theta^B - \theta^A)/2 \tag{15}$$

and, by means of (9), we write them as three quadratic forms:

$$\mathbf{X}^T \mathbf{Q}_i \mathbf{X} = \cos(\theta^B - \theta^A)/2 \quad i = 1, 2, 3 \tag{16}$$

where $\mathbf{X}^T = \left[s_2^A\ s_2^B\ s_2^C\ s_2^D\ c_2^A\ c_2^B\ c_2^C\ c_2^D\ 1\right]$.

Finally, we express that ψ_4^A and ψ_4^B satisfy (1) for the same value of h. We solve (1) for h; since $\alpha_{34}^L = 0$ and $h^L = 0$, we obtain: $h = r_4^L c_4^L + \varepsilon^L ({r_4^L}^2 {c_4^L}^2 - N^L + 2 r_4^L r_5^L s_4^L)^{\frac{1}{2}}$. Thus, the fourth equation is:

$$r_4^A c_4^A - r_4^B c_4^B = \varepsilon^B \sqrt{{r_4^B}^2 {c_4^B}^2 - N^B + 2 r_4^B r_5^B s_4^B} - \varepsilon^A \sqrt{{r_4^A}^2 {c_4^A}^2 - N^A + 2 r_4^A r_5^A s_4^A} \quad (17)$$

After squaring twice, Eq. (17) becomes a second degree equation in s_4^L, c_4^L. Since the PM is rhombic, $\mathbf{k}_3^M \cdot \mathbf{k}_3^N = \cos(2\psi_4^M)$, $\{M, N\} = \{A, C\}, \{B, D\}$. Therefore, ${c_4^M}^2 = (1 + \mathbf{k}_3^M \cdot \mathbf{k}_3^N)/2$ and ${s_4^M}^2 = (1 - \mathbf{k}_3^M \cdot \mathbf{k}_3^N)/2$. Eq. (9) allows to obtain $\mathbf{k}_3^M \cdot \mathbf{k}_3^N$ as a function of θ_2^L.

Eqs. (16–17) are the final system of four equations for the forward kinematics of rhombic PMs.

Square mechanisms. For square PMs, $\theta^A - \theta^B = \pi/2$ and Eqs. (16) are homogeneous. Eq. (17) becomes $\mathbf{k}_3^A \cdot \mathbf{k}_3^C - \mathbf{k}_3^B \cdot \mathbf{k}_3^D = 0$, stating that $\psi_4^A = \psi_4^B$. From Eq. (9), we write this fourth equation as a quadratic form $\mathbf{X}^T \mathbf{Q}_4 \mathbf{X} = 0$. So, the forward kinematics of the square mechanisms is expressed by a system of four second degree, homogeneous equations $\mathbf{X}^T \mathbf{Q}_i \mathbf{X} = 0$, $i = 1, 2, 3, 4$. The matrices $\mathbf{Q}_i$ are given in the Appendix.

Fig. 4 shows a numerical example for a square, cross-platform geometry. This is the problem of placing a "double-compass" tetrapod on four circles of a sphere. The system of the forward kinematics equations is:

$$\begin{cases} -0.1907 s_2^A c_2^B - 0.2892 c_2^A - 0.377 s_2^A + 0.32 c_2^B + 0.8910 s_2^B - 0.0489 + 0.2573 c_2^A c_2^B - 0.0948 c_2^A s_2^B + \\ + 0.0894 s_2^A s_2^B - 0.0981 c_2^C - 0.4620 s_2^C + 0.0567 c_2^B s_2^C + 0.0934 s_2^B c_2^C - 0.3249 c_2^B c_2^C - 0.0589 s_2^B s_2^C \\ -0.1739 c_2^C - 0.9313 s_2^C + 0.0368 s_2^C s_2^D - 0.0586 - 0.042 s_2^C c_2^D + 0.4506 s_2^B + 0.2515 c_2^D + 0.4034 s_2^D + \\ -0.1792 c_2^C s_2^D + 0.1417 c_2^B + 0.2902 c_2^C c_2^D + 0.0567 c_2^B s_2^C + 0.0934 s_2^B c_2^C - 0.3249 c_2^B c_2^C - 0.0589 s_2^B s_2^C \\ -0.1907 s_2^A c_2^B - 0.5963 c_2^A - 0.73 s_2^A + 0.1782 c_2^B + 0.4404 s_2^B + 0.1436 + 0.2573 c_2^A c_2^B - 0.0948 c_2^A s_2^B + \\ + 0.0894 s_2^A s_2^B + 0.2114 c_2^D + 0.4174 s_2^D - 0.2143 c_2^A c_2^D + 0.1538 c_2^A s_2^D + 0.2232 s_2^A c_2^D - 0.061 s_2^A s_2^D \\ -0.22 s_2^A s_2^C - 0.26 s_2^A c_2^C - 0.263 c_2^A s_2^C + 0.224 c_2^A c_2^C + 0.223 s_2^D s_2^B - 0.082 s_2^A + 0.259 c_2^D s_2^B + 0.065 s_2^C + \\ -0.016 c_2^A + 0.0027 + 0.0525 c_2^C + 0.262 s_2^D c_2^B - 0.225 c_2^D c_2^B - 0.045 s_2^D - 0.002 c_2^B + 0.071 s_2^B - 0.055 c_2^D \end{cases}$$

Geometry and configuration parameters are reported in the figure. We have identified the same ten solutions with several different numeric procedures, yet, so far, there is no proof that others do not exist.

5. Conclusions

The paper provides an analytical form solution of the inverse kinematics of a general class of 4-dof PMs. It also presents a relatively simple formulation of the forward kinematics of an interesting mechanism subclass, including some new PMs with special geometries. The proposed mathematical models are fully parametric and free of mathematic singularities, i.e., valid in all configurations. The obtained equations are a natural starting point for any further research on this mechanism class, such as workspace analysis or force transmission, as well for the design and any practical application of this architecture. Challenging open problems, related to the solution of the direct kinematics and the evaluation of the number of possible solutions, remain.

References

Gohberg, T., Lancaster, P., Rodman, L. (1982), *Matrix Polynomials*, New York, Academic Press;

Huang, Z., Li, Q.C. (2002), On the Type Synthesis of Lower-Mobility Parallel Manipulators, *Proc. of the workshop Fundamental Issues and Future Research Directions for Parallel Mechanisms and Manipulators*, Québec City, Québec, Canada;

Saaty, T. L. (1981), *Modern Non-linear Equations*, New York (1967), Dover (1981), McGraw-Hill;

Salmon, G. (1859), *Introductory to the Modern Higher Algebra*, Dublin, Hodges, Smith, and co.;

Zlatanov, D., Gosselin, C. (2001), A family of new parallel architectures with four degrees of freedom, *In F.C. Park C.C. and Iurascu, editor, Computational Kinematics*, pp.57-66, Mai 2001.

Appendix

$$\mathbf{Q}_1=\begin{bmatrix}
0 & \mathbf{b}_c^B\cdot\mathbf{b}_c^A & 0 & -\mathbf{b}_c^D\cdot\mathbf{b}_c^A & 0 & \mathbf{a}_c^B\cdot\mathbf{b}_c^A & 0 & -\mathbf{a}_c^D\cdot\mathbf{b}_c^A & \mathbf{b}_c^A\cdot\mathbf{g}_c^B-\mathbf{b}_c^A\cdot\mathbf{g}_c^D\\
0 & 0 & 0 & 0 & \mathbf{a}_c^A\cdot\mathbf{b}_c^B & 0 & 0 & 0 & \mathbf{b}_c^B\cdot\mathbf{g}_c^A\\
0 & 0 & 0 & 0 & 0 & 0 & 0 & 0 & 0\\
0 & 0 & 0 & 0 & -\mathbf{a}_c^A\cdot\mathbf{b}_c^D & 0 & 0 & 0 & -\mathbf{b}_c^D\cdot\mathbf{g}_c^A\\
0 & 0 & 0 & 0 & 0 & \mathbf{a}_c^B\cdot\mathbf{a}_c^A & 0 & -\mathbf{a}_c^D\cdot\mathbf{a}_c^A & \mathbf{a}_c^A\cdot\mathbf{g}_c^B-\mathbf{a}_c^A\cdot\mathbf{g}_c^D\\
0 & 0 & 0 & 0 & 0 & 0 & 0 & 0 & \mathbf{a}_c^B\cdot\mathbf{g}_c^A\\
0 & 0 & 0 & 0 & 0 & 0 & 0 & 0 & 0\\
0 & 0 & 0 & 0 & 0 & 0 & 0 & 0 & -\mathbf{a}_c^D\cdot\mathbf{g}_c^A\\
0 & 0 & 0 & 0 & 0 & 0 & 0 & 0 & \mathbf{g}_c^B\cdot\mathbf{g}_c^A-\mathbf{g}_c^D\cdot\mathbf{g}_c^A
\end{bmatrix}$$

$$\mathbf{Q}_2=\begin{bmatrix}
0 & \mathbf{b}_c^B\cdot\mathbf{b}_c^A & 0 & 0 & 0 & \mathbf{a}_c^B\cdot\mathbf{b}_c^A & 0 & 0 & \mathbf{b}_c^A\cdot\mathbf{g}_c^B\\
0 & 0 & -\mathbf{b}_c^C\cdot\mathbf{b}_c^B & 0 & \mathbf{a}_c^A\cdot\mathbf{b}_c^B & 0 & -\mathbf{a}_c^C\cdot\mathbf{b}_c^B & 0 & \mathbf{b}_c^B\cdot\mathbf{g}_c^A-\mathbf{b}_c^B\cdot\mathbf{g}_c^C\\
0 & 0 & 0 & 0 & 0 & -\mathbf{a}_c^B\cdot\mathbf{b}_c^C & 0 & 0 & -\mathbf{b}_c^C\cdot\mathbf{g}_c^B\\
0 & 0 & 0 & 0 & 0 & 0 & 0 & 0 & 0\\
0 & 0 & 0 & 0 & 0 & \mathbf{a}_c^B\cdot\mathbf{a}_c^A & 0 & 0 & \mathbf{a}_c^A\cdot\mathbf{g}_c^B\\
0 & 0 & 0 & 0 & 0 & 0 & -\mathbf{a}_c^C\cdot\mathbf{a}_c^B & 0 & \mathbf{a}_c^B\cdot\mathbf{g}_c^A-\mathbf{a}_c^B\cdot\mathbf{g}_c^C\\
0 & 0 & 0 & 0 & 0 & 0 & 0 & 0 & -\mathbf{a}_c^C\cdot\mathbf{g}_c^B\\
0 & 0 & 0 & 0 & 0 & 0 & 0 & 0 & 0\\
0 & 0 & 0 & 0 & 0 & 0 & 0 & 0 & \mathbf{g}_c^B\cdot\mathbf{g}_c^A-\mathbf{g}_c^C\cdot\mathbf{g}_c^B
\end{bmatrix}$$

$$\mathbf{Q}_3=\begin{bmatrix}
0 & 0 & 0 & 0 & 0 & 0 & 0 & 0 & 0\\
0 & 0 & \mathbf{b}_c^C\cdot\mathbf{b}_c^B & 0 & 0 & 0 & \mathbf{a}_c^C\cdot\mathbf{b}_c^B & 0 & \mathbf{b}_c^B\cdot\mathbf{g}_c^C\\
0 & 0 & 0 & -\mathbf{b}_c^D\cdot\mathbf{b}_c^C & 0 & \mathbf{a}_c^B\cdot\mathbf{b}_c^C & 0 & -\mathbf{a}_c^D\cdot\mathbf{b}_c^C & \mathbf{b}_c^C\cdot\mathbf{g}_c^B-\mathbf{b}_c^C\cdot\mathbf{g}_c^D\\
0 & 0 & 0 & 0 & 0 & 0 & -\mathbf{a}_c^C\cdot\mathbf{b}_c^D & 0 & -\mathbf{b}_c^D\cdot\mathbf{g}_c^C\\
0 & 0 & 0 & 0 & 0 & 0 & 0 & 0 & 0\\
0 & 0 & 0 & 0 & 0 & 0 & \mathbf{a}_c^C\cdot\mathbf{a}_c^B & 0 & \mathbf{a}_c^B\cdot\mathbf{g}_c^C\\
0 & 0 & 0 & 0 & 0 & 0 & 0 & -\mathbf{a}_c^D\cdot\mathbf{a}_c^C & \mathbf{a}_c^C\cdot\mathbf{g}_c^B-\mathbf{a}_c^C\cdot\mathbf{g}_c^D\\
0 & 0 & 0 & 0 & 0 & 0 & 0 & 0 & -\mathbf{a}_c^D\cdot\mathbf{g}_c^C\\
0 & 0 & 0 & 0 & 0 & 0 & 0 & 0 & \mathbf{g}_c^C\cdot\mathbf{g}_c^B-\mathbf{g}_c^D\cdot\mathbf{g}_c^C
\end{bmatrix}$$

$$\mathbf{Q}_4=\begin{bmatrix}
0 & 0 & \mathbf{b}_c^C\cdot\mathbf{b}_c^A & 0 & 0 & 0 & \mathbf{a}_c^C\cdot\mathbf{b}_c^A & 0 & \mathbf{b}_c^A\cdot\mathbf{g}_c^C\\
0 & 0 & 0 & -\mathbf{b}_c^D\cdot\mathbf{b}_c^B & 0 & 0 & 0 & -\mathbf{a}_c^D\cdot\mathbf{b}_c^B & -\mathbf{b}_c^B\cdot\mathbf{g}_c^D\\
0 & 0 & 0 & 0 & \mathbf{a}_c^A\cdot\mathbf{b}_c^C & 0 & 0 & 0 & \mathbf{b}_c^C\cdot\mathbf{g}_c^A\\
0 & 0 & 0 & 0 & 0 & -\mathbf{a}_c^B\cdot\mathbf{b}_c^D & 0 & 0 & -\mathbf{b}_c^D\cdot\mathbf{g}_c^B\\
0 & 0 & 0 & 0 & 0 & 0 & \mathbf{a}_c^C\cdot\mathbf{a}_c^A & 0 & \mathbf{a}_c^A\cdot\mathbf{g}_c^C\\
0 & 0 & 0 & 0 & 0 & 0 & 0 & -\mathbf{a}_c^D\cdot\mathbf{a}_c^B & -\mathbf{a}_c^B\cdot\mathbf{g}_c^D\\
0 & 0 & 0 & 0 & 0 & 0 & 0 & 0 & \mathbf{a}_c^C\cdot\mathbf{g}_c^A\\
0 & 0 & 0 & 0 & 0 & 0 & 0 & 0 & -\mathbf{a}_c^D\cdot\mathbf{g}_c^B\\
0 & 0 & 0 & 0 & 0 & 0 & 0 & 0 & \mathbf{g}_c^A\cdot\mathbf{g}_c^C-\mathbf{g}_c^B\cdot\mathbf{g}_c^D
\end{bmatrix}$$

KINEMATIC STRUCTURE OF A PARALLEL ROBOT FOR MEMS FABRICATION

Hagay Bamberger[1,2], Moshe Shoham[2]
[1]*RAFAEL – Armament Development Authority Ltd.*
[2]*Robotics Laboratory*
Department of Mechanical Engineering
Technion – Israel Institute of Technology
hagayb@rafael.co.il
shoham@tx.technion.ac.il

Abstract This investigation proposes a parallel robot structure that can be realized by MEMS technology. The robot consists of linear actuators located at the base and revolute joints only, thus making fabrication using silicon etching technology easier. The kinematic structure contains three single loop sub-mechanisms connected in parallel to the moving platform, enabling generation of three or six degrees-of-freedom mechanisms. The forward and inverse kinematics solutions are presented.

Keywords: direct kinematics, parallel robot, MEMS, micro actuator, micro joint

1. Introduction

Micro-Electro-Mechanical-Systems (MEMS) is a relative new technology, enabling fabrication of miniature mechanical devices coupled to electronics on silicon wafers. Most fabrication methods are based on lithography, which produces two-dimensional shapes by layer etching and deposition. Consequently, MEMS usually have a planar geometry, and it is difficult to achieve structures that are fully three-dimensional.

The first micro planar parallel mechanism utilizing MEMS technology was reported by Mehregany et al., 1988; it realized rotating actuators to provide a triangular moving platform with planar motion. The hinge proposed by Pister et al., 1992, was one of the breakthroughs that enabled production of three-dimensional devices using surface micro-machining. The hinge elements are fabricated in the substrate plane, but they are free to rotate out of the plane, resulting in a three-dimensional structure. A variety of devices were developed using this hinge and other similar ones.

Translational motion of an optical element along three axes is reported in Fan et al., 1997. Another XYZ device is presented in Jokiel et al., 2001, which also present two more devices with three degrees-of-freedom. An interesting walking mechanism is shown in Hollar et al., 2003, in which

J. Lenarčič and C. Galletti (eds.), On Advances in Robot Kinematics, 113-122.

the leg itself is a parallel spatial mechanism with two degrees-of-freedom.

Joint manufacturing on the micro scale is problematic, pointing to a different solution such as flexible joints (Piyawattanametha et al., 1998). A spatial mechanism with linear and revolute flexible joints was also developed by Liu and DeVoe, 2001, using bulk micro-machining and precise bonding of two wafers.

A good acquaintance with and understanding of MEMS fabrication technologies is a precondition for the development of a micro-robot. These considerations are discussed in the next section. Section 3 suggests a new kinematic configuration based on the limitations of micro-machining. The inverse and the direct kinematical problems are solved in sections 4 and 5, respectively, and numerical example is provided in section 6.

2. Design Considerations

In the micro world, parallel robot structures are more suitable for MEMS fabrication technology. Fabrication process limitations force the motors to be located at the silicon substrate and not at the links. Furthermore, implementation of transmissions and wiring to moving motors is very problematic. In addition, only a parallel structure can carry a reasonable external load. All of these advantages of parallel robot structure are at the expense of a smaller work volume.

2.1 Actuators for micro robot

Several physical principles are used in actuating MEMS devices, e.g.: electro-static (Jokiel et al., 2001, Piyawattanametha et al., 1998, Fan et al., 1997, Hollar et al., 2003), thermal (Liu and DeVoe, 2001), magnetic, piezo-electric, and others (Tabib-Azar, 1998). Most of the actuators enable linear motion (Jokiel et al., 2001, Fan et al., 1997, Hollar et al., 2003, Liu and DeVoe, 2001), but a few execute rotating motion (Mehregany et al., 1988). A very common actuation method based on techniques derived from integrated circuit fabrication is the electro-static method. A variety of geometries of electro-static actuators were developed, enabling relatively large displacements. Scratch-drive-actuators (Fan et al., 1997) are compact and provide gross motion, large forces and fine accurate steps. Comb-drives can shift a toothed rack slipping in a groove, actually providing unlimited displacement (Jokiel et al., 2001, Hollar et al., 2003). A suitable design enables bi-directional translation using the same actuators.

Since all actuators developed in MEMS have been built on the silicon substrate and their stators are not lifted off from that plane, it is preferable to use this well-known design to produce actuators that are all linear and are located at the base.

2.2 Joints for micro robot

Designing a robot for the micro scale should take into consideration that the available joints differ from those in the macro world not only in size but also in relative clearance and friction. Joints can be classified according to the following categories: type (flexible or rigid), motion (translation or rotation), and degrees-of-freedom (one, two, or more).

Flexible joints with a variety of geometries enable translation (Liu and DeVoe, 2001) or rotation (Piyawattanametha et al., 1998), with one or more degrees-of-freedom. The challenge is always how to design the shape of the joint so that it can move in the desired direction, while avoiding movements in other directions (Kota, 1999). The advantages of these kinds of joints include smoothness of motion, good repeatability, high resolution, acceptable reliability, and saving of assembly processes. Their main disadvantage is the inherent stress at the joints that limits the travel range and causes a non-linear restitution force or moment. In MEMS this problem is even worse, since the structure has to be lifted off the substrate, causing an initial stress load at the home position.

The main advantages of micro rigid joints are their large achievable motion and the absence of a restitution force. Even though the main disadvantage of the micro rigid joints is clearance, which results in noise, impacts, friction and wear, nevertheless when a reasonable work volume is needed for parallel micro robots, rigid hinges are preferable. The well known hinge shown by Pister et al., 1992, enables a very large angle of revolution, even more than 180°. The axis of that joint is parallel to the substrate plane. Universal joints and spherical joints, having two and three degrees-of-freedom respectively, are much more uncommon. Jokiel et al., 2001, combines several single degree-of-freedom joints in order to achieve complex joints, some of which consist of non-intersecting axes.

3. Kinematic Configurations of Parallel Robots for MEMS Fabrication

Based on the design considerations discussed in the previous section, a new configuration of a six degrees-of-freedom parallel robot was shown in Bamberger and Shoham, 2004. The kinematic configuration that fits MEMS fabrication consists of only revolute hinges, whose axes lie in the substrate plane during the fabrication process, and six linear motors located at the base. Another configuration, which is a degeneration of the previous one, having only three translatorial degrees-of-freedom, is discussed hereinafter. Fig. 1 shows an enlarged model demonstrating the suggested configuration in manufacturing and operating phases, and Fig. 2 shows a perspective skeleton view that clarifies the subsequent definitions.

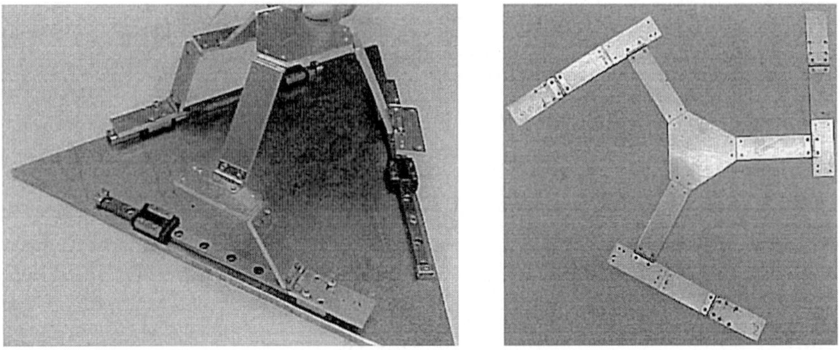

(a) 3-D configuration, after lifting *(b)* manufacturing configuration

Figure 1. Enlarged model of the kinematic configuration

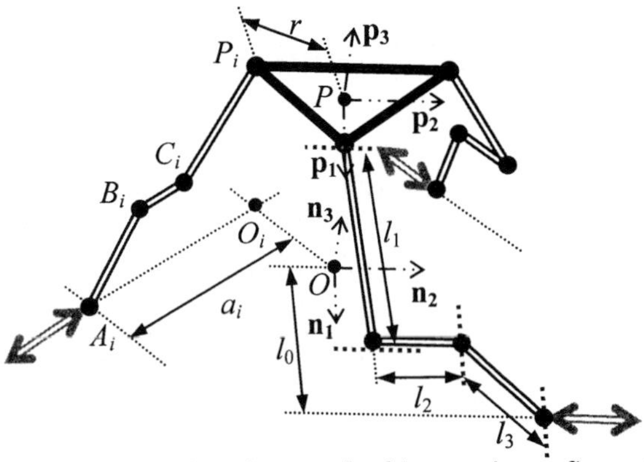

Figure 2. Perspective view on the kinematic configuration

A platform $P_1P_2P_3$ is connected to the base via three legs. Each leg has three arms and one linear motor. The motor is located at the substrate plane and determines the position of point A_i (i=1,2,3). The arm A_iB_i is connected to the motor by an axis, which is perpendicular to the line of action of the motor. The arm B_iC_i is connected to the arm A_iB_i by an axis, which is parallel to the previous one. It will be later proved that the arm B_iC_i must be parallel to the substrate plane. The arm C_iP_i is perpendicular to the arm B_iC_i, and has two parallel axes situated at its ends: the axis at point P_i lies at the platform plane and the axis at point C_i lies along the arm B_iC_i.

The point O, which is situated in the substrate plane, is the origin of the coordinate system n_1-n_2-n_3. The vector n_3 is perpendicular to the substrate plane. The distance from point O to the line of action of the motor is indicated by l_0. Point P is the geometric center of the equilateral triangle $P_1P_2P_3$, and is the origin of the coordinate system p_1-p_2-p_3, which is attached to the platform. Point P is located above the point O at the symmetrical position, when all the motors make the same travel. The distance from point P to point P_i is r. The lengths of the arms C_iP_i, B_iC_i, and A_iB_i are l_1, l_2, and l_3, respectively. The position of point A_i is indicated by a_i, and is measured from the fixed point O_i, defined as the projection of point O on the line of action of the motor.

Next it is proven that the moving platform orientation remains fixed, so it has only three translational degrees-of-freedom.

The directions of the axes situated at the points A_i and P_i (i=1,2,3) are

$$\begin{aligned} \mathbf{e}^{A_i} &= \cos(\psi_i)\mathbf{n}_1 + \sin(\psi_i)\mathbf{n}_2 \\ \mathbf{e}^{P_i} &= \sin(\psi_i)\mathbf{p}_1 - \cos(\psi_i)\mathbf{p}_2 \end{aligned} , \tag{1}$$

where the angle ψ_i is constant and equal to

$$\psi_1 = 0°, \psi_2 = 120°, \psi_3 = 240° \ . \tag{2}$$

The axes situated at points B_i and C_i are parallel to $\mathbf{e}^{A_i}$ and $\mathbf{e}^{P_i}$, respectively, and are perpendicular to each other, thus

$$\mathbf{e}^{A_i} \cdot \mathbf{e}^{P_i} = 0 \ . \tag{3}$$

Substituting Eq. 1 and Eq. 2 yields three equations:

$$\begin{aligned} &R_{21} = 0 \\ &3R_{12} - R_{21} - \sqrt{3}(R_{11} - R_{22}) = 0 \ , \\ &3R_{12} - R_{21} + \sqrt{3}(R_{11} - R_{22}) = 0 \end{aligned} \tag{4}$$

where R_{ij} are the components of the rotation matrix ${}^{P}\mathbf{R}_{N}$. This matrix must fulfill the orthogonality conditions (Tsai, 1999), and in view of Eq. 4, it must be a unit matrix. That is, the coordinate systems n_1-n_2-n_3 and p_1-p_2-p_3 are parallel to each other, so the platform has, indeed, a fixed orientation. Note that the arm B_iC_i must be parallel to the base and to the platform, like in the manipulator of Carricato and Parenti-Castelli, 2003, which also has two pairs of parallel axes in each limb, however, its axes are all perpendicular to the arm B_iC_i, while here it is not the case.

4 Inverse Kinematics

The goal of the inverse kinematic solution is to find the positions of the motors when the position of the moving platform is given. It should be mentioned that the algorithm is calculated under the assumption of no joint clearance, which might be way off at the MEMS relative scale.

The vector from point O to point P defines the position of the moving platform:

$$ {}^{O}\mathbf{r}^{P} = x\mathbf{n}_1 + y\mathbf{n}_2 + z\mathbf{n}_3 \ . \tag{5} $$

The kinematics equations are

$$ z = h_{i1} + h_{i2} \quad (i = 1,2,3) \ , \tag{6} $$

where

$$ \begin{aligned} h_{i1} &= {}^{A_i}\mathbf{r}^{B_i} \cdot \mathbf{n}_3 = \sqrt{l_3^2 - [a_i - l_2 + x\sin(\psi_i) - y\cos(\psi_i)]^2} \\ h_{i2} &= {}^{C_i}\mathbf{r}^{P_i} \cdot \mathbf{n}_3 = \sqrt{l_1^2 - [l_0 - r - x\cos(\psi_i) - y\sin(\psi_i)]^2} \end{aligned} \ . \tag{7} $$

Solving Eq. 6 for the motor locations yields

$$ a_i = l_2 - x\sin(\psi_i) + y\cos(\psi_i) + \sqrt{l_3^2 - (z - h_{i2})^2} \ . \tag{8} $$

The inverse kinematic solution for the six degrees-of-freedom version is shown in Bamberger and Shoham, 2004.

5 Direct Kinematics

The goal of the direct kinematic solution is to find the position of the platform when the positions of the motors are given, i.e. Eq. 6 should be solved for x, y, and z. By squaring Eq. 6 twice, the following polynomials are obtained:

$$ z^4 - 2z^2\left(h_{i1}^2 + h_{i2}^2\right) + \left(h_{i1}^2 - h_{i2}^2\right)^2 = 0 \quad (i = 1,2,3) \ . \tag{9} $$

These are four-degrees polynomial equations in the three unknowns. However, unlike Innocenti and Parenti-Castelli, 1990, and Tahmasebi and Tsai, 1994, where the distances from the base to the platform are known, and in each equation only two unknowns appear, and hence the solution is simpler, in our case the three unknowns appear in all the three equations. The same is true for the configuration of Carricato and Parenti-Castelli, 2003, where they solved the problem by implementing Sylvester's Dialytic method (Salmon, 1964) and obtaining a 28x28 matrix, whose determinant is a 64-degrees polynomial in one unknown.

The fact that z appears only in bi-squared form can be utilized. Eq. 9 can be rewritten as

$$ \begin{pmatrix} 1 & H_{11} & H_{12} \\ 1 & H_{21} & H_{22} \\ 1 & H_{31} & H_{32} \end{pmatrix} \cdot \begin{pmatrix} z^4 \\ z^2 \\ 1 \end{pmatrix} = \begin{pmatrix} 0 \\ 0 \\ 0 \end{pmatrix} , \tag{10} $$

where

$$H_{i1} = -2\left(h_{i1}^2 + h_{i2}^2\right), \quad H_{i2} = \left(h_{i1}^2 - h_{i2}^2\right)^2 . \tag{11}$$

The matrix in Eq. 10 must be singular in order to have a non-trivial solution (Salmon, 1964). Therefore, the determinant of that matrix, which is a five-degrees polynomial in x and y, must be zero:

$$\det\begin{pmatrix} 1 & H_{11} & H_{12} \\ 1 & H_{21} & H_{22} \\ 1 & H_{31} & H_{32} \end{pmatrix} = \sum_{m=0}^{5}\sum_{n=0}^{5} \kappa_{0mn} x^m y^n = 0 . \tag{12}$$

The expressions of κ_{0mn} are too long to be written here, but it should be noted that they are all known.

Nine equations in x and y can be derived by writing two of Eq. 9 with two more equations after multiplying them by z^2:

$$\begin{pmatrix} 0 & 1 & H_{i1} & H_{i2} \\ 0 & 1 & H_{j1} & H_{j2} \\ 1 & H_{m1} & H_{m2} & 0 \\ 1 & H_{n1} & H_{n2} & 0 \end{pmatrix} \cdot \begin{pmatrix} z^6 \\ z^4 \\ z^2 \\ 1 \end{pmatrix} = \begin{pmatrix} 0 \\ 0 \\ 0 \\ 0 \end{pmatrix} \quad \begin{pmatrix} ijmn = 1212, 1223, 1231, \\ 2312, 2323, 2331, \\ 3112, 3123, 3131 \end{pmatrix} . \tag{13}$$

Again, the determinant of the matrix, which now is an eight-degree polynomial, must be zero:

$$\sum_{m=0}^{8}\sum_{n=0}^{8} \kappa_{imn} x^m y^n = 0 \quad (i = 1,..,9) . \tag{14}$$

Multiplying Eq. 12 by y, y^2, and y^3, and arranging them with Eq. 14 in a matrix form yields

$$\begin{pmatrix}
0 & 0 & 0 & \kappa_{005} & \sum_{m=0}^{1}\kappa_{0m4}x^m & \sum_{m=0}^{2}\kappa_{0m3}x^m & \sum_{m=0}^{3}\kappa_{0m2}x^m & \sum_{m=0}^{4}\kappa_{0m1}x^m & \sum_{m=0}^{5}\kappa_{0m0}x^m \\
0 & 0 & \kappa_{005} & \sum_{m=0}^{1}\kappa_{0m4}x^m & \sum_{m=0}^{2}\kappa_{0m3}x^m & \sum_{m=0}^{3}\kappa_{0m2}x^m & \sum_{m=0}^{4}\kappa_{0m1}x^m & \sum_{m=0}^{5}\kappa_{0m0}x^m & 0 \\
0 & \kappa_{005} & \sum_{m=0}^{1}\kappa_{0m4}x^m & \sum_{m=0}^{2}\kappa_{0m3}x^m & \sum_{m=0}^{3}\kappa_{0m2}x^m & \sum_{m=0}^{4}\kappa_{0m1}x^m & \sum_{m=0}^{5}\kappa_{0m0}x^m & 0 & 0 \\
\kappa_{005} & \sum_{m=0}^{1}\kappa_{0m4}x^m & \sum_{m=0}^{2}\kappa_{0m3}x^m & \sum_{m=0}^{3}\kappa_{0m2}x^m & \sum_{m=0}^{4}\kappa_{0m1}x^m & \sum_{m=0}^{5}\kappa_{0m0}x^m & 0 & 0 & 0 \\
\kappa_{108} & \sum_{m=0}^{1}\kappa_{1m7}x^m & \sum_{m=0}^{2}\kappa_{1m6}x^m & \sum_{m=0}^{3}\kappa_{1m5}x^m & \sum_{m=0}^{4}\kappa_{1m4}x^m & \sum_{m=0}^{5}\kappa_{1m3}x^m & \sum_{m=0}^{6}\kappa_{1m2}x^m & \sum_{m=0}^{7}\kappa_{1m1}x^m & \sum_{m=0}^{8}\kappa_{1m0}x^m \\
\vdots & \vdots & \vdots & \vdots & \vdots & \vdots & \vdots & \vdots & \vdots \\
\kappa_{i08} & \sum_{m=0}^{1}\kappa_{im7}x^m & \sum_{m=0}^{2}\kappa_{im6}x^m & \sum_{m=0}^{3}\kappa_{im5}x^m & \sum_{m=0}^{4}\kappa_{im4}x^m & \sum_{m=0}^{5}\kappa_{im3}x^m & \sum_{m=0}^{6}\kappa_{im2}x^m & \sum_{m=0}^{7}\kappa_{im1}x^m & \sum_{m=0}^{8}\kappa_{im0}x^m \\
\vdots & \vdots & \vdots & \vdots & \vdots & \vdots & \vdots & \vdots & \vdots \\
\kappa_{908} & \sum_{m=0}^{1}\kappa_{9m7}x^m & \sum_{m=0}^{2}\kappa_{9m6}x^m & \sum_{m=0}^{3}\kappa_{9m5}x^m & \sum_{m=0}^{4}\kappa_{9m4}x^m & \sum_{m=0}^{5}\kappa_{9m3}x^m & \sum_{m=0}^{6}\kappa_{9m2}x^m & \sum_{m=0}^{7}\kappa_{9m1}x^m & \sum_{m=0}^{8}\kappa_{9m0}x^m
\end{pmatrix} \cdot \begin{pmatrix} y^8 \\ y^7 \\ y^6 \\ y^5 \\ y^4 \\ y^3 \\ y^2 \\ y \\ 1 \end{pmatrix} = \begin{pmatrix} 0 \\ 0 \\ 0 \\ 0 \\ 0 \\ 0 \\ 0 \\ 0 \\ 0 \\ 0 \\ 0 \\ 0 \\ 0 \end{pmatrix} . \tag{15}$$

Although the size of this matrix is 13x9, its rank is only 5, because its rows are linearly dependent. Therefore, one cannot find nine independent rows and write a sub-matrix whose determinant is not identically zero. But, multiplying by y the last ten rows of Eq. 15 gives a matrix, whose size is 23x10 and its rank is 10. Now, one can find ten independent rows and write a sub-matrix, whose rank is also 10. The determinant of this sub-matrix is a 33-degrees polynomial in x, and it must be equal to zero for non-trivial solution.

The 33-degree polynomial can be numerically solved. The real solutions are then back-substituted into Eq. 12, which gives five solutions for y. Every pair of x and y should be substituted three times in Eq. 7 and in Eq. 6, one time per each leg (i=1,2,3). When the three substitutions yield the same height z, the solution is correct.

6 Numerical Example

The inverse kinematical problem and the direct kinematical problem are solved using the algorithms developed in the above sections, with the following dimensions: The platform radius is taken as r =150, and the distances to the lines of action of the motors are l_0 =400. The arm lengths are l_1 =350, l_2 =100, and l_3 =400. All dimensions are in microns – a reasonable size for MEMS.

For the inverse kinematics solution, assume that the required platform location is

$$ {}^{O}\mathbf{r}^{P} = -50\mathbf{n}_1 + 500\mathbf{n}_3 \ . \tag{16} $$

Using Eqs. 2, 7, and 8, the locations of the motors are

$$ a_1 = 340.37, \quad a_2 = 469.22, \quad a_3 = 382.61 \ . \tag{17} $$

To check the direct kinematics, substitute Eq. 17, which now is given, in Eqs. 7, 11, and 12, resulting in the following five-degree polynomial:

$$ \begin{aligned} &-1.9590\cdot 10^{14} + 4.6069\cdot 10^{12}x + 2.0001\cdot 10^{11}x^2 + 5.9017\cdot 10^{8}x^3 \\ &-8.8572\cdot 10^{4}x^4 - 1.6935\cdot 10^{3}x^5 - 6.2464\cdot 10^{12}y - 1.4689\cdot 10^{11}xy \\ &+2.1922\cdot 10^{9}x^2y - 4.7016\cdot 10^{4}x^3y - 6.4952\cdot 10^{2}x^4y - 1.0778\cdot 10^{11}y^2 \\ &-2.8513\cdot 10^{9}xy^2 - 1.1304\cdot 10^{6}x^2y^2 + 4.5723\cdot 10^{3}x^3y^2 - 1.0211\cdot 10^{9}y^3 \\ &+2.4808\cdot 10^{5}xy^3 - 1.2990\cdot 10^{3}x^2y^3 + 4.6146\cdot 10^{5}y^4 + 3.8952\cdot 10^{3}xy^4 \\ &+1.1691\cdot 10^{3}y^5 = 0 \end{aligned} \ . \tag{18} $$

Substituting Eq. 17 in Eqs. 7, 11, and 14 yields nine eight-degree polynomials, so Eq. 15 can be calculated. The 33-degree polynomial has 33 solutions, and in this case 15 of them are real. Every x yields five y's while solving Eq. 18:

Table 1. The real solutions of *x* and all the solutions of *y*

x	Five solutions of *y*				
-706.98	-626.79	-200+428i	-200-428i	390.98	2596.58
-607.00	-568.15	-243+192i	-243-192i	472.99	2209.37
-337.36	-854.02	-271.88	-20.81	702.09	1173.90
-241.82	-938.09	-194.64	-24.04	763.45	804.28
-213.04	-960.16	-171.12	-26.53	717.92	754.97
-211.88	-961.02	-170.18	-26.64	712.40	756.67
-55.17	-1069.56	-66.16	-7.36	120.74	811.43
-54.55	-1069.99	-66.07	-6.59	118.14	811.54
-50.00	-1073.18	-65.49	0.00	98.22	812.34
1.00	-1111.11	-55.19	-25+51i	-25-51i	819.18
9.89	-1118.29	-50+40i	-50-40i	-30.10	819.99
675.14	-2953.25	-315.83	-237.84	104.53	758.36
685.19	-2991.25	-273.46	-244.33	68.95	762.58
699.22	-3044.39	-231.77	-211.64	-5.36	768.90
710.31	-3086.48	-241.99	-104+86i	-104-86i	774.24

Every pair of *x* and *y* is substituted three times in Eq. 7 and in Eq. 6, one time per each leg (i=1,2,3). The eight correct solutions occur when the three substitutions yield the same height *z*. The next figure shows the solutions where the height *z* is positive. There are four more solutions, which are mirrored with respect to the *x-y* plane. It should be noted that if only the positive square roots in Eq. 7 are taken, there is only one solution for the direct kinematical problem, which is written in Eq. 16 and is shown at the upper left side of the next figure:

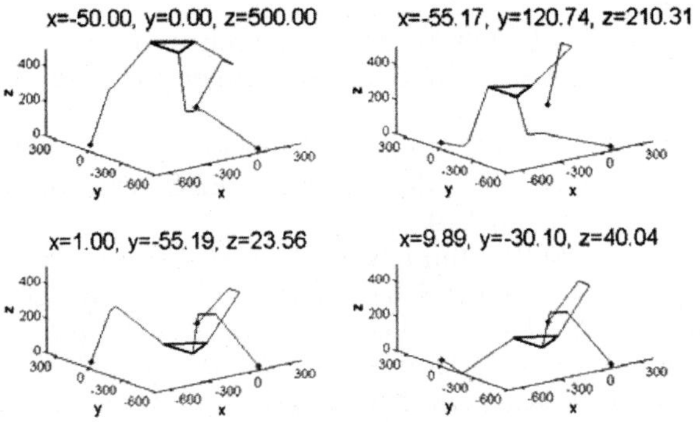

Figure 3. Four of the solutions of the direct kinematical problem

7. Conclusions

This paper presents a structure of a translational three degrees-of-freedom parallel robot that is suitable for MEMS fabrication techniques, namely, all the joints are revolute, all the actuators are linear and both are located at the base during the manufacturing process. The inverse and direct kinematic solutions are given. The direct kinematic solution yields a 33-degrees polynomial in one unknown, and a numerical example with eight solutions is presented for a given motor locations.

References

Bamberger, H., and Shoham, M. (2004), A New Configuration of a Six Degrees-of-Freedom Parallel Robot for MEMS Fabrication, *IEEE Int. Conference on Robotics and Automation (ICRA 2004)*, New Orleans, LA, USA, accepted.

Carricato, M., and Parenti-Castelli, V. (2003), Position Analysis of a New Family of 3-DOF Translational Parallel Manipulators, *ASME Journal of Mechanical Design*, Vol. 125(2), pp. 316-322.

Fan, L., Wu, M.C., Choquette, K.D., and Crawford, M.H. (1997), Self-assembled Microactuated XYZ Stages for Optical Scanning and Alignment, *Proceedings of International Solid State Sensors and Actuators Conference (Transducers'97)*, Chicago, IL, USA, pp. 319-322.

Hollar, S., Bergbreiter, S., and Pister, K.S.J. (2003), Bidirectional Inchworm Motors and Two-DOF Robot Leg Operation, *The 12th International Conference on Solid State Sensors, Actuators and Microsystems (Transducers'03)*, Boston, MA, USA, pp. 262-267.

Innocenti, C., and Parenti-Castelli, V. (1990), Direct Position Analysis of the Stewart Platform Mechanism, *Mechanism and Machine Theory*, Vol. 25, No. 6, pp. 611-621.

Jokiel, B.Jr., Benavides, G.L., Bieg, L.F., and Allen, J.J. (2001), Planar and Spatial 3-Degree-of-Freedom Micro-Stages in Silicon MEMS, *Annual Meeting of the American Society for Precision Engineering*, Crystal City, VA, USA.

Kota, S. (1999), Design of Compliant Mechanisms: Applications to MEMS, *SPIE Conference on Smart Electronics and MEMS*, Newport Beach, CA, USA, Vol. 3673, pp. 45-54.

Liu, Z., and DeVoe, D.L. (2001), Micromechanism Fabrication Using Silicon Fusion Bonding, *Robotics and Computer Integrated Manufacturing*, Vol. 17, No. 1, pp. 131-137.

Mehregany, M., Gabriel, K.J., and Trimmer, W.S.N. (1988), Integrated Fabrication of Polysilicon Mechanisms, *IEEE Transactions of Electron Devices*, Vol. 35, No. 6, pp. 719-723.

Pister, K.S.J., Judy, M.W., Burgett, S.R., and Fearing, R.S. (1992), Micro-fabricated Hinges, *Sensors and Actuators A-Physical*, Vol. 33(3), pp. 249-256.

Piyawattanametha, W., Fan, L., Lee, S.S., Su, J.G.D., and Wu, M.C. (1998), MEMS Technology for Optical Crosslink for Micro/Nano Satellites, *The International Conference on Integrated Nano/Microtechnology for Space Applications*, NASA/Johnson Space Center, Houston, TX, USA.

Salmon, G. (1964), Modern Higher Algebra, *5th edition reprinted by Chelsea Publishing Company*, Bronx, NY, USA.

Tabib-Azar, M. (1998), Microactuators: Electrical, Magnetic, Optical, Mechanical, Chemical & Smart Structures, *Kluwer Academic Publishers*, Norwell, MA, USA.

Tahmasebi, F., and Tsai, L.W. (1994), Closed-Form Direct Kinematics Solution of a New Parallel Minimanipulator, *ASME Journal of Mechanical Design*, Vol. 116, pp. 1141-1147.

Tsai, L.W. (1999), Robot Analysis: The Mechanics of Serial and Parallel Manipulators, *John Wiley & Sons*, New York, NY, USA.

ISOLATING SELF-MOTION MANIFOLDS ON A PLAYSTATION™

J.M. Porta, L. Ros, F. Thomas
Institut de Robòtica i Informàtica Industrial (CSIC-UPC)
Llorens i Artigas 4-6, 08028-Barcelona, Spain
{jporta,llros,fthomas}@iri.upc.es

Abstract In this paper, we show how it is possible to use the specialized hardware of a Play Station 2 (PS2) to speed up the execution of the Cuik algorithm, an interval-based inverse kinematics algorithm. A PS2 includes a RISC CPU and two parallel vector units able to perform four floating point multiplications and four additions in a single cycle. A careful use of these two vector units to implement the basic operations of the Cuik algorithms allows increasing the speed of the algorithm in one order of magnitude. The final implementation in the PS2 is as fast as that in a Pentium 4 at 2.6GHz (that is at least three times more expensive than a plain PS2).

Keywords: Inverse kinematics, hardware implementation of algorithms, parallel algorithms, graphic hardware, Play Station 2.

1. Introduction

To speed up inverse kinematics computations we can either develop better algorithms and heuristics or provide better implementations of already existing ones. One possibility to speed up a given algorithm is to use special hardware to implement its basic operations. This is the line followed, for instance, by Lee and Chang, 1987, using the CORDIC architecture. In this work, the inverse kinematics of a particular mechanism is analytically solved and the resulting equations are efficiently implemented using the special operations provided by the CORDIC processors. A limitation of this work, however, is that it can only be applied to mechanisms whose inverse kinematics can be solved in closed form.

In Porta *et. al*, 2002, we introduced the Cuik algorithm, an interval-based algorithm able to deal with mechanism with an arbitrary number of kinematic loops. The basic operations of this algorithm are homogeneous matrix products. Nowadays, commercially available graphic cards include specialized hardware to perform products of homogeneous matrices. The problem is how to use this hardware to implement our algorithms. In general, the communication between the CPU and the graphic card is slow and unidirectional. Therefore, although we can per-

J. Lenarčič and C. Galletti (eds.), On Advances in Robot Kinematics, 123-132.

form some computations using this specialized hardware, it is hard to move back the results to the main CPU.

Recently, Sony has released a Linux kit for Play Station 2 (PS2) that makes its programming rather easy. The design of the PS2 differs in its principles from that of a PC (Diefendorff and Dubey, 1997). The PC was designed for *static applications*: large programs, such as word processors or spreadsheets, that deal with relatively static data (i.e., the document). PS2 was designed for *media applications*: small applications dealing with a constantly changing data stream. In a graphical context, a PC is designed work with a mainly static environment including a large number of geometric elements. This geometry is send once to the graphic card and, afterwards, many operations (viewpoint changes, etc) are applied to it. On the contrary, a PS2 is designed to repetitively apply the same type of simple operations to a continuously changing geometry. This has many implications in the hardware design. Thus, while the PC has large graphic memories (to store the static geometry) and a relatively slow communication bus between the CPU and the graphical card, the PS2 has a reduced graphic memory and a fast communication bus.

An interval-based algorithm (and the Cuik algorithm in particular) can be considered a *media application*: a relatively simple process (i.e., the box reduction) is repetitively applied to different inputs (the set of boxes and sub-boxes where we recursively look for solutions). Thus, the Cuik algorithm could substantially benefit from being implemented on a PS2.

In section 2, we briefly describe the Cuik algorithm. Next (section 3), we describe the hardware resources of the PS2 and how use them to implement the Cuik algorithm. In section 4, we show the improvements resulting of using the PS2 and we conclude in section 5 summarizing our work and pointing out some directions for further development.

2. The Cuik Algorithm

A kinematic loop yields an equation of the form

$$c(x_1, \ldots, x_n, \theta_1, \ldots, \theta_m) = I, \tag{1}$$

where c is a matrix function, $x_1, \ldots, x_n$ are variables involved in translations, $\theta_1, \ldots, \theta_m$ are rotational variables, and I is the identity matrix. Translations and rotations can be represented using homogeneous transforms. Thus, equation 1 can be seen as

$$K_1 H_1(x_1) \ldots K_n H_n(x_n) L_1 J_1(\theta_1) \ldots L_m J_m(\theta_m) = I, \tag{2}$$

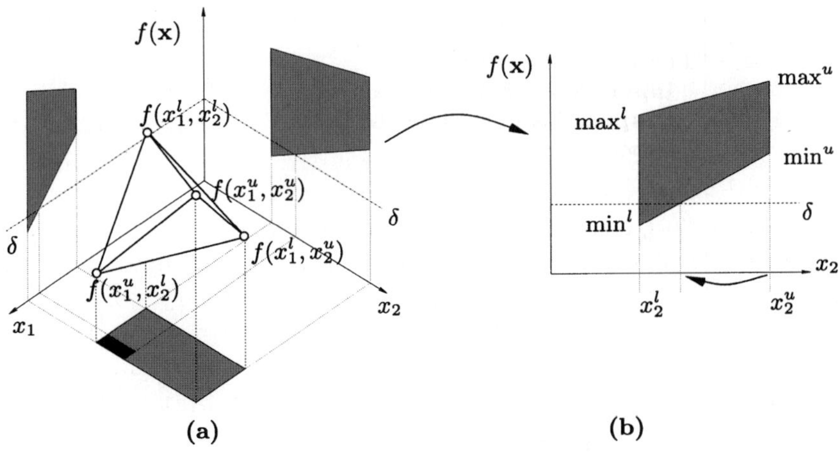

Figure 1. Segment-trapezoid clipping.

with K_i and L_i constant homogeneous transforms (possibly the identity), H_i a translation homogeneous matrix (along X, Y or Z) and J_i us a rotation matrix (around either X, Y or Z).

Each matrix equation leads to 12 scalar equations, one for each row i and column j in the transformation matrix:

$$f_{i,j}(x_1, \ldots, x_n, \theta_1, \ldots, \theta_m) = \delta_{i,j}, \tag{3}$$

where $\delta_{i,j}$ is 1 if i is equal to j, and 0 otherwise.

To avoid trigonometric expressions in the $f_{i,j}$ functions, we use the substitutions $y_k = \cos\theta_k$ and $z_k = \sin\theta_k$, and we add the circle constraint $y_k^2 + z_k^2 = 1$ to the system. The drawback of this variable change is that we have m additional variables and, thus, the search space is considerably enlarged. The advantage is that now all $f_{i,j}$ functions are multilinear, allowing us to use the following procedure to isolate the solutions of the resulting set of equations.

Assume we want to find all solutions of a multilinear equation $f(\mathbf{x}) = \delta$, for $\mathbf{x}$ in the box $\mathcal{B} = [x_1^l, x_1^u] \times \ldots \times [x_n^l, x_n^u] \in \Re^n$. According to Rikun, 1997, the graph of $(\mathbf{x}, f(\mathbf{x}))$ must lie within the convex hull of the 2^n points $\{(\mathbf{x}, f(\mathbf{x}))|\ \mathbf{x} \in \{x_1^l, x_1^u\} \times \ldots \times \{x_n^l, x_n^u\}\}$. To efficiently bound the intersection of this convex hull with the plane $f(\mathbf{x}) = \delta$ we project the hull onto each coordinate plane, as depicted in Fig. 1a for $n = 2$, and we intersect each of the resulting trapezoids with the $f(\mathbf{x}) = \delta$ line, as shown in Fig. 1b. Usually, these segment-trapezoid clippings reduce the ranges of some variables, and the Cartesian product of them all gives a box smaller than $\mathcal{B}$ still bounding the root locations (the black rectangle

```
The Cuik Algorithm
  Input: A set of kinematic equations
  Output: A set of solution boxes (S)
  Process:
    S ← ∅
    L ← Initial list of boxes
    while not empty(L)
        B ← first box(L)
        do
            s ←size(B)
            Reduce_Box(B)
        until empty(B) or size(B)< σ or size(B)/s > ρ
        if not empty(B) then
          if size(B)≤ σ then S ← S ∪ {B}
          else
            Split B into two sub-boxes: B1, B2
            Add B1 and B2 to L
          endif
        endif
    endwhile
```

Figure 2. The Cuik algorithm.

in Fig. 1a). If the range of a variable y_k involved in a circle constraint $y_k^2 + z_k^2 = 1$ is reduced, then the range of z_k can also be reduced using a rectangle-circle clipping operation.

Fig. 2 shows the high level form of the *Cuik* algorithm. To determine the set of solution boxes, each of the initial boxes is reduced as much as possible by the successive application of the segment-trapezoid and rectangle-circle clippings (see Fig. 3). If one of the clippings yields no solutions, then the search can be stopped in the area delimited by the box in process. If a box becomes smaller than a given limit (σ) we consider it a solution. Finally, if a box can not be significantly reduced (its reduction ratio is below a user-provided parameter ρ), we split it and we process the two resulting sub-boxes in a recursive way.

3. Cuik on a PlayStation 2

The *Emotion Engine* (see Fig. 4) (the core controller of the PS2) is equipped with one RISC processor that works at 300MHz and two vector units (Sony, 2001). The RISC processor executes the operating system (Linux in our case). The vector units (VU0 and VU1) have two sub-processors. The first one, called Floating Point Unit (FPU), is

```
Reduce_Box(B)
  Input: A box defined as a set of intervals:
      B = {[x_1^l, x_1^u], ..., [x_n^l, x_n^u]}
  Output: The same box but eventually resized
  Process:
    for each kinematic loop c
      V ← {v_0, ..., v_k} (Set of indices of variables involved in c)
      (Initialize the projection matrices)
      for each v ∈ V
          min_v^l ← +∞
          min_v^u ← +∞
          max_v^l ← −∞
          max_v^u ← −∞
      endfor
      for each x ∈ {x_{v_0}^l, x_{v_0}^u} × ... × {x_{v_k}^l, x_{v_k}^u} (For each corner of box B)
          (Generate a sequence of matrices for this corner)
              S ← c(x)
          (Multiply the matrix sequence)
              s ← eval(S)
          (Project the matrix on the coordinate planes)
              for each v ∈ V
                  if x[v] = x_v^l then
                      min_v^l ← min(min_v^l, s)
                      max_v^l ← max(max_v^l, s)
                  else
                      min_v^u ← min(min_v^u, s)
                      max_v^u ← max(max_v^u, s)
                  endif
              endfor
      endfor
      (Perform the clippings)
      for each v ∈ V
        for each i ∈ [1,3] and j ∈ [1,4]
          if i = j then δ = 1 else δ = 0
          Trapezoid_Clipping(x_v^l, x_v^u, min_v^l(i,j), max_v^l(i,j), min_v^u(i,j), max_v^u(i,j), δ)
          if x_v is involved in a circle equation x_v^2 + x_w^2 = 1 then
            Rectangle_Circle_Clipping(x_v^l, x_v^u, x_w^l, x_w^u)
          endif
        endfor
      endfor
    endfor
```

Figure 3. The **Reduce_Box** function.

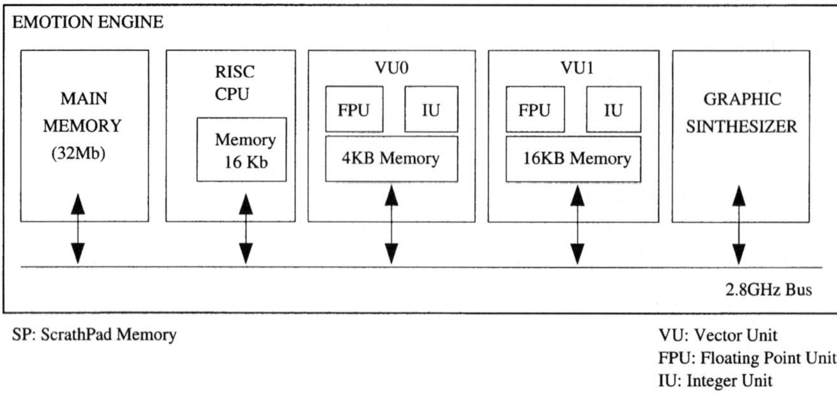

Figure 4. Sketch of the PS2 hardware.

specialized in the operation of vectors of four single-precession floating points elements. For instance, it can perform the element-wise product of two vector of four elements (i.e., four products) and accumulate the result to another vector (i.e., four additions) with a latency of 1 cycle. Using these operations, a full 4×4 matrix multiplication (as that performed in the *matrix sequence multiplication* step of the *reduce box* process) can be performed in just 16 cycles. Other basic operations the FPU can perform are the minimum and maximum of two vectors. We can use these operations to speed up the *matrix projection* step of the *reduce box* algorithm. The second sub-processor of the vector units, the Integer Unit (IU), can be used to feed the FPU with new data so that wait cycles are minimized.

The main memory of the Emotion Engine has 32 Mb. However, each processor has its own fast-access local memory. The bus communicating all the devices in the Emotion Engine operates at 2.8GHz, faster than the AGP graphic bus currently used in the PCs.

The special design of the PS2 enforces a particular programming philosophy. Small programs must be loaded in the VUs (with micro-instructions both to the FPU and to the IU). Then, the a continuous stream of data must be placed into the VU memories at the same speed as the VUs consumes the data.

To efficiently use the three processors of the PS2, we have to divide the Cuik algorithm in three sub-processes. Since, the VUs are specialized in matrix operations, we can use them to perform the matrix multiplications and the matrix projections included in the inner loop of the *reduce box* process (Fig. 3). The cost of both operations is $O(n)$, with n the

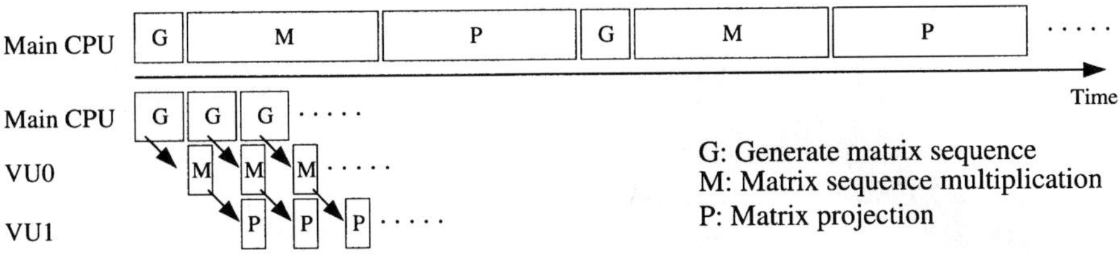

Figure 5. Time arrangement of the basic operation of the reduce-box process in a sequential processor (up) and in the PS2 (down).

number of variables involved in the loop. In this way, the load of the two VUs is balanced. The main RISC processor will take care of the rest of the algorithm: to generate the sequences of matrices to be multiplied (one sequence for each corner of the box in process, $\mathcal{B}$), to coordinate the execution of the vector units (sending data to them when appropriate), and to perform the clippings when the matrix multiplication and projections are done.

Fig. 5 shows a sketch with the arrangement of operations when using a sequential processors and when using the PS2. The use of three processors instead of just one, reduces the total cost of the algorithm at least by a factor of 3. Additionally, the time of the matrix multiplications and the matrix projection is considerably reduced due to the use of the special hardware of the VUs. The final result is a large increase of the speed in the algorithm.

4. Experiments and Results

Fig. 6a represents the Bennett linkage, a 4R spatial closed chain with 1-d.o.f. mobility. Fig. 6b shows the discretization of the one-dimensional solution space provided by the Cuik algorithm with $\sigma = 0.1$ and $\rho = 0.9$. This discretization includes about 300 boxes and it was achieved after applying the *box reduction* process 4150 times to 1575 boxes. Thus, the *box reduction* is applied, in average, 2.6 times to each processed box. The problem has 4 rotational variables but, after the sin/cos substitutions, we work with 8 variables. Consequently, each box has 256 corners and for each one of them we have to apply the *matrix sequence generation*, *matrix multiplication* and *matrix projection* steps.

If we apply the sequential version of the Cuik algorithm to this problem using the main processor of the PS2, the solution manifold of the Bennett linkage is isolated in about 265 seconds.

The parallelized version explained in the previous section, solves the problem in less than 22 seconds. So, using the special hardware of the

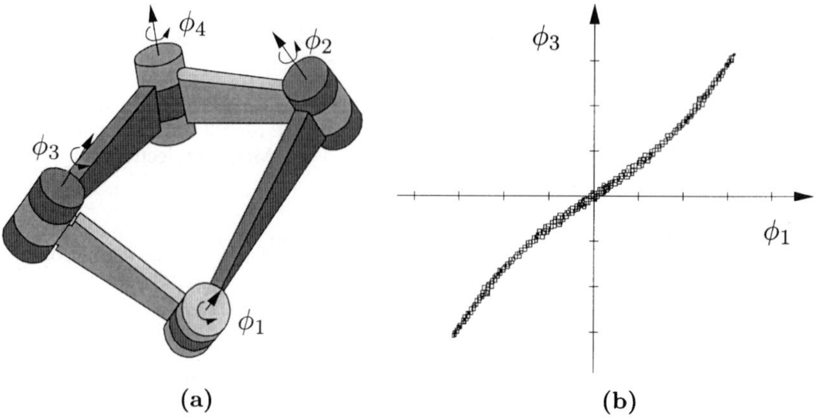

Figure 6. **(a)** The Bennett linkage. **(b)** The discretization of the one-dimensional space of solutions provided by the Cuik algorithm.

PS2 decreases the execution time to 10% of the original cost. The final execution time on the PS2 is roughly the same as that on a Pentium 4 at 2.6GHz, that is a much more expensive machine.

After the parallelization, the bottleneck in the PS2 implementation is the generation of the matrix sequences for each corner of all processed boxes. This takes about 97% of the final execution time. This process is that slow since it is executed by the RISC CPU of the PS2 that only operates at 300MHz. So, what in the sequential version of the algorithm was non-relevant (less than 10% of the execution time) becomes the most expensive step in the parallelized version. Therefore, any improvement in the matrix sequence generation procedure would result in a direct improvement of the global performance of the system.

The matrix multiplications and projections for the 4150 processed boxes are performed in less than 1 second. Thus, the PS2 achieves 1.8GFlops. This is less than the PS2 theoretical maximum (4.8GFlops) since not all instructions in the VUs perform floating point operations. (initializations, branches, etc are performed by the IU while the FPU is stopped).

5. Conclusions

Currently available graphic hardware offer many opportunities to speed up kinematic algorithms. In this paper, we describe the use of a PS2 to implement Cuik, a kinematic problems solver. Previously existing works on the use of hardware devices for inverse kinematic computations where limited to particular mechanism while the Cuik algorithm can deal with (almost) general mechanisms.

The internal controller of the PS2 with three CPU's, two of them specialized in vector operations, happened to be perfectly tailored for the structure of the Cuik algorithm.

The use of the PS2 resulted in the parallelization of our algorithm at two different levels of granularity (Henrich *et al*, 1997). The first one is the in-box level: the different operations applied to each box can be implemented in a different processor of the PS2. The second level of parallelism we achieved is the operation-level: the basic floating-point vector operations are performed in parallel in the vector units of the PS2.

Additionally, the interval-based algorithms, such as Cuik, are inherently parallelizable at the box-level (different boxes can be processed by different computers). Thus, one possible extension of this work is the set up of a grid of PS2s to cooperatively solve large problems.

The version of the Cuik algorithm described in this paper is the most basic one. A future improvement is to incorporate to the PS2 version the heuristics and improvements we have already developed for other versions of Cuik. This would result in a even larger increase of the speed of the algorithm.

Another possibility we want to explore is the use of PC graphic cards to implement Cuik since they are more easily available in research labs than PS2.

Acknowledgments

This research has been partially supported by the Spanish CICYT under contract TIC2003-03396.

References

Diefendorff K. and Dubey P.K. (1997), *How Multimedia Workloads Will Change Processor Design*, IEEE Computer, Vol. 30, No. 9, pp. 43-45.

Henrich D., Karl J. and Wörn H. (1997) *A Review of Parallel Processing Approaches to Robot Kinematics and Jacobian* Technical Report 10/97, Computer Science Department, University of Karlsruhe, ISSN 1432-7864.

Lee C.S.G. and Chang P.R. (1987), *A Maximum Pipelined CORDIC Architecture for Inverse Kinematic Position Computation.* IEEE Journal of Robotics and Automation. Vol. 3, No. 5, pp. 445–548.

Porta J.M., Ros L., Thomas F. and Torras C. (2002), *Solving Multi-Loop Linkages by Iterating 2D Clippings*, Advances in Robot Kinematics. Kluwer Academic Publishers, pp.255–264.

Rikun, A.D. (1997), A Convex Envelope Formula for Multilinear Functions, *J. of Global Optimization*, Vol. 10, pp. 425-437.

Sony Computer Entertainment Inc. (2001) *The Emotion Engine User's Manual.*

MOBILITY OF SPATIAL MULTI-LOOP AND MIXED LINKAGES

José M. Rico
Departamento de Ingeniería Mecánica, Instituto Tecnológico de Celaya
Celaya, Gto. 38000, México
mrico@itc.mx

Bahram Ravani
Department of Mechanical and Aeronautical Engineering, University of California-Davis
Davis, California 95616, U.S.A.
bravani@ucdavis.edu

Abstract The paper develops a method for systematic calculation of mobility of spatial multi-loop and mixed linkages. It uses the Lie algebra $se(3)$ of the Euclidean group $SE(3)$ and builds upon the previous work of the authors on developing a group theoretic mobility criterion for single loop linkages. The resulting method can correctly calculate the degree of freedom of multi loop linkages even for cases where the linkage is a mixed linkage or when some of the loops are over-constrained of exceptional type. The case of linkages with paradoxical loops still remains unsolved. Several examples are used to illustrate the theory.

Keywords: Mobility, Multi-Loop Linkages, Mixed Linkages

1. Introduction

It is well known that the mobility of linkages can be obtained by applying the Kutzbach-Grübler criterion. One form of this criterion (see for example Hunt, 1990) is

$$F = \lambda(n - r - 1) + \sum_{i=1}^{r} f_i. \tag{1}$$

where, F is the number of degrees of freedom, λ is the dimension of the tangent space of the Lie subgroup of the Euclidean group that contains all the relative displacements among arbitrary links of the linkage, n is the number of links of the linkage including the fixed link, r is the number of kinematic pairs of the linkage, and f_i is the number of degrees of freedom of the relative movement allowed by the i-th kinematic pair.

It is also known that there are many single-loop linkages whose mobility is not correctly computed by the Kutzbach-Grübler criterion. These

J. Lenarčič and C. Galletti (eds.), On Advances in Robot Kinematics, 133-142.

linkages are loosely denoted as "overconstraint", they owe their mobility to their special geometry that introduces unexpected symetries. What is lesser known is the existence of multi-loop linkages whose geometry has nothing in particular; but, their mobility can not be computed by the Kutzbach-Grübler criterion. The reason is the existence of loops whose links have relative motions associated with different subgroups of the Euclidean group. Hence, it is not possible to use a unique value of λ in equation (1). These linkages are called "mixed linkages".

Wittenbauer, 1923, presented several mixed linkages that included loops with prismatic and revolute pairs and included a mobility equation for these linkages. It was Boden, 1962, who introduced the term of mixed linkages, *gemischt Getriebe.* Freudenstein and Alizade, 1975, in a review of the mobility of mechanisms, defined λ in equation (1) as the **mobility number**, a characteristic of an independent loop of a mechanism. Further, they characterized a **mechanism with mixed mobility** as a mechanism with loops of different mobility number. Liu and Yu, 1995, indicated that different selection of independent loops may yield different results for the degrees of freedom of multi-loop linkage. In all these contributions, the authors analyzed only planar linkages on the basis that the determination of the mobility number for spatial linkages involved geometrical restrictions. Fayet, 1995a and b, analyzed multi-loop linkages, it appears that the focus of the contribution was the infinitesimal mobility of such linkages; unlike these, the focus of this contribution is the finite mobility of multi-loop linkages.

In this paper, we present a systematic method for computing the degree of freedom or mobility of a large class of spatial mixed and multi-loop linkages. The method presented handles all classes of trivial and exceptional loops, but it cannot handle paradoxical loops. The method is based on a new group theoretic and Lie algebra based mobility analysis method developed by the authors, Rico and Ravani, 2003 and Rico *et al*, 2003.

2. Fundamentals

The method illustrated in this contribution is based on the determination of the mobility of single-loop spatial linkages as developed in Rico and Ravani, 2003, and Rico *et al*, 2003. The analysis is based on the Lie algebra, $e(3)$, of the Euclidean group, $E(3)$. Here, only a summary of the most necessary results for computing the mobility are included.

Given a pair of links, i and j, in a single-loop closed kinematic chain, it is possible to define a **clockwise and counterclockwise infinitesimal mechanical liaisons** between i and j, denoted as $V_c(i,j)$ and

$V_{cc}(i,j)$ respectively. Simply speaking, the infinitesimal mechanical liaisons are the column spaces of the Jacobian matrices whose columns are the screw representing the kinematic pairs that join the links, from link i to link j following a clockwise or counterclockwise path. It is also possible to define the **clockwise and counterclockwise closure subalgebras** between i and j, denoted by $A_c(i,j)$ and $A_{cc}(i,j)$ respectively. These subalgebras are the algebras generated by $V_c(i,j)$ and $V_{cc}(i,j)$ respectively by requesting that the subspaces be also closed under the Lie product. Then, one can compute the **absolute closure algebra** as $A_a(i,j) \equiv A_c(i,j) \cap A_{cc}(i,j)$. Once the absolute closure algebra is known it is possible to apply the following Table.

Table I. Mobility Conditions for Kinematic Chains

1 **Chains with Partitioned Mobility** $A_a(i,j) = \{\vec{0}\}$.

2 **Trivial, Exceptional and Paradoxical Chains** $A_a(i,j) \neq \{\vec{0}\}$.

(a) **Exceptional and Paradoxical Chains**

$$A_a(i,j) \neq A_c(i,j) \quad \text{and} \quad A_a(i,j) \neq A_{cc}(i,j).$$

i **Exceptional Chains**

$$V_c(i,j) = A_c(i,j) \quad \text{and} \quad V_{cc}(i,j) = A_{cc}(i,j).$$

$$F = \sum f_i - dim(A_c(i,j)) \quad - \quad dim(A_{cc}(i,j)) + dim(A_a(i,j)).$$

ii **Paradoxical Chains Class I**

$$V_c(i,j) \neq A_c(i,j) \quad \text{or} \quad V_{cc}(i,j) \neq A_{cc}(i,j).$$

(b) **Trivial and Paradoxical Chains**

$$A_a(i,j) = A_c(i,j) \quad \text{or} \quad A_a(i,j) = A_{cc}(i,j).$$

i **Trivial Chains**

$$\text{If} \quad A_a(i,j) = A_{cc}(i,j) \quad \text{then} \quad V_c(i,j) = A_c(i,j), \quad \text{or}$$
$$\text{If} \quad A_a(i,j) = A_c(i,j) \quad \text{then} \quad V_{cc}(i,j) = A_{cc}(i,j).$$

$$F = \sum f_i - dim(A_G), \quad \text{where} \quad A_G = A_c(i,j) \quad \text{or} \quad A_G = A_{cc}(i,j).$$

ii **Paradoxical Chains Class II**

$$\text{If} \quad A_a(i,j) = A_{cc}(i,j) \quad \text{then} \quad V_c(i,j) \neq A_c(i,j), \quad \text{or}$$
$$\text{If} \quad A_a(i,j) = A_c(i,j) \quad \text{then} \quad V_{cc}(i,j) \neq A_{cc}(i,j),$$

As indicated in Table I, if a single-loop linkage is not paradoxical, then it is always possible to determine its mobility using this approach. This is the starting point of the original contribution of the present paper.

3. Mobility of Multi-Loop Linkages

Consider a multi-loop spatial linkage and assume that the linkage has a finite number of loops and none of the loops of the linkage is paradoxical. Then, provided that for any loops that contain non-binary links, the mobility of the loop can be determined by evaluating the relative displacements between non-binary links, the mobility of the multi-loop linkage can be computed as follows:

1 Determine the mobility of each and everyone of the loops of the multi-loop linkage as if there were single-loop linkages, the mobility must be computed by following the approach indicated in Table I.

2 If a loop of the spatial linkage is a structure; namely it has zero mobility, the joints are locked to produce a unique rigid link. This step may require a renumbering of the links and pairs as well as computing again the mobility of the remaning loops.

3 If a loop of the spatial linkage has partitioned mobility, it will assumed that the pair of links that partition the mobility of a linkage are regarded as one. Furthermore the mobility of the clockwise and counterclockwise "loops", formed by the pair of links that partition the mobility of the chain, are computed and regarded as the mobility of these new loops.

4 From the set of loops of freedom greater than zero, we then choose a maximal set of independent loops. Assume that there is k independent loops in the maximal set.

5 Then the total mobility of the multi-loop chain, F_t, is given by

$$F_t = \sum_{i=1}^{k} F_i - \sum f_{rp}, \tag{2}$$

where F_i is the computed freedom for the $i-th$ loop of the maximal set, and f_{rp} is the degrees of freedom of the "repeated pairs"; namely the degrees of freedom of those pairs that appear in the mobility computation of different loops of the maximal set of independent loops.

3.1 Example of Wittenbauer

Consider the multi-loop and mixed linkage with cross links presented by Wittenbauer, 1923, (Figure 1). The ternary links are link numbers $3, 1, 5$ and 8, (link 9 is assumed to be connected to 8 and not to 7). The selected independent loops are: Loop 1 which is the loop 1-2-3-9-8, Loop 2 which is the loop 9-3-4-5-6-7 and Loop 3 which is the loop 1-2-3-4-5-10.

The first loop contains the ternary links 3 and 8, they will be employed to compute the mobility of the loop. The infinitesimal mechanical liaison

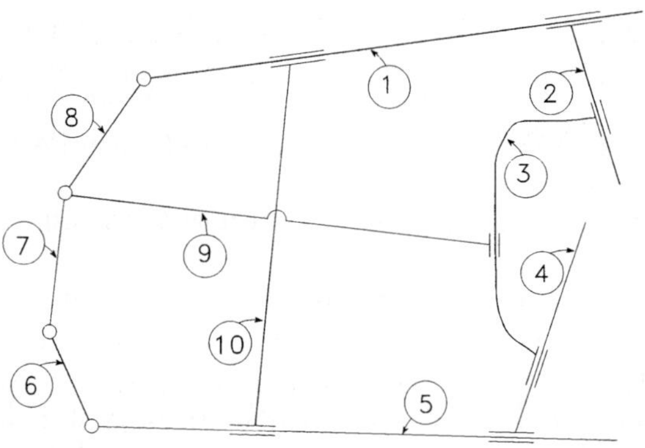

Figure 1. A Mixed Linkage with a Crossing Link Presented by Wittenbauer.

in the clockwise direction is:

$$V_c(8,3) = A_c(8,3) = pl_{\hat{u}},$$

where $\hat{u}$ is a unit vector perpendicular to the plane of the paper. Then $dim(A_c(8,3)) = 3$. Similarly, in the counterclockwise direction is:

$$V_{cc}(8,3) = V(8,9) + V(9,3) \quad \text{and} \quad A_{cc}(8,3) = V(8,9) \odot V(9,3) = pl_{\hat{u}}.$$

However, $dim(V_{cc}(8,3)) = 2 \neq 3 = dim(A_{cc}(8,3))$. Therefore $V_{cc}(8,3) \neq A_{cc}(8,3)$. The absolute closure algebra is given by

$$A_a(8,3) = A_c(8,3) \cap A_{cc}(8,3) = pl_{\hat{u}}$$

and $dim(A_a(8,3)) = 3$. It must be noted that this loop is a trivial one and its mobility can be computed from

$$F_1 = \sum f_i - dim(A_G) = 5 - 3 = 2.$$

where $A_G = A_c(8,3) = A_{cc}(8,3) = pl_{\hat{u}}$.

The second loop 9-3-4-5-6-7 contains the ternary link 3; therefore, links 3 and 7 will be employed to compute the mobility of the loop. The clockwise mechanical liaison is:

$$V_c(3,7) = A_c(3,7) = pl_{\hat{u}},$$

where $\hat{u}$ is a unit vector perpendicular to the plane of the paper. Furthermore $dim(A_c(8,3)) = 3$. Similarly in the counterclockwise direction

$$V_{cc}(3,7) = V(3,9) + V(9,7) \quad \text{and} \quad A_{cc}(3,7) = V(3,9) \odot V(9,7) = pl_{\hat{u}}.$$

However, $dim(V_{cc}(3,7)) = 2 \neq 3 = dim(A_{cc}(3,7))$. Therefore $V_{cc}(3,7) \neq A_{cc}(3,7)$, and the resulting absolute closure algebra is given by

$$A_a(3,7) = A_c(3,7) \cap A_{cc}(3,7) = pl_{\hat{u}} \quad \text{and} \quad dim(A_a(3,7)) = 3.$$

This loop is trivial, and its mobility is computed from

$$F_2 = \sum f_i - dim(A_G) = 6 - 3 = 3.$$

where $A_G = A_c(3,7) = A_{cc}(3,7) = pl_{\hat{u}}$.

The third loop 1-2-3-4-5-10 contains the ternary link 3; therefore, the links 3 and 10 will be employed to compute the mobility of the loop, as follows

$$V_c(3,10) = A_c(3,10) = p2, \quad \text{and} \quad dim(A_c(3,10)) = 2,$$

$$V_{cc}(3,10) = A_{cc}(3,10) = p2, \quad \text{and} \quad dim(A_{cc}(3,10)) = 2.$$

Moreover

$$A_a(3,10) = A_c(3,10) \cap A_{cc}(3,10) = p2 \quad \text{and} \quad dim(A_a(3,10)) = 2.$$

Again, the loop is trivial and its mobility is given by

$$F_3 = \sum f_i - dim(A_G) = 6 - 2 = 4.$$

where $A_G = A_c(3,10) = A_{cc}(3,10) = p2$.

Finally, in order to find the mobility of the entire multi-loop linkage, the kinematic pairs that appear in more than one loop are counted. These are $(1,2), (2,3), (3,4), (3,9)$ and $(4,5)$. Hence, the degree of freedom or the mobility of the multi-loop linkage is then:

$$F_t = \sum_{i=1}^{2} F_i - \sum f_{rp} = (2+3+4) - 5 = 4. \qquad (3)$$

3.2 A Multi-Loop Linkage with an Exceptional Loop

Consider the multi-loop linkage with an exceptional loop shown in Figure 2. Note that the loop formed by links 1-2-3-4-5-6 is a Sarrus linkage that is an exceptional single-loop linkage. The implication of this is that none of the existing reported methods would be able to correctly compute its degrees of freedom. The multi-loop linkage has three loops only two of which are independent. The loops that are selected for our analysis are loops 1-2-3-4-5-6 and 1-6-5-4-7-8-9-10. The linkage has two

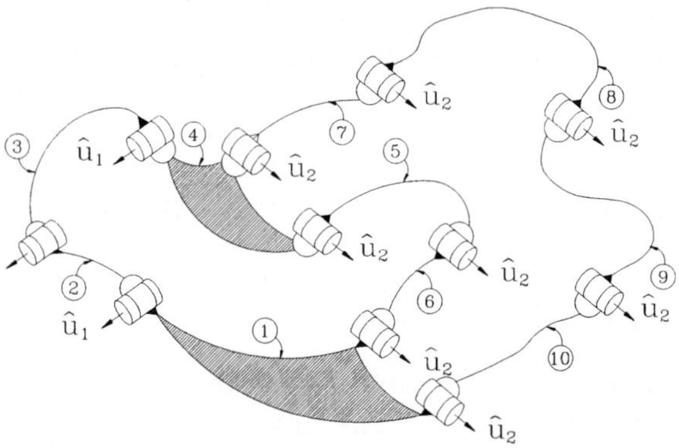

Figure 2. A Multi-Loop Linkage with a Sarrus Linkage Embedded.

ternary links 1 and 4. These links are selected to compute the mobility of the loops.

For the first loop 1-2-3-4-5-6, we have

$$V_c(1,4) = A_c(1,4) = pl_{\hat{u}_1}, \quad \text{and} \quad dim(A_c(1,4)) = 3,$$

$$V_{cc}(1,4) = A_{cc}(1,4) = pl_{\hat{u}_2}, \quad \text{and} \quad dim(A_{cc}(1,4)) = 3.$$

Moreover, the absolute closure algebra is

$$A_a(1,4) = A_c(1,4) \cap A_{cc}(1,4) = t_{\hat{u}_1 \times \hat{u}_2} \quad \text{and} \quad dim(A_a(1,4)) = 1.$$

The is exceptional loop and its mobility is computed from

$$\begin{aligned} F_1 &= \sum f_i - dim(A_c(1,4)) - dim(A_{cc}(1,4)) + dim(A_a(1,4)) \\ &= 6 - 3 - 3 + 1 = 1. \end{aligned}$$

For the loop 1-6-5-4-7-8-9-10:

$$V_c(1,4)A_{cc}(1,4) = pl_{\hat{u}_2}, \quad \text{and} \quad dim(A_c(1,4)) = 3,$$

$$V_{cc}(1,4) = A_{cc}(1,4) = pl_{\hat{u}_2} \quad \text{and} \quad dim(A_{cc}(1,4)) = 3.$$

Moreover, the absolute closure algebra is given by

$$A_a(1,4) = A_c(1,4) \cap A_{cc}(1,4) = pl_{\hat{u}_2} \quad \text{and} \quad dim(A_a(1,4)) = 3.$$

The loop is trivial and its mobility is computed from

$$F_2 = \sum f_i - dim(A_G) = 8 - 3 = 5.$$

where $A_G = A_c(1,4) = A_{cc}(1,4) = pl_{\hat{u}_2}$.

Finally, in order to find the mobility of the multi-loop linkage, the kinematic pairs that appear in more than one loop are counted. These are $(4,5), (5,6)$ and $(6,1)$. Hence the degree-of-freedom of the multi-loop linkage is:

$$F_t = \sum_{i=1}^{2} F_i - \sum f_{rp} = (1+5) - 3 = 3. \tag{4}$$

3.3 Example of Liu and Yu

Consider the planar multi-loop and mixed linkage studied in Liu and Yu, 1995, and shown in Figure 4. The linkage has three loops and any two of them are independent. Furthermore, the linkage has a pair of ternary links, namely links 1 and 3. For the mobility analysis we will employ the loops 1-4-3-5 and 1-2-3-5.

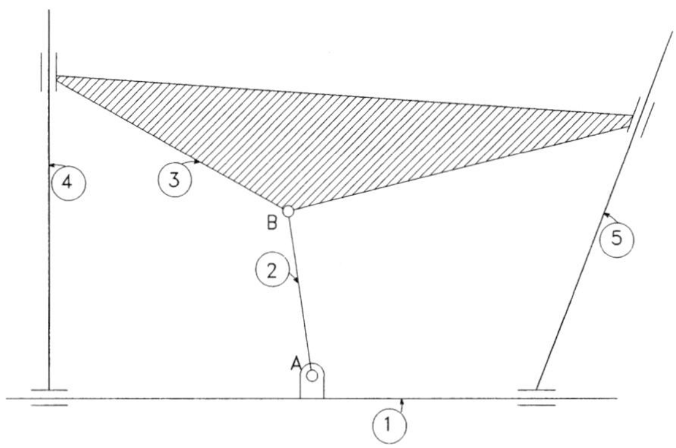

Figure 3. A Mixed Parallel Linkage Presented by Liu and Yu.

The first loop 1-4-3-5 contains the ternary links 1 and 3, they will be employed to compute the mobility of the loop as follows:

$$V_c(1,3) = V(1,4) + V(4,3) = V(1,4) \odot V(4,3) = A_c(1,3) = p2,$$

where $p2$ is the algebra associated with plane translations. Furthermore $dim(A_c(8,3)) = 2$. Similarly

$$V_{cc}(1,3) = V(1,5) + V(5,3) = V(1,5) \odot V(5,3) = A_{cc}(1,3) = p2.$$

Moreover

$$A_a(1,3) = A_c(1,3) \cap A_{cc}(1,3) = p2 \quad \text{and} \quad dim(A_a(1,3)) = 2.$$

The loop is trivial and its mobility can be computed as

$$F_1 = \sum f_i - dim(A_G) = 4 - 2 = 2.$$

where $A_G = A_c(1,3) = A_{cc}(1,3) = p2$.

The second loop 1-2-3-5 contains the ternary links 1 and 3. These are employed to compute the mobility of the loop.

$$V_c(1,3) = V(1,2) + V(2,3) \neq A_c(1,3) = V(1,2) \odot V(2,3) = pl_{\hat{u}},$$

where $pl_{\hat{u}}$ is the algebra associated with plane motions in the plane of the paper. Since $dim(V_c(1,3)) = 2 \neq 3 = dim(A_c(1,3))$. It follows that $V_c(1,3) \neq A_c(1,3)$. Similarly

$$V_{cc}(1,3) = V(1,5) + V(5,3) = V(1,5) \odot V(5,3) = A_{cc}(1,3) = t_2.$$

Moreover

$$A_a(1,3) = A_c(1,3) \cap A_{cc}(1,3) = A_{cc}(1,3) = t_2$$

However, because $V_c(1,3) \neq A_c(1,3)$, the loop is paradoxical, when analyzing the mobility of the loop by employing links 1 and 3.

The method presented so far by the authors does not apply to paradoxical linkages. The problem, however, can be overcome by noting that since $A_a(1,3) = p2$, the sum of the rotation angles associated with the revolute pairs must be zero; i.e.

$$\theta_{12} + \theta_{23} = 0, \quad \text{therefore} \quad \omega_{12} + \omega_{23} = 0.$$

The clockwise infinitesimal mechanical liaison and clockwise closure algebra are recomputed as follows

$$\begin{aligned} V_c(1,3)_n &= \omega_{12}{}^1\$^2 + \omega_{23}{}^2\$^3 = \omega_{12}\begin{bmatrix} \hat{u} \\ \vec{r}_A \times \hat{u} \end{bmatrix} + \omega_{23}\begin{bmatrix} \hat{u} \\ \vec{r}_B \times \hat{u} \end{bmatrix} \\ &= \begin{bmatrix} (\omega_{12} + \omega_{23})\hat{u} \\ \omega_{12}(\vec{r}_A - \vec{r}_B) \times \hat{u} \end{bmatrix} = \begin{bmatrix} \vec{0} \\ \omega_{12}(\vec{r}_A - \vec{r}_B) \times \hat{u} \end{bmatrix} = A_c(1,3)_n \end{aligned}$$

The recomputed infinitesimal mechanical liaison is equal to the closure algebra. This is the algebra associated with a line translation perpendicular to the unit vector $\hat{u}$ and the vector $\vec{r}_A - \vec{r}_B$. Further details are given in Rico and Ravani, 2004. Thus $dim(A_c(1,3)_n) = 1$. Under the new recomputed closure algebra,

$$A_a(1,3)_n = A_c(1,3)_n \cap A_{cc}(1,3) = A_c(1,3)_n = t_{\hat{u}\times(\vec{r}_A - \vec{r}_B)}.$$

Thus, under these new conditions, the loop is trivial and its mobility is

$$F_2 = \sum f_i - dim(A_G) - N_{ac} = 4 - 2 - 1 = 1$$

where N_{ac} is the number of additional constraints necessary to compute the mobility of the loop, and $A_G = A_{cc}(1,3) = p2$.

Finally, to obtain the mobility of the multi-loop linkage, it is necessary to determine the "repeated pairs" which are, in this case, the pairs $(1,5)$ and $(5,3)$. Thus

$$F_t = \sum_{i=1}^{2} F_i - \sum f_{rp} = (2+1) - 2 = 1. \tag{5}$$

4. Conclusions

This contribution has presented a method for correctly determining the mobility of multi-loop and mixed linkages. Unlike previously reported techniques, the method presented here can be applied to spatial multi-loop linkages where the loops may not only be trivial but also exceptional, the only limitation is that none of the loops can be paradoxical.

5. Acknowledgments

The first author thanks Prof. Bahram Ravani, from the Mechanical and Aeronautical Engineering Department of the University of California-Davis and the California Department of Transportation (CalTrans) for their support during a summer stay in the year 2003. The first author also thanks Concyteg and Cosnet for its continuous support.

References

Boden, H. (1962), Zum Zwanglauf gemischt räumliche-ebener Getriebe, *Maschinenbautechnik*, Vol. 11, pp. 612–615.

Wittenbauer, F., (1923), *Graphische Dynamik*, Springer Verlag: Berlin.

Freudenstein, F. and Alizade, R. (1975), On the Degree of Freedom of Mechanisms with Variable General Constraint, *Proceedings of the Fourth World Congress on the Theory of Machines and Mechanisms*, pp. 51–56.

Liu, T. and Yu, C. H. (1995), Identification and Classification of Multi-Degree-of-Freedom and Multi-Loop Mechanisms, *Journal of Mechanical Design*, Vol. 117, pp. 104–111.

Fayet, M. (1995a), Mécanismes Multi-Boucles–I, *Mechanism and Machine Theory*, Vol. 30, pp. 201–217.

Fayet, M., (1995b), Mécanismes Multi-Boucles–III, *Mechanism and Machine Theory*, Vol. 30, pp. 233–252.

Hunt, K. H., (1990), *Kinematic Geometry of Mechanisms*, Oxford University Press: Oxford.

Rico, J.M. and Ravani, B., (2003), On Mobility Analysis of Linkage Using Group Theory, *ASME Journal of Mechanical Design*, Vol. 125, pp. 70–80.

Rico, J.M. and Ravani, B., (2004), On Calculating the Degrees-of-Freedom or Mobility of Overconstrained Linkages, Submitted to ASME-DETC 2004.

Rico, J.M., Gallardo, J, and Ravani, B. (2003), Lie Algebra and the Mobility of Kinematic Chains, *Journal of Robotic Systems*. Vol. 20, pp. 477–499.

Humanoids and Biomedical Applications

HUMAN-LIKE MOTION FROM PHYSIOLOGICALLY-BASED POTENTIAL ENERGIES

O. Khatib, J. Warren, V. De Sapio, L. Sentis
Robotics Laboratory, Department of Computer Science
Stanford University, Stanford, California 94305
{khatib, warren, vdesap, lsentis}@robotics.stanford.edu

Abstract Generating coordinated natural motion in human-like robotic structures has proved to be a challenging task. Given that humans easily solve this problem, we propose a methodology to devise the underlying strategies of human movement and apply them for robotic control. We use this approach to examine how humans utilize their muscles while performing positioning tasks. This analysis suggests an effort potential that is shown to characterize human postural motion. By applying this methodology to other criteria, we seek to establish a basis of human motion characteristics.

Keywords: human motion behaviors, task-level control, muscle kinematics and dynamics, task/posture decomposition

1. Introduction

The challenge of synthesizing motion behaviors is a long-standing problem within the robotics community. With the recent advent of complex humanoid systems, this challenge grows ever demanding. Due to their anthropomorphic design, humanoids should move in a human-like manner to facilitate movement within man-made environments and to promote interaction with their biological counterparts. Common strategies involve generating joint space trajectories (Kuroki, 2003) or learning specific motions (Ijspeert, 2002), but these approaches require off-line computations and do not generalize well to related tasks. There is a pressing need for a framework where natural motion is generated in real-time for a large range of tasks.

From observations of human motion, it is apparent that people perform many different behaviors while executing a task. These secondary criteria are the essence of natural human motion, affecting not *whether* but *how* the task is achieved. Fig. 1 contrasts poses of two humanoids performing a hands positioning task. By simultaneously controlling the balance and maximizing mobility, the left humanoid visibly exhibits a more human-like behavior. With this as motivation, we have developed

J. Lenarčič and C. Galletti (eds.), On Advances in Robot Kinematics, 145-154.

Figure 1. Comparison of configurations while performing a hand positioning task. By maintaining balance and maximizing mobility, the left humanoid has a markedly more natural posture than the one on the right.

a prioritized multiple task controller that in real-time dynamically decouples each task.

Equipped with this control framework, we propose a general methodology to discover the innate subtasks of human motion. These behaviors are initially identified by hypothesis or direct observation. By designing targeted motion capture experiments that excite these behaviors, we then confirm or refute their presence and ascertain their importance relative to other tasks. After identification, these behaviors are then incorporated into our robotic controller, incrementally creating a basis of tasks that characterize human-like movement.

We will illustrate this methodology by examining the muscular effort required to perform positioning tasks. In general terms, it is evident that humans avoid overexertion in commonplace motions. However, this concept is not easily quantified, nor is it clear of its relative importance to other behaviors. By utilizing biomechanical models for neuromuscular dynamics and control (Zajac, 1993; Hogan, 1991), we demonstrate that humans seek to minimize the muscular effort (normalized by torque capacity) needed to compensate for gravity while performing positioning tasks. While there are additional behaviors that characterize human motion, humans use much of the available redundancy for gravity effort minimization.

2. Prioritized Task-Level Control Framework

For any human movement, a person satisfies various criteria while performing the task at hand. These tasks have different priorities - maintaining balance is typically more vital than minimizing the required effort - but all are addressed subject to this hierarchy. With this model in mind, we have previously developed a task-level control framework that performs multiple behaviors in a prioritized manner (Khatib, 2004).

To illustrate this framework, we examine the scenario involving two tasks. We begin by describing the dynamics of the robot in terms of its joints coordinates, $\boldsymbol{q}$,

$$\boldsymbol{A}(\boldsymbol{q})\ddot{\boldsymbol{q}} + \boldsymbol{b}(\boldsymbol{q}, \dot{\boldsymbol{q}}) + \boldsymbol{g}(\boldsymbol{q}) = \boldsymbol{\Gamma} \tag{1}$$

For this system of n equations, $\boldsymbol{\Gamma}$ is the set of generalized joint torques, $\boldsymbol{A}(\boldsymbol{q})$ is the inertia matrix, $\boldsymbol{b}(\boldsymbol{q}, \dot{\boldsymbol{q}})$ represents the centrifugal and Coriolis forces and $\boldsymbol{g}(\boldsymbol{q})$ is the gravity effect.

We denote the m parameters of the primary task as $\mathbf{x}_t$, with $J_t(\boldsymbol{q})$ representing its joint space Jacobian. The operational space framework (Khatib, 1987) first introduced the dynamically consistent inverse of the Jacobian,

$$\bar{\boldsymbol{J}}_t = \boldsymbol{A}^{-1}\boldsymbol{J}_t^T \left[\boldsymbol{J}_t\boldsymbol{A}^{-1}\boldsymbol{J}_t^T\right]^{-1} \tag{2}$$

and demonstrated that its use provides a task dynamic behavior model. More explicitly, $\bar{\boldsymbol{J}}_t^T$ projects the joint space dynamics (Eq. 1) into task space,

$$\bar{\boldsymbol{J}}_t^T [\boldsymbol{A}\ddot{\boldsymbol{q}} + \boldsymbol{b} + \boldsymbol{g} = \boldsymbol{\Gamma}] \Longrightarrow \boldsymbol{\Lambda}_t\ddot{\mathbf{x}}_t + \boldsymbol{\mu}_t + \boldsymbol{p}_t = \boldsymbol{F}_t \tag{3}$$

In this space, $\boldsymbol{\Lambda}_t$ is the $m \times m$ task inertia matrix, and $\boldsymbol{\mu}_t$, $\boldsymbol{p}_t$, and $\boldsymbol{F}_t$ are respectively the centrifugal and Coriolis forces, gravity effect, and generalized force.

By the principle of virtual work, forces in the task space, $\boldsymbol{F}_t$, are transformed into a joint torque, $\boldsymbol{\Gamma}_{\text{task}}$, by the relationship,

$$\boldsymbol{\Gamma}_{\text{task}} = \boldsymbol{J}_t^T\boldsymbol{F}_t \tag{4}$$

This torque/force relationship allows for the decomposition of the total torque into two dynamically decoupled vectors: the torque controlling the task behavior of $\mathbf{x}_t$ and a torque that does not affect the task but controls the robotic posture (Khatib, 1995),

$$\boldsymbol{\Gamma} = \boldsymbol{\Gamma}_{\text{task}} + \boldsymbol{\Gamma}_{\text{posture}} \tag{5}$$

Torques are projected into the posture space by the operator,

$$\boldsymbol{N}_t^T = [\boldsymbol{I} - \boldsymbol{J}_t^T\bar{\boldsymbol{J}}_t^T] \tag{6}$$

and it is within this space that we will address our secondary task.

Since it will be controlled within the posture space of the primary task, we choose $\mathbf{x}_p$ to denote the coordinates of the subtask and $\boldsymbol{J}_p$ as its Jacobian. The task-consistent posture Jacobian was introduced to incorporate the constraints of $\mathbf{x}_t$ (Khatib, 2004),

$$\boldsymbol{J}_{p|t} = \boldsymbol{J}_p \boldsymbol{N}_t \tag{7}$$

The subtask may not be achievable if it conflicts with the primary task, but this Jacobian identifies the controllable directions. Moreover, the task-consistent posture Jacobian provides the dynamic behavior model of the subtask in this posture space,

$$\bar{\boldsymbol{J}}_{p|t}^T[\boldsymbol{A}\ddot{\boldsymbol{q}} + \boldsymbol{b} + \boldsymbol{g} = \boldsymbol{\Gamma}] \Longrightarrow \boldsymbol{\Lambda}_{p|t}\ddot{\mathbf{x}}_{p|t} + \boldsymbol{\mu}_{p|t} + \boldsymbol{p}_{p|t} = \boldsymbol{F}_{p|t} \tag{8}$$

and the force/torque relationship,

$$\boldsymbol{\Gamma}_{\text{posture}} = \boldsymbol{J}_{p|t}^T \boldsymbol{F}_{p|t} \tag{9}$$

With these dynamic behavior models we can formulate a dynamically decoupled control to perform both tasks. For each task, we have a desired unit inertial behavior $\ddot{\mathbf{x}}_t = \boldsymbol{F}_t^\star$ and $\ddot{\mathbf{x}}_p = \boldsymbol{F}_p^\star$. To compensate for the dynamics within their respective spaces, we select control forces $\boldsymbol{F}_t$ and $\boldsymbol{F}_{p|t}$ by,

$$\begin{aligned} \boldsymbol{F}_t &= \widehat{\boldsymbol{\Lambda}}_t \boldsymbol{F}_t^\star + \widehat{\boldsymbol{\mu}}_t + \widehat{\boldsymbol{p}}_t \\ \boldsymbol{F}_{p|t} &= \widehat{\boldsymbol{\Lambda}}_{p|t} \boldsymbol{F}_{p|t}^\star + \widehat{\boldsymbol{\mu}}_{p|t} + \widehat{\boldsymbol{p}}_{p|t} \end{aligned} \tag{10}$$

where $\widehat{\cdot}$ denotes the estimates of the components of the dynamic models. (The modification $\boldsymbol{F}_{p|t}^\star = \boldsymbol{F}_p^\star - \ddot{\mathbf{x}}_{\overline{p|t}}$ is needed since $\boldsymbol{\Gamma}_{\text{task}}$ induces a bias acceleration on $\mathbf{x}_p$.) Using the task/posture decomposition of torque and the force/torque relationships, the resulting control torque $\boldsymbol{\Gamma}$ is determined by,

$$\boldsymbol{\Gamma} = \boldsymbol{\Gamma}_{\text{task}} + \boldsymbol{\Gamma}_{\text{posture}} = \boldsymbol{J}_t^T \boldsymbol{F}_t + \boldsymbol{J}_{p|t}^T \boldsymbol{F}_{p|t} \tag{11}$$

This control framework allows for a conceptual partitioning of motion. The task can be controlled by a task field U_t that determines the desired behavior by its gradient: $\boldsymbol{F}_t^\star = -\nabla_{\mathbf{x}_t} U_t$. In a similar manner, the posture behavior $\boldsymbol{F}_p^\star$ can be specified by a separate posture field U_p. We can therefore incorporate human motion behaviors by expressing these strategies as energy potentials. In the following sections we examine how humans address muscular effort and construct a posture field to capture this behavior.

3. Human Model

To examine muscular effort within the prioritized task control framework we have constructed a musculoskeletal model in conjunction with a dynamic simulation environment. The SAI environment (Khatib, 2002) uses an $\mathcal{O}(n)$ recursive algorithm for computing the dynamics and control of an n-joint branching, articulated robotic mechanism.

Our human model consists of 25 joints and 144 musculo-tendon actuators (Fig. 2). The musculoskeletal data used in this model has been derived from SIMM models (Delp and Loan, 1995). The skeleton is modelled as a multibody system within SAI and each musculo-tendon unit is modelled using a three element Hill-type model (Schutte, 1992). In this model all musculo-tendon lengths, $\boldsymbol{l}$, can be uniquely determined from the skeletal configuration. From kinematics, differential changes in $\boldsymbol{l}$ are given by $d\boldsymbol{l} = \boldsymbol{L}\, d\boldsymbol{q}$, where $\boldsymbol{L}(\boldsymbol{q})$ is the muscle Jacobian. From the principle of virtual work we conclude,

$$\boldsymbol{\Gamma} = \boldsymbol{L}^T \boldsymbol{f} \tag{12}$$

where $\boldsymbol{f}$ is the vector of net musculo-tendon forces (with an appropriate sign convention adopted).

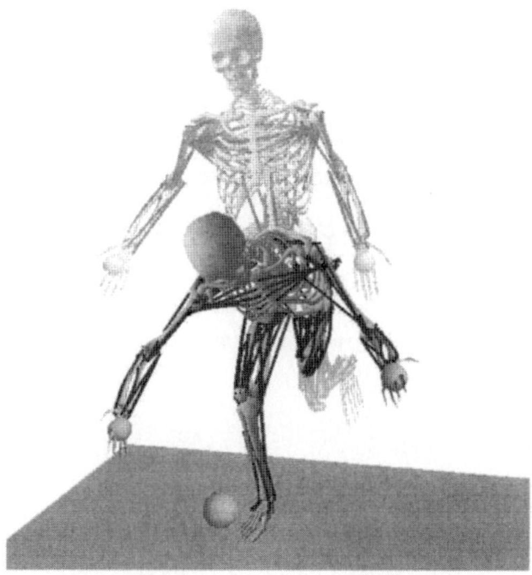

Figure 2. A human musculoskeletal model simulated in the SAI environment. This model consists of 25 joints and 144 musculo-tendon actuators.

The contraction dynamics of each musculo-tendon unit is modelled as depicted in Fig. 3a. The force-length-velocity relationship of the muscle at full activation is depicted in Fig. 3b. The active force component of this surface linearly scales by activation ($a \in [0, 1]$).

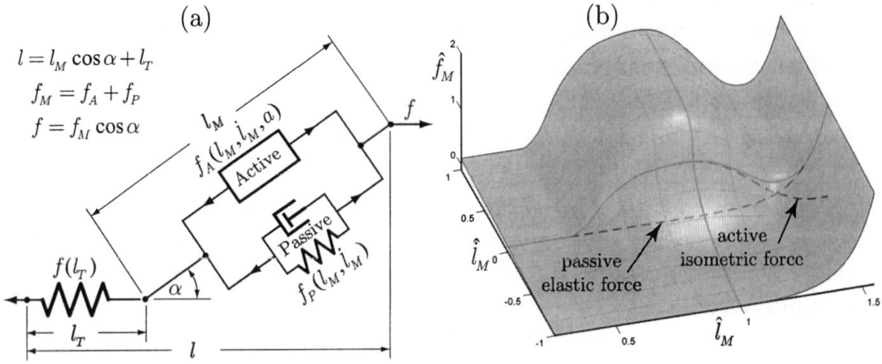

Figure 3. (a). Muscle model consisting of a parallel-series arrangement of active and passive elements. (b). Force-length-velocity relationship showing normalized muscle force, $\widehat{f}_M$, as a function of muscle length, $\widehat{l}_M$, and rate of contraction, $\widehat{\dot{l}}_M$.

This overall model can be simplified if we take the tendon to be infinitely stiff. In this case the musculo-tendon force, f, can be computed as a function of $\boldsymbol{q}$, $\dot{\boldsymbol{q}}$, and a. From this we can compute feasible ranges for the tendon forces (between zero and full activation). Using Eq. 12, in conjunction with the feasible ranges of tendon forces, muscle induced torque boundaries, Γ_B, can be determined.

4. Muscular Effort Induced Postures

With the premise that musculoskeletal physiology plays an important role in the nature of human motion, we conducted preliminary exper-

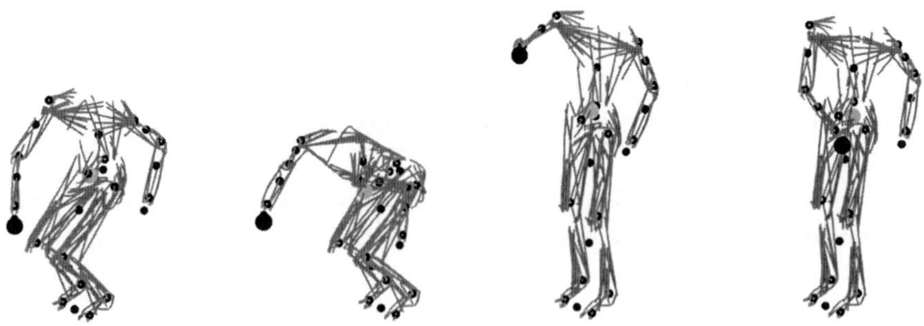

Figure 4. Captured skeletal configurations associated with a subject reaching to four different target locations with weight in hand. Muscle paths were overlayed based on the muscle model.

iments to investigate the manner in which humans address muscular effort while performing tasks. Three subjects were tracked using a motion capture system while performing whole-body reaching movements. This set of tasks required the subject to move a normalized weight to visual goal locations while keeping both feet grounded. Upon reaching a goal location, the subject maintained this task for 5 seconds while adjusting his posture to minimize discomfort. Fig. 4 depicts frames from the motion capture of a subject moving to separate target locations. Computational muscle models were used to overlay muscle paths over the configuration data.

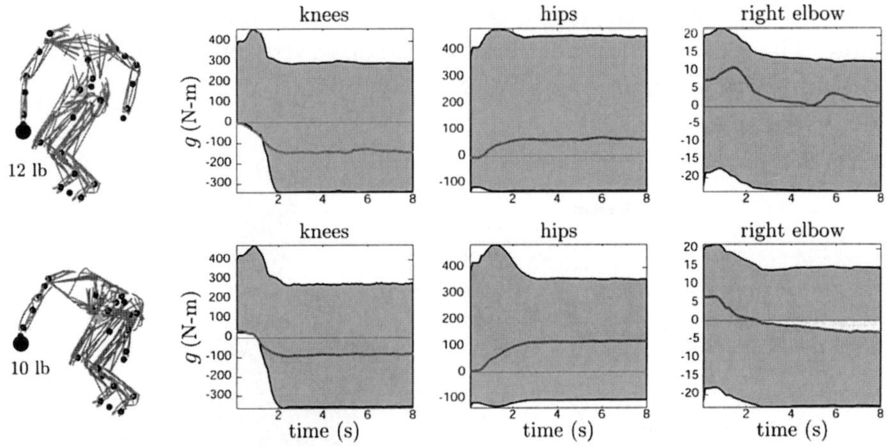

Figure 5. Captured skeletal configurations for reaching tasks, with selected gravity compensation torques shown. Boundaries represent muscle induced torque capacities.

From the joint space trajectories and our musculoskeletal model we can determine the muscular effort required to compensate for gravity. Fig. 5 shows samples of the processed data for two tests. We observe a strong trend that the gravity torques are balanced over the various joints in a manner that is correlated to the available torque capacity at each joint. Based on these observations, and past use of muscle-level weighted norm criteria in the biomechanics community (Anderson, 2001; Crowninshield, 1981), we conjecture that postural motion involves the minimization of a muscle effort potential, $U(\boldsymbol{q})$, of the form,

$$U(\boldsymbol{q}) = \sum_{i=1}^{n} w_i \frac{g_i(\boldsymbol{q})^2}{\Gamma_{B_i}(\boldsymbol{q})^2} \tag{13}$$

where g_i is the gravity torque about joint i, Γ_{B_i} is the muscle induced boundary torque (upper or lower boundary depending on the sign of g_i), and w_i is a weighting term.

We analyzed the suitability of the muscle effort potential in predicting quasi-static postures. Fig. 6 shows muscle effort computed over null space motion of arm configuration. A one dimensional null space was chosen for illustrative purposes. The variable q_p spans the task consistent manifold in configuration space (self motion manifold) and, in this case, represents the inclination angle of the plane of the arm. The configuration $\boldsymbol{q}_{p_o}$ is the observed steady state configuration from the experimental data (third configuration in Fig. 4). This observed configuration is 0.13 radians (7.5°) from the configuration where the computed minimum occurs. The muscle effort computed at the observed configuration exceeds the minimum computed muscle effort by 1.6%.

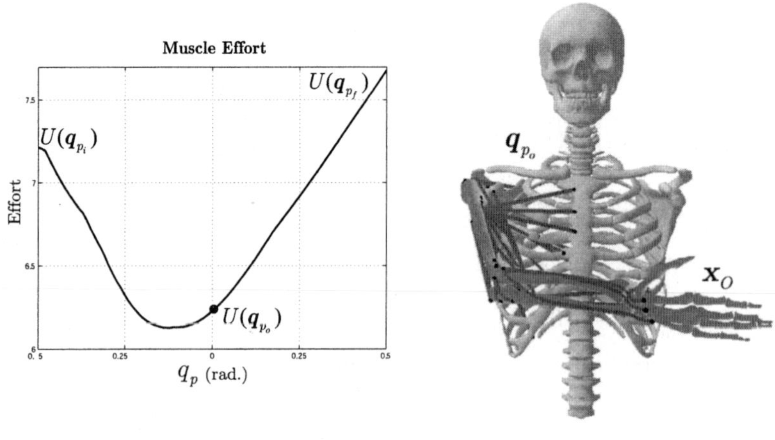

Figure 6. Plot showing muscle effort over null-space motion of arm configuration. Observed configuration, $\boldsymbol{q}_{p_o}$, is 0.13 radians (7.5°) from the configuration where the computed minimum occurs.

We also computed muscle effort over null space motion in whole body configuration. This is shown in Fig. 7. In addition to a task point at the right hand, $\mathbf{x}_O$, balancing was imposed with a task point at the center of mass of the system, $\mathbf{x}_{cm}$. Joints above the chest were locked to limit the dimension of the null space to one. The observed steady state configuration (second configuration in Fig. 4) is 0.27 radians (15.5°) from the configuration where the computed minimum occurs. The muscle effort computed at the observed configuration exceeds the minimum computed muscle effort by 3.6%.

For both the arm and whole body, uniform weighting was applied in computing the muscle effort. The efforts computed at the observed configurations exceed the minimum computed muscle efforts by less than 4%. This suggests strong correlation between task consistent human postures and a muscular effort criteria. While there are a number of sources of variability, including experimental error, model inaccuracies, and intrinsic variability in human motion (e.g. motion deadband); the results suggest a robust relationship between human motion and task-level muscular effort minimization.

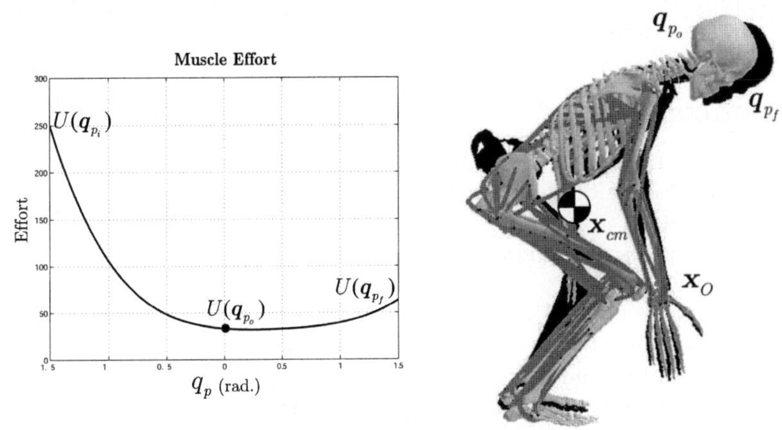

Figure 7. Plot showing muscle effort over null-space motion in whole body posture. Observed configuration, q_{p_o}, is 0.27 radians (15.5°) from the configuration where the computed minimum occurs.

The identification of this important postural behavior is encouraging for our research to map human motion behaviors for humanoid robot control. By using this potential to define the posture field U_p, we can utilize the posture space to minimize effort in real-time while performing any primary task $\mathbf{x}_t$. Moreover, the flexibility of our control framework will easily incorporate other behaviors upon identification. By establishing this basis, we expect to capture the characterstics that drive most of natural human motion.

5. Conclusions

Having constructed a dynamic musculoskeletal model of a human, we have proposed a methodology for identifying physiological characteristics and constraints that shape human motion. These characteristics are then mapped for robotic control by incorporating potentials into a prioritized task-level control framework. We investigated the muscular effort required for reaching tasks using this approach, and our analysis suggests that minimization of muscular effort is a vital objective of postural motion. By applying this methodology to other motion characteristics, we seek to incrementally construct a basis of human motion strategies.

6. Acknowledgements

The authors would like to acknowledge Scott Delp for his assistance and for providing musculoskeletal data. Additionally, the motion capture contributions of Chris Dyrby and Jaeheung Park are gratefully acknowledged.

References

Anderson, F.C., and Pandy, M.G. (2001), Static and dynamic optimization solutions for gait are practically equivalent, *Journal of Biomechanics*, vol. 34, pp. 153–161.

Crowninshield, R.D., and Brand, R.A. (1981), A physiologically based criterion of muscle force prediction in locomotion, *Journal of Biomechanics*, vol. 14, pp. 793–801.

Delp, S., and Loan, P. (1995), A software system to develop and analyze models of musculoskeletal structures, *Computers in Biology and Medicine*, vol. 25, pp. 21–34.

Hogan, N. (1991), Mechanical Impedance of Single- and Multi-Articular Systems. In *Multiple Muscle Systems*, New York, Springer Verlag.

Ijspeert, A., Nakanishi, J., and Schaal, S. (2002), Movement imitation with nonlinear dynamical systems in humanoid robots, *Proceedings of the International Conference on Robotics and Automation*, pp. 1398-1403.

Khatib, O. (1987), A unified approach to motion and force control of robot manipulators: The operational space formulation, *International Journal of Robotics Research*, vol. 3, no. 1, pp. 43–53.

Khatib, O. (1995), Inertial properties in robotic manipulation: An object level framework, *International Journal of Robotics Research*, vol. 14, no. 1, pp. 19–36.

Khatib, O., Brock, O., Chang, K-S., Conti, F., Ruspini, D., Sentis, L. (2002), Robotics and Interactive Simulation, *Communications of the ACM*, vol. 45, no. 3, pp. 46–51.

Khatib, O., Sentis, L., Park, J., and Warren, J. (2004), Whole body dynamic behavior and control of human-like robots, *International Journal of Humanoid Robotics.*

Kuroki, Y., Blank, B., Mikami., T., Mayeux, P., Miyamoto, A., Playter, R., Nagasaka, K., Raibert, M., Nagano, M., and Yamaguchi, J. (2003), Motion creating system for a small biped entertainment robot, *Proceedings of the International Conference on Intelligent Robots and Systems, pp. 1394–1399.*

Schutte, L.M. (1992), *Using musculoskeletal models to explore strategies for improving performance in electrical stimulation-induced leg cycle ergometry*, PhD thesis, Stanford University.

Zajac, F.E. (1993), Muscle coordination of movement: a perspective, *Journal of Biomechanics*, vol. 26, pp. 109–124.

ROLE OF BIARTICULAR MUSCLES IN HUMAN JUMP

Jan Babič

Jožef Stefan Institute, Department of Automatics, Biocybernetics and Robotics
Ljubljana, Slovenia
jan.babic@ijs.si

Jadran Lenarčič

Jožef Stefan Institute, Department of Automatics, Biocybernetics and Robotics
Ljubljana, Slovenia
jadran.lenarcic@ijs.si

Abstract The purpose of the study was to build an efficient biomechanical model and to evaluate the influence of the timing of gastrocnemius muscle activation and its elastic properties on vertical jump performance. A series of countermovement vertical jump simulations have been performed. First with a stiff biarticular link where the influence of the muscle activation timing has been demonstrated. Afterwards, the stiff biarticular link was replaced by an elastic biarticular link to evaluate the role of elasticity and resonance in human jump.

Keywords: human jump, biarticular muscle, elasticity.

1. Introduction

Biarticular muscles are muscles that are passing two joints. Their functions during human jumping have been studied extensively by many researchers but these functions still remain unclear. One of such functions is the transportation of mechanical energy from proximal to distal joints. It is believed that this transportation causes an effective transformation of rotational motion of body segments into translation of the body center of gravity (Ingen Schenau, 1989). The mechanism by which this is achieved is the timely activation of biarticular muscles before the end of the push-off phase of the jump. The activation of the biarticular gastrocnemius muscle prior to the end of the push-off enables the transportation of the power generated by the knee extensors from the knee to the ankle joint. This transfer of mechanical energy by gastrocnemius can be explained using the following example. During the push-off phase of the human jump, the knee joint is rapidly extended as a result of the positive work done by the knee extensor muscles. If the

J. Lenarčič and C. Galletti (eds.), On Advances in Robot Kinematics, 155-162.

biarticular gastrocnemius muscle contracts isometrically (its length does not change), the additional mechanical work is done at the ankle joint because of the gastrocnemius muscle, which contributes no mechanical work by itself. A part of the energy generated by the knee extensors appears as mechanical work at the ankle joint and the height of the jump is significantly increased. This is because, as the jump proceeds and the knee straightens, the angular position changes of the knee have progressively less effect on vertical velocity of the jumper center of gravity. By gastrocnemius muscle activation, a rapid extension of the foot is produced. This extension has a greater effect on the vertical velocity than the extension of the almost straightened knee. The energy is more effectively translated into vertical velocity and a greater height of the jump is achieved. However, the timing of the gastrocnemius muscle activation is critical to obtain a maximum effect. This was demonstrated by an articulated physical model of the vertical jump by Bobbert et al., 1987.

Besides biarticularity, the gastrocnemius muscle has one more interesting feature. It is connected to the foot by an elastic tendon. The elasticity in the muscle fibers and tendons plays an important role in enhancing the effectiveness and the efficiency of human performance. An enhanced performance of human motion has been most effectively demonstrated for jumping (Cavagna, 1970; Bobbert et al., 1996) and running (Shorten, 1985). An important feature of elastic tissues is the ability to store elastic energy when stretched and to recoil this energy afterwards as a mechanical work (Asmussen and Bonde-Petersen, 1974). Beside this feature, oscillatory movements performed at the natural frequency of muscle-tendon complex could maximize the performance. A countermovement vertical jump can be treated as one period of oscillatory movement and from this point of view the natural frequency, as well as parameters that define the natural frequency, can be determined. For a simple mass-spring system, these parameters can easily be analytically determined. However, for a complex multibody elastic system, such as the human body, an analytical solution is often impossible.

The purpose of the present study was to build an efficient biomechanical model and to evaluate the influence of the timing of gastrocnemius muscle activation and its elastic properties on the vertical jump performance.

2. Basic mechanism

We built a biomechanical model of the human body as a planar, four-segment, articulated linkage, with links joined together by frictionless hinges. To achieve the balance during the jump, the motion of the body was constrained by a virtual vertical groove. Segmental anthropometric parameters, such as masses, moments of inertia about the transverse axes, lengths and locations of the centers of gravity (COG), were estimated using regression equations (Winter, 1990). The model parameters are specified in Tb. 1.

Table 1. Segmental anthropometric parameters used in biomechanical model of human body

	Mass (kg)	Length (m)	Joint-COG (m)	Inertia moment (kg/m^2)
Foot	2.43	0.16	0.0830	0.0154
Shank	7.85	0.43	0.0245	0.1340
Thigh	16.86	0.43	0.0245	0.3330
Body	56.00	0.54	0.2740	4.3960

The interaction between the foot and the floor was modelled by a highly damped, stiff and nonlinear spring at the toes and the heel. However, this approach introduces the pseudo-stiffness into the equations of motion which demands a decrease in the integration step of the simulation. The mechanical behavior of the knee extensor muscle was described by a Hill-type contractile element which models its force-length-velocity characteristics. The torque in the knee joint, exerted by the muscle, is given by a form of Hill's equation (Woledge et al., 1985)

$$T = T_0 \frac{\dot{\alpha}_0 - \dot{\alpha}}{\dot{\alpha}_0 + G\dot{\alpha}}. \tag{1}$$

Here, T_0 is the isometric torque, $\dot{\alpha}$ is the velocity of the knee extension, $\dot{\alpha}_0$ is the unloaded value of $\dot{\alpha}$, and G is a constant that describes the curvature of the relationship between the torque T and the velocity of the knee extension $\dot{\alpha}$. It is assumed that the moment arm of the muscle is independent of the knee joint angle. Therefore, the angle of the knee joint α does not appear in the equation.

The moments in the hip, knee and ankle due to passive elastic structures at the joints were modelled as simple nonlinear springs (Audu and Davy, 1985) which provide strong recurrent joint moments near the extremes of the joint motion.

Simulations of the counter movement vertical jumps were performed. At the beginning of the simulation, the legs were fully straightened. A counter movement was performed by the flexion of the knee until the desired minimal knee angle was reached. The flexion of the knee was such that the center of gravity of the jumper dropped with the acceleration of the gravity. Lifting of the foot from the floor would occur if faster flexions of the knee were applied. Flexion was immediately followed by a rapid extension of the knee which provided the energy for the jump.

Simulations of the counter movement vertical jump with the model configuration described above (Fig. 1) yielded the jump height of 0.27m regarding to the straightened position at the beginning of the simulation.

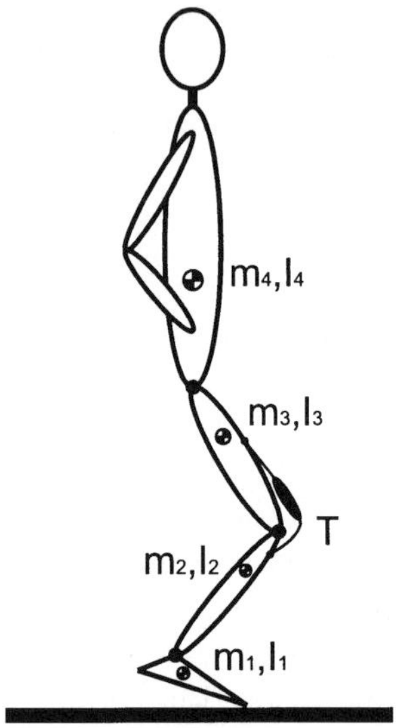

Figure 1. Schematic representation of basic musculoskeletal model of human body

3. Stiff biarticular link

To study the influence of the timing of the gastrocnemius muscle activation on the height of the human jump, a stiff biarticular link has been added to the basic musculoskeletal model of the human body as shown in Fig. 2. Insertions of the biarticular link on the foot (T_2) and thigh (T_1) have been determined from the muscle data provided by Brand et al., 1982. In contrast to the real gastrocnemius muscle, biarticular link can not contract. It can only be enabled or disabled at different moments during the push-off phase of the jump. High pennation angle of gastrocnemius muscle fibers suggest that the role of gastrocnemius muscle is not in its contraction but in its ability to bear high forces and to enable the energy transportation from the knee to the ankle (Legreneur et al., 1997). Therefore, a biarticular link that can not contract and can only be enabled or disabled is considered as an appropriate model of the gastrocnemius muscle.

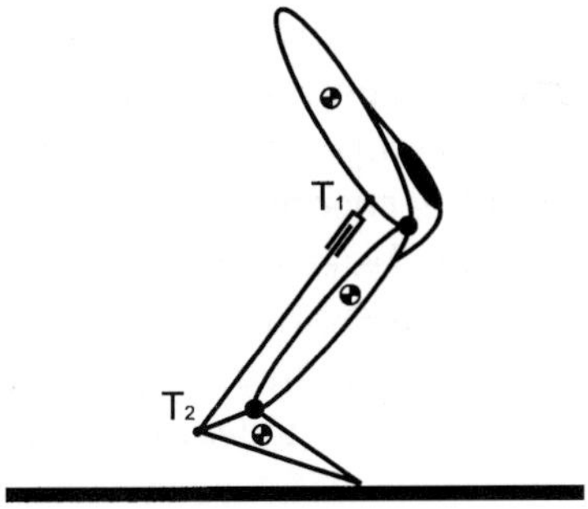

Figure 2. Schematic representation of added stiff biarticular link

A series of the countermovement vertical jump simulations have been performed, each with a different timing of the biarticular link activation. The moment of biarticular link activation was varied in order to find the optimum timing that results in the highest vertical jump. The relationship of the jump height and the knee angle when the biarticular link is activated is presented in Fig. 3.

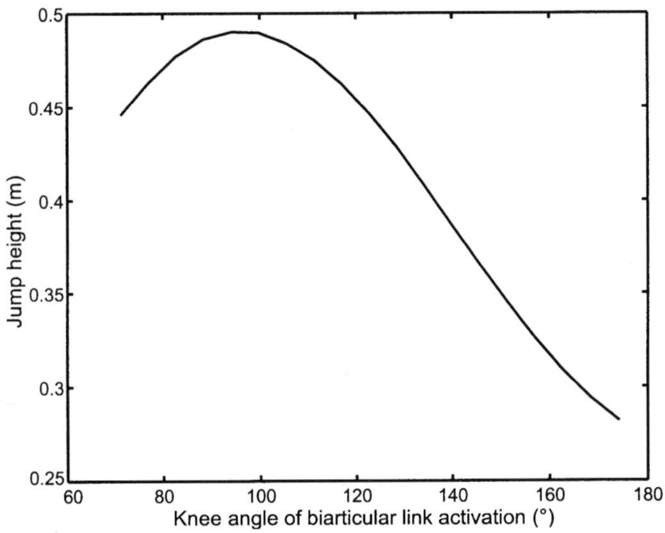

Figure 3. Jump height as function of knee angle - stiff biarticular link is activated

The highest jump has been achieved when the biarticular link was activated in the moment when the knee angle was approximately 95°. With this optimal timing, the model jumped almost twice as high as when no biarticular link was included. If the biarticular link activates too early in the push-off phase, the jumper looses the ground contact too early. If the biarticular link activates too late in the push-off phase, the rotational energy of the thigh and shank increases uselessly. The optimal timing represents a compromise between these two cases.

The results suggest that the coactivation of the knee extensor and the biarticular gastrocnemius can be highly effective in explosive push-offs of the human body. By measuring the activation of the muscles during the vertical jump we found that human subjects indeed use coactivation of the knee extensors and the gastrocnemius.

4. Elastic biarticular link

Gastrocnemius muscle is connected to the foot by an elastic Achilles tendon. During the countermovement of the jump this elastic tendon is pre-stretched and temporarily stores the work of stretching as elastic energy. This energy can later be reutilized in the push-off phase of the jump. This reutilization of the energy increases the effective energy contributing to the jump height. The portion of the energy reutilized depends on the stiffness of the tendon and velocity of the countermovement and the push-off. The motion of the jumper during the countermovement jump can be treated as one period of the oscillatory movement. If this motion coincides with the natural frequency of the human body elastic system, the highest portion of stored energy is reutilized and the highest jump is achieved. To demonstrate this explanation, the stiff biarticular link from the previous section has been replaced by an elastic biarticular link as shown in Fig. 4.

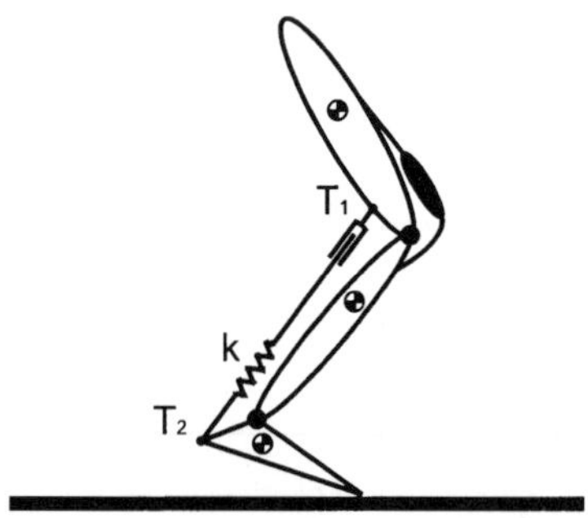

Figure 4. Schematic representation of added elastic biarticular link with stiffness k

A series of the countermovement vertical jump simulations have been performed, each with a different stiffness of the biarticular link. For each and every stiffness of the biarticular link the optimal timing has been determined as described in the previous section. The relationship of the jump height and the biarticular link stiffness is presented in Fig. 5.

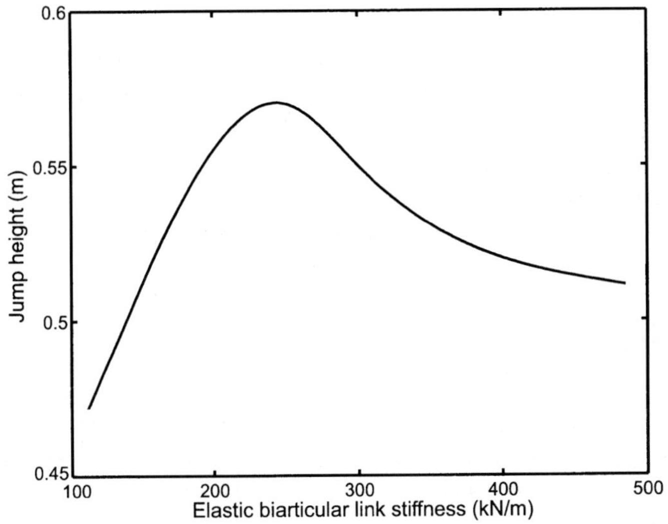

Figure 5. Jump height as function of biarticular link stiffness

The highest jump has been achieved when the stiffness of the elastic biarticular link was approximately 245 kN/m. This value is comparable to real human Achilles tendon stiffness reported in the past (Cavagna, 1970; Shorten, 1987). The highest jump with the elastic biarticular link was 16% higher than the highest jump with the stiff biarticular link. We can thus conclude that the reutilization of the elastic energy stored during the countermovement contributes 16% to the height of the vertical jump.

5. Conclusions

By building an efficient biomechanical model of the human body with stiff and elastic biarticular link and by simulation of the vertical jump we have demonstrated that timing of the biarticular link activation and stiffness of the biarticular link considerably influence the height of the human jump.

The methodology and the obtained results of our study offer an effective tool for improvement of the human jump performance. The timing of muscle activation and the viscoelastic properties of muscles and tendons can easily be experimentally determined. By adjusting the proposed biomechanical model to the subject's body and his or her performance, improved training methods for jump improvement can be suggested. However, it has to be considered that this study deals only with one biarticular link. Although the biarticular muscle we included in our study has the most distinctive elastic properties among all biarticular muscles of the human leg, the influence of other biarticular muscles should also be studied. A special challenge would be to demonstrate the joint effect of all biarticular muscles together.

6. Acknowledgement

This investigation was supported by the Slovenian Ministry of Education, Science and Sport.

References

Asmussen, E., Bonde-Petersen, F. (1974), Storage of elastic energy in skeletal muscles in man, *Acta Physiologica Scandinavica*, vol. 91, pp. 385–392.

Audu, M.L., Davy, D.T. (1985), The influence of muscle model complexity in musculoskeletal motion modeling, *Journal of Biomechanical Engineering*, vol. 107, pp. 147–157.

Bobbert, M.F., Gerritsen, K.G.M., Litjens, M.C.A., Soest, A.J. van (1996), Why is countermovement jump height greater than squat jump height?, *Medicine and Science in Sports and Exercise*, vol. 28, pp. 1402–1412.

Bobbert, M.F., Hoek, E., Ingen Schenau, G.J. van, Sargeant, A.J., Schreurs, A.W. (1987), A model to demonstrate the power transporting role of biarticular muscles, *ASME Journal of Physiology*, vol. 387, pp. 24.

Brand, R.A., Crowninshield, R.D., Wittstock, C.E., Pedersen, D.R., Clark, C.R., Krieken, F.M. van (1982), A model of lower extremity muscular anatomy, *ASME Journal of Biomechanical Engineering*, vol. 104, pp. 304–310.

Cavagna, G.A. (1970), Elastic bounce of the body, *Journal of Physiology*, vol. 29, pp. 279–282.

Ingen Schenau, G.J. van (1989), From rotation to translation: constraints on multijoint movements and the unique action of bi-articular muscles, *Human Movement Science*, vol. 8, pp. 301–337.

Legreneur, P., Morlon, B., Hoecke, J. van (1997), Joined effects of pennation angle and tendon compliance on fibre length in isometric contractions: a simulation study, *J. Archives of Physiology and Biochemistry*, vol. 105, pp. 450–455.

Shorten, M.R. (1985), Mechanical energy changes and elastic energy storage during treadmill running *Biomechanics IXB*, Champaign, Human Kinetics Publications.

Shorten, M.R. (1987), Muscle elasticity and human performance, *Medicine in Sports and Science*, vol. 25, pp. 1–18.

Winter, D.A. (1990), *Biomechanics and Motor Control of Human Movement*, New York, John Wiley and Sons.

Woledge, R.C, Curtin, N.A., Homster, E. (1985), *Energetic Aspects of Muscle Contraction*, Bologna, London, Academic Press.

HUMAN HAND MOVEMENT KINEMATICS AND KINESTHESIA

J. P. Schmiedeler
The Ohio State University, Department of Mechanical Engineering
Columbus, OH USA
schmiedeler.2@osu.edu

J. J. Stephens*, C. R. Peterson+, and W. G. Darling*+
*University of Iowa, *Dept. of Mech. & Ind. Eng. and +Dept. of Exercise Science*
Iowa City, IA USA
joshua-stephens@uiowa.edu, clayton-peterson@uiowa.edu, and warren-darling@uiowa.edu

Abstract For reaching tasks, the human kinesthetic sensory system uses a trunk-fixed anterior/posterior (a-p) axis as a frame of reference in the horizontal plane. Hand motion is aligned more precisely to the a-p axis than to visually presented axes, and precision degrades with angular displacement from the a-p axis. The central nervous system's need to perform coordinate transformations between the vision and kinesthetic systems explains this result. Also, kinesthetic perception of hand motion is more precise in the ipsilateral than in the contralateral hemifield, likely due to higher precision in sensing elbow flexion than shoulder rotation.

Keywords: Kinesthetic perception, arm kinematics, coordinate transformations

1. Introduction

Kinesthesia is the human sense of body segment position and movement. Mechanoreceptors in the joints, skin, and muscles provide body segment orientation information to the central nervous system (CNS) just as encoders provide joint position information to a robot control system. The receptors do not in general, though, provide information directly related to individual joint angles. For example, length sensitive muscle spindles produce signals related to muscle fiber length, while fibers commonly act across multiple joints (Cordo et al., 2000). It is not a priori known how kinesthetic information is represented in the CNS.

Reaching tasks require the CNS to solve the inverse kinematics problem for the arms, which have redundant degrees of freedom. Target position and orientation are defined by the vision system, and the CNS determines, typically at a subconscious level, the limb segment orientations needed to acquire the target. If the target is represented in the

J. Lenarčič and C. Galletti (eds.), On Advances in Robot Kinematics, 163-170.

vision system relative to a coordinate system different from the one relative to which the kinesthetic sensations are interpreted, the CNS must perform a transformation between the two (Darling & Miller, 1993).

Static arm positioning without visual feedback has been investigated to determine the coordinate system in which kinesthetic perception is coded. Results indicate that upper limb segment orientations are specified using frames of reference of an earth-fixed vertical axis in the vertical plane (Darling & Hondzinski, 1999) and a trunk-fixed anterior/posterior (a-p) axis in the horizontal plane (Darling & Williams, 1997). Experiments in which vision was employed to statically position external objects have shown that the vision system can use these same axes as frames of reference (Darling et al., 1996, Darling & Hondzinski, 1997). A recent study has shown that human subjects can align their forearms with equal accuracy to a trunk-fixed a-p axis and to the visually specified orientation of an external object (Darling & Bartelt, 2003).

The present work seeks to determine what effects the kinematics of horizontal plane hand motion have on kinesthetic perception. It is theorized that the kinesthetic system specifies motion relative to the same frame of reference employed for specifying static orientation of the upper limb segments. This theory stems from work indicating that the vision system defines the direction of motion relative to an earth-fixed vertical axis and an external horizontal axis when one is available, or a trunk-fixed a-p axis when trunk orientation is neutral (Darling & Pizzimenti, 2001). The same extension of static to dynamic frames of reference is anticipated for kinesthetic perception. The present work furthermore theorizes that the precision of kinesthetic perception is at least in part dependent upon joint positions during hand motion.

2. Experimental Apparatus

A table-top revolute-prismatic (R-P) mechanism, shown in Fig. 1, was constructed to investigate kinesthetic perception of passive arm motion in the horizontal plane. A linear slide with 370 mm of travel provides the translational degree of freedom and is driven through a ballscrew by a brushless DC motor. The rotational degree of freedom is driven through a gearhead also by a brushless DC motor, and optical encoders on each motor provide position feedback. Two cam followers whose axes of rotation intersect at the axis of the mechanism's revolute joint provide rolling contact for the slide on the table. Hard stops limit the angular rotation to approximately ± 85 degrees from a zero position perpendicular to the table's long axis. A cylindrical handle is mounted to a six-axis force sensor, which is in turn mounted to the linear slide. The subject

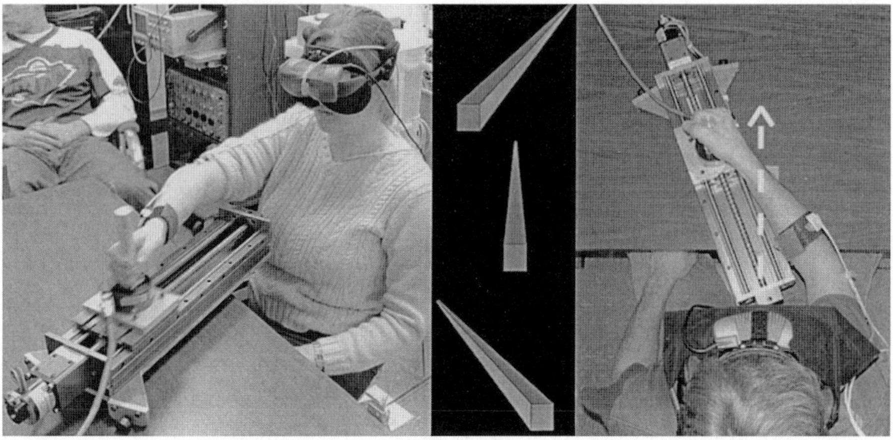

Figure 1. Two views of subjects and the apparatus that imposes hand motion. Subjects align their linear hand motion parallel to a visually displayed perspective line, examples of which are shown between the two views, and to the trunk-fixed anterior-posterior axis, indicated by the dashed white line in the top view.

grasps the handle with one hand during the experiment, and the two actuators are driven to impose hand motion. The force sensor measures the subject's active resistance to the imposed motion, although in all of these experiments, subjects were instructed not to resist the motion.

3. Experimental Procedure

Each subject sat comfortably close to the mechanism's revolute joint centered on his/her chest and with his/her trunk-fixed a-p axis perpendicular to the table's long axis. (One subject's trunk-fixed a-p axis is shown as a dashed white line in the right side of Fig. 1.) The subject's extended arm length was measured, and the linear slide's commanded travel was scaled to impose motion near a maximal reach. The subject was outfitted with 3-D, projection goggles that eliminate vision of the surroundings. Although the subject was instructed not to rotate his/her chest or head, no constraints were imposed, so magnetic position/orientation sensors were attached to the top of the sternum and goggles to measure the chest and head orientations. Additional sensors were attached on the acromion, on the humerus near the elbow, and on the forearm near the wrist to measure the joint positions.

Each trial began with the subject grasping the handle in his/her right hand and the revolute joint driven to an initial angle. The repeated forward/backward translation of the linear slide began while a perspective line, drawn as a long, rectangular parallelepiped, was projected in the goggles. (Three perspective line examples are shown in the center of Fig. 1.) The subject was instructed to align the orientation of his/her linear hand motion with the projected line. By depressing the right/left buttons of a standard computer mouse held in his/her left hand, the subject drove the revolute joint clockwise/counterclockwise. Alignment to the projected line was indicated by depressing the middle mouse button, and three seconds of sensor data was then collected to record the relevant positions and orientations. The translation of the linear slide was stopped, the revolute joint was driven to a new initial angle, a new perspective line was presented in the goggles, and the process was repeated.

Seven healthy, young adult subjects participated in the experiments. Six trials were performed for each of the presented target angles of 0, ± 20, ± 30, and ± 40 degrees. The initial mechanism orientations were selected from this same set, with the exception of 0 degrees. The initial mechanism orientation and the projected line angle were varied in a pseudo-random fashion, with the order known to the experimenter, but appearing random to the subject. Six additional trials were conducted in which the subject was instructed to alter the initial orientation of hand motion to align it with his/her trunk-fixed a-p axis.

4. Data Analysis

Since kinesthetic perception is subject to drift over time, the acuity of kinesthetic hand motion orientation is determined by the level of noise in the system (van Beers et al., 1998). Variable errors provide a measure of noise, so the standard deviation in alignment to each target angle was calculated for each subject. The actual angle of hand motion was recorded from the mechanism's revolute joint encoder. The effective target angle was calculated by combining the angle of the projected line with the head orientation recorded from the magnetic sensor on the goggles. Although subjects were instructed to keep their heads still, all demonstrated some head rotation. Error in alignment was calculated as the difference between the actual hand motion angle and the effective target angle. For the six trials of alignment to the trunk-fixed a-p axis, the error was calculated as the difference between the actual hand motion angle and the trunk orientation recorded from the magnetic sensor on the sternum. The vectors of the humerus and forearm at both maximum and minimum reach were calculated from the recorded positions of the magnetic sensors attached at the joints. The elbow angular displacement was calculated as the change in angle between the forearm and humerus from minimum to maximum reach. The shoulder angular displacement was calculated as the angle between the two humerus vectors.

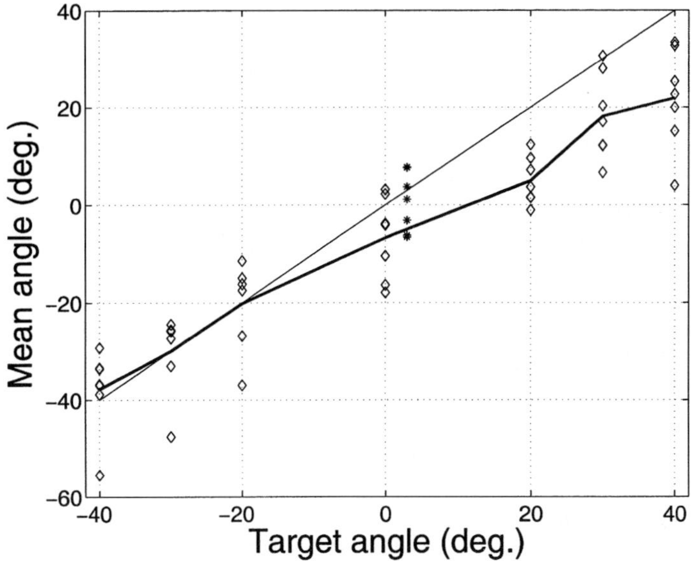

Figure 2. Mean final angles at visually presented target angles for all seven subjects. The heavy linear plot tracks the mean of all subjects at each angle, and the lighter line with a slope of unity indicates what would be ideal alignment. The * data points indicate the mean angle in aligning to the trunk-fixed a-p axis.

5. Results

Fig. 2 plots the average final angle of each subject's hand motion alignment to the seven target angles (◇ data points) and to his/her trunk-fixed a-p axis (* data points shown offset from 0 degrees for clarity). Individual subjects demonstrated different biases, but the notable trend is that positive target angles were consistently undershot, possibly due to a visual perception bias. Fig. 3 plots the standard deviation of each subject's hand motion alignment. In general, subjects demonstrated the highest precision in kinesthetically aligning hand motion to a trunk-fixed a-p axis and to a visually presented line parallel to this axis (target angle 0 degrees). Precision generally degraded with increasing difference in orientation of the visually presented line from this axis.

These results support the theory that like static positioning, kinesthetic perception of hand motion is represented relative to a trunk-fixed a-p axis rather than a visually specified axis. The superior kinesthetic alignment of hand motion to the trunk-fixed a-p axis was expected since the axis itself is kinesthetically defined. Thus, no transformation between coordinate systems is likely required in the CNS to achieve the alignment. To kinesthetically align hand motion to a visually defined

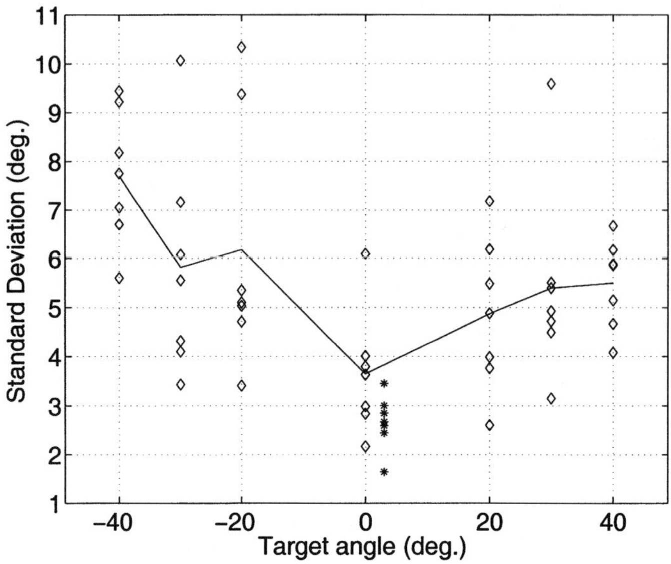

Figure 3. Standard deviation at visually presented target angles for all seven subjects. The linear plot tracks the mean standard deviation at each angle, and the * data points indicate the standard deviation in aligning to the trunk-fixed a-p axis.

axis, a coordinate transformation is required. As the orientation of that visually defined axis differs from the trunk-fixed a-p axis, noise in the kinesthetic system leads to increasing imprecision in the alignment.

The data in Fig. 3 also displays a slightly less dramatic trend. Five of the seven subjects demonstrated more precise alignment for positive target angles than for negative target angles of the same magnitude. Since each subject aligned the motion of his/her right hand, this corresponds to higher precision in the ipsilateral hemifield than in the contralateral hemifield. The trend is consistent with findings of better position sense accuracy in the ipsilateral hemifield during imperceptibly slow movements (Cordo et al., 2000). Other work has shown that hand localization in the radial direction with respect to the shoulder is more precise than that in the azimuthal direction (van Beers et al., 1998). This can be explained in part by considering the manipulability ellipsoid (Yoshikawa, 1985) of the arm. For a nearly extended arm, noise in perception of both elbow and shoulder angles results in uncertainty in movement in the azimuthal direction. Uncertainty in the radial direction is small because relatively large joint motions generate relatively small hand motions.

Because hand motion was predominantly in the radial direction with respect to the shoulder in the present work, consideration of the individual joint displacements offers one explanation for the asymmetry in

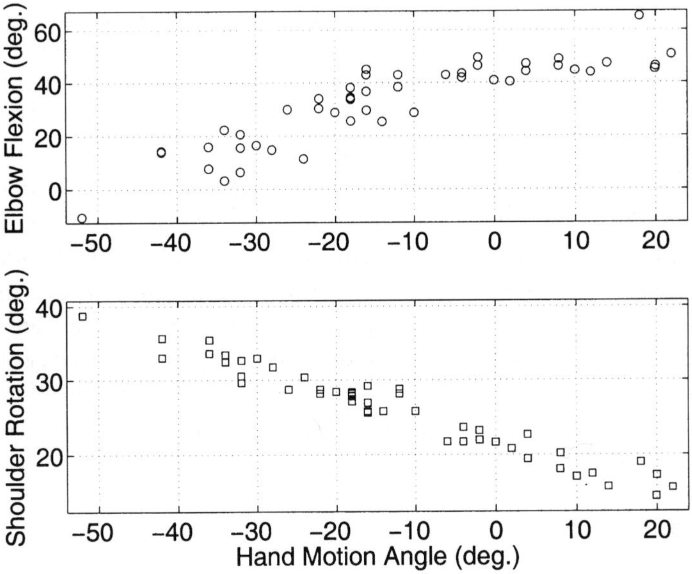

Figure 4. Elbow and shoulder angular displacement versus actual hand motion angle for an individual subject.

kinesthetic perception. Fig. 4 plots elbow flexion and shoulder rotation as functions of actual hand motion angle for an individual subject. Other subjects displayed similar joint angular displacements. The actual hand motion angles do not match the target angles due to head rotation, but data points toward the left side of each plot correspond to hand motion in the contralateral hemifield. Similarly, data points toward the right side correspond to hand motion in the ipsilateral hemifield. In the former case, shoulder rotation dominates the motion, while in the latter, elbow flexion dominates. The greater precision in the ipsilateral hemifield suggests superior kinesthetic perception of elbow flexion compared to shoulder rotation, which may be due to the greater complexity of the shoulder joint. For these experiments, the elbow is fundamentally a one-degree-of-freedom joint, while the shoulder remains a complex system producing spatial pointing of and roll orientation about the humerus (Lenarčič & Stanišić, 2003). With its complexity, noise in the kinesthetic sensations at the shoulder likely generates greater uncertainty in shoulder angular displacement compared to elbow flexion.

6. Conclusions

This work shows that the kinesthetic sensory system uses a trunk-fixed a-p axis as a frame of reference in the horizontal plane for upper limb reaching tasks. The same axis has been shown in previous work to be a frame of reference for static positioning, so it is concluded that arm motion does not cause a change in reference frame. Linear hand motion is aligned more precisely to a trunk-fixed a-p axis than to visually presented axes when the trunk has a neutral orientation, and precision degrades with angular displacement of the motion from the trunk-fixed a-p axis. This can be explained by the need for the CNS to perform a coordinate transformation between the vision and kinesthetic systems. Additionally, kinesthetic perception of hand motion in the ipsilateral hemifield is more precise than that in the contralateral hemifield. A likely explanation for this perception asymmetry is greater precision in sensing elbow flexion than shoulder rotation. Due to the relative complexity of the shoulder, noise in the kinesthetic sensations results in larger uncertainty in the joint position.

References

Beers, R.J. van, Sittig, A.C., and Denier van der Gon, J.J. (1998), The precision of proprioceptive position sense, *Experimental Brain Research*, vol. 122, pp. 367-377.

Cordo, P.J., Gurfinkel, V.S., and Levik, Y. (2000), Position sense during imperceptibly slow movements, *Experimental Brain Research*, vol. 132, pp. 1-9.

Darling, W.G. (1991), Perception of forearm angles in 3-dimensional space, *Experimental Brain Research*, vol. 87, pp. 445-456.

Darling, W.G., and Bartelt, R. (2003), Kinesthetic perception of visually specified axes, *Experimental Brain Research*, vol. 149, pp. 40-47.

Darling, W.G., Butler, A.J., and Williams, T.E. (1996), Visual perceptions of head-fixed and trunk-fixed anterior/posterior axes, *Experimental Brain Research*, vol. 112, pp. 127-134.

Darling, W.G., and Hondzinski, J.M. (1997), Visual perceptions of vertical and intrinsic longitudinal axes, *Experimental Brain Research*, vol. 116, pp. 485-492.

Darling, W.G., and Hondzinski, J.M. (1999), Kinesthetic perceptions of earth- and body-fixed axes, *Experimental Brain Research*, vol. 126, pp. 417-430.

Darling, W.G., and Miller, G.F. (1993), Transformations between visual and kinesthetic coordinate systems in reaches to remembered object locations and orientations, *Experimental Brain Research*, vol. 93, pp. 534-547.

Darling, W.G., and Pizzimenti, M.A. (2001), A coordinate system for visual motion perception, *Experimental Brain Research*, vol. 141, pp. 174-183.

Darling, W.G., and Williams, T.E. (1997), Kinesthetic perceptions of intrinsic anterior-posterior axes, *Experimental Brain Research*, vol. 117, pp. 465-471.

Lenarčič, J., and Stanišić, M. (2003), A humanoid shoulder complex and the humeral pointing kinematics, *IEEE Transactions on Robotics and Automation*, vol. 19, no. 3, pp. 499-506.

Yoshikawa, T. (1985), Manipulability of robotic mechanisms, *International Journal of Robotics Research*, vol. 4, pp. 3-9.

KINEMATIC SYNTHESIS OF AVATAR SKELETONS FROM VISUAL DATA

Mari Cruz Villa-Uriol, Falko Kuester, Nader Bagherzadeh
Visualization and Interactive Systems Group,
University of California, Irvine, CA 92697
mvillaur, fkuester, nader@uci.edu

Alba Perez, J. Michael McCarthy
Robotics and Automation Laboratory,
University of California, Irvine, CA 92697
maperez, jmmccart@uci.edu

Abstract The recovery of 3D models from visual data generally results in a geometric skeleton that is a simplification of the shape of the captured figure, called an avatar. In this paper, we introduce a method called hierarchical kinematic synthesis that identifies an articulated skeleton, called kinematic skeleton, which provides a compact representation of the movement of the tracked subject.

In this work, the kinematic skeleton of a human is computed from a finite number of key poses captured from full body articulated movements of an arbitrary subject, and provides location of joints, length and twist angle of links that form the limbs of the 3D avatar. We use an approximation to the human skeleton which consists of five serial chains constructed from revolute and spherical joints. To recover the kinematic skeleton, a hierarchical approximate finite-position synthesis methodology determines the dimensions of these chains limb by limb.

We show that this technique effectively recovers the kinematic skeleton for several synthetically generated datasets, and that the identification of the kinematic skeleton improves pose estimation for 3D data while simplifying the generation of avatar movement.

Keywords: Kinematic synthesis, Avatar, Skeleton.

1. Introduction

Diverse techniques have been studied for recovery of posture, motion analysis and motion tracking of articulated subjects from video sequences or sets of static images (Kakadiaris and Metaxas, 1998; Bregler and Malik, 1998). This has applications in the study of gesture, biomechanics, anthropometry, human gait analysis, human-computer interaction, realistic computer animation and control of virtual characters.

J. Lenarčič and C. Galletti (eds.), On Advances in Robot Kinematics, 171-180.

Body pose recovery methods from images frequently use geometrically defined models of humans as a reference for pose estimation (e.g. Cheung and Kanade, 2000; Weik and Liedtke, 2001; Kakadiaris and Metaxas, 1998).

Previous research has focused on creating a *geometric skeleton* that represents a simplification of the subject's geometric properties as a set of solids roughly aligned to the medial axis of each limb. In this paper, we determine a *kinematic skeleton*, which is a compact representation of the movement of the subject. The geometric skeleton locates a set of rigid bodies in space, while the kinematic skeleton adds to that the location of joints that allow the movement of the rigid bodies that form the limbs.

Animation of an articulated model to achieve human-like movement is easier using a kinematic skeleton, where configuration changes are obtained by specifying angles at each joint. We find that anthropomorphic movements at the level of a complete avatar are easily obtained with simplified skeletons (for detailed modeling of human joints see for instance Lenarcic, Stanisic and Schearer, 2002).

The presented approach uses *kinematic synthesis* to find the kinematic skeleton from the movement of the figure. In particular, the input data is a series of positions of the limbs of the captured subject. Synthetically generated movements of a simplified human skeleton were used for testing the algorithm. We define the synthesis problem using dual quaternions to represent the chains and spatial displacements (Perez, 2003). The result is a compact formulation which suggests it will be feasible to be used in real time applications.

2. Data Acquisition

The actual data consists of synthetically generated movements resembling human motion, represented as a sequence of 4x4 homogeneous matrix transformations for each limb. In the future, video images will be used to obtain the motion information characterizing the subject. For each of the acquired image frames, a set of silhouettes of the actor and their corresponding camera calibration parameters are obtained, as shown in Figure 1. The visual hull (Laurentini et al., 1994) containing the actor is calculated by intersecting the camera frustums, and subdivided into voxels with a resolution defined by the user. This volumetric reconstruction can be described as the maximal shape that generates the same silhouette for every reference view outside of the object's convex hull.

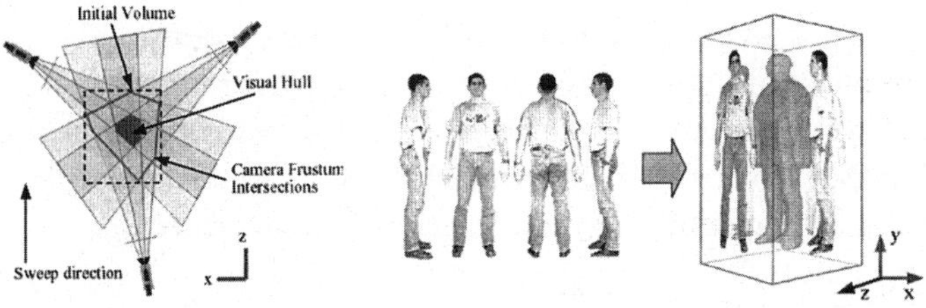

Figure 1. Visual Hull: concept (left) and example obtained from 4 images (right).

3. Synthesis of Kinematic Chains

The finite-position synthesis of a kinematic chain seeks to find locations and directions of each of its joint axes from a finite set of positions defined as rigid transformations. In approximate synthesis, the kinematic chain is dimensioned such that its final element passes approximately through this set of positions. Approximate synthesis is used to solve this problem since it is not possible to model the movement of real figures as rigid motions.

In the presented approach, we use the robot *kinematics equations* or *forward kinematics* to define the movement of a skeleton topology (Craig, 1989). We then equate the avatar kinematics equations to the input data for the limb positions from a reference configuration (Mavroidis et al., 2001; Tsai, 1999), formulated using dual quaternion algebra (Perez, 2003). The dimensions of the avatar limbs are obtained by minimizing the distance between the avatar and goal configurations.

3.1 The Kinematics Equations

The kinematics equations of a robot equate the 4×4 homogeneous matrix transformation $[D]$ between the last element of the kinematic chain and the base frame to the sequence of local coordinate transformations along the m joints of the chain,

$$\begin{aligned}[D] = [G][Z(\theta_1, d_1)][X(\alpha_{12}, a_{12})][Z(\theta_2, d_2)]\dots \\ [X(\alpha_{m-1,m}, a_{m-1,m})][Z(\theta_m, d_m)][H]. \qquad (1)\end{aligned}$$

The parameters (θ, d) and (α, a) are known as the *Denavit-Hartenberg parameters.* The transformation $[G]$ defines the position of the first joint of the chain relative to a fixed frame, and $[H]$ locates the last limb relative to the last joint.

We use a dual quaternion representation of the displacements instead of the homogeneous matrix representation. In order to create the dual quaternion kinematics equations, the expression in Eq.(1) is transformed into successive screw displacements by choosing a reference position $[D_0]$. Let $[D_i]$ be the homogeneous matrix describing the transformation from the fixed frame to a moving frame M_i, then we can compute $[D_{0i}] = [D_i][D_0]^{-1}$, and Equation (1) can be transformed into

$$[D_{0i}] = [T(\Delta\theta_1, \mathsf{S}_1)][T(\Delta\theta_2, \mathsf{S}_2)] \ldots [T(\Delta\theta_m, \mathsf{S}_m)], \quad (2)$$

where $\Delta\theta_j = \theta_j^i - \theta_j^0$ or $\Delta\theta_j = d_j^i - d_j^0$ depending on whether the joint is revolute or prismatic, respectively. The displacements $[T(\Delta\theta_i, \mathsf{S}_i)]$ define the rotations about and translations along the joint axes S_i measured in the fixed frame relative to the reference configuration $[D_0]$.

3.2 Dual Quaternion Kinematics Equations

The kinematics equations of the serial chain are expressed as elements of the Clifford algebra of dual quaternions. A spatial displacement consisting of a rotation of θ and translation of d around and along a screw axis S is written as the dual quaternion $\hat{S}(\hat{\theta}) = \sin(\frac{\hat{\theta}}{2})\mathsf{S} + \cos(\frac{\hat{\theta}}{2})$, where $\hat{\theta} = \theta + \epsilon d$, and $\mathsf{S} = \mathbf{S} + \epsilon\mathbf{p} \times \mathbf{S}$ is the expression of the screw axis as a dual vector using Plücker coordinates. The dual unit ϵ is defined so that $\epsilon^2 = 0$ (Bottema and Roth, 1979).

Notice that the dual quaternion expression encodes the same information as the 4×4 screw displacement matrix $[T(\Delta\theta, \mathsf{S})]$. Thus, we can define the dual quaternion kinematics equations for the serial chain by simply replacing the 4×4 matrices in (2) by their dual quaternion equivalents to obtain

$$\hat{D}^i = \hat{S}_1(\Delta\hat{\theta}_1^i)\hat{S}_2(\Delta\hat{\theta}_2^i) \ldots \hat{S}_m(\Delta\hat{\theta}_m^i). \quad (3)$$

The Plücker coordinates of the screw axes S_i, $i = 1, \ldots, m$, are defined in the base frame F.

3.3 Design Equations

For the solution of our problem, we are given a series of positions (location and orientation) of the subject at different time frames. Let the n positions be defined as the 4×4 transforms $[P_i]$, $i = 1, \ldots, n$. Construct the $n - 1$ relative transformation matrices $[P_{1i}] = [P_i][P_1]^{-1}$ and their associated dual quaternions $\hat{P}_{1i}$, $i = 2, \ldots, n$. Equating the task dual quaternions $\hat{P}_{1i}$ to the kinematics equations (3), yields the design equations

$$\mathcal{Q}_i : \quad \hat{S}_1(\Delta\hat{\theta}_1^i)\hat{S}_2(\Delta\hat{\theta}_2^i) \ldots \hat{S}_m(\Delta\hat{\theta}_m^i) - \hat{P}_{1i} = 0, \quad i = 2, \ldots, n. \quad (4)$$

The solution of the total set of design equations provides with the location and orientation of the joint axes in the reference configuration. In addition, the solution provides also with the inverse kinematics, that is, the joint angles needed to reach a given position of the limbs.

4. Kinematic Skeleton Synthesis

The synthesis of a kinematic skeleton begins with the definition of the skeleton topology and the input data for each limb. From this information, the design equations for each of the five chains are stated. A hierarchical solving process is used to synthesize an approximated solution, including the successive averaging of solutions to obtain an acceptable final skeleton.

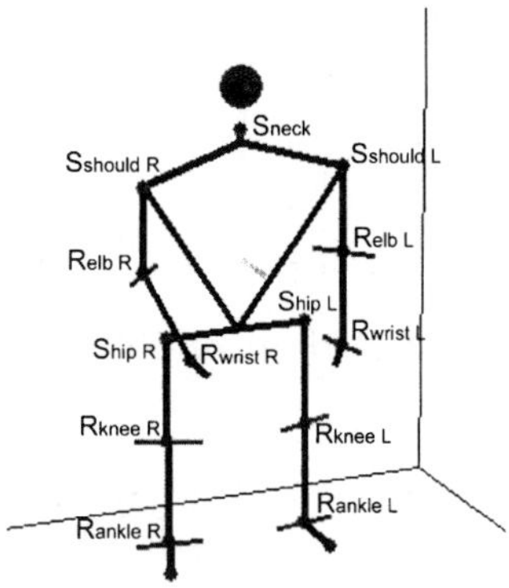

Figure 2. The topology of the kinematic skeleton.

4.1 The topology of the kinematic skeleton

Our approach to the kinematic human skeleton consists of five serial chains starting from the center of the chest: a chain for the neck and head, a chain for each of the arms and a chain for each of the legs. The five chains that form the skeleton, presented in Figure 2, are defined as follows. *Head* goes from the chest to the center of the head, and consists of a spherical joint (S) that performs the movements of the neck. *Left arm* goes from the chest to the left hand. It is composed of a spherical joint in the shoulder, a revolute joint in the elbow, and a revolute joint in the wrist. We denote the topology of this chain as S-R-R. *Right arm* has the same S-R-R structure as the left arm. *Left leg* consists of a spherical joint in the hip, a revolute joint in the knee, and a revolute joint in the ankle. It is hence also a S-R-R chain, and the same holds for *Right leg.* In Figure 2, spherical joints are represented as spheres and revolute joints as lines located at the joints.

4.2 Design equations for the kinematic skeleton

The kinematics equations in dual quaternion form for each of the limbs are

$$\begin{aligned}\hat{Q}_{head} &= \hat{S}_{neck}(\theta_{h1}, \theta_{h2}, \theta_{h3}),\\ \hat{Q}_{arm} &= \hat{S}_{should}(\theta_{a1}, \theta_{a2}, \theta_{a3})\hat{R}_{elb}(\theta_{a4})\hat{R}_{wrist}(\theta_{a5}),\\ \hat{Q}_{leg} &= \hat{S}_{hip}(\theta_{l1}, \theta_{l2}, \theta_{l3})\hat{R}_{knee}(\theta_{l4})\hat{R}_{ankle}(\theta_{l5}).\end{aligned} \tag{5}$$

The dual quaternion expression of rotations of θ around an axis of Plücker coordinates $\mathsf{S} = (s_x, s_y, s_z) + \epsilon(s_x^0, s_y^0, s_z^0)$, is

$$\hat{S}(\theta) = \begin{Bmatrix} \sin\frac{\theta}{2}s_x \\ \sin\frac{\theta}{2}s_y \\ \sin\frac{\theta}{2}s_z \\ \cos\frac{\theta}{2} \end{Bmatrix} + \epsilon \begin{Bmatrix} \sin\frac{\theta}{2}s_x^0 \\ \sin\frac{\theta}{2}s_y^0 \\ \sin\frac{\theta}{2}s_z^0 \\ 0 \end{Bmatrix}. \tag{6}$$

For the spherical joint, the dual quaternion expression for the displacements is given by

$$\hat{S}(\theta_1, \theta_2, \theta_3) = \begin{Bmatrix} \alpha_1\mathbf{g}_1 + \alpha_2\mathbf{g}_2 + \alpha_3\mathbf{g}_3 \\ \alpha_4 \end{Bmatrix} + \epsilon \begin{Bmatrix} \alpha_1\mathbf{g}_1^0 + \alpha_2\mathbf{g}_2^0 + \alpha_3\mathbf{g}_3^0 \\ 0 \end{Bmatrix}, \tag{7}$$

where $\mathbf{g}_1 + \epsilon\mathbf{g}_1^0$, $\mathbf{g}_2 + \epsilon\mathbf{g}_2^0$ and $\mathbf{g}_3 + \epsilon\mathbf{g}_3^0$ are the Plücker coordinates of three perpendicular axes intersecting at a point, and the α_i appear as combinations of the joint variables,

$$\begin{aligned}\alpha_1 &= \sin\frac{\theta_1}{2}\cos\frac{\theta_2}{2}\cos\frac{\theta_3}{2} + \cos\frac{\theta_1}{2}\sin\frac{\theta_2}{2}\sin\frac{\theta_3}{2},\\ \alpha_2 &= \cos\frac{\theta_1}{2}\sin\frac{\theta_2}{2}\cos\frac{\theta_3}{2} - \sin\frac{\theta_1}{2}\cos\frac{\theta_2}{2}\sin\frac{\theta_3}{2},\\ \alpha_3 &= \sin\frac{\theta_1}{2}\sin\frac{\theta_2}{2}\cos\frac{\theta_3}{2} + \cos\frac{\theta_1}{2}\cos\frac{\theta_2}{2}\sin\frac{\theta_3}{2},\\ \alpha_4 &= \cos\frac{\theta_1}{2}\cos\frac{\theta_2}{2}\cos\frac{\theta_3}{2} - \sin\frac{\theta_1}{2}\sin\frac{\theta_2}{2}\sin\frac{\theta_3}{2}.\end{aligned} \tag{8}$$

In order to create the design equations, we equate the expressions in Eq.(5) to the input data obtained from the images, $\hat{P}^i$, for as many time frames i as are needed to solve for the variables,

$$\begin{aligned}&\hat{Q}_{head} = \hat{P}^j_{head}, \quad j = 2, \ldots n,\\ &\hat{Q}_{arm} = \hat{P}^i_{arm}, \quad \hat{Q}_{leg} = \hat{P}^i_{leg}, \quad i = 2, \ldots m,\end{aligned} \tag{9}$$

and solve individually for each of the chains that form the kinematic skeleton.

4.3 Solution process

In order to synthesize the kinematic skeleton efficiently, we use a hierarchical kinematic synthesis approach. It consists of solving the design equations body segment by body segment, starting from the closest to the base or fixed frame and working sequentially forward along the chain. In the process, we define the location of each segment at each instant of time. The resulting equations can be easily solved.

For instance, for the kinematic chain corresponding to one of the arms in Eq.(5), we solve first for the shoulder using data from three frames (two frames and the reference or initial frame), by stating the set of equations

$$\hat{S}_{should}(\theta^i_{a1}, \theta^i_{a2}, \theta^i_{a3}) = \hat{P}^i_{upper}, \quad i = 2, 3, \tag{10}$$

that are solved numerically to obtain the coordinates of the point defining the spherical joint, **c**, and the joint angles $\theta^i_{a1}, \theta^i_{a2}, \theta^i_{a3}$ that most closely reach positions 2 and 3. These data allow us to create the dual quaternion that approximates the movement of the upper arm, namely $\tilde{S}_{should}(\theta_{a1}, \theta_{a2}, \theta_{a3})$.

To solve for the elbow, the input data of the lower arm must be equated to the series of transformations from the base to the lower arm,

$$\hat{S}_{should}(\theta^i_{a1}, \theta^i_{a2}, \theta^i_{a3})\hat{R}_{elb}(\theta^i_{a4}) = \hat{P}^i_{lower}, \quad i = 2, ..., n. \tag{11}$$

To isolate the dual quaternion corresponding to the elbow, Eq. (11) is left-multiplied by the inverse of the solution for the shoulder using the dual quaternion conjugate,

$$\hat{R}_{elb}(\theta^i_{a4}) = \tilde{S}^*_{should}(\theta^i_{a1}, \theta^i_{a2}, \theta^i_{a3})\hat{P}^i_{lower}, \quad i = 2, ..., n, \tag{12}$$

where the right part of the equality consists of known data, and $n = 2$ for the case of a revolute joint. We obtain the Plücker coordinates of the joint axis and the joint rotation for the elbow, which can be used to create the dual quaternion of the joint $\tilde{R}_{elb}(\theta_{a4})$.

We use data from the movement of the hand to solve for the last joint of the chain. The design equations for the wrist are, after pre-multiplying by the solved joints,

$$\hat{R}_{wrist}(\theta^i_{a5}) = \tilde{R}^*_{elb}(\theta^i_{a4})\tilde{S}^*_{should}(\theta^i_{a1}, \theta^i_{a2}, \theta^i_{a3})\hat{P}^i_{hand}, \quad i = 2, ..., n, \tag{13}$$

where again, $n = 2$ and we obtain the solution $\tilde{R}_{wrist}(\theta_{a5})$.

A similar operation is performed independently for each of the five chains that define the kinematic skeleton.

To obtain a kinematic skeleton with good accuracy, we need to repeat this process several times, due both to the lack of accuracy of the input data, and the disparity between a model made of rigid rotations and the real movement of the human body. Also the input data has to be general enough to show the movement of all joints. We perform the synthesis for a number of steps and average the solutions for each joint, until the error between solutions is small enough, according to the user needs.

4.4 Minimization and measuring

Approximate synthesis is performed by minimizing the distance between the position obtained from the kinematics equations and the input data in Eq.(9). The derivative of the 8-dimensional distance vector,

$$d_{norm} = (\hat{Q}_{arm} - \hat{P}^i_{arm}) \cdot (\hat{Q}_{arm} - \hat{P}^i_{arm}) \tag{14}$$

is set equal to zero to create a system of equations

$$[\frac{\partial Q_{arm}}{\partial s_k}]_i (\hat{Q}_{arm} - \hat{P}^i_{arm}) = 0 \tag{15}$$

that are solved for the parameters s_k of the joints.

The length of the difference vector is a metric for spatial displacements that, as any other metric that we can define, is not bi-invariant, see Martinez and Duffy, 1995. However, since the reference frame is defined in the center of the body, the displacements of all five chains are close to the standard dimensions of the human body, thus the reference frame error can be bounded. We expect to explore this in more detail with real subject data.

5. Results

The present results are based on synthetically generated data and are used to test the procedure. In Figure 3 we present the results obtained for a specific subject in 8 different poses, extracted from a sequence of 80 frames, which are the input data for the synthesis method. The first row shows the input synthetic data at frames $i = 41$ to 48, the second row the corresponding recovered pose of the synthesized skeletons approximating the task configuration and the third row shows the kinematic skeleton at the reference position. We can see that a good match is obtained when using synthetic data. For this dataset, 15 frames were needed to obtain a good approximation of the whole skeleton.

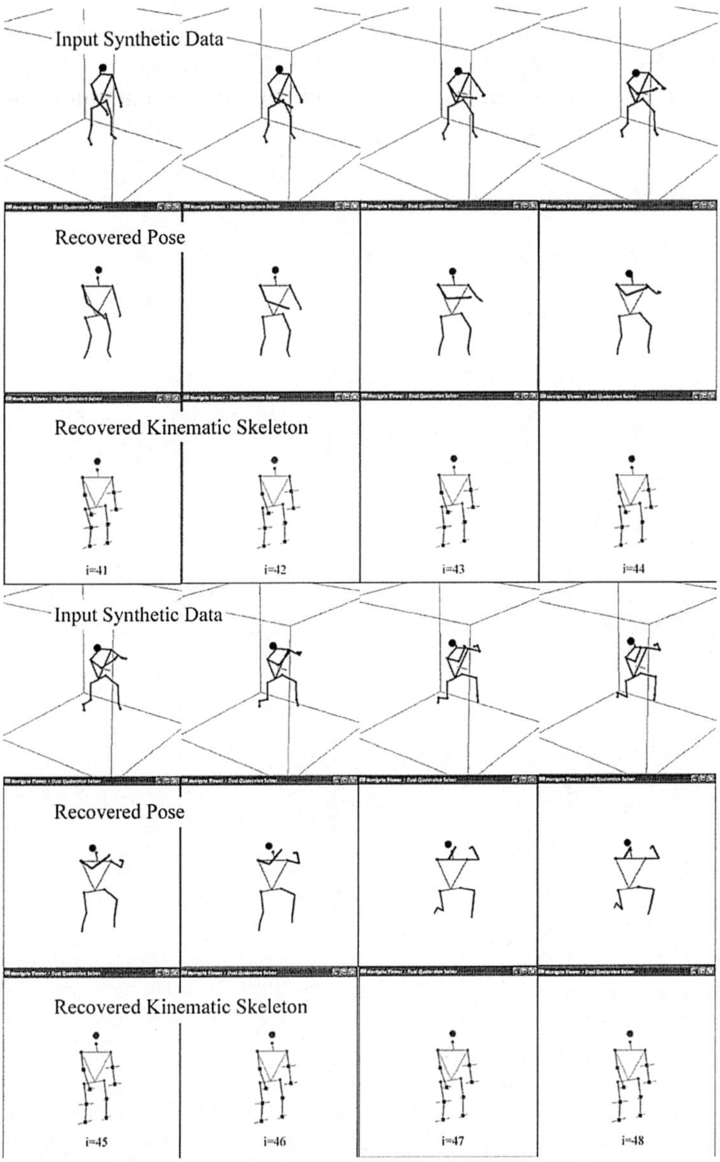

Figure 3. Synthesized kinematic skeleton at frames $i = 41$ to 48.

6. Conclusions

This paper presents a new method for determining a kinematic skeleton and poses of an arbitrary human subject. Finite-position kinematic synthesis is used to fit a kinematic skeleton defined in dual quaternion coordinates to the data that defines movement of the subject. A major advantage of this approach is that it accepts arbitrary poses of human subjects. In addition, this technique eliminates the need for manual intervention commonly found in geometric skeleton identification methods.

7. Acknowledgements

This research was supported in part by the National Science Foundation under contracts CCR-0083080, EIA-0203528 and National Science Foundation award DMII 0218285.

References

Bottema, O. and Roth, B., *Theoretical Kinematics.* NY: North Holland Press, 1979.

Bregler, C. and Malik, J. "Tracking people with twists and exponential maps", in *IEEE Conference on Computer Vision and Pattern Recognition*, pp. 8–15, 1998.

Cheung, K. M., Kanade, T., Bouguet, J.-Y. and Holler, M., "A real time system for robust 3d voxel reconstruction of human motions", in *IEEE Conference on Computer Vision and Pattern Recognition*, vol. 2, pp. 714–720, 2000.

Craig, J., *Introduction to Robotics, Mechanics and Control*, Ed. Addison Wesley Publ. Co, 1989.

Kakadiaris,I. A. and Metaxas, D. N., "Three-dimensional human body model acquisition from multiple views", *International Journal of Computer Vision*, vol. 30, no. 3, pp. 191–218, 1998.

Laurentini, A., "The visual hull concept for silhouette-based image understanding", *IEEE Transactions on Pattern Analysis and Machine Intelligence*, vol. 16, pp. 150–162, 1994.

Lenarčič, J., Stanišić, M. and Schearer, E., "Humanoid Humeral Pointing Kinematics", *Advances in Robot Kinematics*, eds. J. Lenarcic and F. Thomas, Kluwer Academic Publishers, 2002.

Martinez,J. M. R. and Duffy, J., "On the metrics of rigid body displacements for infinite and finite bodies", *ASME Journal of Mechanical Design*, vol. 117, pp. 41-47, 1995.

Mavroidis, C., Lee, E. and Alam, M., "A new polynomial solution to the geometric design problem of spatial rr robot manipulators using the denavit-hartenberg parameters", *ASME J. Mechanical Design*, vol. 123, no. 1, pp. 58–67, 2001.

Perez, A., "Dual quaternion synthesis of constrained robotic systems", Ph.D. dissertation, Department of Mechanical and Aerospace Engineering, University of California, Irvine, September 2003.

Tsai, L. W., *Robot Analysis: The Mechanics of Serial and Parallel Manipulators.* New York, NY: John Wiley and Sons, 1999.

Weik, S. and Liedtke, C.-E., "Hierarchical 3d pose estimation for articulated human body models from a sequence of volume data", in *International Workshop Robot Vision*, ser. Lecture Notes in Computer Science, vol. 1998. Springer-Verlag, pp. 27–34, 2001.

KINEMATICS AND MODELLING OF A SYSTEM FOR ROBOTIC SURGERY

Hermann Mayer, István Nagy and Alois Knoll
Robotics and Embedded Systems
Technische Universität München
{mayerh|nagy|knoll}@in.tum.de

Abstract We have developed an open experimental platform for robot assisted minimally invasive heart surgery with haptic feedback. The manipulator set-up is composed of two low-payload robots carrying delicate surgical instruments. One of the main requirements is full Cartesian control of the end effectors to enable the surgeon to precisely control the position and orientation of the instruments. This must be realized in the presence of severe mechanical constraints imposed by the small ports ("key holes") in the patient's body which drastically restrict the motion space of the instrument. In this paper we present the complete set-up with the underlying theory and the software simulator that allows us to test operations before applying them to the real system.

Keywords: robotic heart surgery, minimally invasive surgery, trocar kinematics

1. Introduction

Minimally invasive surgery (MIS) has become a promising option for a great number of medical interventions (like coronary heart surgery). However, apart from obvious advantages to the patient (including reduced tissue trauma and shorter recovery times), using this technique entails additional difficulties. The basic limitations are reduced sight and manipulability. This gap is closed by application of robotic systems. Available systems like the *daVinci* workstation (cf. G. Guthart and J. Salisbury, 2000) or the *ZEUS* system provide the surgeon with stereo vision of the operating environment and restore full control of the instruments. While having pioneered the field of endoscopic surgery, these systems suffer from several deficiencies: they are telemanipulators with no direct position control (the control loop is implicitly closed by visual servoing of the surgeon) and they provide no possibility for force feedback. Both features are important in order to move the surgeon up in the control hierarchy, i.e. to implement "partial autonomy". There is a number of research projects that aim at remedying these deficiencies. At the University of California, Berkeley, a robotic system was developed, which has already been used to perform certain surgi-

J. Lenarčič and C. Galletti (eds.), On Advances in Robot Kinematics, 181-190.

cal tasks like suturing and knot-tying (M. Cavusoglu et al., 2003). The Korean Advanced Institute of Science and Technology has developed a micro-telerobot system that also provides force feedback (D. Kwon et al., 1998). In Germany two systems for robotic surgery were built at the Research Facility in Karlsruhe (U. Voges et al. 1997) and at the DLR in Oberpfaffenhofen (R. Konietschke et al., 2003). While the first system provides no force feedback, the latter system is equipped with *PHANToM* devices for haptic display.

The goal of our prototypical system is three-fold: (i) implementing full Cartesian control of the combination of robot and articulated instrument along with software facilities for realistic simulation, (ii) meeting all the requirements for sensitive force feedback enabling complex and complicated surgical procedures like knot-tying, and (iii) providing an open experimental platform for researchers that have no access to proprietary software interfaces of the other systems. In the scope of this paper we will concentrate on (i) and describe the mechanical properties of our platform and sketch the simulator.

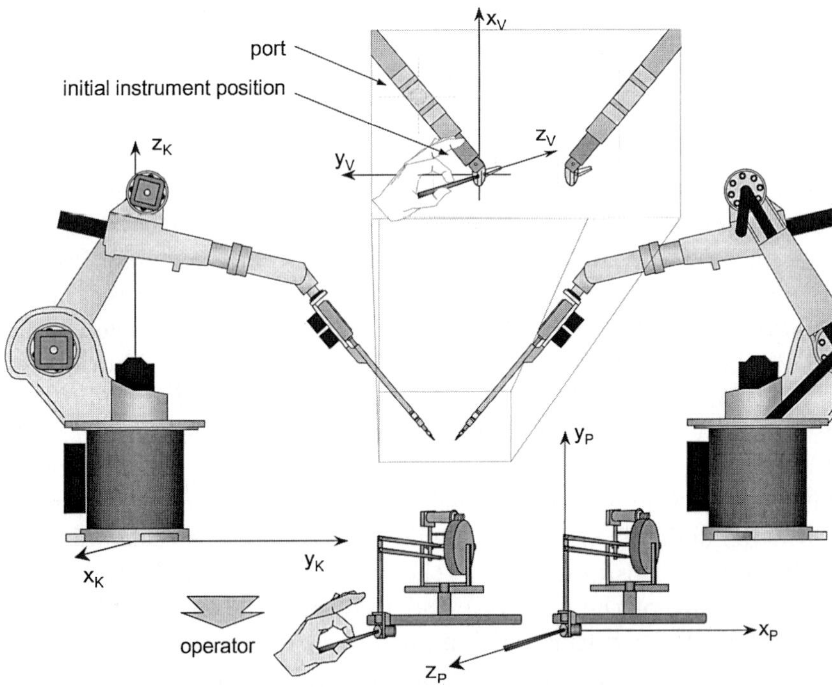

Figure 1. System Overview

2. System Setup and Virtual Instrument

The motor part of our system consists mainly of two instruments for minimally invasive surgery, which are mounted on industrial-grade low-payload robots (Fig. 1). To address the aspect of intuitive operability of the user interface, we apply the concept of so-called *trocar kinematics*: the manipulator has to pass through a fixed hole ("port") in the the patient's body. This restricts the degrees of freedom of the instrument. Feed (translation) and rotation axes must always intersect with the fixed port. Given the position and rotation of the end effector, we have to calculate all joint angles of our eight degrees of freedom (DOF) system. The resulting angles can be directly applied to the robotic system – or they can first be evaluated in a simulation environment.

Because the system's working space is mostly determined by the working space of the robots, we have chosen their base coordinate system K as our base system. In our setup, the surgeon (handling two haptic input devices of type *PHANToM*) and the observation camera, respectively, are placed in front of the robots: the x_K-axis points in the direction of the user, while the z_K-axis points upward to the ceiling (Fig. 1). The force-feedback styluses are placed in front of the user and, therefore, the z_P-axis is collinear with the x_K-axis of the robot system. The y_P-axis points up to the ceiling, while the x_P-axis points to the right hand side (Fig. 1).

The specification of the coordinate system for the minimally invasive instrument is crucial for all further considerations. As mentioned above, we want to make the control of the instrument appear to the surgeon as "natural" or "intuitive". We therefore modelled the kinematic structure on the observation of the human hand: if humans perform very precise manual tasks (e.g. a surgeon making a cut), we turn the hand about a rotation center that lies near the first link of the fingers (Fig. 2, left). If we want to mimic this behavior in our instrument control – while preserving mechanical feasibility – then a good compromise is a setting as shown in Fig. 2, center. We call the coordinate system spanned by the axes x_V, y_V and z_V the virtual instrument system. The mechanical rotation axes of the real instrument are named x_M, y_M and z_M (Fig. 2, right).

3. Trocar Kinematics

As mentioned above, all instruments have to pass through the incisions in the patient's body (ports). In most cases three ports are needed (see Fig. 3); two ports for the instruments and one port for the endoscopic stereo camera. Clearly, the possible movement of the instrument's shaft

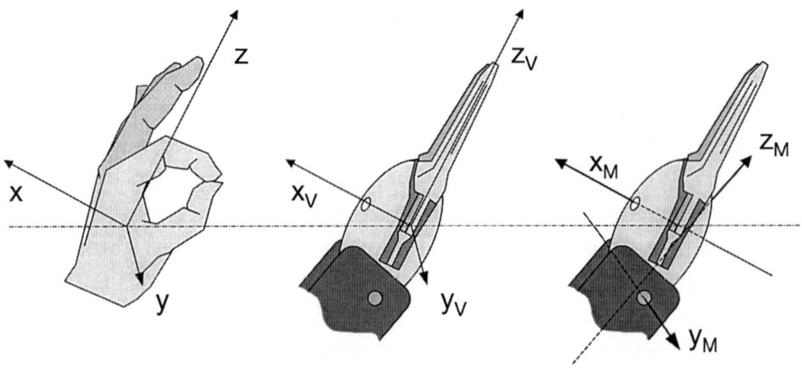

Figure 2. Virtual Instrument Definition

is restricted to insertion, retraction and rotation about the center axis of the corresponding port P.

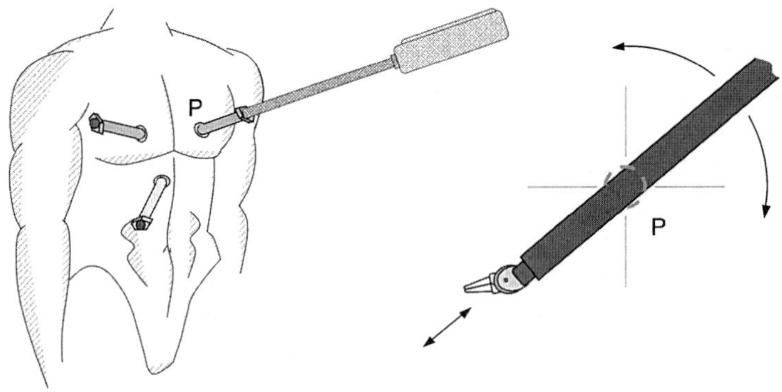

Figure 3. Location of the Instrument and Camera Port

The most comfortable way of moving a surgical instrument inside the body would be Cartesian control. The homogeneous transformation matrix ${}^K_V T$ describing the pose of the virtual instrument is given by the position of the input stylus of the *PHANToM* haptic devices.

We now determine the transformation matrix for an *initial* instrument system I. This system is the initial virtual instrument system V with all instrument angles set to zero. The basic idea is that we can easily calculate the homogenous transformation of this system in relation to the robot system. Then, the system I can be transferred to the virtual instrument system V by subsequent rotations of shaft, wrist and fingers.

First, we describe the rotation of system I, relative to the base (robot) system, by Z-Y-X-Euler angles:

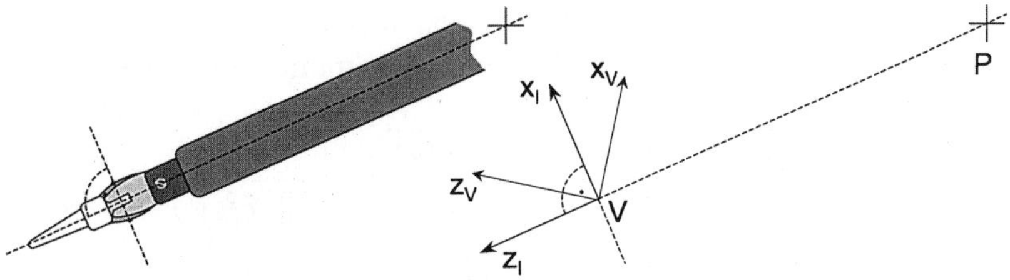

Figure 4. Initial Position of the Virtual Instrument

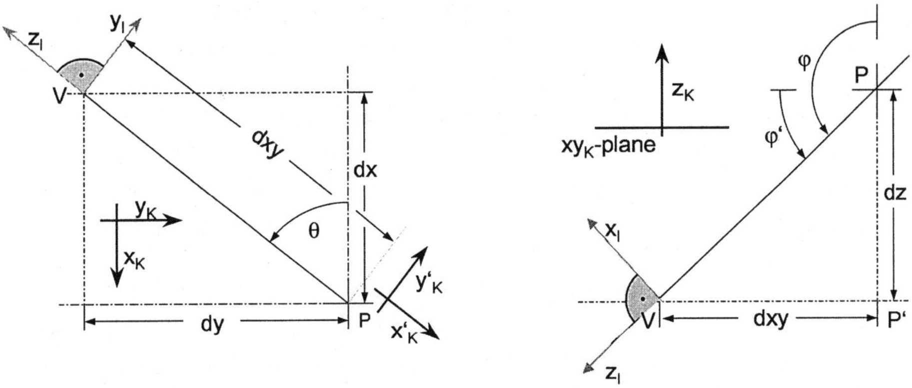

Figure 5. Z-Y-(X)-Euler Angles of the Virtual Instrument System

Because we know the position of the port P and the instrument V, we can derive the Z-Y-Euler angles θ and φ from the geometry shown in Fig. 5. Rotation about the x-axis is set to zero (therefore y_I is parallel to the xy_K plane. We have:

$$\theta = atan2(dy, dx) \tag{1}$$

$$\varphi = atan2(dz, dxy) + 90^\circ \tag{2}$$

With these angles we can now arrange the homogenous rotation matrix ${}^K_I T$ that transforms the robot base system to the initial position of the instrument. The derivation of this relation is straightforward and well-known (e.g. J. Craig, 1986 or T. Yoshikawa, 1990). Note, however, that the position of the origin is the same for both the virtual instrument system and the initial instrument system. Nevertheless, the frames differ in their orientations. Therefore, in a second step, we extract the corresponding rotation angles, which lead to this difference. To do this, we need to know the transformation from ${}^K_I T$ (initial position) to ${}^K_V T$ (desired position). In other words: we have to calculate ${}^I_V T$. This is done by taking the inverse matrix of the description of the initial position: ${}^K_I T^{-1} = {}^I_K T$. Now we get ${}^I_V T$ by matrix multiplication:

${}^{I}_{V}T = {}^{I}_{K}T \cdot {}^{K}_{V}T$. Based on this result, we can determine the X-Y-Z Yaw-Pitch-Roll rotation angles for the instrument as follows:

$$\alpha_V = atan2({}^{I}_{V}T_{21}, {}^{I}_{V}T_{22}) \tag{3}$$

$$\beta_V = atan2(-{}^{I}_{V}T_{20}, \sqrt{{}^{I}_{V}T_{00}^2 + {}^{I}_{V}T_{10}^2}) \tag{4}$$

$$\gamma_V = atan2({}^{I}_{V}T_{10}, {}^{I}_{V}T_{00}) \tag{5}$$

We now have the rotation angles of the virtual instrument from its initial position to the desired pose. In the definition of V (see above) we assumed that all rotation axes of the virtual instrument intersect in one point.

We now have to consider how to simulate this behavior with the real instrument and its mechanical axes. The mechanical x_M-axis and the virtual x_V-axis are identical by definition. That means we can directly apply the x_V-rotation of the virtual instrument to its mechanical counterpart, the yaw of the fingers ($\alpha_M = \alpha_V$).

Because the x_M and y_M-axis of the instrument do not intersect, we cannot apply the y_V-rotation without modifications. We have to calculate it from known geometric properties of the set-up: (i) the rotation angle about the y_V-axis of the virtual instrument (β_V) is known, (ii) the shaft of the instrument always passes through the port P and (iii) though not necessarily identical, virtual and mechanical y-axis are always parallel. Using these properties we can compute the rotation about the mechanical y_M-axis (β_M). Note that the position of V remains unchanged but the shaft of the instrument is tilted about the port. As can

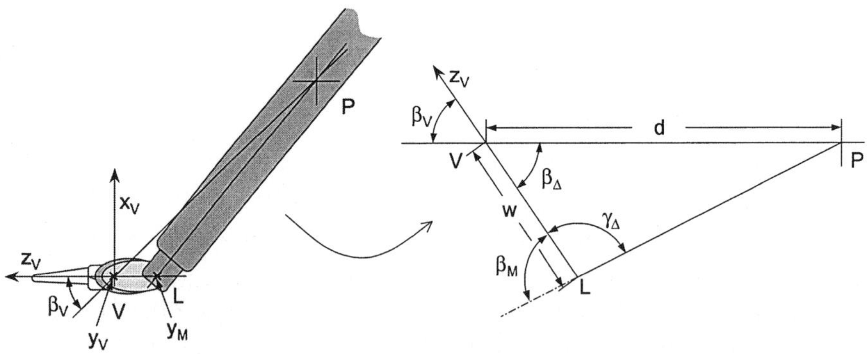

Figure 6. Determination of β_M

be seen in Fig. 6, two sides of a triangle are known, with w being the length of the instrument's wrist and d being the distance between the port and the position of the virtual instrument system. Additionally we are given the rotation angle $\beta_\triangle = \beta_V$ which is constituted by w and d. Therefore the following holds:

$$\tan\left(\frac{\gamma_\triangle - \alpha_\triangle}{2}\right) = \frac{d-w}{d+w}\cot\left(\frac{\beta_\triangle}{2}\right) \tag{6}$$

$$\frac{\gamma_\triangle + \alpha_\triangle}{2} = 90^\circ - \frac{1}{2}\beta_\triangle; \tag{7}$$

By applying the atan-function to Eq. (6) and adding (7), we have:

$$\gamma_\triangle = \tan^{-1}\left(\frac{d-w}{d+w}\cot\left(\frac{\beta_\triangle}{2}\right)\right) + 90^\circ - \frac{1}{2}\beta_\triangle \tag{8}$$

Moreover, $\beta_M = 180^\circ - \gamma_\triangle$.

The last angle to determine is the mechanical rotation of the instrument's shaft. The initial transformation of the shaft is set by the transformation of the robot flange F. In order to reach the desired position of the virtual instrument, we perform a rotation about the z_I-axis (see derivation of transform ${}^I_V T$). On the other hand, we only can apply rotations about the real z_M-axis, which is identical with the z_F-axis (see Fig. 7). It is clear from picture 6 that the longitudinal axis of the shaft z_M ($\overrightarrow{PL}$) is not collinear with the z_I-axis ($\overrightarrow{PV}$) when β_M is not zero. In this case, we have to determine γ_M by an additional calculation. To this end we need to find the positions of the mechanical instrument axes (esp. the shaft axis $z_M = z_F$) relative to the robot base system. Our chosen convention for the y_F-axis is that it is initially parallel to the xy_K-plane. Note that every other convention would naturally change γ_M, but will not influence the final position of the instrument. In Fig. 7, the point V^+ is the position of the virtual instrument if γ_M were zero. We thus have to rotate the shaft until V^+ overlaps with the actual position of the virtual instrument $V = {}^K_V T_t$. In other words: we have to find the rotation transforming x_F, which is parallel to a plane spanned by z_K and $\overrightarrow{PL}$, to x_I, which is parallel to a plane spanned by $\overrightarrow{PL}$ and $\overrightarrow{PV}$ (see Fig. 7). Therefore we first have to find the position of the inflexion point L (also see Fig. 6). We know the transformation of the virtual instrument ${}^K_V T$. Since the x_V- and the x_M-axis are identical, we can get the orientation of the wrist ${}^K_W T_r$, by back-rotating ${}^K_V T$ by $-\alpha_M$. Because the z-axis of the resulting system points in the direction of the wrist, we can move back along it the known length of the wrist w. Thus, altogether we get:

$$\vec{L} = {}^K_V T \cdot {}^V_W T \cdot \begin{pmatrix} 0 \\ 0 \\ -w \\ 1 \end{pmatrix} \tag{9}$$

Given all the points, we can now estimate γ_M. The best way to do this is

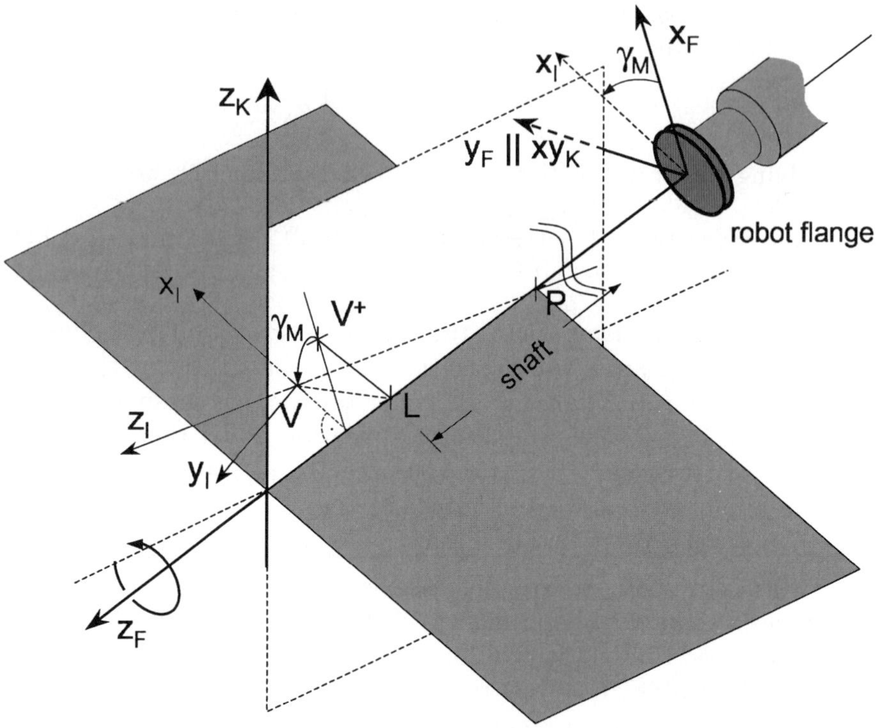

Figure 7. Calculation of γ_M

calculating the intersection angle of the plane spanned by V^+, L and P, and the plane spanned by V, L and P. We get this angle by intersecting the normals of these planes. Now we have all angles necessary to control the instrument.

The final step is to determine the position and rotation of the flange of the robot (${}^K_F T$), which the instrument attached to. Since we know the position of the port P and the inflexion point L (Fig. 6), we can determine the Z-Y-Z Euler-angles α_K, β_K and γ_K in a similar fashion we used for the rotation of the initial instrument (Fig. 5). Because we required parallelism of the y_i-axis of the initial virtual instrument system (cf. Fig. 4) with the xy_K-plane of the robot system, we set the rotation of the flange to zero. We have:

$$\alpha_K = atan2(P_y - L_y, P_x - L_x) \quad (10)$$

$$\beta_K = atan2(P_z - L_z, \sqrt{(P_x - L_x)^2 + (P_y - L_y)^2} + 90^\circ \quad (11)$$

$$\gamma_K = 0 \quad (12)$$

The position of the shaft can be found by shifting back L along $\overrightarrow{PL}$ for the length of the instrument. With orientation and position we can arrange the homogenous transform matrix of the flange. Given the transformation, we can derive the joint angles of the robot by standard inverse kinematics.

4. Simulation

In order to check certain operation sequences (e.g. the complicated procedure of knot-tying) before applying them to the real world, we have developed a realistic simulation of our system. Since the model has the same geometry as the real system, all joint angles obtained from the inverse kinematics can be directly applied to it. The model is displayed in an *Open Inventor*-GUI. Input data can be recorded to a data base for subsequent use with the simulation or the real system. This simulation was especially useful to detect some unusual motion sequences that could lead to failures of the real system. For example, the robot tends to move too fast if the instrument tips approach come too close to the port. The simulation can also be used in parallel with real manipulations. This can be very helpful if the remote user has no full sight of the operation environment (e.g. if instruments are occluded by other objects).

5. Conclusion

We have presented an experimental system for robot assisted minimally invasive surgery with haptic feedback. It enables full Cartesian control of the instruments by implementing trocar kinematics. This is an important aspect for increasing the acceptance of such systems with surgeons. We are currently working on further improvements of both, the real system and the simulation environment. The robotic system will be continued to be developed towards a reliable and stable surgical workstation. The future plans for the simulation environment include inclusion of haptic feedback by means of a realistic tissue model, thread modelling and implementation of augmented reality techniques.

References

Cavasoglu M. et al. (2003), Robotics for Telesurgery: Second Generation Berkeley/UCSF Laparoscopic Telesurgical Workstation and Looking towards the Future Applications, *Industrial Robot, Special Issues on Medical Robotics*, Vol. 30, no. 1

Craig, John J. (1986), Introduction to Robotics (Mechanics and Control), Addison Wesley, Massachusetts, USA.

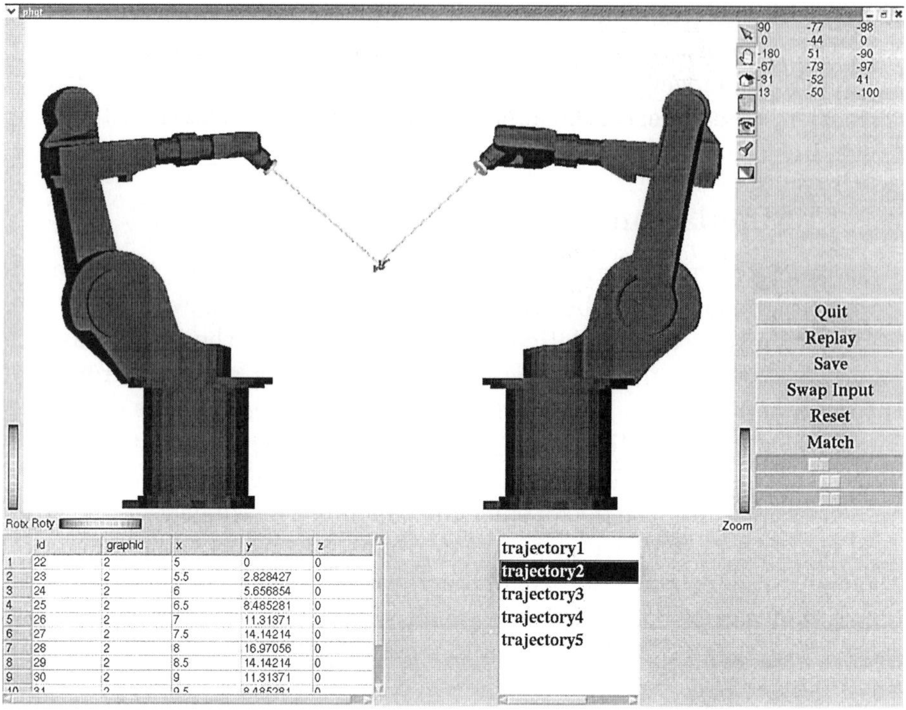

Figure 8. Screenshot of the Simulation Environment

Guthart, Gary S., and Salisbury, J. K. (2000), The Intuitive^{TM} Telesurgery System: Overview and Application, *IEEE International Conference on Robotics and Automation*, San Francisco CA, USA.

R. Konietschke et al. (2003), Optimal Design of a Medical Robot for Minimally Invasive Surgery, *2. Jahrestagung der Deutschen Gesellschaft fuer Computer- und Roboterassistierte Chirurgie (CURAC)*, Nuernberg, Germany

Kwon D. et al. (1998), Microsurgical Telerobot System, *IEEE/RSJ International Conference on Intelligent Robots and Systems*, pp. 945-950

Massie, Thomas. H. and Salisbury, J. Kenneth (1994), The PHANTOM Haptic Interface: A Device for Probing Virtual Objects, *Proceedings of the ASME Winter Annual Meeting, Symposium on Haptic Interfaces for Virtual Environment and Teleoperator Systems*, Chicago, IL, USA.

Voges U. et al., Evaluation of ARTEMIS: the Advanced Robotics and Telemanipulator System for Minimally Invasive Surgery, *Proceedings IARP 2nd Workshop on Medical Robotics*, pp. 137-148

Yoshikawa, Tsuneo (1990), Foundations of Robotics: Analysis and Control, MIT, Massachusetts, USA.

MANIPULABILITY AND ACCURACY MEASURES FOR A MEDICAL ROBOT IN MINIMALLY INVASIVE SURGERY

Rainer Konietschke[1], Tobias Ortmaier, Holger Weiss, Gerd Hirzinger
German Aerospace Center, Institute of Robotics and Mechatronics
P.O. Box 1116, D-82230 Weßling, Germany
[1]Rainer.Konietschke@dlr.de

Robert Engelke
Munich University of Technology, Institute of Applied Mechanics
D-85747 Garching, Germany

Abstract This paper presents measures for manipulability and accuracy that are specifically adapted to the conditions found in robotically assisted minimally invasive surgery. The considered robot consists of 9 active joints, thus allowing for full manipulability at the tool tip as well as null-space motion. The presented manipulability and positioning accuracy measures are based on an "inverse Jacobian" approach since the constraints at the entry point into the human body forbid a classic formulation. High significance of the measures is reached by including robot design parameters such as encoder resolution and maximum joint velocity.

Keywords: manipulability, positioning accuracy, robotic assistance, minimally invasive surgery, optimisation, robot design

1. Introduction

In minimally invasive surgery (MIS), the field of operation is reached through small incisions into the human body with the use of thin cylindrical instruments. In the case of robotic assistance, these instruments are tele-operated by systems such as the daVinci^{TM} or ZEUS^{TM} robot. The movement of the instrument is constrained in two directions at the entry point into the human body as shown in Fig. 1. Since the operating room is an unpredictable and overcrowded environment, it is desirable to have a medical robot with a redundant kinematic structure, thus permitting the robot to avoid collisions without changing the position and orientation of the tool tip (null-space motion).

The optimal positioning of both the robot base(s) and the entry point(s) into the human body are determined in a preoperational planning step, where different base positions and entry points are evaluated

J. Lenarčič and C. Galletti (eds.), On Advances in Robot Kinematics, 191-198.

according to predefined optimisation criteria such as e.g. manipulability, accuracy, geometrical considerations, insensitiveness of the robot setup or overall size of the robot.

Among these, manipulability and positioning accuracy are of great interest due to the following reasons:

- Singular configurations which could cause system failures must be avoided.
- A certain minimal velocity of the instrument tip both in terms of translation and rotation must be realised all over the considered operating field. This is important to assure that motions commanded by the surgeon can be performed by the robot system.
- The demanded accuracy of the positioning of the instrument tip must be guaranteed. This is of great interest if very fine structures (e.g. blood vessels) are manipulated.

Previous work concerning preoperational planning of MIS procedures (Adhami, 2002; Chiu et al., 2000; Engel et al., 2003; Lehmann et al., 2001; Selha et al., 2003; Tabaie et al., 1999) does however not take into account manipulability and positioning accuracy sufficiently. Instead, the used approaches *tool dexterity* (Adhami, 2002; Lehmann et al., 2001) as well as *magic pyramid* (Chiu et al., 2000) evaluate the orientation of the instruments, the endoscope, and the surface normal of the considered operating field to each other. Engel et al., 2003 consider the distance of the joint angles of the robot from its maximum values. Experimental trials (Adhami, 2002) suggest the conclusion that the use of more sophisticated descriptions of manipulability and accuracy measures is reasonable. Since their classic formulations are not applicable to the MIS setup due to the restriction at the entry point, this paper presents a suitable formulation.

In the next section, the underlying problem is described. The derivation of measures for manipulability and accuracy is shown in Sect. 3 and 4. The results are discussed in Sect. 5 and a conclusion is given in Sect 6.

2. Problem Statement

A robot with 7 degrees of freedom (DoFs) as shown in Fig. 1 is considered. Two additional DoFs (Θ_8 and Θ_9) are added by the actuated instrument. The notation Θ_i stands both for the joint i and the axis direction parallel to the rotation axis of that same joint. Taking into account the loss of 2 DoFs due to the kinematic restriction at the entry

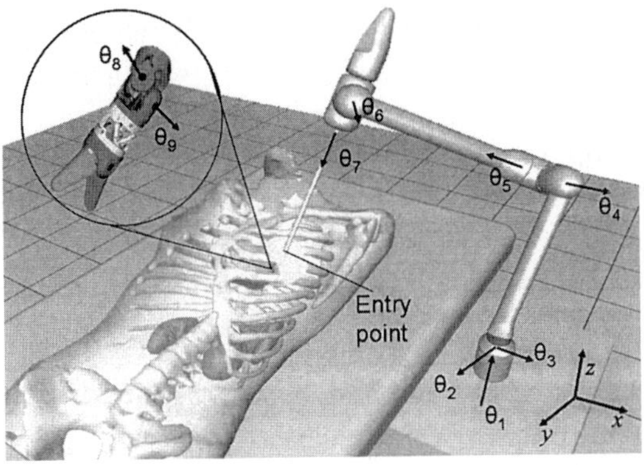

Figure 1. Surgical robot with actuated instrument.

point into the human body, full 6 DoFs remain for the manipulation of the instrument tip and an extra degree of freedom serves to accomplish null-space motion.

Dynamics are not included in the modelling since typical motions in MIS are too small to induce significant forces.

3. Manipulability

The classic formulation of manipulability based on YOSHIKAWA, 1990 investigates the singular values of the Jacobian matrix $\mathbf{P} \in \mathbb{R}^{6\times 9}$ in Eq. 1:

$$\mathbf{v} = \mathbf{P}\dot{\mathbf{\Theta}}\,, \tag{1}$$

$$\text{with} \quad \dot{\mathbf{\Theta}} = [\dot{\Theta}_1, \ldots, \dot{\Theta}_9]^{\mathrm{T}} \quad \text{the joint velocities} \tag{2}$$

$$\text{and} \quad \mathbf{v} = [v_x, v_y, v_z, \omega_x, \omega_y, \omega_z]^{\mathrm{T}} \tag{3}$$

the translational velocity of the instrument tip and the angular velocity of the last segment of the instrument expressed in the inertial coordinate system $[x, y, z]$ as depicted in Fig. 1. However, in the considered case Eq. 1 cannot be used to address manipulability because it does not take into account the motion restriction due to the entry and the possibility of null-space motion. At the entry point, the following kinematical constraint must be satisfied (Ortmaier and Hirzinger, 2000):

$$\mathbf{C}_1\dot{\mathbf{\Theta}} = \mathbf{0}\,, \mathbf{C}_1 \in \mathbb{R}^{2\times 9}. \tag{4}$$

Null-space motion implies that a given tool tip position and orientation can be realised with an infinite number of joint sets. To calculate the most appropriate joint set, another constraint has to be introduced which represents some kind of optimisation among the possible joint sets:

$$\mathbf{C}_2\dot{\mathbf{\Theta}} = \mathbf{0}\,, \mathbf{C}_2 \in \mathbb{R}^{1\times 9}, \tag{5}$$

resulting in

$$\mathbf{C}\dot{\mathbf{\Theta}} = \mathbf{0} \quad \text{with} \quad \mathbf{C} = \begin{bmatrix} \mathbf{C}_1 \\ \mathbf{C}_2 \end{bmatrix} \in \mathbb{R}^{3\times 9}. \tag{6}$$

If the vector of joint velocities $\dot{\mathbf{\Theta}}$ is divided arbitrarily into a dependent part $\dot{\mathbf{\Theta}}_\mathrm{d} \in \mathbb{R}^{3\times 1}$ and an independent part $\dot{\mathbf{\Theta}}_\mathrm{i} \in \mathbb{R}^{6\times 1}$, it can be reordered as follows:

$$\dot{\mathbf{\Theta}} = \begin{bmatrix} \dot{\mathbf{\Theta}}_\mathrm{d} \\ \dot{\mathbf{\Theta}}_\mathrm{i} \end{bmatrix}. \tag{7}$$

The equations 1 and 6 can then be rewritten as (Ortmaier and Hirzinger, 2000):

$$\mathbf{v} = \mathbf{P}_\mathrm{d}\dot{\mathbf{\Theta}}_\mathrm{d} + \mathbf{P}_\mathrm{i}\dot{\mathbf{\Theta}}_\mathrm{i} \quad \text{with} \quad \mathbf{P} = [\mathbf{P}_\mathrm{d}\ \ \mathbf{P}_\mathrm{i}] \tag{8}$$

$$\text{and} \quad \dot{\mathbf{\Theta}}_\mathrm{d} = \mathbf{B}\dot{\mathbf{\Theta}}_\mathrm{i} \quad \text{with} \quad \mathbf{B} \in \mathbb{R}^{3\times 6}. \tag{9}$$

After insertion of the kinematical constraint (Eq. 6), Eq. 8 has form

$$\mathbf{v} = \mathbf{P}'\dot{\mathbf{\Theta}}_\mathrm{i} \quad \text{with} \quad \mathbf{P}' = \mathbf{P}_\mathrm{d}\mathbf{B} + \mathbf{P}_\mathrm{i} \in \mathbb{R}^{6\times 6}, \tag{10}$$

where only 6 of the 9 joint velocities occur. Depending on which joint velocities are chosen as dependent, the Jacobian matrix $\mathbf{P}'$ has different elements and different singular values. Therefore, it is not useful for the formulation of manipulability. A more suitable formulation is achieved if the inverse correlation is used. This can be done by solving Eq. 10 for $\dot{\mathbf{\Theta}}_\mathrm{i}$ and combining the result with Eq. 9 (Konietschke, 2001):

$$\dot{\mathbf{\Theta}} = \mathbf{D}\mathbf{v} \quad \text{with} \quad \mathbf{D} = \begin{bmatrix} \mathbf{B}\mathbf{P}'^{-1} \\ \mathbf{P}'^{-1} \end{bmatrix}. \tag{11}$$

This equation includes all joint velocities. It relates a given instrument velocity $\mathbf{v}$ to the joint velocities $\dot{\mathbf{\Theta}}$ in a non-ambiguous way for every non-singular robot configuration by consideration of the kinematic constraint due to the entry point. Alternatively to the above described calculations, matrix $\mathbf{D}$ can also be obtained by numerical differentiation. This can be done by applying difference quotients to the inverse kinematics which expresses the joint angles as function of the instrument tip position and orientation and the position of the entry point.

To define a measure for manipulability, a connection between the maximum joint velocity $\dot{\Theta}_{\max}$ and the desired minimal translational velocity $\dot{x}_{\min}$ and angular velocity $\dot{\alpha}_{\min}$ of the instrument tip has to be established. To do so, the matrix $\mathbf{D}$ in Eq. 11 is split into two components

$$\mathbf{D} = [\mathbf{E}\,\mathbf{F}] \quad \text{with} \quad \mathbf{E} = \begin{bmatrix} e_{11} & & e_{13} \\ \vdots & \ddots & \vdots \\ e_{81} & & e_{83} \end{bmatrix}, \mathbf{F} = \begin{bmatrix} f_{11} & & f_{13} \\ \vdots & \ddots & \vdots \\ f_{81} & & f_{83} \end{bmatrix}, \tag{12}$$

and Eq. 11 is rewritten:

$$\dot{\mathbf{\Theta}} = \mathbf{E} \begin{bmatrix} v_x \\ v_y \\ v_z \end{bmatrix} + \mathbf{F} \begin{bmatrix} \omega_x \\ \omega_y \\ \omega_z \end{bmatrix}. \tag{13}$$

For each joint $i \in [1, ..., 8]$, the maximum velocity $\dot{\Theta}_{i,\,\max}$ is computed by solving the optimisation problem (14):

$$\dot{\Theta}_{i,\max} = \left\{ |\dot{\Theta}_i(\dot{\mathbf{x}})| \xrightarrow{\text{opt}} \max \right\} = \left\{ |e_{i1}\dot{x} + e_{i2}\dot{y} + e_{i3}\dot{z} \right.$$
$$\left. + f_{i1}\dot{\alpha}_1 + f_{i2}\dot{\alpha}_2 + f_{i3}\dot{\alpha}_3| \xrightarrow{\text{opt}} \max \right\} \tag{14}$$

under the constraints

$$\sqrt{v_x^2 + v_y^2 + v_z^2} - \dot{x}_{\min} \le 0 \quad \text{and} \quad \sqrt{\omega_x^2 + \omega_y^2 + \omega_z^2} - \dot{\alpha}_{\min} \le 0\,. \tag{15}$$

Solution is done using the Lagrange function and yields (see Konietschke, 2001 for further details):

$$\dot{\Theta}_{i,\,\max} \quad = \quad \sqrt{e_{i1}^2 + e_{i2}^2 + e_{i3}^2}\; \dot{x}_{\min} \;+\; \sqrt{f_{i1}^2 + f_{i2}^2 + f_{i3}^2}\; \dot{\alpha}_{\min}\,. \tag{16}$$

Thus the maximum joint velocity $\dot{\Theta}_{\max} = \max(\dot{\Theta}_{i,\,\max})$ is determined which can be used as a reciprocal manipulability measure:

$$w_{\text{man}} = \frac{1}{\dot{\Theta}_{\max}}\,. \tag{17}$$

For a given, desired minimal velocity at the tool tip, low values of $\dot{\Theta}_{\max}$ denote that the joint velocities remain low. This indicates good manipulability.

4. Positioning accuracy

The following question is considered to define an (in)accuracy measure:

> How far can the instrument tip be moved at maximum with all the changes in the articular space remaining below the encoder resolution $\Delta\Theta_{\min}$?

This describes the maximum translational and rotational movement $\Delta x_{\min}$ and $\Delta\alpha_{\min}$ of the instrument tip that is not detectable by the robot control system and thus provides a worst case estimation of the positioning accuracy that can be commanded.

For small changes $\Delta\mathbf{\Theta}$ and $\mathbf{u}$, where $\mathbf{u} = \begin{bmatrix} \mathbf{u}_{\text{trans}} \\ \mathbf{u}_{\text{rot}} \end{bmatrix}$ describes small displacements of the instrument tip and small changes in the orientation of the instrument, the following relation approximately holds:

$$\Delta\mathbf{\Theta} = \mathbf{D}\mathbf{u}\,. \tag{18}$$

Since the vector $\mathbf{u}$ contains both translational and rotational components, normalisation is applied:

$$\Delta\mathbf{\Theta} = \tilde{\mathbf{D}}\tilde{\mathbf{u}} \quad \text{with} \quad \tilde{\mathbf{u}} = \begin{bmatrix} \mathbf{u}_{\text{trans}}/\Delta x_{\min} \\ \mathbf{u}_{\text{rot}}/\Delta\alpha_{\min} \end{bmatrix}, \; \tilde{\mathbf{D}} = [\Delta x_{\min}\mathbf{E} \;\; \Delta\alpha_{\min}\mathbf{F}]\,. \tag{19}$$

Thus, calculating the smallest singular value $\tilde{\sigma}_{\min}$ of $\tilde{\mathbf{D}}$ yields (Konietschke, 2001):

$$\|\Delta\mathbf{\Theta}\|_2 \geq \tilde{\sigma}_{\min}\,\|\tilde{\mathbf{u}}\|_2\,. \tag{20}$$

One is interested in the maximum change of one of the joint angles which is synonymous to the maximum norm $\|\Delta\mathbf{\Theta}\|_\infty$ and not to the Euclidean norm as appearing in Eq 20. Therefore, the following estimation is used:

$$\|\Delta\mathbf{\Theta}\|_2 \geq \|\Delta\mathbf{\Theta}\|_\infty \cdot \sqrt{\dim(\Delta\mathbf{\Theta})}\,. \tag{21}$$

With movements $\|\mathbf{u}_{\text{trans}}\|_2 \leq \Delta x_{\min}$ and $\|\mathbf{u}_{\text{rot}}\|_2 \leq \Delta\alpha_{\min}$, Eq. 22 holds:

$$\|\tilde{\mathbf{u}}\|_2 = \sqrt{\|\mathbf{u}_{\text{trans}}\|_2^2/\Delta x_{\min}^2 + \|\mathbf{u}_{\text{rot}}\|_2^2/\Delta\alpha_{\min}^2} \leq \sqrt{2}\,. \tag{22}$$

Thus, the following measure w_{pos} for positioning accuracy can be established:

$$w_{\text{pos}} = \Delta\Theta_{\min} = \frac{\tilde{\sigma}_{\min}}{\sqrt{\frac{1}{2}\dim(\Delta\Theta)}}\,, \tag{23}$$

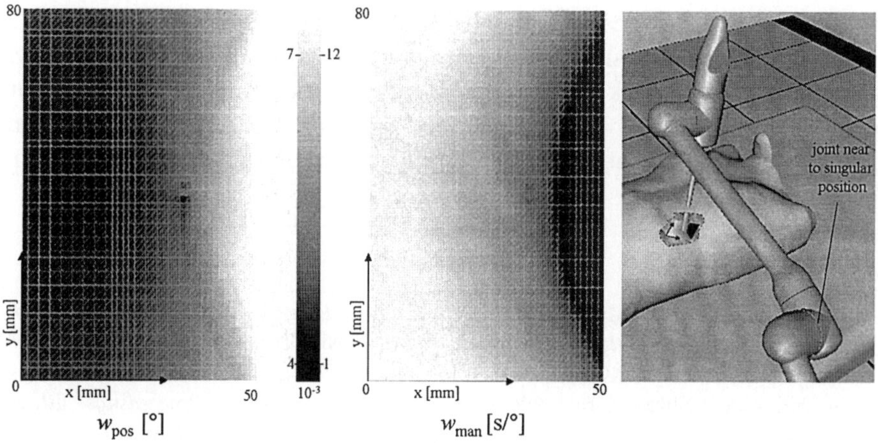

Figure 2. The measures w_{man} and w_{pos} in the vicinity of a singular configuration.

where $\Delta\Theta_{\mathrm{min}}$ is the necessary encoder resolution to detect movements greater than Δx_{min} resp. $\Delta\alpha_{\mathrm{min}}$.

5. Results

Fig. 2 shows the manipulability measure w_{man} and the positioning accuracy w_{pos} for different instrument tip positions in the vicinity of a singularity. Since w_{man} decreases when approaching the border line to singularity, it can be used to omit singular configurations.

The measure for positioning accuracy w_{pos} mainly shows a reverse behavior compared to w_{man} what is understandable since high manipulability is defined to allow fast motion of the tool tip with low joint velocities whereas high positioning accuracy means small displacements with large joint rotations. Concerning the validation of a robot configuration, it has to be found a tradeoff between high manipulability and high positioning accuracy.

6. Conclusion

In this paper, modified measures for manipulability and accuracy are presented to overcome the problems such as unexpected singularities or poor manipulability that were encountered with previous approaches in the field of robotically assisted MIS. The presented measures are very descriptive since they comprise the robotic design parameters *encoder resolution* and *maximum joint velocity.* In order to establish quantitative information whether a certain value for w_{man} or w_{pos} is sufficient to accomplish robotically assisted MIS, the requirements concerning desired velocity and positioning accuracy of the tool tip have to be known. In this context, work remains to be done since previous publications (Riviere and Jensen, 2000; Singh and Riviere, 2002; Hotraphinyo and Riviere, 2001; Cao et al., 1996) only cover analyses of translational aspects.

References

Adhami, L. (2002), *An Architecture for Computer Integrated Mini-Invasive Robotic Surgery: Focus on Optimal Planning*, Ph.D. Thesis, Ecole des Mines de Paris.

Cao, C. G. L., C. L. MacKenzie, and S. Payandeh (1996), Task and Motion Analyses in Endoscopic Surgery, In: *1996 ASME IMECE Conference Proceedings: 5th Annual Symposium on Haptic Interfaces for Virtual Environment and Teleoperator Systems.* Atlanta, Georgia, USA.

Chiu, A. M., D. Dey, M. Drangova, D. Boyd, and T. M. Peters (2000), 3-D Image Guidance for Minimally Invasive Robotic Coronary Artery Bypass, *The Heart Surgery Forum* **3**(3), 224–231.

Engel, D., W. Korb, J. Raczkowsky, S. Hassfeld, and H. Woern (2003), Location Decision for a Robot Milling Complex Trajectories in Craniofacial Surgery, In: *Proceedings of the 17th International Congress and Exhibition CARS 2003.* London.

Hotraphinyo, L. and C. Riviere (2001), Three-Dimensional Accuracy Assessment of Eye Surgeons, In: *Proc. 23rd Annual Intl. Conf. IEEE Engineering in Medicine and Biology Society.* pp. 3458–3461.

Konietschke, R. (2001), *Aufbauoptimierung für Roboter in medizinischen Anwendungen*, Diploma Thesis: Munich University of Technology, Germany.

Lehmann, G., A. Chiu, D. Gobbi, Y. Starreveld, D. Boyd, M. Drangova, and T. M. Peters (2001), Towards dynamic planning and guidance of minimally Invasive Robotic Cardiac Bypass Surgical Procedures, In: *Proc MICCAI 2001. 4th International Conference.* Utrecht, The Netherlands, pp. 368–375.

Ortmaier, T. and G. Hirzinger (2000), Cartesian Control Issues for Minimally Invasive Robot Surgery, In: *Proc. of the IEEE/RSJ International Conference on Intelligent Robots and Systems IROS 2000.* Takamatsu, Japan.

Riviere, C. and P. Jensen (2000), A Study of Instrument Motion in Vitreoretinal Microsurgery, In: *Proceedings of the 22nd International Conference of the IEEE Engineering in Medicine and Biology Society.*

Selha, S., P. Dupont, R. Howe, and D. Torchiana (2003), Dexterity Optimization by Port Placement in Robot-Assisted Minimally Invasive Surgery, In: *Proc. of the 2001 SPIE Int. Symp. on Intelligent Systems and Advanced Manufacturing.*

Singh, S. and C. Riviere (2002), Physiological Tremor Amplitude During Vitreoretinal Microsurgery, In: *Proceedings of the IEEE 28th Annual Northeast Bioengineering Conference.* 445 Hoes Lane, Piscataway, NJ 08855-1331, pp. 171–172, IEEE.

Tabaie, H. A., J. A. Reinbolt, W. P. Graper, T. F. Kelly, and M. A. Connor (1999), Endoscopic Coronary Artery Bypass Graft (ECABG) Procedure with Robotic Assistance, *The Heart Surgery Forum* **2**(4), 310–317.

Yoshikawa, T. (1990), *Foundations of Robotics: Analysis and Control*, The MIT Press.

HELIX-HELIX PACKING IN PROTEINS: THE KINEMATICS OF INTERACTING LINE SEGMENTS

Sangyoon Lee

School of Mechanical and Aerospace Engineering, Konkuk University

Seoul, Korea

slee@konkuk.ac.kr

Gregory S. Chirikjian

Department of Mechanical Engineering, Johns Hopkins University

Baltimore, Maryland, USA

gregc@jhu.edu

Abstract In the structural bioinformatics literature, a number of works address the geometry of how alpha-helices (one of the major components of proteins) pack against each other. Typically such studies involve isolating the axis of each helix, measuring the angle between helices that are judged to be in contact, and generating statistics of the inter-helical angle over many contacting pairs. Using techniques from theoretical kinematics, we show that previous efforts to normalize angle distribution data did not include important effects. Namely, helices can interact with each other in three distinct ways which we refer to as "line on line", "end point to line" and "endpoint to endpoint," and each of these interactions has its own geometric effects which must be included in the proper normalization of statistical data. This problem is one of many in the area of structural bioinformatics which can benefit from kinematics techniques.

Keywords: Alpha-helix, angular distribution, statistical bias, empirical potential

1. Introduction

Proteins are biological macromolecules which form much of the structure and function of living organisms. Proteins are composed of 20 different amino acid residues which are connected sequentially by covalent chemical bonds. These polypeptide chains twist and writhe in space to form helical and sheet-like structures which then pack against each other in various ways to form the final protein. Whereas the prediction of these so-called "secondary structures" themselves is relatively reliable, methods for predicting how these structures assemble to form the resulting protein are not yet reliable enough to declare the

J. Lenarčič and C. Galletti (eds.), On Advances in Robot Kinematics, 199-208.

protein folding problem solved. For this reason, a number of researchers over the years have developed algorithms which search known protein structures in an attempt to find patterns in the way secondary structures pack against each other. Since the alpha-helix is perhaps the most common secondary structure in proteins, helix-helix packing has received the most attention.

In structural bioinformatics, the protein data bank is searched, and protein structures that have been determined at high resolution and have little resemblance to each other are used to examine preferred packing motifs [Chothia et al., 1977, Chothia et al., 1981, Crick, 1953, Finkelstein and Ptitsyn, 1987, Murzin and Finkelstein 1988, Richmond and Richards, 1978, Wang and Dunbrack, 2003]. In the context of helix-helix interactions, there are significant geometric effects which bias these statistics. In this short paper, we use methods of theoretical kinematics to derive these biases, and determine the unbiased angular preferences that alpha-helices have for packing against each other.

Since alpha-helices are effectively rigid bodies, kinematics is the appropriate tool for describing their spatial relationships. We review rigid-body kinematics in Section 2, and present our methodology for classifying and analyzing helix-helix pairs in Section 3. In Section 3, the appropriate volume elements for normalizing randomly oriented helices is derived. This is required in order to be able to normalize statistical data correctly. Section 4 presents results that we have obtained with a very large number of helix-helix pairs, and Section 5 presents our conclusions. Many of the kinematic tools used here were developed in the kinematics literature [Angeles, 1988, Basavaraj and Duffy, 1993, Bottema and Roth, 1990, Chirikjian and Kyatkin, 2000, McCarthy, 1990, McCarthy and Roth, 1982, Nayak and Roth, 1981, Ravani and Roth, 1984, Roth, 1967].

2. Review of Kinematics

Recall that for counter-clockwise rotations about the $\mathbf{e}_3$, $\mathbf{e}_2$, and $\mathbf{e}_1$ axes can be written as the matrix exponential $R_i(\phi) = \exp(\phi\, E_i)$ where each E_i is the unique skew-symmetric matrix defined to satisfy

$$E_i \mathbf{x} = \mathbf{e}_i \times \mathbf{x}$$

This relationship is described using the notation $\mathbf{e}_i = \text{vect}\,(E_i)$, and since the "vect" operation is linear, it can be used to relate any three-dimensional vector and any 3×3 matrix.

Note that $E_i \mathbf{e}_i = 0$ and $E_1 \mathbf{e}_2 = \mathbf{e}_3$, $E_3 \mathbf{e}_1 = \mathbf{e}_2$, $E_2 \mathbf{e}_3 = \mathbf{e}_1$. In contrast, $R_i(\phi)\mathbf{e}_i = \mathbf{e}_i$. Another property which was used multiple times to obtain the results presented below is that for any 3×3 rotation matrix R and any 3×3 skew symmetric matrix Ω,

$$\text{vect}\,(R\,\Omega\,R^{\mathrm{T}}) = R\,\text{vect}\,(\Omega)$$

The relative position and orientation between rigid bodies can be expressed with a pair $(A, \mathbf{a})$ where A is a rotation matrix and $\mathbf{a}$ is a translation vector. To be more precise, assume two bodies have reference frames attached to them in some canonical way. We define the relative motion that will take the frame of reference attached to body 1 into the frame of reference attached to body 2 to be $(A, \mathbf{a})$. Then from the perspective of body 2, the relative motion that it would need to undergo for its frame to become coincident with frame 1 would be $(A^{\mathrm{T}}, -A^{\mathrm{T}}\mathbf{a})$. The set of all such rigid-body motions forms a manifold, and the operation of composing two rigid-body motions endows this manifold with the structure of a Lie group. It is convenient to think of this Lie group as the set of all 4×4 *homogeneous transformation* matrices of the form

$$H = \begin{pmatrix} A & \mathbf{a} \\ \mathbf{0}^{\mathrm{T}} & 1 \end{pmatrix},$$

in which case matrix multiplication corresponds to the group operation and

$$H[A_1, \mathbf{a}_1]\,H[A_2, \mathbf{a}_2] = H[(A_1, \mathbf{a}_1) \circ (A_2, \mathbf{a}_2)] = H[(A_1 A_2, A_1 \mathbf{a}_2 + \mathbf{a}_1)].$$

Two special kinds of homogeneous transformations are pure rotations and pure translations along the axes of local coordinate systems:

$$rot(\mathbf{e}_i, \theta) = \begin{pmatrix} R_i(\theta) & \mathbf{0} \\ \mathbf{0}^{\mathrm{T}} & 1 \end{pmatrix} \quad \text{and} \quad trans(\mathbf{e}_i, x) = \begin{pmatrix} \mathrm{I} & x\mathbf{e}_i \\ \mathbf{0}^{\mathrm{T}} & 1 \end{pmatrix}.$$

For "small" rigid-body motions,

$$H \approx \mathrm{I} + \begin{pmatrix} \Omega & \mathbf{v} \\ \mathbf{0}^{\mathrm{T}} & 0 \end{pmatrix} \Delta t$$

where $\Omega = -\Omega^{\mathrm{T}}$. The angular velocity vector can be extracted from the skew-symmetric matrix Ω as $\text{vect}(\Omega) = \boldsymbol{\omega}$ to describe the rotational part of the displacement. Since the second term in the above approximation to H consists mostly of zeros, it is common to extract the information necessary to describe the motion as

$$\begin{pmatrix} \Omega & \mathbf{v} \\ \mathbf{0}^{\mathrm{T}} & 0 \end{pmatrix}^{\vee} = \begin{pmatrix} \boldsymbol{\omega} \\ \mathbf{v} \end{pmatrix}.$$

This six-dimensional vector is called an *infinitesimal* screw motion or *infinitesimal* twist.

3. Methods

We examine 1290 protein structures which have been resolved to 2.0 Å or better and possess less than 20 percent of their sequences in common [Wang and Dunbrack, 2003]. Within these structures there are 12,207 helices (the vast majority of which are alpha-helices) and there are 90,438 helix-helix pairs (many of which do not represent helices in direct contact). We examine the distribution of helix-helix angle over all of these pairs. The full angular distribution is broken down according to inter-helical distance and the number of residues that separate helices from each other in sequence.

The angle distributions examined here are over the range [-180,180] degrees, corresponding to helix axes with directionality. We number residues and helices sequentially in the usual way, starting at the N terminus of each protein. We assign a unit vector along the helix axis (choosing the direction representing an increase in sequence number over the direction representing decrease in sequence number). The angle between two helices is considered to be positive if helix 2 is rotated counterclockwise by an angle between 0 and 180 degrees about the unique line segment pointing from the axis of helix 1 to that of helix 2 and intersecting both axes at right angles. This convention is well known in many fields and does not depend on the numbering of the helices after the direction of the helix unit vector is established. Such a line of closest approach will always exist for infinitely long helix axes. In contrast, alpha-helices have finite length, and so a distinction must be made between the axis of the helix and that finite part of the axis which lies inside the helix. We call the ideal case an "infinite axis" and the actual case a "finite axis."

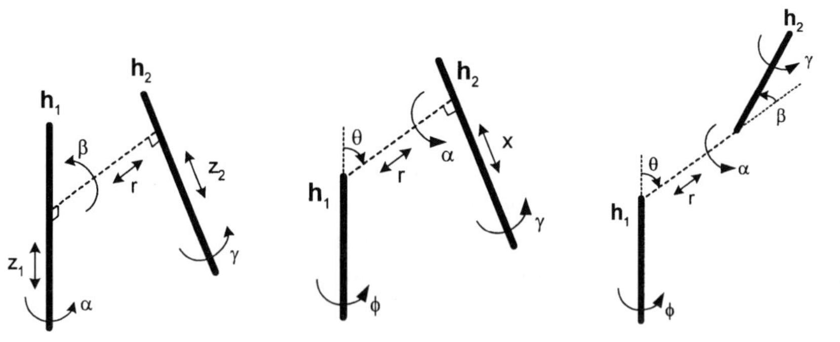

Figure 1 (a) Line to line; (b) Endpoint to line; (c) Endpoint to endpoint

There are three very different scenarios that are possible in the finite case: (1) the line of closest contact between infinite axes intersects both finite axes at right angles in which case a "line on line" contact is made; (2) the line of closest contact between two finite axes meets one helix at its end and the other at a right angle, in which case an "endpoint to line" contact results; (3) both helices interact only at their ends in which case an "end to end" contact results. These case are all displayed in Fig. 1 together with variables which parameterize them.

An implicit assumption in the study of helix-helix interactions is that alpha-helices are essentially rigid objects, and in each of the three cases shown above six parameters describe the spatial relationships between the finite helices (which are represented as line segments). Given a homogeneous transform consisting of motions that are not necessarily small,

$$H(\mathbf{q}) = \begin{pmatrix} A(\mathbf{q}) & \mathbf{a}(\mathbf{q}) \\ \mathbf{0}^{\mathrm{T}} & 0 \end{pmatrix}$$

parameterized with coordinates $(q_1, ..., q_6)$, which we write as a vector $\mathbf{q} \in \mathrm{R}^6$, one can express the homogeneous transform corresponding to a slightly changed set of parameters as the truncated Taylor series

$$H(\mathbf{q} + \delta\mathbf{q}) = H(\mathbf{q}) + \sum_{i=1}^{6} \Delta q_i \frac{\partial H}{\partial q_i}(\mathbf{q}) .$$

This result can be shifted to the identity transformation by multiplying on the left by H^{-1} to define an equivalent relative infinitesimal motion. In this case we write

$$\begin{pmatrix} \boldsymbol{\omega}_R \\ \mathbf{v}_R \end{pmatrix} = \vartheta_R(\mathbf{q})\dot{\mathbf{q}} \quad \text{where} \quad \vartheta_R(\mathbf{q}) = \left[\left(H^{-1} \frac{\partial H}{\partial q_1} \right)^{\vee}, ..., \left(H^{-1} \frac{\partial H}{\partial q_6} \right)^{\vee} \right].$$

Here

$$\mathbf{v}_R = A^{\mathrm{T}} \dot{\mathbf{a}} .$$

The volume element for integrating over rigid-body motions in the coordinates $q_1, ..., q_6$ is:

$$dH = \left| \det \vartheta_R \right| dq_1 ... dq_6 .$$

In cases where the two rigid bodies have symmetries, it makes sense to use a parameterization which captures this fact, and then integrate out all such coordinates. In this way marginal probability densities on a space of reduced-dimension can be examined. Below, the form of the volume elements is given for the three cases shown in Fig. 1.

3.1 Case 1: Line to Line Interaction

In Fig. 1(a), the series of rigid-body motions that result in the frame attached at the base of helix 1 being moved to the base of helix 2 parameterize the homogeneous transformation:

$$H(\alpha, z_1, \beta, r, \gamma, z_2) = rot(\mathbf{e}_3, \alpha)\, trans(\mathbf{e}_3, z_1)\, rot(\mathbf{e}_1, \beta)\, trans(\mathbf{e}_1, r)\, rot(\mathbf{e}_3, \gamma)\, trans(\mathbf{e}_3, z_2)$$

Performing the multiplications above,

$$H = \begin{pmatrix} R_3(\alpha)R_1(\beta)R_3(\gamma) & z_2 R_3(\alpha)R_1(\beta)\mathbf{e}_3 + r\, R_3(\alpha)\mathbf{e}_1 + z_1\, \mathbf{e}_3 \\ \mathbf{0}^{\mathrm{T}} & 1 \end{pmatrix}.$$

Unlike the other two cases below, in this case three variables which appear in the translation part of the homogeneous transformation matrix do not appear in the rotation part. This gives a block structure to the Jacobian matrix, and makes the determinant easy to compute.

In particular, if we group the variables as $\mathbf{q}_1 = (\alpha, \beta, \gamma)$ and $\mathbf{q}_2 = (z_1, r, z_2)$, then in this case the Jacobian will have the form

$$\mathcal{J}_R = \begin{pmatrix} J_R & \mathbf{0}_{3\times 3} \\ A^{\mathrm{T}} \dfrac{\partial \mathbf{a}}{\partial \mathbf{q}_1} & A^{\mathrm{T}} \dfrac{\partial \mathbf{a}}{\partial \mathbf{q}_2} \end{pmatrix}.$$

Here the matrix J_R is [Chirikjian and Kyatkin, 2000]:

$$J_R(A) = \left[\mathrm{vect}\left(A^{\mathrm{T}} \frac{\partial A}{\partial \alpha} \right), \mathrm{vect}\left(A^{\mathrm{T}} \frac{\partial A}{\partial \beta} \right), \mathrm{vect}\left(A^{\mathrm{T}} \frac{\partial A}{\partial \gamma} \right) \right] = \begin{pmatrix} \sin\beta \sin\gamma & \cos\gamma & 0 \\ \sin\beta \cos\gamma & -\sin\gamma & 0 \\ \cos\beta & 0 & 1 \end{pmatrix}.$$

Due to the block lower diagonal form of this matrix, and the fact that A is a rotation matrix (and therefore $\det A = +1$), it is clear that

$$|\det \mathcal{J}_R| = |\det J_R| \left| \det \frac{\partial \mathbf{a}}{\partial \mathbf{q}_2} \right|.$$

From the above expression for H it is clear that

$$\frac{\partial \mathbf{a}}{\partial z_1} = \mathbf{e}_3; \qquad \frac{\partial \mathbf{a}}{\partial r} = R_3(\alpha)\mathbf{e}_1; \qquad \frac{\partial \mathbf{a}}{\partial z_2} = R_3(\alpha)R_1(\beta)\mathbf{e}_3.$$

It is interesting to note that these are the columns of J_R, and since it is known that $|\det J_R| = \sin\beta$, it follows that

$$|\det \mathcal{J}_R| = \sin^2 \beta.$$

3.2 Case 2: End to Line Interaction

Observing Fig. 1(b), it is clear that the rigid-body motion taking frame 1 into frame 2 is of the form:

$$H(\phi,\theta,r,\alpha,x,\gamma) = rot(\mathbf{e}_3,\phi)\, rot(\mathbf{e}_1,\theta)\, trans(\mathbf{e}_3,r)\, rot(\mathbf{e}_3,\alpha)\, trans(\mathbf{e}_1,x)\, rot(\mathbf{e}_1,\gamma)$$

Performing the multiplications above one finds

$$H = \begin{pmatrix} R_3(\phi)R_1(\theta)R_3(\alpha)R_1(\gamma) & r\,R_3(\phi)R_1(\theta)\mathbf{e}_3 + xR_3(\phi)R_1(\theta)R_3(\alpha)\mathbf{e}_1 \\ \mathbf{0}^{\mathrm{T}} & 1 \end{pmatrix}$$

Analytically computing the corresponding Jacobian and taking the determinant, one finds

$$\left|\det \mathcal{G}_R\right| = r\sin\theta\,.$$

3.3 Case 3: End to End Interaction

Looking at Fig. 1(c), the sequence of concatenated rigid-body motions that takes frame 1 to frame 2 is:

$$H(\phi,\theta,r,\alpha,\beta,\gamma) = rot(\mathbf{e}_3,\phi)\, rot(\mathbf{e}_1,\theta)\, trans(\mathbf{e}_3,r)\, rot(\mathbf{e}_3,\alpha)\, rot(\mathbf{e}_1,\beta)\, rot(\mathbf{e}_3,\gamma)\,.$$

Performing these multiplications, one finds that

$$H = \begin{pmatrix} R_3(\phi)R_1(\theta)R_3(\alpha)R_1(\beta)R_3(\gamma) & r\,R_3(\phi)R_1(\theta)\mathbf{e}_3 \\ \mathbf{0}^{\mathrm{T}} & 1 \end{pmatrix}$$

Analytically computing the corresponding Jacobian and taking the determinant, one finds

$$\left|\det \mathcal{G}_R\right| = r^2 \sin\beta \sin\theta\,.$$

4. Results

1,290 protein structures with resolution of 2.0 Å better and sequence identity of less than 20 % were examined. There were 12,207 helices and 90,438 helix pairs. The helix pairs were divided into the three cases, and the parameters describing the spatial relationship between each helix pair were recorded. A contour plot of the relative frequency of occurrence of these parameters was then formed to capture this scattered discrete data.

In these contour plots, the magnitude of each occurrence of a particular spatial relationship between alpha-helices in contact is normalized by the appropriate Jacobian factor calculated in the previous section. These are shown in Fig. 2, 3, and 4 below. In these figures, closer contours (as well as red color) indicates greater preference in parameter values. The fact that "clumps" are observed in all cases indicates that alpha-helices do not pack against each other randomly, but rather there are distinct packing preferences.

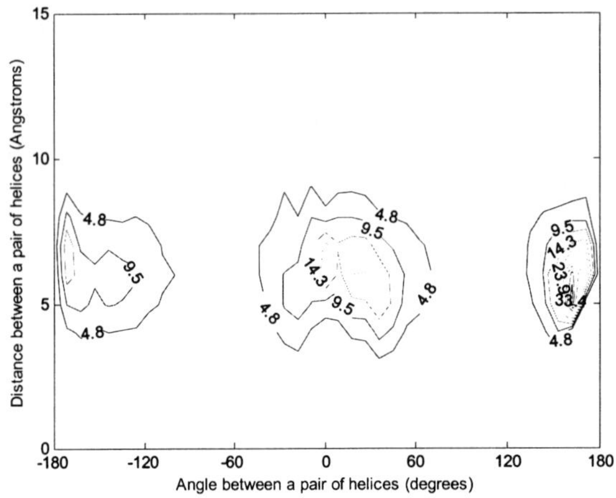

Figure 2. Case 1: Line on line contact (5534 pairs)

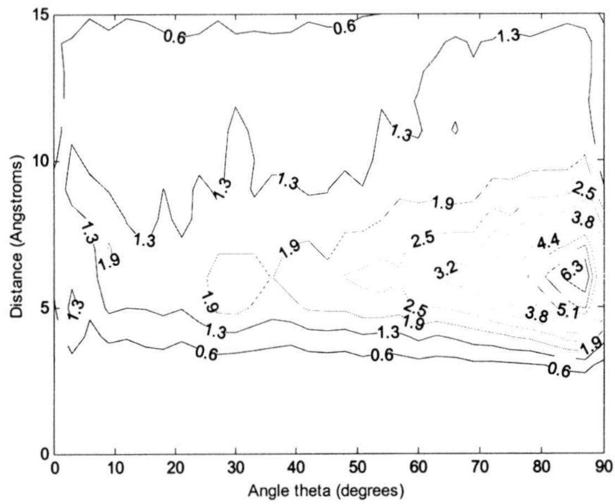

Figure 3. Case 2: End point to line contact (13,983 pairs)

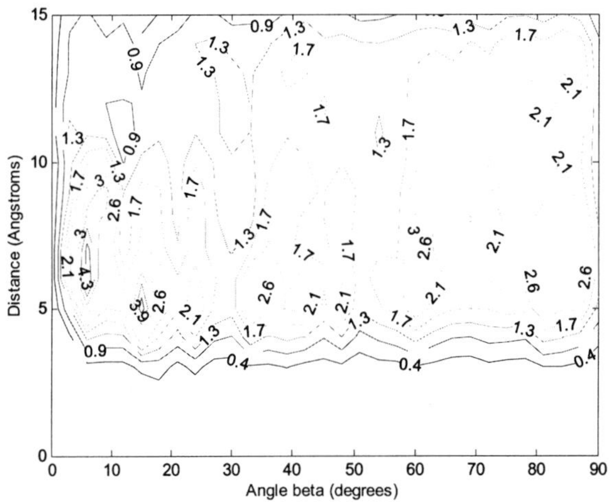

Figure 4. Case 3: End point to end point contact (8847 pairs)

5. Conclusions

Principles of theoretical kinematics have been used to derive the correct normalization factors for statistical analysis of inter-helical angle distributions in proteins. Using these normalization factors, it is clear that there are preferred orientational preferences between helices in contact. We believe that empirically obtained helix-helix potentials obtained from these distributions may be useful for incorporation in protein folding algorithms. This paper represents one of the many problems in structural bioinformatics that can be addressed using techniques from theoretical kinematics.

References

Angeles, J. (1988), *Rational Kinematics*, New York, Springer-Verlag.

Basavaraj, U., and Duffy, J. (1993), End-effector Motion Capabilities of Serial Manipulators, *International Journal of Robotics Research*, no. 2, vol. 12, pp.132-145.

Bottema, O., and Roth, B. (1990), *Theoretical Kinematics*, New York, Dover Publications, Inc.

Chirikjian, G.S., and Kyatkin, A.B. (2000), *Engineering Applications of Noncommutative Harmonic Analysis*, Boca Raton, CRC Press.

Chothia, C., Levitt, M., and Richardson, D. (1977), Structure of proteins: Packing of α-helices and pleated sheets, *Proc. Natl. Acad. Sci. USA*, no. 10, vol. 74, pp.4130-4134.

Chothia, C., Levitt, M., and Richardson, D. (1981), Helix to helix packing in proteins, *J. Mol. Biol.*, vol. 145, pp.215-250.

Crick, F. (1953), The packing of α-helices: simple coiled coils, *Acta Crystallogr.*, vol. 6, pp.689-697.

Finkelstein, A.V., and Ptitsyn, O.B. (1987), Why Do Globular Proteins Fit the Limited Set of Folding Patterns?, *Prog. Biophys. Molec. Biol.*, vol. 50, pp.171-190.

McCarthy, J.M. (1990), *Introduction to Theoretical Kinematics*, Cambridge, MIT Press.

McCarthy, J.M., and Roth, B. (1982), Instantaneous Properties of Trajectories Generated by Planar, Spherical, and Spatial Rigid Body Motions, *Journal of Mechanical Design-Transactions of the ASME*, no. 1, vol. 104, pp.39-51.

Murzin, A.G., and Finkelstein, A.V. (1988), General Architecture of the α-Helical Globule, *J. Mol. Biol.*, vol. 204, pp.749-769.

Nayak, J.H., and Roth, B. (1981), Instantaneous Kinematics of Multiple-Degree-of-Freedom Motions, *Journal of Mechanical Design-Transactions of the ASME*, no. 3, vol. 103, pp.608-620.

Ravani, B., and Roth, B (1984), Mappings of Spatial Kinematics, *Journal of Mechanisms Transmissions and Automation in Design-Transactions of the ASME*, no. 3, vol. 106, pp.341-347.

Richmond, T.J., and Richards, F.M. (1978), Packing of α-helices: geometrical constraints and contact areas, *J. Mol. Biol.*, vol. 119, pp.537-555.

Roth, B. (1967), On Screw Axes and Other Special Lines Associated with Spatial Displacements of a Rigid Body, *Journal of Engineering for Industry*, no. 1, vol. 89, pp.102-?.

Wang, G., and Dunbrack, R.L. (2003), *Culling the PDB by Resolution and Identity*, http://www.fccc.edu/reseach/labs/dunbrack/pisces/culledpdb.html.

Workspace and Isotropy

ANALYSIS OF THE INFLUENCE OF WIRES INTERFERENCE ON THE WORKSPACE OF WIRE ROBOTS

J-P. Merlet
INRIA Sophia-Antipolis, France
Jean-Pierre.Merlet@sophia.inria.fr

Abstract The workspace of wire robots is limited due to three factors: minimal and maximal lengths of the wires, static equilibrium that imposes to have a positive only force in the wires and wires interference (either intersection between 2 wires or between a wire and the end-effector). The limitation by the two first factors may be reduced by having wires with a large coiling ability and by increasing the number of wires. We propose in this paper algorithms that allow to study the influence of wire/wire and wire/end-effector interference on the workspace, assuming a fixed orientation of the end-effector. These algorithms allow one to determine exactly which region of the workspace is interference free.

Keywords: workspace analysis,self-collision,wire robots,parallel robots

1. Introduction

Wire robots are constituted of an end-effector connected to the ground by a set of wires whose lengths can be modified by coiling devices. Modifying the length of the wires allows one to control the pose of the end-effector. Recently wires robots have attracted the attention of many researchers as they have some interesting features:

- *high speed*: the moving parts of wire robots have a very low mass and fast coiling devices are available. Hence high speed wire robots have been developed such as the Falcon robot (Kawamura, 2000)
- *large workspace*: compared to their rigid link equivalent the actuators of wire robots have a larger possible change in lengths which allows to obtain a larger workspace
- *force sensing*: like all parallel structures a carefully instrumented wire robots may be used both as a positioning device and also as a force sensor

Wire robots have also drawbacks:

J. Lenarčič and C. Galletti (eds.), On Advances in Robot Kinematics, 211-218.

- *lower accuracy*: the errors in the positioning of the end-effector is clearly related to the accuracy in the measurement of the wire lengths. Here the larger change in length of the actuators will lead to a larger error in the length measurements for a given accuracy of the sensors. Furthermore the deformation of the wires (which are due to the mass of the wires) are difficult to model: hence their straight-line model is only an approximation
- *control problem*: wires can only pull and are not rigid. Hence control should address the problems of vibrations in the wire together with force control

Wire robots have been used in multiple applications: robotics crane such as the Robocrane (Albus, 1993; Bostelman, 1996) or the device proposed in (Arai, 1992; Higuchi, 1988; Viscomi, 1994), haptic devices (Baumann, 1997), pose measurements (Geng, 1994; Jeong, 1999), medical assistance devices (Homma, 1994; Wendlandt, 1994), sport training (Morizono, 1997), telescope supporting system (Su, 2001).

We have started recently the development of a modular wire robot that will be used in virtual reality application in conjunction with a *workbench* which allows a 3D rendering of large virtual scene. This system uses a magnetic sensing device that measure the position/orientation of the head of the user, an information that is used to update the scene presented on the screen. Although this system is effective it has two main drawbacks: no metallic object should be added in the surrounding of the workbench otherwise a tedious calibration has to be performed and no tactile feedback is provided The use of a wire robot in place of the current sensing system will allow to get a similar pose measurement ability while providing a minimal interference with the screen view and may also provide some force-feedback ability that will be useful in many applications such as medical training. Another specificity of our system is its modularity: we aim to provide modular coiling devices whose location on the ground can be easily adjusted to adapt the wires robot to the task (optimal location of the coiling devices will be determined using algorithms such as the one described in (Merlet, 2003)).

Previous works on wires robot have focused on kinematics (Maier, 1998; Yamamoto, 2000), planar robots (Ming, 1994-1; Ming, 1994-2; Osumi, 2000) and force controllable workspace (Verhoeven, 1998; Verhoeven, 2000). Clearly modularity should address first the workspace problem. Three elements play a role on the limitation of the workspace of wires robots:

- *limitation due to the wire lengths*: here there is no differences between a wires robot and a classical parallel robot and this subject is now mastered using for example the algorithms described in (Gosselin, 1990) when the orientation is constant or in (Merlet, 1999) in the general case

- *limitation due to the force in the wires*: forces in the wires should always be positive
- *limitation due to the interference between wires and between the wires and the end-effector*

The purpose of this paper is to address the last point which, to the best of the author knowledge, has never been addressed for wires robot.

2. Wires interference

In this paper we consider a wires robot with n wires whose coiling devices are located at point A_i on the ground and at point B_i on the end-effector. A reference frame is attached to the ground so that the coordinates of A_i are xa_i, ya_i, za_i and a mobile frame is attached to the end-effector so that the coordinates of B_i are xb_i, yb_i, zb_i. The location of the end-effector is defined by the coordinates X, Y, Z in the reference frame of a specific point C of the end-effector while its orientation is defined by the rotation matrix R that allows to express in the reference frame the components of a vector whose components in the mobile frame are known.

Two wires i, j will interfere if they have a common point M such that $\mathbf{A_iM} = l_i\mathbf{A_iB_i}$ and $\mathbf{A_jM} = l_j\mathbf{A_jB_j}$ with l_i, l_j in [0,1]. We get

$$\mathbf{A_iA_j} = \mathbf{A_iM} - \mathbf{A_jM} = l_i\mathbf{A_iB_i} - l_j\mathbf{A_jB_j} \tag{1}$$

Equation (1) is a system of 3 equations which are linear in the unknowns l_i, l_j. Two equations of this system are chosen to determine l_i, l_j and using these values in the third equation allows one to get an interference condition that is free of l_i, l_j. Geometrically speaking this interference condition is equivalent to state that the points A_i, A_j, B_i, B_j *lie in the same plane.*

3. Interference for a fixed orientation of the end-effector

3.1 Interference condition and constraints planes

When the orientation is fixed the interference condition implies that C lie in a plane $\mathcal{P}$. An interesting property of the interference condition is that when two wires i, j have the same z coordinates and the orientation of the end-effector is such that the z components Bz_i, Bz_j of the vectors $\mathbf{CB_i}, \mathbf{CB_j}$ are identical, then the interference condition is written as:

$$U\ (Z - Bz_i - za_i) = 0 \tag{2}$$

where U is not a function of X, Y, Z and cancel only if B_iB_j is parallel to A_iA_j. Hence interference may occur only if B_iB_j is parallel to A_iA_j or if the points B_i, B_j lie in the same horizontal plane than A_i, A_j.

The values of l_i, l_j may be written as:

$$l_i = \frac{N(l_i)}{D(l_i)} \quad l_j = \frac{N(l_j)}{D(l_j)}$$

The region of the workspace in which interference will occur is bounded by geometrical elements for which l_i or l_j are exactly 0 or 1. Hence it is necessary to investigate the constraint conditions $N(l_i) = 0, D(l_i) = 0, N(l_i) = D(l_i)$ and $N(l_j) = 0, D(l_j) = 0, N(l_j) = D(l_j)$. All these constraint conditions are also equations of planes for C that will be called the *constraints planes*. Note that we have $D(l_i) = D(l_j)$ and hence only 5 constraints planes that will be denoted $\mathcal{C}^{N(l_i)=0}$, $\mathcal{C}^{N(l_j)=0}$, $\mathcal{C}^{D(l_i)=0}$, $\mathcal{C}^{N(l_i)=D(l_i)}$ and $\mathcal{C}^{N(l_j)=D(l_j)}$. Interference between the wires will thus happen when C belong to a region of $\mathcal{P}$ that is bounded by the intersection of $\mathcal{P}$ with the constraints planes.

The constraints planes have interesting properties: the planes obtained for $N(l_i) = 0$ and $N(l_j) = 0$ are parallel as the coefficients in X, Y, Z of these equations are identical and similarly the planes obtained for $N(l_i) = D(l_i)$ and $N(l_j) = D(l_j)$ are parallel.

A direct consequence is that one of the half space bounded by $N(l_i) = 0$ or $N(l_j) = 0$ will be such that $N(l_i) \geq (\leq)0$ **and** $N(l_j) \geq (\leq)0$. Similarly one of the half space bounded by $N(l_i) = D(l_i)$ or $N(l_j) = D(l_j)$ will be such that $N(l_i) \leq D(l_i)$ **and** $N(l_j) \leq D(l_j)$. Combined with the constraints on $D(l_i)$ this will allow us to determine elements on which either l_i or l_j is exactly 0 (1) while on one side of the element we will have $l_i(l_j) > 0(1)$ and $l_i(l_j) < 0(1)$ on the other side.

Hence only three constraints planes have to be calculated: the planes separating the half-space where $N(l_i) \geq (\leq)0$ **and** $N(l_j) \geq (\leq)0$ and $N(l_i) \leq D(l_i)$ **and** $N(l_j) \leq D(l_j)$ together with the plane $D(l_i) = 0$. These three constraints plane will be called the *separating planes* $\mathcal{S}_1, \mathcal{S}_2, \mathcal{S}_3$. Determining the two first separating planes among the five constraints planes is easy. We select an arbitrary point on $\mathcal{C}^{N(l_i)=0}$ and we calculate $N(l_j)$ for this point: if $N(l_j) \geq 0$ then we get $\mathcal{S}_1 = \mathcal{C}^{N(l_i)=0}$ otherwise $\mathcal{S}_1 = \mathcal{C}^{N(l_j)=0}$ and a similar procedure is used for determining $\mathcal{S}_2$ among $\mathcal{C}^{N(l_i)=D(l_i)}$ and $\mathcal{C}^{N(l_j)=D(l_j)}$.

3.2 Algorithm for calculating the interference region

The *interference region* for the wires i, j is the region of $\mathcal{P}$ for which each point satisfies $l_i, l_j \in [0,1]$. This region can be calculated as follows:

1 calculate the intersecting lines D_1, D_2, D_3 between $\mathcal{P}$ and $\mathcal{S}_1$, $\mathcal{S}_2$ and $\mathcal{S}_3$

2 compute the intersection points between D_1, D_2, D_3

3 for each D_i:

(a) order the set $\{S_0^i, \ldots, S_k^i\}$ of $k+1$ intersection points that belong to D_i along this line

(b) for each segment $S_l^i S_{l+1}^i$, $l \in [0, k-1]$ takes the mid-point M of the segment and then 2 points on each side of the segment close to M. Calculate the value of l_i, l_j for these two points. If for one point we have $0 \leq l_i \leq 1, 0 \leq l_j \leq 1$ while for the other point one of these constraints is violated, then the segment is part of the border of the interference region

4 connect all the parts of the segments that has been determined as part of the border of the interference region

A similar algorithm may be derived for computing the intersection of the interference region with a given plane (for example for a given Z) that are called the *interference lines.* Both algorithms have been implemented in Maple and their computation time is less than 10 s. Consider for example the 6-wired robot whose A_i, B_i are:

A_1	0 200 150	A_2	173.205081 -100 150
A_3	-173.205081 -100 150	A_4	-173.205081 100 -150
A_5	173.205081 100 -150	A_6	0 -200 -150
B_1	-25.980762 15 0	B_2	25.980762 15 0
B_3	0 -30 0	B_4	-25.980762 -15 0
B_5	0 30 0	B_6	25.980762 -15 0

with a minimal wire length of 100 and a maximal wire length of 500. Figure 1 present a cross-section of the workspace for $Z = 0$ and an orientation of the end-effector defined by the Euler angles $\psi = 70°$, $\theta = 30°$, $\phi = 0°$ as derived from the limitation of the wires lengths together with the interference lines. It may be seen that the wires interference will play a role for trajectory planning as some interferences lines are inside the potential workspace of the wires robot.

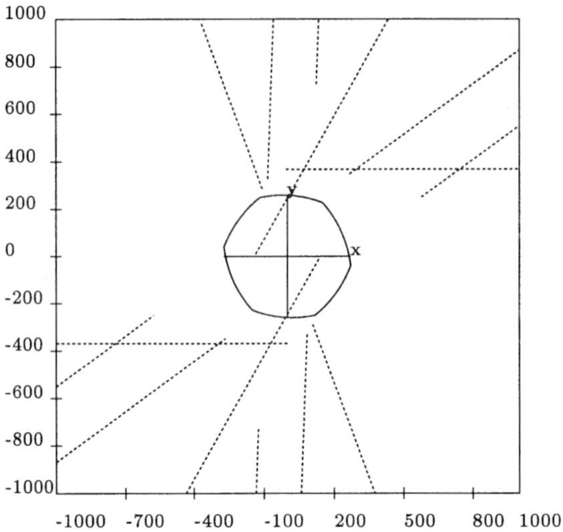

Figure 1. Cross-section of the workspace at $Z = 0$ for $\psi = 70$ degree, $\theta = 30$ degree and $\phi = 0$. The interference lines for all pair of wires are represented in dashed lines

4. Interference between the end-effector and the wires

Another factor that may limit the workspace is the interference between the wires and the end-effector. We will assume here that the end-effector is a polyhedral object with N planar facets $\mathcal{F}_i$. Each facet i is defined in the mobile frame by a set of m^i ordered vertices $\{V_1^i, \ldots, V_m^i\}$ and the exterior of the end-effector is defined by a unit vector $\mathbf{n}^i$ that is perpendicular to the facet plane. Interference will occur when a wire cross the interior of the end-effector. The purpose of this section is to determine the region of the workspace, called the *end-effector interference region*, for which such interference will occur, assuming that the orientation of the end-effector is fixed.

The border of the end-effector interference region will be obtained when one wire lie exactly in a facet plane and has an intersection with a facet edge. A necessary condition for a wire j to be in the i-th facet plane is:

$$\mathbf{n^i}.\mathbf{A_jB_j} = 0 \tag{3}$$

Wire j will have an intersection with a facet edge $V_k^i V_{k+1}^i$ if there is a point M that is common to A_jB_j and $V_k^i V_{k+1}^i$. We have

$$\mathbf{A_jV_k^i} = \mathbf{A_jM} - \mathbf{V_k^iM} = l_l \mathbf{A_jB_j} - l_m \mathbf{V_k^iV_{k+1}^i} \tag{4}$$

in which it is necessary to have l_l, l_m in the range [0,1]. A point of the border of the end-effector interference region will be such that either l_l or l_m is exactly 0 or 1. Equation (4) is a system of 3 linear equations in l_l, l_m: two equations are used to determine l_l, l_m and we get a constraint equation with the third one. This equation is linear in X, Y, Z as are the constraint equation (3), the numerators and the denominator of l_l, l_m.

The border of the end-effector interference region is hence obtained by dealing with constraint planes as in the previous section: the same algorithm may thus be used. For each facet and each edge we have 4 constraints planes (the 2 numerators and the denominator of l_l, l_m and the constraints plane issued from equation (4) and for each facet we have another constraints plane issued from equation (3). Hence we have to deal with a total of $n \sum_{k=1}^{k=N}(4m^k + 1)$ constraints planes.

5. Conclusion

Limitation to the workspace due to interference is a problem that has rarely been addressed. In this paper we have proposed algorithms that allows one to determine exactly the location of the end-effector (its orientation being fixed) where interference between 2 wires or one wire and a polyhedral end-effector will occur. We have shown on an example that indeed interference is a factor that may limit the workspace of wires robots.

References

Albus J., Bostelman R., and Dagalakis N. The NIST ROBOCRANE. *J. of Robotic Systems*, 10(5):709–724, July 1993.

Arai T. and others . A hybrid drive parallel robot for heavy material handling. *IEEE Robotics and Automation Magazine*, 9(1):45–54, March 2002.

Baumann R., Maeder W., Glauser D., and Clavel R. The Pantoscope: a spherical remote-center-of-motion parallel manipulators for force reflection. In *IEEE Int. Conf. on Robotics and Automation*, pages 718–723, Albuquerque, April, 21-28, 1997.

Bostelman R., Albus J., Dagalakis N., and Jacoff A. RoboCrane project: an advanced concept for large scale manufacturing. In *Proceedings Association for Unmanned vehicle systems International*, pages 509–521, Orlando, July, 15-19, 1996.

Geng Z. and Haynes L.S. A 3-2-1 kinematic configuration of a Stewart platform and its application to six degree of freedom pose measurements. *Robotics & Computer-Integrated Manufacturing*, 11(1):23–34, March 1994.

Gosselin C. Determination of the workspace of 6-dof parallel manipulators. *ASME J. of Mechanical Design*, 112(3):331–336, September 1990.

Higuchi T., Ming A., and Jiang-Yu J. Application of multi-dimensional wire crane in construction. In *5th Int. Symp. on Robotics in Construction*, pages 661–668, Tokyo, June, 6-8, 1988.

Homma K. and Arai T. Upper limb motion assist system with parallel mechanisms. In *2nd Japan-France Congress on Mechatronics*, pages 388–391, Takamatsu, November, 1-3, 1994.

Jeong J.W., Kim S.H., and Kwak Y.K. Kinematics and workspace analysis of a parallel wire mechanism for measuring a robot pose. *Mechanism and Machine Theory*, 34(6):825–841, August 1999.

Kawamura S. and others . High-speed manipulation by using parallel wire-driven robots. *Robotica*, 18(1):13–21, 2000.

Maier T. and Woernle C. Inverse kinematics for an underconstrained cable suspension manipulator. In *ARK*, pages 97–104, Strobl, June 29- July 4, 1998.

Merlet J-P. Determination of the optimal geometry of modular parallel robots. In *IEEE Int. Conf. on Robotics and Automation*, Taipei, September, 14-19, 2003.

Merlet J-P. Determination of 6d workspaces of Gough-type parallel manipulator and comparison between different geometries. *Int. J. of Robotics Research*, 18(9):902–916, October 1999.

Ming A. and Higuchi T. Study on multiple degree of freedom positioning mechanisms using wires, Part 1, Concept, Design and Control. *Int. J. Japan Soc. Prec. Eng.*, 28(2):131–138, June 1994.

Ming A. and Higuchi T. Study on multiple degree of freedom positioning mechanisms using wires, Part 2, Development of a planar completely restrained positioning mechanism. *Int. J. Japan Soc. Prec. Eng.*, 28(3):235–242, September 1994.

Morizono T., Kurahashi K., and Kawamura S. Realization of a virtual sports training system with parallel wire mechanism. In *IEEE Int. Conf. on Robotics and Automation*, pages 3025–3030, Albuquerque, April, 21-28, 1997.

Osumi H. and others . Development of a manipulator suspended to parallel wire structure. In *IEEE Int. Workshop on Intelligent Robots and Systems (IROS)*, Takamatsu, Japan, October 30- November 5, 2000.

Su Y.X. and others . Development of a large parallel-cable manipulator for the feed-supporting system of a next-generation large radio telescope. *J. of Robotic Systems*, 18(11):633–643, 2001.

Verhoeven R., Hiller M., and Tadoroko S. Workspace, stiffness, singularities and classification of tendon driven Stewart platforms. In *ARK*, pages 105–114, Strobl, June 29- July 4, 1998.

Verhoeven R. and Miller M. Estimating the controllable workspace of tendon-based Stewart platforms. In *ARK*, pages 277–284, Piran, June, 25-29, 2000.

Viscomi B.V., Michalerya W.D., and Lu L-W. Automated construction in the ATLSS integrated building systems. *Automation in Construction*, 3(1):35–43, May 1994.

Wendlandt J.M. and Sastry S.S. Design and control of a simplified Stewart platform for endoscopy. In *33nd Conf. on Decision and Control*, pages 357–362, Lake Buena Vista, December, 14-16, 1994.

Yamamoto M. and Mohri A. Inverse kinematic analysis for imcompletely restrained parallel wire mechanisms. In *IEEE Int. Workshop on Intelligent Robots and Systems (IROS)*, Takamatsu, October 30- November 5, 2000.

WORKSPACES CLASSIFICATION OF 3R ORTHOGONAL MANIPULATORS

Philippe Wenger, Maher Baili and Damien Chablat
Institut de Recherche en Communications et Cybernétique de Nantes, UMR CNRS 6597
1 rue de la Noë, BP 92101, 44321 Nantes Cedex 3, France
Philippe.Wenger@irccyn.ec-nantes.fr

Abstract A classification of a family of 3-revolute (3R) positioning manipulators is established. This classification is based on the topology of their workspace. The workspace is characterized in a half-cross section by the singular curves of the manipulator. The workspace topology is defined by the number of cusps and nodes that appear on these singular curves. The design parameters space is shown to be partitioned into nine subspaces of distinct workspace topologies. Each separating surface is given as an explicit expression in the DH-parameters.

Keywords: Classification, Workspace, Singularity, Cusp, Node, Orthogonal Manipulator, Computer Algebra.

1. Introduction

This paper focuses on positioning 3R manipulators with orthogonal joint axes, referred to as *orthogonal manipulators*. Orthogonal manipulators may have different global kinematic properties according to their link lengths and joint offsets. Unlike usual industrial manipulators, orthogonal manipulators may be cuspidal, that is, they can change their posture without meeting a singularity (Parenti and Innocenti, 1988, Burdick, 1988). This property was unknown before 1988 (Borrel and Liégeois, 1988). Several years later, some conditions for a manipulator to be noncuspidal were provided, which include simplifying geometric conditions like parallel and intersecting joint axes (Burdick, 1995) and also nonintuitive conditions (Wenger, 1997). A general necessary and sufficient condition for a 3-DOF manipulator to be cuspidal was established in (El Omri and Wenger, 1995), namely, the existence of at least one point in the workspace where the inverse kinematics admits three equal solutions. The word "cuspidal manipulator" was defined in accordance to this condition because a point with three equal IKS forms a cusp in a cross section of the workspace (Arnold, 1981). The categorization of all generic 3R manipulators was established in (Wenger, 1998) based on the homotopy class of the

J. Lenarčič and C. Galletti (eds.), On Advances in Robot Kinematics, 219-228.

singular curves in the joint space. Wenger, 1999 proposed a procedure to take into account the cuspidality property in the design process of new manipulators. More recently, Corvez and Rouillier, 2002 applied efficient algebraic tools to the classification of 3R orthogonal manipulators with no offset on their last joint. Five surfaces were found to divide the parameters space into 105 cells with the same number of cusps in the workspace. The equations of these five surfaces were derived as polynomials in the DH-parameters using Groebner Bases. A kinematic interpretation of this theoretical work showed that, in fact, only five different domains exist : two domains of noncuspidal manipulators, one domain where manipulators have two cusps and two domains where they have four cusps (Baili et al, 2003). However, the authors did not provide the equations of the true separating surfaces in the parameters space. On the other hand, they did not take into account the occurrence of nodes, which play an important role for analyzing the number of IKS in the workspace.

The purpose of this work is to classify a family of 3R positioning manipulators according to the topology of their workspace, which is defined by the number of cusps and nodes that appear on the singular curves. The design parameters space is shown to be divided into nine subspaces of distinct workspace topologies. In each subspace, the distribution of the number of IKS is the same. This study is of interest for the design of new manipulators.

The rest of this article is organized as follows. Next section presents the manipulators under study and recalls some preliminary results. The classification is established in section 3. Section 4 synthesizes the results and section 5 concludes this paper.

2. Preliminaries

2.1 Manipulators under study

The manipulators studied in this paper are orthogonal with their last joint offset equal to zero. The remaining lengths parameters are referred to as d_2, d_3, d_4, and r_2. The angle parameters α_2 and α_3 are set to –90° and 90°, respectively. The joint variables θ_1, θ_2 and θ_3 will be assumed unlimited. Figure 1 shows the kinematic architecture of the manipulators under study in the zero configuration. The position of the end-tip (or wrist center) is defined by the three Cartesian coordinates x, y and z of the operation point P with respect to a reference frame (O, **x**, **y**, **z**) attached to the manipulator base as shown in Fig. 1.

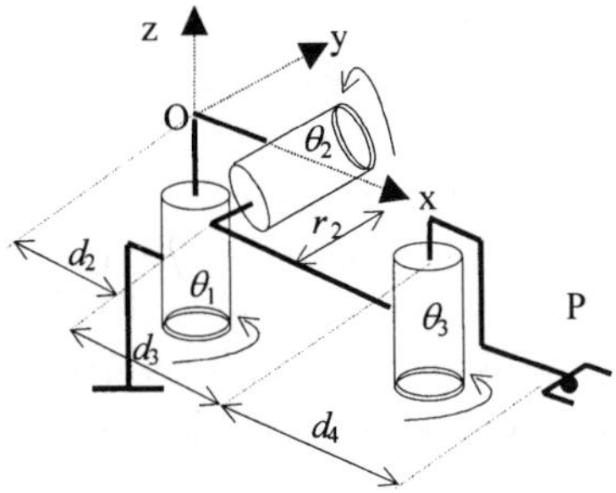

Figure 1. Orthogonal manipulators under study.

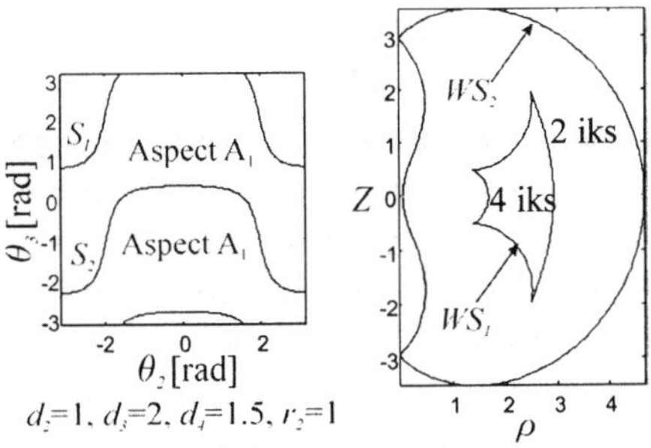

Figure 2. Singularity curves in joint space (left) and workspace (right)

2.2 Singularities and aspects

The determinant of the Jacobian matrix of the orthogonal manipulators under study is det(**J**) = $(d_3 + c_3d_4)(s_3d_2 + c_2(s_3d_3 - c_3r_2))$, where c_i=cos(θ_i) and s_i=sin(θ_i). A singularity occurs when det(**J**)=0. The contour plot of det(**J**)=0 forms two or four separate curves in $-\pi \leq \theta_2 < \pi, -\pi \leq \theta_3 < \pi$, according to whether $d_3>d_4$ or not.

If $d_3>d_4$, the first factor of det(**J**) cannot vanish and the singularities form two distinct curves S_1 and S_2 in the joint space (El Omri, 1996), which divide the joint space into two singularity-free open sets A_1 and A_2 called *aspects* (Borrel and Liégeois, 1988). The singularities can also be displayed in the Cartesian space (Kholi and Hsu, 1987, Ceccarelli, 1996). Thanks to their symmetry about the first joint axis, a 2-dimensional representation in a half cross-section of the workspace is sufficient. The singularities form two disjoint sets of curves in the workspace. These two sets define the internal boundary WS_1 and the external boundary WS_2, respectively, with WS_1=f(S_1) and WS_2=f(S_2). Figure 2 shows the singularity curves when d_2=1, d_3=2, d_4=1.5 and r_2=1. This manipulator has four cusp points on its the internal boundary WS_1 and no node. WS_1 divides the workspace into one region with two IKS (the outer region) and one region with four IKS (the inner region).

If $d_3 \leq d_4$, the operation point can meet the second joint axis whenever θ_3=±arccos($-d_3/d_4$) and two horizontal lines appear in the joint space $-\pi \leq \theta_2 < \pi, -\pi \leq \theta_3 < \pi$, which may intersect S_1 and S_2 depending on d_2, d_3, d_4 and r_2 (El Omri, 1996). The number of aspects depends on these intersections. Note that if $d_3<d_4$, no additional curve appears in the workspace cross-section but only two points. This is because, since the operation point meets the second joint axis when θ_3=±arccos($-d_3/d_4$), the location of the operation point does not change when θ_2 is rotated.

3. Workspaces classification

3.1 Classification criteria

The classification is conducted on the basis of the topology of the singular curves in the workspace, which we characterize by (i) the number of cusps and (ii) the number of nodes. A cusp (resp. a node) is associated with one point with three equal IKS (resp. with two pairs of equal IKS). These singular points are interesting features for characterizing the workspace shape, the distribution of the number of IKS and the existence of holes in the workspace (Ceccarelli, 1996).

3.2 Number of cusps

For now on and without loss of generality, d_2 is set to 1. Thus, only three parameters d_3, d_4 and r_2 need to be handled. Baili et al, 2003 showed that one or more surfaces among the five ones found by Corvez and Rouillier, 2002, are not relevant. However, they did not try to find which surfaces are really separating. To derive the equations of the separating surfaces, we investigate the transitions between the five domains found in Baili et al, 2003. First, we recall the 5 different manipulator types associated with these 5 domains.

The first type is a binary manipulator (i.e. it has only two IKS), with a hole (Fig. 3). All other types are quaternary (i.e. with four IKS).

The second type is a quaternary manipulator (like the remaining three types) with four cusps on the internal boundary. There are three instances of type 2 according to the number of nodes, as will be shown in section 3.3. The first one is shown in Fig. 4 with two nodes. The second one was shown in Fig. 2 with no node. The last one is such that $d_3 \leq d_4$ i.e. with the two singular lines $\theta_3 = \pm\arccos(-d_3/d_4)$. It will be shown in Fig. 10 (section 3.3).

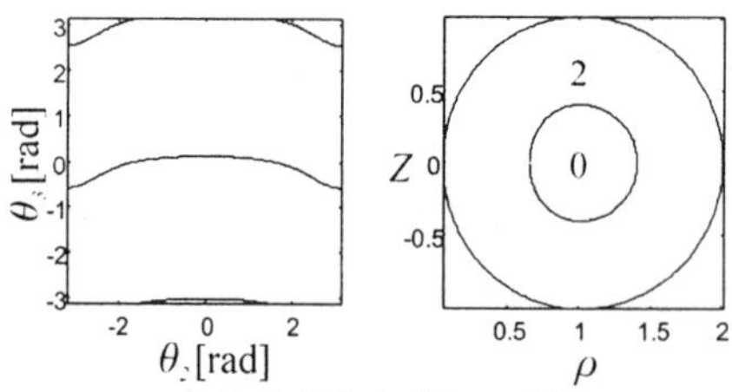

$d_2=1,\ d_3=0.7,\ d_4=0.3,\ r_2=0.2$

Figure 3. Type 1 manipulator.

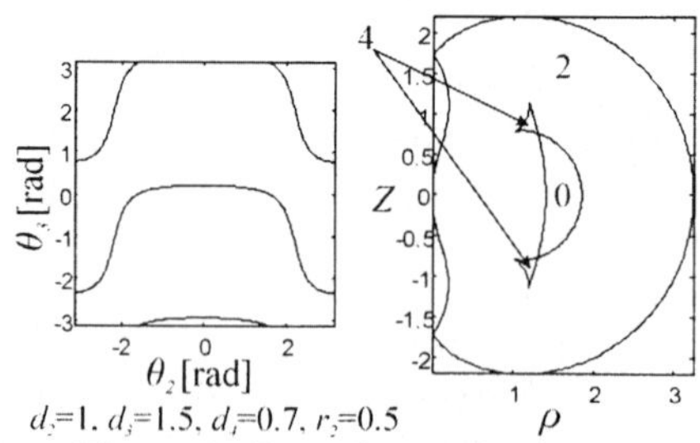

$d_2=1,\ d_3=1.5,\ d_4=0.7,\ r_2=0.5$

Figure 4. Type 2 manipulator.

Transition between type 1 and type 2 is a manipulator having a pair of points with four equal IKS, where two nodes and one cusp coincide. Proof of this result can be obtained as a straightforward consequence of transitions in quartics root patterns (Baili, 2003). Deriving the condition for the inverse kinematic polynomial to have four equal roots yields the equation of the separating surface (Baili, 2003)

$$d_4 = \sqrt{\frac{1}{2}\left(d_3^{\,2} + r_2^{\,2} - \frac{(d_3^{\,2} + r_2^{\,2})^2 - d_3^{\,2} + r_2^{\,2}}{AB}\right)} \tag{1}$$

where $A = \sqrt{(d_3+1)^2 + r_2^{\,2}}$ and $B = \sqrt{(d_3-1)^2 + r_2^{\,2}}$.

The third type is a manipulator with only two cusps on the internal boundary, which looks like a fish with one tail (Fig. 5).

As shown by Baili, 2003, transition between type 2 and type 3 is characterized by a manipulator for which the singular line given by θ_3=–arccos(-d_3/d_4) is tangent to the singularity curve S_1. Expressing this condition yields the equation of the separating surface

$$d_4 = \frac{d_3}{1+d_3}\cdot A \tag{2}$$

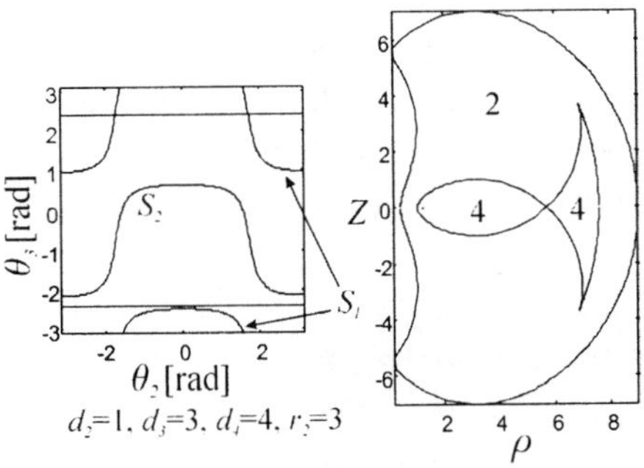

Figure 5. Manipulator of type 3.

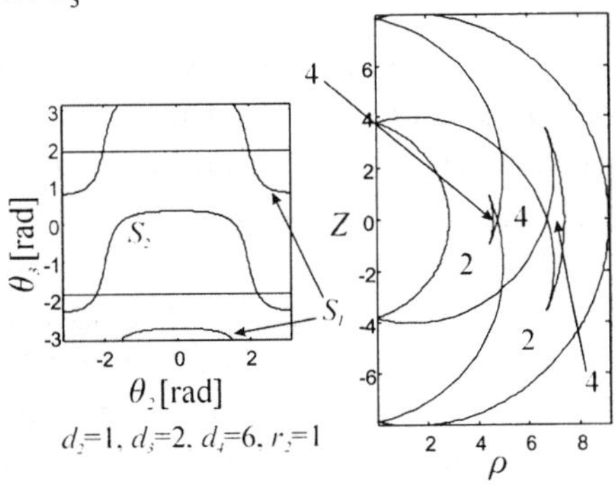

Figure 6. Manipulator of type 4.

The fourth type is a manipulator with four cusps. Unlike type 2, the cusps are not located on the same boundary (Fig. 6).

Transition between type 3 and type 4 is characterized by a manipulator for which the singular line given by θ_3=–arccos(-d_3/d_4) is tangent to the singularity curve S_2 (Baili, 2003). Expressing this condition yields the equation of the separating surface:

$$d_4 = \frac{d_3}{d_3-1}\cdot B \ \text{ and } \ d_3 > 1 \tag{3}$$

Last type is a manipulator with no cusp (Fig. 7). Unlike type 1, the internal boundary does not bound a hole but a region with 4 IKS. The two isolated singular points inside the inner region are associated with the two singularity lines. Transition between type 4 and type 5 is characterized by a manipulator for which the singular line given by θ_3=+arccos(-d_3/d_4) is tangent to the singularity curve S_1 (Baili, 2003).

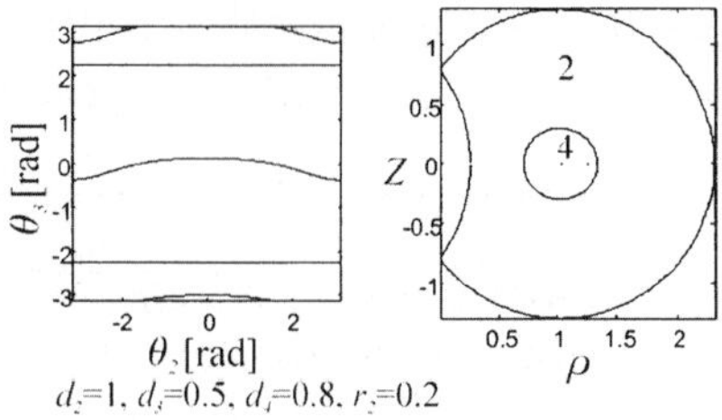

Figure 7. Manipulator of type 5.

Expressing this condition yields the equation of the separating surface:

$$d_4 = \frac{d_3}{1-d_3} \cdot B \text{ and } d_3 < 1 \tag{4}$$

We have provided the equations of four surfaces that divide the parameters space into five domains where the number of cusps is constant. Figure 8 shows the plots of these surfaces in a section (d_3, d_4) of the parameter space for r_2=1. Domains 1, 2, 3, 4 and 5 are associated with manipulators of type 1, 2, 3, 4 and 5, respectively. C_1, C_2, C_3 and C_4 are the right hand side of (1), (2), (3) and (4), respectively.

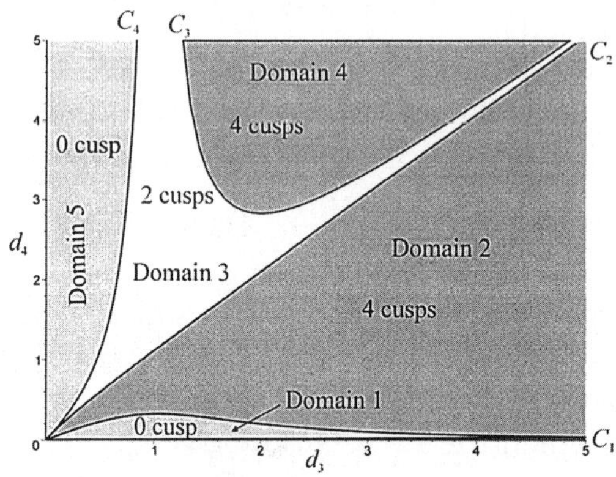

Figure 8. Plots of the four separating surfaces in a section (d_3, d_4) of the parameter space for r_2=1.

It is interesting to see the correspondence between the equations found with pure algebraic reasoning in (Corvez and Rouillier, 2002) and those provided in this paper. The five equations found by Corvez and Rouillier are:

$$-d_3 + d_4 r_2^2 + d_4 = 0 \tag{5}$$

$$d_3^2 - d_4^2 + r_2^2 = 0 \tag{6}$$

$$d_4^2d_3^6 - d_4^4d_3^4 + 3d_4^2d_3^4r_2^2 - 2d_4^2d_3^4 + 2d_4^4d_3^2 - 2d_4^4d_3^2r_2^2 + d_4^2d_3^2 + 3d_4^2d_3^2r_2^4$$
$$-d_3^2r_2^2 - 2d_4^4r_2^2 - d_4^4r_2^4 - d_4^4 + d_4^2r_2^6 + d_4^2r_2^2 + 2d_4^2r_2^4 = 0 \quad (7)$$

$$d_3^2r_2^2 + d_3^2 - 2d_3^3 + d_3^4 - d_4^2 + 2d_3d_4^2 - d_3^2d_4^2 = 0 \quad (8)$$

$$d_3^2r_2^2 + d_3^2 + 2d_3^3 + d_3^4 - d_4^2 - 2d_3d_4^2 - d_3^2d_4^2 = 0 \quad (9)$$

Equation (7) is a second-degree polynomial in d_4^2. Solving this quadratics for d_4 shows that (7) has the following two branches:

$$d_4 = \sqrt{\frac{1}{2}\left(d_3^2 + r_2^2 - \frac{(d_3^2+r_2^2)^2 - d_3^2 + r_2^2}{AB}\right)} \text{ or } d_4 = \sqrt{\frac{1}{2}\left(d_3^2 + r_2^2 + \frac{(d_3^2+r_2^2)^2 - d_3^2 + r_2^2}{AB}\right)}$$

The first branch is the separating surface $d_4=C_1$ between domains 1 and 2.

Equation (8) is a second-degree polynomial in d_4. Solving this quadratics for d_4 and assuming strictly positive values for d_4 and r_2 yields the following two branches for (8):

$$(d_4 = \frac{d_3}{d_3 - 1} \cdot B \text{ and } d_3 > 1) \text{ or } (d_4 = \frac{d_3}{1 - d_3} \cdot B \text{ and } d_3 < 1)$$

These two branches are the separating surfaces $d_4=C_3$ and $d_4=C_4$, respectively. Also, (9) can be shown to be rewritten as $d_4=C_2$.

In conclusion, (5) and (6) do not define separating surfaces and only one branch of (7) defines a separating surface.

3.3 Number of nodes

In this section, we investigate each domain according to the number of nodes in the workspace.

Domain 1: Since all manipulators in this domain are binary, they cannot have any node in their workspace. Thus, all manipulators in domain 1 have the same workspace topology, namely, 0 node, 0 cusp and a hole inside their workspace. This workspace topology is referred to as WT_1 (Workspace Topology #1).

Domain 2: Figures 4 and 2 show two distinct workspace topologies of manipulators in domain 2, which feature 2 nodes and 0 node and which we call WT_2 and WT_3, respectively. Transition between these two workspace topologies is one such that the two lateral segments of the internal boundary meet tangentially (Fig. 9). Equation of this transition can be derived geometrically and the following equation is found (Baili, 2003):

$$d_4 = (A - B)/2 \quad (10)$$

As noted in section 3.2, a third topology exists in this domain, where the internal boundary exhibits a '2-tail fish'.

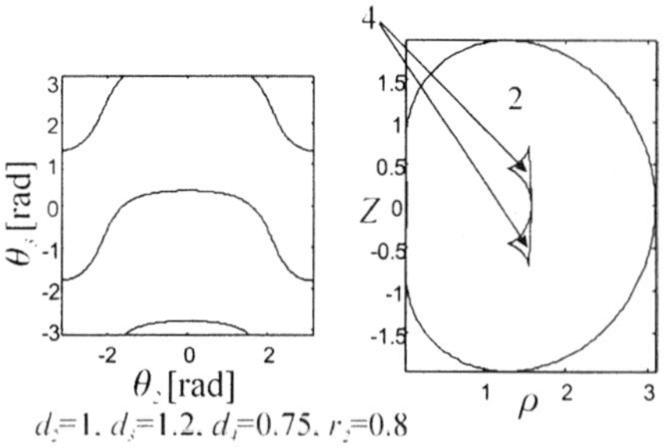

Figure 9. Transition WT_2-WT_3.

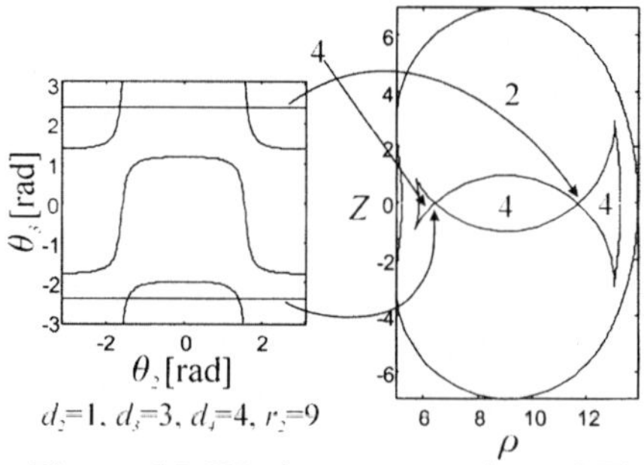

Figure 10. Workspace topology WT_4.

This workspace topology, which we call WT_4, features two nodes like in Fig. 4, but these nodes do not play the same role. They coincide with two isolated singular points, which are associated with the two singularity lines defined by $\theta_3 = \pm\arccos(-d_3/d_4)$ (the operation point lies on the second joint axis and the inverse kinematics admits infinitely many solutions). Also, the nodes do not bound a hole like in Fig. 4 but a region with four IKS (Fig. 10).

Transition between WT_3 and WT_4 is a workspace topology such that the upper and lower segments of the internal boundary meet tangentially (Fig. 11).

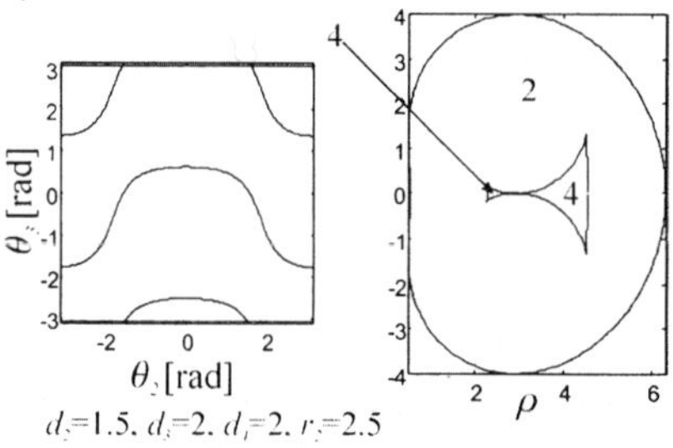

Figure 11. Transition WT_3-WT_4.

As shown in (Baili, 2003), this transition is the occurrence of the additional singularity $d_3 + c_3 d_4 = 0$. This transition is defined by:

$$d_4 = d_3 \tag{11}$$

Domains 3 and 5: The internal boundary has either 2 cusps (domain 3) or 0 cusp (domain 5). This boundary is either fully inside the external boundary (like in Figs 5 and 7), or it can cross the external boundary, yielding two nodes like in Fig. 12. Thus, domain 3 (resp. domain 5) contains two distinct workspace topologies, which we call WT_5 (2 nodes) and WT_6 (resp. WT_8 and WT_9).

Transition between WT_5 and WT_6 and transition between WT_8 and WT_9 are such that the internal boundary meets the external boundary tangentially (Fig. 13). This transition can be derived geometrically and the following equation is found (Baili, 2003):

$$d_4 = \frac{1}{2}(A+B) \qquad (12)$$

Domains 4: Manipulators in domain 4 have four cusps and four nodes. No subcase exists in this domain (Baili, 2003). Such topologies are referred to as WT_7.

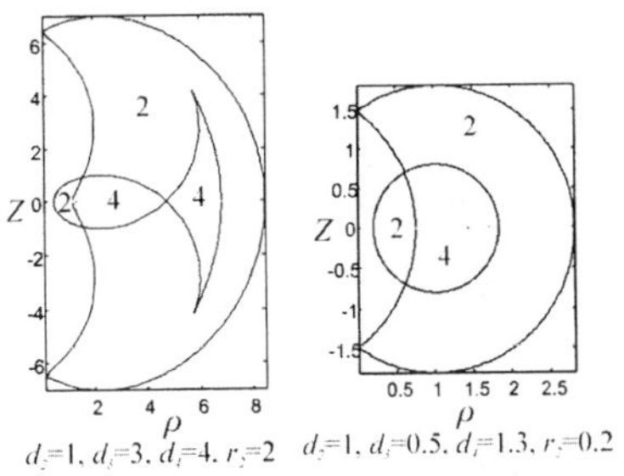

Figure 12. Workspace topologies WT_6 (left) and WT_9 (right)

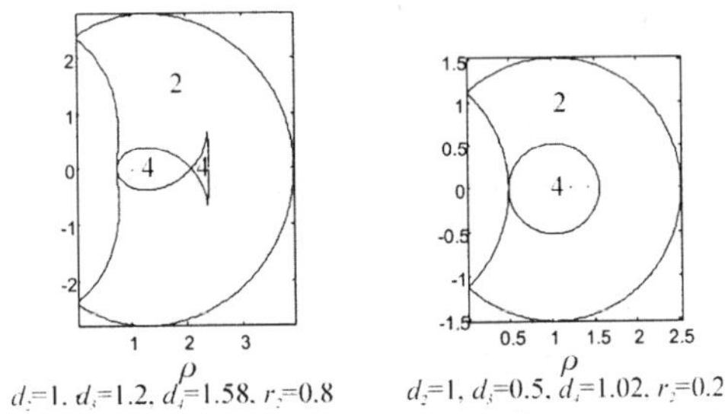

Figure 13. Transition WT_5-WT_6 (left) and WT_8-WT_9 (right).

4. Results synthesis

The partition with cusps and nodes is shown in a section (d_3, d_4) of the parameter space for r_2=1(Fig. 14), where E_1, E_2 and E_3 are the right hand side of (10), (11) and (12), respectively. Plots of the separating curves in sections for different values of r_2 show that they deform smoothly with the same intersections when r_2 varies. The areas of WT_1, WT_2, WT_7 and WT_9 increase when r_2 decreases, whereas those of WT_3, WT_4, WT_5 and WT_6 decrease. The area of WT_4 is very narrow when r_2<1 and almost disappears when r_2<0.1.

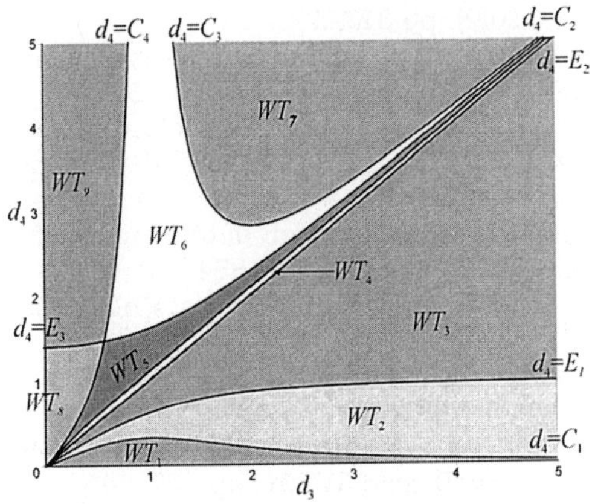

Figure 14. Parameter space partition in a section r_2=1.

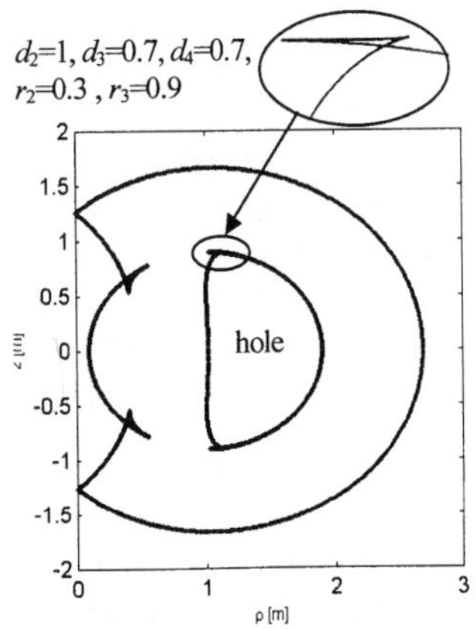

Figure 15. A manipulator with 8 cusps when $r_3 \neq 0$.

5. Conclusions

A family of 3R manipulators was classified according to the topology of the workspace, which was defined as the number of cusps and nodes. The design parameters space was shown to be partitioned into nine subspaces of distinct workspace topologies. Each separating surface was given as an explicit expression in the DH-parameters. This study is being extended to manipulators with $r_3 \neq 0$. First results show that some may have 6 or 8 cusps (Fig. 15). But for small values of r_3, the partition sections look lile in Fig. 14. The subspace WT_4 does not exist any more. It is replaced by two adjacent tiny subspaces with 6 and 8 cusps, located near $d_4 = d_3$. For high values of r_3, the partition gets complicated with more topologies.

6. References

Parenti C.V. and Innocenti C. (1988), Position Analysis of Robot Manipulators: Regions and Sub-regions, in *Proc. Int. Conf. on Advances in Robot Kinematics*, pp 150-158.

Burdick J.(1988), Kinematic analysis and design of redundant manipulators, PhD Dissertation, Stanford.

Borrel P. and Liegeois A. (1986), A study of manipulator inverse kinematic solutions with application to trajectory planning and workspace determination, in *Proc. IEEE Int. Conf. Rob. and Aut.*, pp 1180-1185.

Burdick J. (1995), A classification of 3R regional manipulator singularities and geometries, *Mech. and Machine Theory*, Vol 30(1), pp 71-89.

Wenger P. (1997), Design of cuspidal and noncuspidal manipulators, in *Proc. IEEE Int. Conf. on Rob. and Aut.*, pp 2172-2177.

El Omri J. and Wenger P. (1995), How to recognize simply a non-singular posture changing 3-DOF manipulator, *Proc. 7th ICAR* , p. 215-222.

V.I. Arnold, *Singularity Theory*, Cambridge University Press, 1981.

Wenger P. (1998), Classification of 3R positioning manipulators , *ASME Journal of Mechanical Design*, Vol. 120(2), pp 327-332.

Wenger P. (1999), Some guidelines for the kinematic design of new Manipulators, *Mechanisms and Machine Theory*, Vol 35(3), pp 437-449.

Corvez S. and Rouillier F. (2002), Using computer algebra tools to classify serial manipulators, in *Proc. Fourth International Workshop on Automated Deduction in Geometry*, Linz.

Baili M., Wenger P. and Chablat D. (2003) , Classification of one family of 3R positioning manipulators, in *Proc. 11th ICAR*, pp 1849-1854.

El Omri J., (1996), Kinematic analysis of robotic manipulators, PhD Thesis, University of Nantes (*in french*).

Kohli D. and Hsu (1987), The Jacobian analysis of workspaces of mechanical manipulators, *Mech. & Machine Theory*, Vol. 22(3), pp 265-275.

Ceccarelli M. (1996) , A formulation for the workspace boundary of general n-revolute manipulators, *Mech. & Mach. Theory*, Vol 31, pp 637-646.

Baili M. (2003), Classification of 3R Orthogonal positioning manipulators, technical report, University of Nantes, September 2003.

WORKSPACE CLASSIFICATION OF STEWART-GOUGH MANIPULATORS WITH PLANAR BASE AND PLATFORM

Friedrich Pernkopf
Technical University of Graz
Institute of Mechanics and Mechanisms
friedrich.pernkopf@tugraz.at

Manfred L. Husty
University of Innsbruck
Institute of Engineering Mathematics, Geometry and Computer Science
manfred.husty@uibk.ac.at

Abstract In this paper we present a general workspace classification of a class of spatial Stewart-Gough manipulators (SGP), namely those having the anchor points of platform and base in planes (SGPP). The notion *pseudo isotropy* is introduced as isotropic orientation capability within an interval $[0..\alpha]$, where α is less than 2π.

Keywords: Workspace analysis, Isotropy.

1. Introduction

The main difficulty of workspace analysis of Stewart-Gough platforms (SGP, see Merlet, 2000) is that a complete representation of the workspace needs a 6-dimensional space for which no direct graphical illustration is possible. Therefore only lower dimensional subsets of the workspace can be displayed. So far most investigations have focused on the 3-dimensional constant orientation workspace, which describes all possible locations of an arbitrary point C of the moving system with a constant orientation of the moving body. But many other types of workspace are also of interest, for example the reachable workspace (all positions that can be reached by C in some orientation), the orientation workspace (all possible orientations of the end-effector (EE) about all axis through C for a given position), the inclusive orientation workspace (all the locations that can be reached by C with at least one orientation of a defined set) or the total orientation workspace (all positions that can be reached with all orientations of a given interval $[\varphi_{min}, \varphi_{max}]$ about all axis through C). If the interval of the total orientation workspace is $[0, 2\pi]$ then the workspace is called dextrous (Merlet, 2000).

J. Lenarčič and C. Galletti (eds.), On Advances in Robot Kinematics, 229-236.

Within this paper workspace analysis is performed taking into account singularities, active and passive joint limits. The dimensionality problem of representation of the workspace is overcome by the following procedure:

- Orientation and position are at first considered separately.
- Computation of the orientation WS taking into account the above mentioned constraints for all points of the Cartesian Space.
- Applying Workspace conditions to the ori WS.
- Computing and displaying the set of Cartesian points where these conditions hold.

Note that with this algorithm each point of position WS has obtained additional orientation information and that the position WS is dependent on which type of Ori-Ws was taken.

After a brief introduction to the mathematical tools like Study parameterization, canonical placement of the coordinate frames and singularity-freeness in Section 2, workspace analysis of the defined class of manipulators is contained in Section 3. Important part of this section is the visualization of the workspaces to give fast information to the designer. In Section 4 classification of the workspace is performed and pseudo isotropy is defined.

2. Mathematical and Kinematical Framework

In this section we introduce briefly the mathematical tools used for the workspace analysis. A detailed description can be found in Pernkopf, 2003, Pernkopf and Husty, 2002.

2.1 Study Parameters

Kinematic mapping was introduced by Study, 1912 to map displacements of the three-dimensional Euclidean space $\mathbb{E}^3$ into a seven-dimensional image space (Husty et al., 1997, Bottema and Roth, 1990). The group of Euclidean transformations ($\mathbf{SE}_3$) is parameterized by Study parameters (or biquaternions). The transformation of the platform points ${}^{\mathcal{M}}\mathbf{p}_i = [1, a, b, c]^T$ in the platform frame (moving frame) to the base frame $\mathcal{B}$ is given by:

$$\begin{bmatrix} 1 \\ A \\ B \\ C \end{bmatrix} = \begin{bmatrix} x_0^2+x_1^2+x_2^2+x_3^2 & 0 & 0 & 0 \\ X & x_0^2+x_1^2-x_2^2-x_3^2 & 2(x_1x_2+x_0x_3) & 2(x_1x_3-x_0x_2) \\ Y & 2(x_1x_2-x_0x_3) & x_0^2-x_1^2+x_2^2-x_3^2 & 2(x_2x_3+x_0x_1) \\ Z & 2(x_1x_3+x_0x_2) & 2(x_2x_3-x_0x_1) & x_0^2-x_1^2-x_2^2+x_3^2 \end{bmatrix} \begin{bmatrix} 1 \\ a \\ b \\ c \end{bmatrix}, \quad (1)$$

where

$$X = 2(y_1x_0 - y_0x_1 + y_3x_2 - y_2x_3),$$
$$Y = 2(y_2x_0 - y_0x_2 + y_1x_3 - y_3x_1), \quad (2)$$
$$Z = 2(y_3x_0 - y_0x_3 + y_2x_1 - y_1x_2).$$

Note that vector $[X, Y, Z]$ is the position of the origin of the moving frame in the base frame. As the eight coordinates x_i, y_i are homogenous $(x_0 : x_1 : x_2 : x_3 : y_0 : y_1 : y_2 : y_3)$ they can be interpreted as coordinates of a point in a seven dimensional projective space $\mathbb{P}^7$. Only coordinates fulfilling

$$x_0y_0 + x_1y_1 + x_2y_2 + x_3y_3 = 0, \quad (3)$$

represent valid poses of the platform coordinate system $(\mathcal{M})$ in the base coordinate system $(\mathcal{B})$. The zeros of Eq.3 define a six dimensional quadric S_6^2 usually called Study-quadric. The x_i are the Euler parameters[1] and whenever convenient we shall use the normalizing condition

$$x_0^2 + x_1^2 + x_2^2 + x_3^2 = 1, \quad (4)$$

Furthermore we set $x_0 \geq 0$ to get rid of the double covering of the Euclidean transformation group by Study parameters (Blaschke, 1960).

2.2 Canonical Coordinate System

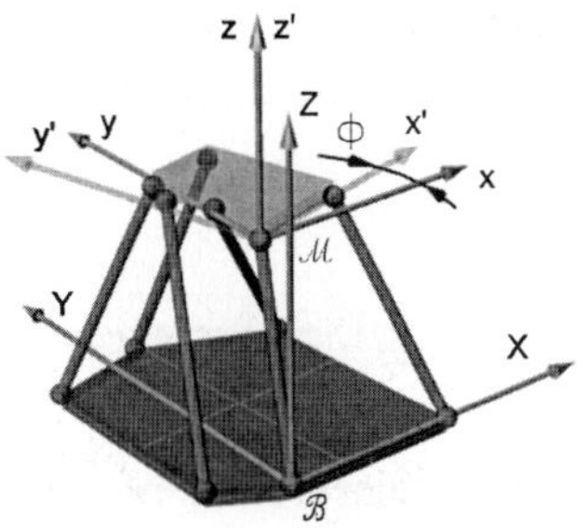

Figure 1. Coordinate frames on base and platform of an SGPP

Figure 2. Pos-WS for home orientation of a 6-6 SGPP design

In Pernkopf and Husty, 2002, canonical coordinate systems (see Fig.1) were described in which the singularity surface of SGPP has the simplest mathematical description[2]. The equation of the singularity surface (i.e. the algebraic variety belonging to the zeros of the manipulator's Jacobian determinant[3]) factorizes into parts which all have significant geometric(and kinematic) meaning. It can be shown, that for every point in the Pos-WS there exists a singularity free void about the origin in Euler-parameter space iff the defined position does not have zero elevation (platform and base do not coincide) and iff the design is not architecture singular (=regular). The canonical home pose is now defined such that platform and base are in parallel planes, having a non zero distance and the coordinate frame on the platform is chosen in such a way, that the home orientation is exactly in the middle of the two orientation singularities which occur if the platform rotates about the z-axis. It was shown that independent of the regular design the two singular orientations are 180° apart. Note that these general singularity properties of SGPP do not change when performing a pure translation.

3. Workspace Analysis

We will use the well defined canonical home pose, which determines the assembly mode and guarantees a singularity-free void around the origin of the Euler-parameter space (Pernkopf and Husty, 2002).

Taking into account the limits of active (link length) and passive joints (spherical, U-joints, U-joints with skew axes) we are interested in the orientability of the EE in a discrete set of positions in Pos-WS of a given SGPP (Pernkopf, 2003). The algorithm for workspace analysis can be described as follows:

- Computation of the Pos-WS of a point C for the canonical home orientation taking into account the limits of active and passive joints (Fig. 2).(Intersection of all feasible regions for every link. The Pos-WS is guaranteed singularity free for the canonical home orientation.)
- Discretization of the Pos-WS.
- Computation of the orientation singularity surface for every point of this grid (wire frame surface in Fig. 3).
- Setting up a bundle of lines representing a discrete set of rotation axes in every point of the Pos-WS-grid.
- Intersection of each axis with the orientation singularity surface. The intersection closest to the origin gives the maximal possible singularity free rotation about this axis.
- Testing each singularity free rotation for active and passive joint limits by applying a bisection routine on the singularity free interval of rotation angle.

This algorithm returns for each discrete point of the Pos-WS a solid in the Euler parameter space, whose center is the apex of the bundle of

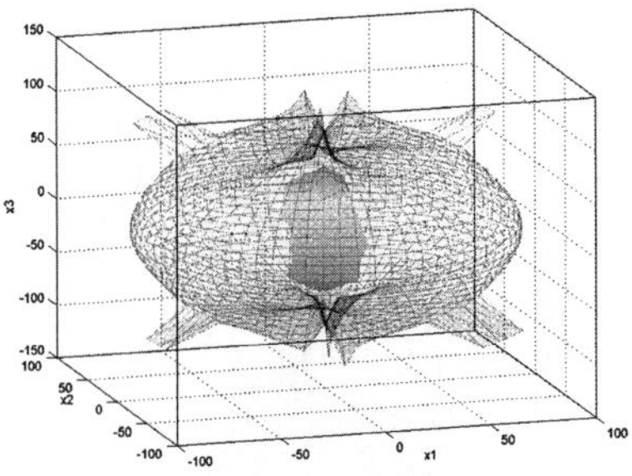

Figure 3. Orientation singularity surface closest to the origin (grid) and Ori-WS (solid) for one given position of an SGPP

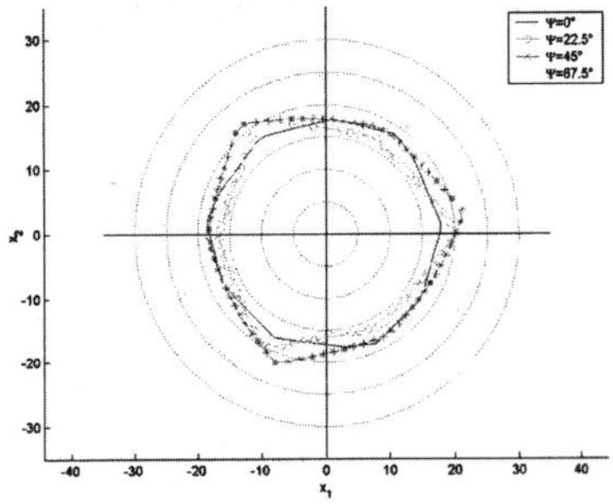

Figure 4. Representation of the orientation part to build up a 6d workspace representation of a 6-6 SGPP design

rotation axes. Each point on an axis represents a rotation about this axis with an angle that corresponds[4] to the distance of the point from the apex. The solid represents for the chosen position the set of all rotations that are possible about axes centered in $\mathcal{C}$ (Fig. 3, inner solid). The volume of the inner solid could be used as a measure of the orientation capability of the manipulator in the chosen position. Performing the algorithm for a set of points in a horizontal cross section of the Pos-WS and plotting the boundary of the intersection of the solids (Fig. 3) with the horizontal plane in the Euler-parameter-space (this represents all horizontal rotation axes) yields the 5D workspace representation given in Fig. 5 for a 6-6 SGPP (Fig. 6).

This workspace representation can be completed by adding 6d information as shown in Fig. 4. In this figure, the orientation capability of the EE about axes, which are tilted with a given elevation Ψ, is depicted. This yields essentially an augmented contour plot of the solid in Fig. 3.

4. Workspace Classification, Pseudo Isotropy

The orientation capability can be defined as a closed interval $\mathcal{O} = [\alpha, \beta]$ in which the rotation angle φ about any axis is included[5]. With this definition a classification of several Pos-WS types is possible. The set of all points of the 3d cartesian space (positions) with following properties of the associated Ori-WS is called:

- position workspace for constant home orientation: The Ori-WS consists of the origin of the Euler parameter space which is associated to the canonical home pose.

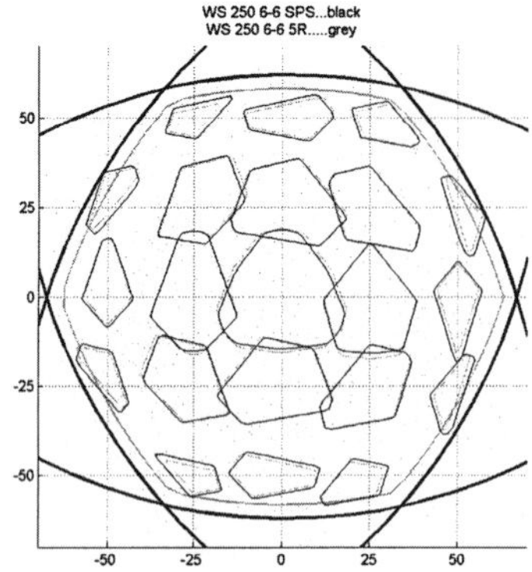

Figure 5. 5D workspace representation of a 6-6 SGPP design at Z=255mm. The black curves illustrate this workspace for SPS- links, whereas the gray curves show the limits for using 5R-links (two U-joints with skew axes)

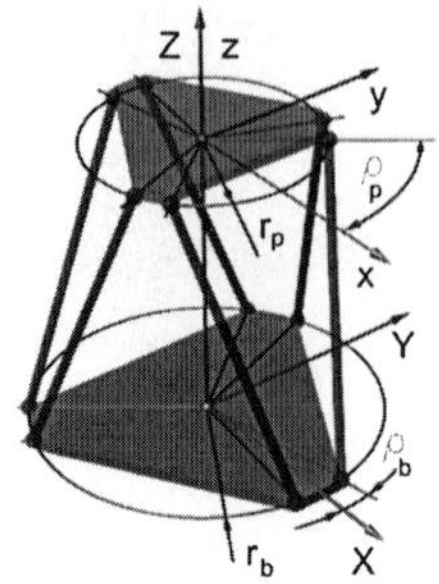

Figure 6. 6-6 SGPP design from PI

- position workspace for constant orientation: The Ori-WS consists of one point of the Euler parameter space.
- dextrous workspace: The Ori-WS is the whole unitsphere ($\mathcal{O} = [0, 2\pi]$). The result of Pernkopf and Husty, 2002 is, that no singularity free dextrous workspace is possible for SGPP.
- total orientation workspace: The possible orientations about every axis are the whole interval $\mathcal{O} = [0, \beta], 0 < \beta < 2\pi$.[6]
- reachable workspace: The set of all positions that can be reached with any orientation. On the boundary of this WS type the Ori-WS is collapsed to one point.
- inclusive orientation workspace: At least one possible orientation about one axis must lie in a certain interval $\mathcal{O} = [\alpha, \beta]$.

With this definitions, the singularity-free balls from Pernkopf and Husty, 2002, can be interpreted as total orientation singularity free workspace without limits to joints and without link interference. By adding these constraints, the total orientation workspace of an SGPP manipulator is obtained by computation of the biggest balls inside the Ori-WS in all reachable positions. Computation of the biggest bal is done by transformation of the Ori-WS (e.g. inner solid of Fig. 3) to spherical coordinates and searching for the for the global minimum of the resulting function. The radius of the biggest ball gives the (isotropic) orientation capability

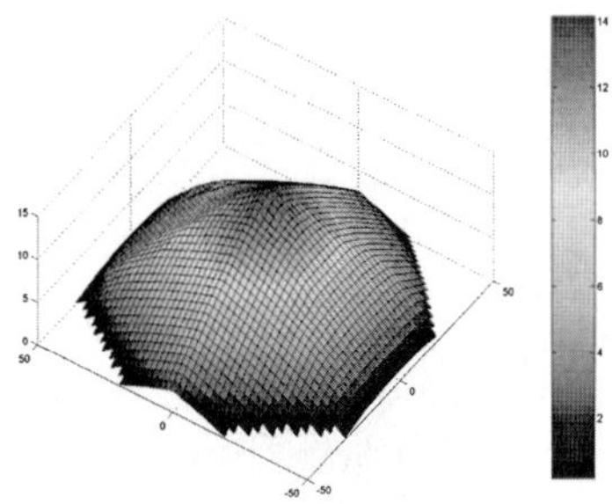

Figure 7. Distribution of the maximal pseudo-isotropic orientation capability over a horizontal cross-section of a 6-6 SGPP manipulator (the measures are in degree)

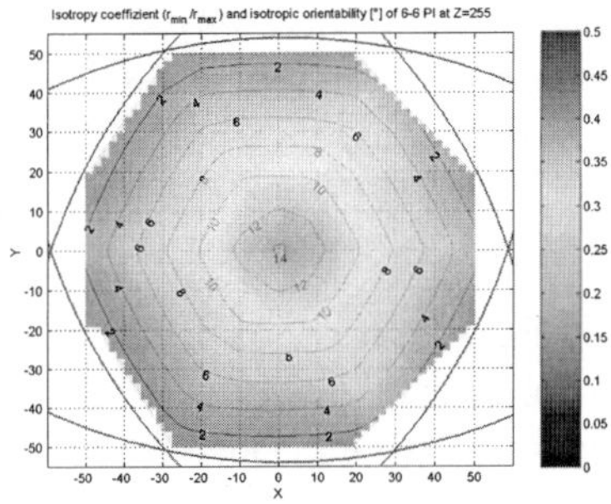

Figure 8. Contour map from Fig. 7 over a kind of isotropy coefficient

for one position. Fig. 7 shows this orientation capabilities over a cross section of the Pos-WS, where the radius of the biggest ball from above is plotted orthogonally to the intersection plane. This figure shows the decreasing isotropic orientation capability from the center (15°) to the boundary of the Pos-WS of constant orientation.

Finally, a special kind of orientability, called *pseudo-isotropic orientability* of the EE in a horizontal cross section of Pos-WS is presented. In Fig. 8 the contour lines show the minimal orientability about every possible axis. This is represented by the largest possible sphere which can be inscribed into the Ori-WS (the solid in the Euler-parameter-space shown in Fig. 3 for one position) for a cross section of a 6-6 SGPP. The underlying gray-scale (shading) gives a kind of isotropy index. It represents the similarity of the Ori-WS to a sphere, namely the division of the shortest possible orientability (the radius of the void in spherical coordinates) by the longest. The interpretation is as following: In home pose (the *"middle"* of Pos-WS) the worst orientability is about a horizontal axis and it is half of the best orientability about the vertical axis. The orientation capability therefore would be better represented by an ellipsiod of revolution with axes ratio 0.5. Moving the manipulator from the central position to the border of Pos-WS, this measure decreases. The *Pseudo-Isotropic-Orientation-Surface* is a "smooth" surface, as it can be seen by the nearly constant distance of the contour lines as well as in Fig. 7, therefore it can be used to give a quick information about the orientability for a regular hexapod design.

All shown figures are computed for an existing manipulator. It is a 6-6 SGPP design, manufactured by PI (Physikinstrumente) in Karlsruhe.

This design is characterized by a base radius r_b=148 mm, a platform radius r_p=93.5 mm, a base delta-angle $\varrho_b = 9°$ and a platform delta-angle $\varrho_p = 11°$.

5. Conclusions

A classification of different position workspace types via the description of special properties of the orientation workspace of all contained points was given. Because of impossibility of singularity freeness of dextrous workspace of SGPP, an definition of pseudo isotropy was introduced.

Acknowledgements

The work reported here was supported by Grant No. P14986 FWF.

Notes

1. These homogenous parameters describe all rotations about a fixed point (parameterization of the group $\mathbf{SO}_3^+$). Using the normalizing condition each rotation can be interpreted as a point in a solid ball centered in the origin of a 3-D Euler parameter space. This interpretation will be used in the paper to discuss the Ori-WS and to distinguish different Pos-WS types.
2. This means that the equation describing the singularity surface in the Study parameter space has canonical form
3. This algebraic variety describes all positions of the platform in which the six legs are in a linear complex.
4. The correspondence is of course nonlinear: $x_0 = cos\frac{\varphi}{2}$ and $[x_1, x_2, x_3]^T$ is the direction of the rotation axis. Note: Eq. 4 must be fulfilled to get the proper metric.
5. This means, the orientation capability is determined by two spheres with its center in the origin of Euler parameter space and with radii α and β.
6. All points(=orientations) within a sphere smaller than the unit sphere in Study/Euler parameter space are reachable for all points within this Pos-WS.

References

Blaschke, W., 1960, *Kinematik und Quaternionen*, VEB Deutscher Verlag der Wissenschaften, Berlin, Germany.

Bottema,O., Roth,B., 1990, *Theoretical Kinematics*, Dover Publishing.

Husty, M.L., Karger, A., Sachs, H., Steinhilper, W., 1997, *Kinematik und Robotik*, Springer-Verlag, Berlin, Germany.

Merlet, J.P., 2000, *Parallel Robots*, Kluwer Academic Publishers, Dordrecht.

Pernkopf, F., 2003, *Workspace Analysis of Stewart-Gough-Platforms*, Phd-thesis, University of Innsbruck, Austria.

Pernkopf, F., Husty, M. L., 2002, *Singularity-Analysis of spatial Stewart-Gough-Platforms with planar base and platform*, DETC2002/MECH-34267, 2002 ASME Design Engineering Technical Conferences, Montreal, Canada.

Study, E., 1912, Grundlagen und Ziele der analytischen Kinematik, Sitzungsberichte der Berliner Mathematischen Gesellschaft, 104. Sitzung.

THE ORTHOGLIDE: KINEMATICS AND WORKSPACE ANALYSIS

A. Pashkevich

Robotic Laboratory, Belarusian State University of Informatics and Radioelectronics

6 P. Brovka St., Minsk, 220027, Belarus

E-mail: pap@ bsuir.unibel.by

D. Chablat, P. Wenger

Institut de Recherche en Communications et Cybernétique de Nantes

1 rue de la Noë, BP 92 101, 44321 Nantes, France

E-mails:Damien.Chablat@irccyn.ec-nantes.fr, Philippe.Wenger@irccyn.ec-nantes.fr

Abstract The paper addresses kinematic and geometrical aspects of the Orthoglide, a three-DOF parallel mechanism. This machine consists of three fixed linear joints, which are mounted orthogonally, three identical legs and a mobile platform, which moves in the Cartesian x-y-z space with fixed orientation. New solutions to solve inverse/direct kinematics are proposed and a detailed workspace analysis is performed taking into account specific joint limit constraints.

Keywords: Parallel manipulators; Workspace; Inverse and direct kinematics.

1. Introduction

For two decades, parallel manipulators attract the attention of more and more researchers who consider them as valuable alternative design for robotic mechanisms (Asada et al, 1986, Fu et al., 1987, Craig, 1989). As stated by a number of authors (Tsai, 1999), conventional serial kinematic machines have already reached their dynamic performance limits, which are bounded by high stiffness of the machine components required to support sequential joints, links and actuators. Thus, while having good operating characteristics (large workspace and high flexibility), serial manipulators have disadvantages of low precision, low stiffness and low power. Also, they are generally operated at low speed to avoid excessive vibration and deflection.

Conversely, parallel kinematic machines offer essential advantages over their serial counterparts (lower moving masses and higher rigidity) that obviously should lead to higher dynamic capabilities. However, most existing parallel manipulators have limited and

J. Lenarčič and C. Galletti (eds.), On Advances in Robot Kinematics, 237-246.

complicated workspace with singularities, and highly non-isotropic input/output relations (Angeles, 2002). Hence, the performances may significantly vary over the workspace and depend on the direction of the motion, which is a serious disadvantage for machining applications. Research in the field of parallel manipulators began with the Stewart-platform used in flight simulators (Stewart, 1965). Many such structures have been investigated since then, which are composed of six linearly actuated legs with different combinations of link-to-platform connections (Merlet, 2000). In recent years, several new kinematic structures have been proposed that possess higher isotropy. In particular, a 3-dof translational mechanism with gliding foot points was found in three separate works to be fully isotropic throughout the Cartesian workspace (Carricato et all, 2002 and Kong et all, 2002). It consists of a mobile platform, which is connected to three orthogonal linear drives through three identical planar 3-revolute jointed serial chains. Although this manipulator behaves like a conventional Cartesian machine, bulky legs are required to assure stiffness because these legs are subject to bending.

In this paper, the Orthoglide manipulator proposed by Wenger et all, 2000, is studied. As follows from previous research, this manipulator has good kinetostatic performances and some technological advantages, such as (i) symmetrical design consisting of similar 1-d.o.f. joints; (ii) regular workspace shape properties with bounded velocity amplification factor; and (iii) low inertia effects (Chablat et all, 2003). This article analyses the kinematics and the workspace of the Orthoglide. Section 2 describes the Orthoglide geometry. Section 3 proposes new solutions for its inverse and direct kinematics. Sections 4, 5 present a detailed analysis of the workspace and jointspace respectively. Finally, Section 6 summarises the main contributions of the paper.

2. Manipulator geometry

The kinematic architecture of the Orthoglide is shown in Fig. 1. It consists of three identical kinematic chains that are formally described as PRP_aR, where P, R and P_a denote the prismatic, revolute, and parallelogram joints respectively. The mechanism input is made up by three actuated orthogonal prismatic joints. The output body is connected to the prismatic joints through a set of three kinematic chains. Each chain includes a parallelogram, so that the output body is restricted to translational movements. To get the Orthoglide kinematic equations, let us locate the reference frame at the intersection of the prismatic joint axes and align the coordinate axis with them (Fig. 2), following the "right-hand" rule. Let us also denote the input vector of the prismatic joints variables as $\boldsymbol{\rho}=(\rho_x, \rho_y, \rho_z)$ and the output position vector of the tool centre point as $\mathbf{p}=(p_x, p_y, p_z)$. Taking into account obvious properties of the parallelograms, the Orthoglide geometrical model can be presented in a simplified form, which consists of three bar links connected by spherical joints to the tool centre point at one side and to the corresponding prismatic joints at another side.

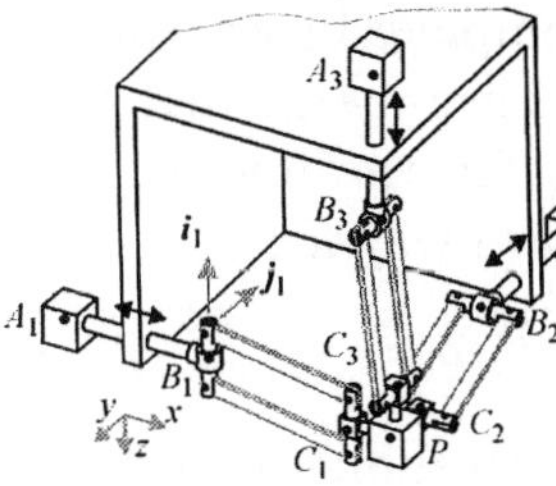

Fig. 1. Orthoglide architecture

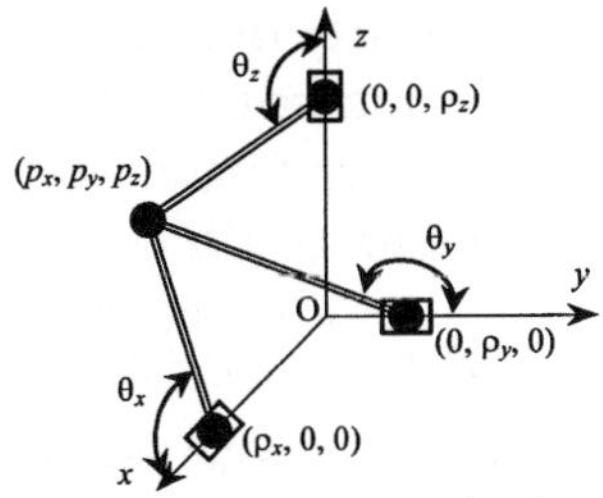

Fig. 2. Orthoglide geometrical model

Using this notation, the kinematic equations of the Orthoglide can be written as follows

$$(p_x-\rho_x)^2+p_y^2+p_z^2=L^2, p_x^2+(p_y-\rho_y)^2+p_z^2=L^2, p_x^2+p_y^2+(p_z-\rho_z)^2=L^2 \quad (1)$$

where L is the length of the parallelogram principal links and the "zero" position $\mathbf{p}_0=(0,0,0)$ corresponds to the joints variables $\boldsymbol{\rho}_0=(L,L,L)$. It should be stressed that the Orthoglide geometry and relevant manufacturing technology impose the following constraints on the joint variables

$$0<\rho_x\le 2L \;;\; 0<\rho_y\le 2L \;;\; 0<\rho_z\le 2L, \quad (2)$$

which essentially influence on the workspace shape. While the upper bound $(\rho\le 2L)$ is implicit and obvious, the lower one $(\rho>0)$ is caused by practical reasons, since safe mechanical design encourages avoiding risk of simultaneous location of prismatic joints in the same point of the Cartesian workspace (here and in the following sections, while referring to symmetrical constraints are subscript omitted, *i.e.* $\rho\in\{\rho_x,\rho_y,\rho_z\}$).

3. Orthoglide Kinematics

3.1 Inverse kinematics

For the inverse kinematics, the position of the end-point (p_x, p_y, p_z) is treated as known and the goal is to find the joint variables (ρ_x, ρ_y, ρ_z) that yield the given location of the tool. Since in the general case the inverse kinematics can produce several solutions corresponding to the same tool location, the solutions must be distinguished with respect to the algorithm "branch". For instance, if the aim is to generate a sequence of points to move the tool along an arc, care must be taken to avoid branch switching during motion, which may cause inefficient (or even impossible) manipulator motions. Moreover, leg singularities may occur at which the manipulator loses degrees of freedom and the joint variables become linearly dependent. Hence, the complete investigation of the Orthoglide kinematics must cover all the above-mentioned topics.

From the Orthoglide geometrical model (1), the inverse kinematic equations can be derived in a straightforward way as:

$$\rho_x = p_x + s_x\sqrt{L^2 - p_y^2 - p_z^2} \quad \rho_y = p_y + s_y\sqrt{L^2 - p_x^2 - p_z^2} \quad \rho_z = p_z + s_z\sqrt{L^2 - p_x^2 - p_y^2} \qquad (3)$$

where s_x, s_y, s_z are the branch (or configuration) indices that are equal to ± 1. It is obvious that (3) yields eight different branches of the inverse kinematic algorithm, which will be further referred to as *PPP*, *MPP*...*MMM* following the sign of the corresponding index (i.e. the notation *MPP* corresponds to the indices $s_x = -1;\ s_y = +1;\ s_z = +1$). The geometrical meaning of these indices is illustrated by Fig. 2, where θ_x, θ_y, θ_z are the angles between the bar links and the corresponding prismatic joint axes. It can be proved that $s = 1$ if $\theta \in (90^\circ, 180^\circ)$ and $s = -1$ if $\theta \in (0^\circ, 90^\circ)$. The branch transition ($\theta = 90^\circ$) corresponds to the serial singularity (where the leg is orthogonal to the relevant translational axis and the input joint motion does not produce the end-point displacement). It is obvious that if the inverse kinematic solution exists, then the target point (p_x, p_y, p_z) belongs to a volume bounded by the intersection of three cylinders

$$C_L = \left\{ \mathbf{p} \mid p_x^2 + p_y^2 \le L^2;\ p_x^2 + p_z^2 \le L^2;\ p_y^2 + p_z^2 \le L^2 \right\} \qquad (4)$$

that guarantees non-negative values under the square roots in (3). However, it is not sufficient, since the lower joint limits (2) impose the following additional constraints

$$p_x > -s_x\sqrt{L^2 - p_y^2 - p_z^2}\ ;\ p_y > -s_y\sqrt{L^2 - p_x^2 - p_z^2}\ ;\ p_z > -s_z\sqrt{L^2 - p_x^2 - p_y^2} \qquad (5)$$

which reduce a potential solution set. For example, it can be easily computed that for the "zero" workspace point $\mathbf{p}_0 = (0, 0, 0)$, the inverse kinematic equations (3) give eight solutions $\rho = (\pm L, \pm L, \pm L)$ but only one of them is feasible. To analyse in details the influence of the joint constraints impact, let us start from separate a study of the inequalities (5) and then summarise results for all possible combinations of the three configuration indices. If $s_x = 1$, then consideration of two cases, $p_x > 0$ and $p_x \le 0$, yields the following workspace set satisfying the constraint $\rho_x > 0$

$$W_L^{+x} = \{\mathbf{p} \in C_L \mid p_x > 0\} \cup \{\mathbf{p} \in C_L \mid p_x \le 0; p_x^2 + p_y^2 + p_z^2 < L^2\} \tag{6}$$

which consists of two fractions (½ of the cylinder intersection denoted C_L and ½ of the sphere whose geometric center is (0,0,0) and radius is L). If $s_x = -1$, then the second case $p_x < 0$ does not give any solution and the joint constraint $\rho_x > 0$ is expressed in the workspace as

$$W_L^{-x} = \{\mathbf{p} \in C_L \mid p_x \ge 0; p_x^2 + p_y^2 + p_z^2 > L^2\}. \tag{7}$$

The latter defines a solid bounded by three cylindrical surfaces and the sphere. The remaining constrains $\rho_y > 0$ and $\rho_z > 0$ can be derived similarly, which differ from (6), (7) by subscripts only.

Then, there can be found intersection of the obtained sets for different combinations of the configuration indices. It can be easily proved that the case "PPP" yields

$$W_L^{PPP} = \{\mathbf{p} \in C_L \mid p_x, p_y, p_z > 0\} \cup \{\mathbf{p} \in C_L \mid p_x^2 + p_y^2 + p_z^2 < L^2\} \tag{8}$$

while the remaining cases give

$$W_L^{MPP} = \dots = W_L^{MMM} = \{\mathbf{p} \in C_L \mid p_x, p_y, p_z > 0\} \cap \{\mathbf{p} \in C_L \mid p_x^2 + p_y^2 + p_z^2 > L^2\} \tag{9}$$

Expressions (8) and (9) can be put in the form

$$W_L^{PPP} = S_L \cup G_L; \qquad W_L^{MPP} = \dots = W_L^{MMM} = G_L \tag{10}$$

where $S_L = \{\mathbf{p} \in C_L \mid p_x^2 + p_y^2 + p_z^2 < L^2\}$;
$G_L = \{\mathbf{p} \in C_L \mid p_x, p_y, p_z > 0; \; p_x^2 + p_y^2 + p_z^2 > L^2\}$; $S_L \cap G_L = \varnothing$.

Therefore, for the considered positive joint limits (2), the existence of the inverse kinematic solutions may be summarised as follows (i) inside the sphere S_L there exist exactly one inverse kinematic solution *PPP* with positive configuration indices s_x, s_y, s_z, (ii) outside the sphere S_L, but within the positive part of the cylinder intersection C_L, there exist 8 solutions of the inverse kinematics (*PPP*, *MPP*, ... *MMM*) corresponding to all possible combinations of the configuration indices s_x, s_y, s_z. These conclusions may be illustrated when $L = 1$ by numerical examples. If the target point $\mathbf{p}$=(-0.5, 0.4, 0.3) is within the sphere S_L, then the joint coordinates must be taken from the sets $\rho_x \in \{0.37, -1.37\}$, $\rho_x \in \{1.21, -0.41\}$, $\rho_x \in \{1.07, -0.47\}$, which allow only one positive combination. In contrast, for the target point $\mathbf{p}$=(0.7, 0.7, 0.7), which is outside the sphere, the inverse kinematics yields solutions with two positive values

$\rho_x, \rho_y, \rho_z \in \{0.84, 0.56\}$ that allow 8 positive combinations of the joint variables. An interesting feature is that intermediate cases (with 2 or 4 feasible solutions) are not possible.

3.2 Direct kinematics

For the direct kinematics, the values of the joint variables (ρ_x, ρ_y, ρ_z) are known and the goal is to find the tool centre point location (p_x, p_y, p_z) that corresponds to the given joint positions. While, in general, the inverse kinematics of parallel mechanisms is straightforward, the direct kinematics is usually very complex. The Orthoglide has the advantage leave an analytical direct kinematics. Like for the previous section, the solutions must be distinguished with respect to the algorithm "branch" that should be also defined both geometrically and algebraically, via a configuration index.

To solve the system (1) for p_x, p_y, p_z, first, let us derive linear relations between the unknowns. By subtracting three possible pairs of the equations (1), we leave

$$2\rho_x p_x - 2\rho_y p_y = \rho_x^2 - \rho_y^2 \;,\; 2\rho_x p_x - 2\rho_z p_z = \rho_x^2 - \rho_z^2 \;,\; 2\rho_y p_y - 2\rho_z p_z = \rho_y^2 - \rho_z^2 \qquad (11)$$

As follows from these expressions, the relation between p_x, p_y, p_z may be presented as

$$p_x = \rho_x/2 + t/\rho_x; \;\; p_y = \rho_y/2 + t/\rho_y; \;\; p_z = \rho_z/2 + t/\rho_z \,, \qquad (12)$$

where t is an auxiliary scalar parameter. From a geometrical point of view, the expression (12) defines the set of equidistant points for the prismatic joint centres (Fig. 6). Also, it can be easily proved that the full set of equidistant points is the line perpendicular to Π and passing through $(\rho_x, \rho_y, \rho_z)/2$, where.

$$\Pi = \{\, \mathbf{p} \mid p_x/\rho_x + p_y/\rho_y + p_z/\rho_z = 1 \,\} \qquad (13)$$

After substituting (12) into any of the equations (1), the direct kinematic problem is reduced to the solution of a quadratic equation in the auxiliary variable t,

$$At^2 + Bt + C = 0 \,, \qquad (14)$$

where $A = (\rho_x \rho_y)^2 + (\rho_x \rho_z)^2 + (\rho_y \rho_z)^2$, $B = (\rho_x \rho_y \rho_z)^2$, $C = (\rho_x^2 + \rho_y^2 + \rho_z^2 - 4L^2)B/4$.

The quadratic formula yields two solutions

$$t = (-B + m\sqrt{B^2 - 4AC})/(2A); \;\; m = \pm 1 \qquad (15)$$

that geometrically correspond to different locations of the target point P (see Fig. 6) with respect to the plane passing through the prismatic joint centres (it should be noted that the intersection point of the plane and the set of equidistant point corresponds to $t_0 = -B/(2A)$). Hence, the Orthoglide direct kinematics is solved analytically, via the quadratic formula (14) for the auxiliary variable t and its substitution into expressions (12). The direct kinematic solution exists if and only if the joint variables satisfy the inequality $B^2 \geq 4AC$, which defines a closed region in the joint variable space

$$\Re_L = \left\{ \rho \mid \left(\rho_x^2 + \rho_y^2 + \rho_z^2 - 4L^2 \right)\left(\rho_x^{-2} + \rho_y^{-2} + \rho_z^{-2} \right) \leq 1 \right\} \tag{16}$$

Taking into account the joint limits (2), the feasible joint space may be presented as $\Re_L^+ = \{ \rho \in \Re_L \mid \rho_x, \rho_y, \rho_z > 0 \}$. Therefore, for the considered positive joint limits (2), the existence of the direct kinematic solutions may be summarised as follows: (i) inside the region $\Re_L^+$, there exist exactly two direct kinematic solutions, which differ by the target point location relative to the plane Π (Fig. 7a). (ii) On the border of the region $\Re_L^+$ located inside the first octant, there exist a single direct kinematic solution, which corresponds to the "flat" manipulator configuration, where both the target point and prismatic joint centres belong to the plane Π (Fig. 7b).

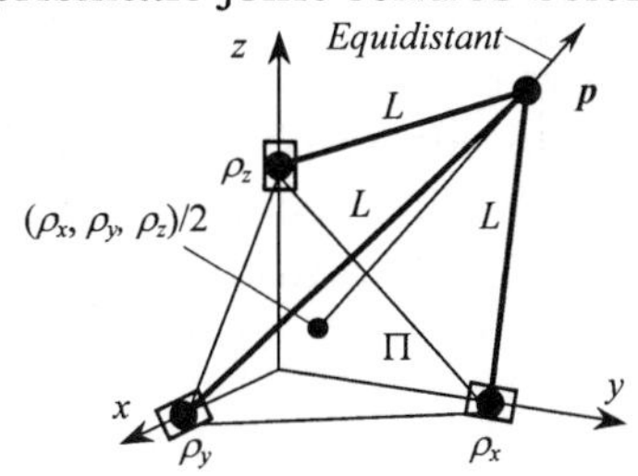

Fig. 6. Geometrical solution of the direct kinematics

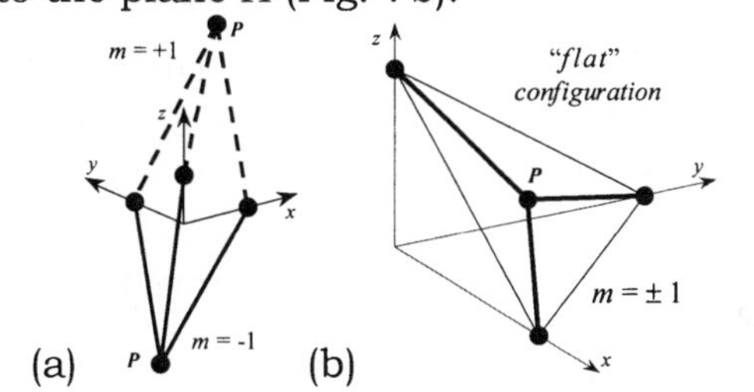

Fig. 7. Double (a) and single (b) solutions of the direct kinematics

These conclusions may be illustrated by the following numerical examples (for the "unit" manipulator, L=1). Since the joint variables $\rho_x = \rho_y = \rho_z = 0.3$ are within $\Re_L^+$, then the end-point coordinates are either $p_x = p_y = p_z = -0.46$ or $p_x = p_y = p_z = -0.66$. In contrast, for the joint variables $\rho_x = \rho_y = \rho_z = \sqrt{1.5}$, which are exactly on the surface $\Re_L^+$, the direct kinematics yields a single solution $p_x = p_y = p_z = \sqrt{1/6}$ corresponding to the "flat" configuration (see Fig. 7b).

3.3 Configuration indices

As follows from the previous sub-sections, both the inverse and direct kinematics of the Orthoglide may produce several solutions. The problem is how to define numerically the *configuration indices*, which allow choosing among the corresponding algorithm branches.

For the inverse kinematics, when the configuration is defined by the angle between the leg and the corresponding prismatic joint axis, the decision equations for the configuration indices are trivial:

$$s_x = \operatorname{sgn}(\rho_x - p_x);\; s_y = \operatorname{sgn}(\rho_y - p_y);\; s_z = \operatorname{sgn}(\rho_z - p_z)$$

Geometrically, $s > 0$ means that (see Fig. 2), $\theta_x, \theta_y, \theta_z \in \left]\pi/2 \;\; 3\pi/2\right[$. For the direct kinematics, the configuration is defined by the end-point location relative to the plane that passes through the prismatic joint centres (see Figs. 6-7). Hence, the decision equation may be derived by analysing the dot-product of the plane normal vector $(\rho_x^{-1},\, \rho_y^{-1},\, \rho_z^{-1})$ and the vector directed along any of the bar links (for instance, $(p_x - \rho_x,\, p_y,\, p_z)$ for the first link: $m = \operatorname{sgn}\left(p_x/\rho_x + p_y/\rho_y + p_z/\rho_z - 1\right)$ which is equivalent to $m = \operatorname{sgn}\left(p_x\rho_y\rho_z + \rho_x p_y \rho_z + \rho_x\rho_y p_z - \rho_x\rho_y\rho_z\right)$ for the positive joint limits.

It should be stressed that the feasible solutions for the inverse/direct kinematics, located in the neighbourhood of the "zero" point, have the configuration indices $s_x = s_y = s_z = +1$ and $m = -1$.

4. Workspace analysis

As follows from Eq. (10), Orthoglide workspace W_L is composed of two fractions: (i) the sphere S_L of radius L and centre point (0, 0, 0), and (ii) the thin non-convex solid G_L, which is located in the first octant and bounded by the surfaces of the sphere S_L and the cylinder intersection C_L. It can be proved that the volume of C_L, S_L and W_L is defined by the expressions

$$Vol(C_L) = 8\left(2-\sqrt{2}\right)L^3,\; Vol(G_L) = \left(2-\sqrt{2}-\pi/6\right)L^3,\; Vol(W_L) = \left(2+7\pi/6-\sqrt{2}\right)L^3 \quad (17)$$

As follows from (17), the Orthoglide with the joint limits (2) uses about 53% of the workspace V_{PPP} of its serial counterpart (a Cartesian PPP machine with $2L \times 2L \times 2L$ workspace). Also, the volume of G_L ($0.062\,L^3$) is insignificant in comparison to the volume of sphere S_L ($4.189\,L^3$), which is equal to 52% of V_{PPP}. On the other hand, releasing the lower joint limit ($\rho > 0$) leads to an increases the workspace volume of up to 59% of V_{PPP} only, since the volume of the workspace is, then, equal to C_L. The mutual location of G_L and S_L (and their size ratio) may be also evaluated by the intersection points of the first octant bisector. In particular, for the sphere S_L the bisector intersection point is located at distance $1/\sqrt{3} \approx 0.58$ from the origin, while for the solid G_L the corresponding distance is $1/\sqrt{2} \approx 0.71$ (assuming that L=1). Moreover, G_L touches the sphere by its circular edges, which are located on the borders of the first octant.

5. Joint Space Analysis

The properties of the *feasible* jointspace are essential for the Orthoglide control, in order to avoid impossible combinations of the prismatic joint variables ρ_x, ρ_y, ρ_z, which are generated by the control system and are followed by the actuators. For serial manipulators, this problem does not usually exist because the jointspace is bounded by a parallelepiped and mechanical limitations of the joint values may be verified easily and independently. For parallel manipulators, however, we needs to check both (i) separate input coordinates (to satisfy the joint limits), and (ii) their combinations that must be feasible to produce a direct kinematic solution.

As follows from Sub-Section 3.3, the Orthoglide jointspace $\Re_L^+$ is located within the first octant and is bounded by a surface, which corresponds to a single solution of the direct kinematics. Therefore, the jointspace boundary is defined by the relation $B^2 = 4AC$ (see equation (14)), which may be rewritten as

$$\left(\rho_x^2 + \rho_y^2 + \rho_z^2 - 4L^2\right)\left(\rho_x^{-2} + \rho_y^{-2} + \rho_z^{-2}\right) = 1 \tag{18}$$

and solved for ρ_x assuming that ρ_y, ρ_z are known, $D\rho_x^4 + DE\rho_x^2 + E = 0$, where $D = \rho_y^{-2} + \rho_z^{-2}$; $E = \rho_y^2 + \rho_z^2 - 4L^2$. However, this equation is non-symmetrical with respect to ρ_x, ρ_y, ρ_z and, therefore, is not convenient the real-time control. An alternative way to obtain the jointspace boundary, which is more computationally efficient, is based on the conversion from Cartesian to spherical coordinates, $\rho_x = e_x t$, $\rho_y = e_y t$, $\rho_z = e_z t$, where $t \geq 0$ is the length of the vector ρ, and (e_x, e_y, e_z) are the components of the unit direction vector, which are expressed via two angles φ, θ with $e_x = \cos\varphi\cos\theta$, $e_y = \cos\varphi\sin\theta$, $e_z = \sin\varphi$, where $\varphi, \theta \in \left]0, \pi/2\right]$. For such a notation, the original equation (18) is transformed into a linear equation for t^2, $(F-1)t^2 = 4L^2F$, $F = e_x^{-2} + e_y^{-2} + e_z^{-2}$, with an obvious solution $t = 2L\sqrt{F/(F-1)}$. As follows from its analyses, the bounding surface is close to the 1/8th of the sphere S_{2L}. At the edges, which are exactly quarters of the circles of the radius $2L$, the surface touches the sphere. However, in the middle, the surface is located out of the sphere. In particular, the intersection point of the first octant bisector is located at the distance $\sqrt{3/2} \approx 1.22$ from the coordinate system origin for the jointspace border and at the distance $2/\sqrt{3} \approx 1.15$ for the sphere S_{2L} (assuming L=1).

6. Conclusions

This article focuses on the kinematics and workspace analysis of the Orthoglide, a 3-DOF parallel mechanism with a kinematic behaviour close to the conventional Cartesian machine taking into account the specific manufacturing constraints in the joint variables. We proposed a formal definition of the configuration indices and developed new simple analytical expressions for the Orthoglide inverse/direct kinematics. It was proved that, for the considered joint limits, the Orthoglide workspace is composed of two fractions, the sphere and a thin non-convex solid in which there are 1 and 8 inverse kinematic solutions, respectively. The total workspace volume comprises about 53% of the corresponding serial machine workspace, where over 52% belongs to the sphere (for comparison, releasing of the joint limits yields to an increase of up to 59% in the workspace volume). It was also shown, that the Orthoglide jointspace is bounded by surface with circular edges, which is more convex than the sphere but is rather close to it. These results can be further used for the optimisation of the Orthoglide parameters, which is the subject of our future research work.

7. References

Asada, H. and Slotine, J.J., (1986),"*Robot Analysis and Control,*" John Wiley & Sons.

Fu, K.S., Gonzalez, R. and Lee, C.S.G., (1987), "*Robotics: Control, Sensing, Vision, and Intelligence,*" McGraw-Hill.

Craig, J.J., (1989), "*Introduction to Robotics: Mechanics and Control,*" Addison Wesley.

Tsai, L.W., (1999), "*Robot Analysis, The Mechanics of Serial and Parallel Manipulators,*" John Wiley & Sons.

Angeles, J., (2002), "*Fundamentals of Robotic Mechanical Systems,*" Springer-Verlag.

Stewart, D., (1965), "A Platform with Six Degrees of freedom," *Proc. of the Institution of Mechanical Enginners,* Vol. 180/1, No. 15, 371-386.

Merlet, J.-P., (2000), "*Parallel Robots,*" Kluwer Academic Publishers.

Carricato, M. and Parenti-Castelli, V., (2002), "Singularity-Free Fully-Isotropic Translational Parallel Mechanisms," *The International Journal of Robotics Research* Vol. 21, No. 2, 161-174.

Kong, X., and Gosselin, C.M., (2002), "A Class of 3-DOF Translational Parallel Manipulators with Linear I-O Equations," *Workshop on Fundamental Issues and Future Research Directions for Parallel Mechanisms and Manipulators,* Quebec, Canada, pp. 161-174.

Wenger, Ph. and Chablat, D., (2002), "Kinematic Analysis of a New Parallel Machine Tool: the Orthoglide," *7th International Symposium on Advances in Robot Kinematics,* Piran-Portoroz, Slovenia, pp. 305-314.

Chablat, D. and Wenger, Ph., (2003), "Architecture Optimization of a 3-DOF Parallel Mechanism for Machining Applications, the Orthoglide," *IEEE Trans. on Robotics and Automation,* Vol. 19/3, pp. 403-410.

A CARTESIAN REPRESENTATION FOR THE BOUNDARY WORKSPACE OF 3R MANIPULATORS

Erika Ottaviano*, Manfred Husty** and Marco Ceccarelli*

* *LARM: Laboratory of Robotics and Mechatronics – DiMSAT – University of Cassino Via Di Biasio 43 - 03043 Cassino (Fr), Italy*
{ ottaviano,ceccarelli }@unicas.it

** *Institute for Engineering Mathematics Geometry and Computer Science University of Innsbruck, Technikerstr. 13 A-6020 Innsbruck, Austria*
manfred.husty@uibk.ac.at

Abstract In this paper an algebraic formulation is presented for the boundary workspace of 3R manipulators in Cartesian Space. It is shown that the cross-section boundary curve can be described by a 16-th order polynomial as function of radial and axial reaches. The cross-section boundary curve and workspace boundary surface are fully cyclic. Geometric singularities of the curve are identified and characterized. Numerical examples are presented to show the usefulness of the proposed investigation.

Keywords: Kinematics, Serial Manipulators, Workspace, Void, Geometric Singularities.

1. Introduction

The computation of the workspace and its boundary is of significant interest because of their impact on manipulator design and trajectory planning. An algebraic formulation for determining the workspace of 3R manipulators has been presented in (Ceccarelli, 1989) and then generalized for nR manipulators in (Ceccarelli, 1996). Furthermore, the proposed formulation has been used for design purposes in (Ceccarelli, 1995), and optimal design procedure in (Lanni et al., 2002). Most of the work has been done on 3R manipulators for either positioning (Freudenstein and Primrose, 1984; or orienting tasks (Angeles, 1988). The determination of the workspace boundary in Cartesian Space has been attempted also by (Smith and Lipkin, 1993). Other papers are related to the singularity of the Jacobian matrix that is usually expressed in the Joint Space. Regions that are free of singularities in the Joint Space have been named C-sheets, (Burdick, 1995). In C-sheets it is possible to change posture without passing through singularities (Parenti-Castelli and Innocenti, 1988). Manipulators that can change

J. Lenarčič and C. Galletti (eds.), On Advances in Robot Kinematics, 247-254.

posture without meeting a singularity have been named cuspidal manipulators in (Wenger, 2000). Although there are many reports in the field, there are still open problems in the characterization of the workspace boundary. It is of practical interest to detect the presence of a void in the workspace, since it is a region of unreachable points for the end-effector of the manipulator. A ring void is a void that is buried within the workspace volume and has ring topology, (Ceccarelli, 1996). The presence of cusps and void can be detected by analyzing the geometric singularities of the cross-section boundary curve (Ottaviano et al., 1999). Analysis of cusps in the workspace boundary has been proposed also in (Burdick, 1995; Saramago et al., 2002).

In this paper an algebraic formulation is presented for the representation of the boundary workspace of 3R manipulators in Cartesian Space. Characteristic points and geometric singularities of the cross-section of the curve of the workspace are identified.

2. An algebraic formulation for the workspace boundary of 3R manipulators

A general 3R manipulator is sketched in Fig.1, in which the kinematic parameters are denoted by the standard Hartenberg and Denavit (H-D) notation. Without loss of generality the base frame is assumed to be coincident with $X_1Y_1Z_1$ frame when $\theta_1= 0$, $a_0=0$ and $d_1=0$. Point H is placed on the X_3 axis at a distance a_3 from O_3, as shown in Fig.1. The general 3R manipulator is described by the H-D parameters a_1, a_2, d_2, d_3, α_1 and α_2, and θ_i, for $(i = 1,...,3)$, as shown in Fig.1. r is the distance of point H from the Z_1-axis and z is the axial reach, both are expressed in H-D parameters. The position workspace of the 3R manipulator can be obtained by a θ_1 rotation of the generating torus that is traced from H by full revolution of θ_2 and θ_3.

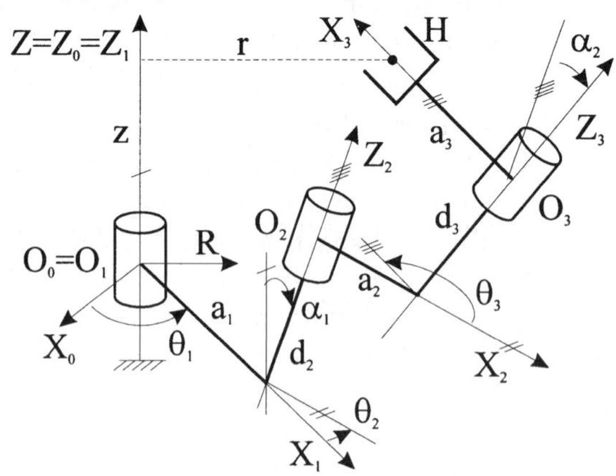

Figure1. A kinematic scheme of a 3R manipulator.

Alternatively, the boundary of position workspace can be determined by considering the envelope surface of the torus family that is traced as function of θ_3, when a torus of the family is obtained by full revolution of θ_1 and θ_2, as outlined in (Ceccarelli 1989, and 1996). In this case, the equation of the workspace boundary of point H can be expressed in the cross-section plane R-Z of the base frame when $\alpha_1 \neq 0$, $C \neq 0$ and $E \neq 0$, in the form of so-called ring equation (Ceccarelli 1989, 1996),

$$z = \frac{-L \pm Q^{1/2}}{KC} - \frac{D}{C}, \quad r = \left[A - z^2 + \frac{(Cz + D)G + F}{E} \right]^{1/2},$$
$$L = FG, \quad K = G^2 + E^2, \quad Q = L^2 - K\left(F^2 + BE^2\right). \tag{1}$$

in which A, B,C,D, E, F and G are named as architecture coefficients.

The sign ambiguity in z expression can be solved to give two envelope branches that together give the envelope of the torus family.

Architecture coefficients are function of H-D parameters in the form

$$A = a_1^2 + r_2 + \left(z_2 + d_2\right)^2, \; B = -4a_2^2 r_2, \; C = \frac{2a_1}{s\alpha_1}, \; D = -2a_1\left(z_2 + d_2\right)\frac{c\alpha_1}{s\alpha_1},$$
$$E = -2a_3\left(a_2 s\theta_3 + d_2 s\alpha_2 c\theta_3\right), G = \frac{2a_1 a_3 c\theta_3 s\alpha_2 c\alpha_1}{s\alpha_1}, \tag{2}$$
$$F = 4a_1^2 a_3\left(a_2 s\theta_3 + a_3 s\theta_3 c\theta_3 \left(s\alpha_2\right)^2 - d_3 c\alpha_2 s\alpha_2 c\theta_3\right).$$

in which

$$r_2 = \left(a_3 c\theta_3 + a_2\right)^2 + \left(a_3 s\theta_3 c\alpha_2 + d_3 s\alpha_2\right)^2, \qquad z_2 = d_3 c\alpha_2 - a_3 s\theta_3 s\alpha_2. \tag{3}$$

where $c\alpha_i$, $s\alpha_i$, $c\theta_i$ and $s\theta_i$ denote $\cos\alpha_i$, $\sin\alpha_i$, $\cos\theta_i$ and $\sin\theta_i$, respectively.

The cross-section boundary curve f of the 3R manipulator workspace in R-Z plane can be thought as the envelope of the θ_3-family curves. A generating torus of the enveloping θ_3-family can be expressed as

$$\left(r^2 + z^2 - A\right)^2 + \left(Cz + D\right)^2 + B = 0. \tag{4}$$

In order to obtain a closed-form expression in R-Z coordinates from the ring equation (1), it is convenient to manipulate Eq.(4) by performing the half-tangent substitution $u = \tan(\theta_3/2)$. This yields

$$k_4 u^4 + k_3 u^3 + k_2 u^2 + k_1 u + k_0. \tag{5}$$

Parameters k_i ($i = 0,\ldots,4$) of Eq.(5) are functions of the H-D parameters, z and r. They can be expressed in the form

$$k_i = K_{i40}\, r^4 + K_{i20}\, r^2 + K_{i00} + K_{i04}\, z^4 + K_{i02}\, z^2 + K_{i22}\, r^2 z^2 + K_{i01}\, z\,. \tag{6}$$

As the coefficients k_i are functions of the parameters r, z, Eq. (5) represents a one-parameter family of curves. The envelope of this family is obtained by eliminating u from Eq. (5) and its derivative with respect u. This yields to the function f expressing the cross-section boundary curve in the form

$$f = \sum_{i=0}^{4} c_{2i} \left(r^2 + z^2\right)^{2i} + \sum_{i=1}^{4} c_{2i-1}\, z \left(r^2 + z^2\right)^{2i-1} = 0\,. \tag{7}$$

Coefficients c_{2i} and c_{2i-1} are non-linear expressions of the H-D parameters and can be obtained from k_i coefficients in Eq. (6) together with Eqs. (2-3). For example, c_8 coefficient can be expressed as

$$c_8 = 65{,}536\, a_3^4 a_1^4 (s\alpha_1)^8 (s\alpha_2)^4\,. \tag{8}$$

Equation (7) represents the Cartesian expression in R-Z coordinates for the cross-section boundary curve for a 3R manipulator. Its expression is function of the H-D parameters through c_i coefficients.

3. Characteristics of the cross-section boundary curve

The cross-section boundary curve f in Eq.(7) is fully cyclic and of 16-th degree; it has 2 singular points at infinity with multiplicity 8 each, that are the 2 circle points. A curve that has the imaginary circular points as double, triple,... points is said to have circularity 2, 3, ... (Hunt, 1978). Therefore, the cross-section boundary curve has circularity 8.

A descriptive proof is reported in Fig.2, which shows the cross-section boundary curve as an envelope with an inner branch, which may give the boundary of a ring void, and an outer branch, which represents the boundary of the bulk workspace ring.

The boundary surface of the 3R manipulator workspace is fully cyclic too and its degree is 16. This can be proved since r^2 in Eq.(7) is expressed as (x^2+y^2), in which x and y are components of the position vector of point H with respect to the base frame. Furthermore, f is symmetric with respect to the Z axis. This can be proved since Eq.(7) contains only even powers of r.

The degree of the curve does not change under rotation because of its symmetry.

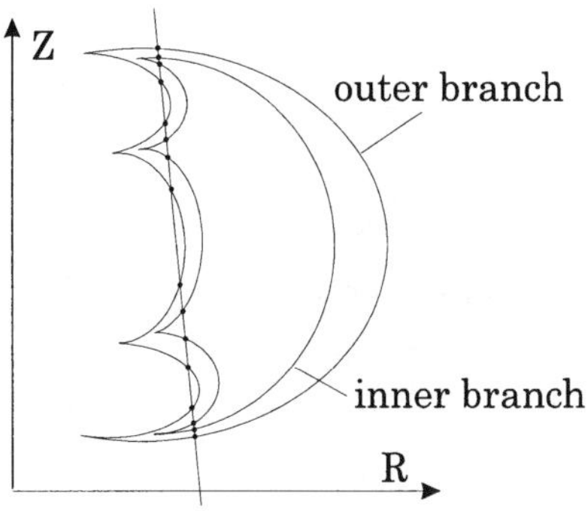

Figure 2. A descriptive representation of the 16-th degree of the cross-section boundary curve.

3.1 Special cases

The polynomial expression of Eq.(7) has the degree lower than 16 iff:
- a_3=0. This means that the operation point H coincides with the origin O_3 of the $X_3Y_3Z_3$ reference frame. In such case the θ_3 rotation has no effect on the workspace and the ring workspace degenerates in a 2R chain workspace.
- α_1=0. This is not allowed since the formulation implies that $\alpha_1 \neq 0$.
- α_2=0. This means that Z_2 and Z_3 axes are parallel. The outcome of the computation is that the cross-section boundary curve is of degree 12.
- a_1=0. This is not allowed since the formulation implies that $a_1 \neq 0$.

By analyzing Eq.(8) one can also note that c_8 coefficient, that gives the full degree of the curve, does not depend on a_2, d_2 and d_3 parameters.

3.2 Geometric singularities

Real geometric singularities of f can be found by introducing the homogeneous coordinate w in Eq.(7). Thus, one can consider Eq.(7) as function of r, z and w, and furthermore its partial derivatives f_r f_z and f_w with respect to r, z, and the homogeneous coordinate w, respectively. The zeros of the set of equations: f = 0; f_r = 0; f_z = 0; and f_w = 0 gives the geometric singularities of the cross-section boundary curve f. The points belonging to these zeros are denoted by C_i, and D_i, where C_i indicates cusps and D_i double points, referring to a geometric interpretation that is shown in Fig. 3. They can be classified by considering the second partial derivatives g of f in the form (Gibson, 1998)

$$g = f_{rz}^2 - f_{rr} f_{zz} . \tag{9}$$

The functions g and f_r f_z and f_w can be useful to fully characterize real geometric singularities in the workspace of the 3R manipulators.

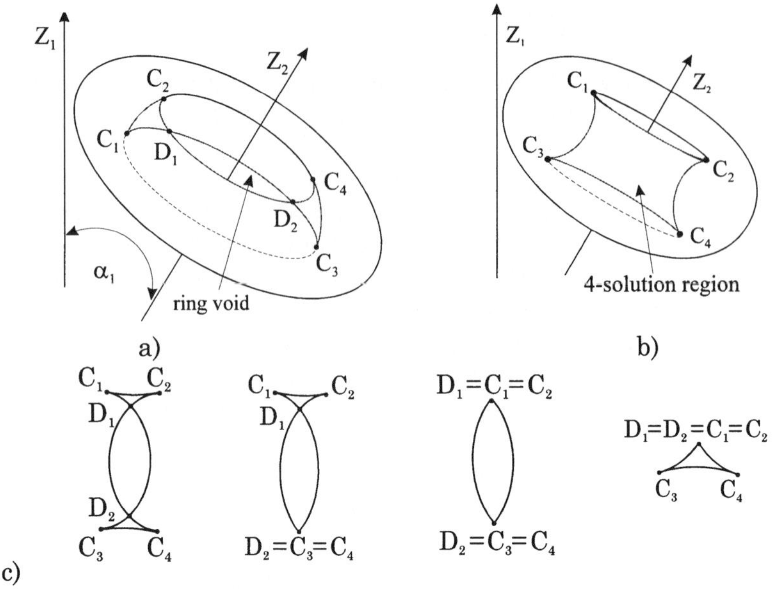

Figure 3. Characteristic points in the workspace boundary of 3R manipulators: a) a general case; b) a cuspidal manipulator; c) envelopes for void determination.

A point D_i, whose coordinates are (rd_i,zd_i), is a double point iff Eq.(9) is greater than zero. A point C_i, whose coordinates are (rc_i,zc_i), is a cusp iff Eq.(9) is equal to zero. A condition on Eq.(9) is useful to characterize the cross-section boundary curve and ring void of the workspace of 3R manipulators through the points shown in Fig. 3.

An enumeration of all possible types of ring void has been presented in (Ottaviano et al., 1999) by analyzing the internal branch of the cross-section boundary envelope curve, as shown in the examples of Fig.3c). Internal branch of the boundary envelope curve in the cross section R-Z shows generally 3 loops. The middle loop delimits the ring void and it is a part of the boundary curve; the others are related to 4-solution regions for the Inverse Kinematics problem. A ring void is characterized by the presence of 2 double points, which have been named as D_1 and D_2 in Fig. 3. Points D_1 and D_2 are double points, i.e. intersections of the inner branch. If there are no double points but there are cusps there is not a void inside the workspace. In this case, the manipulator belongs to the class of the cuspidal manipulators, Fig.3b). Double points indicate the presence of a void that corresponds to a 0-solution region for the Inverse Kinematics problem. Cusp points C_1 to C_4 can be considered characteristic points for the cross-section boundary curve since they delimit special workspace regions with multiple configuration reachability.

4. Numerical examples

Figure 4a) shows the cross-section boundary curve of a general case like in Fig. 3. The inner loop is characterized by the presence of 2 double points and represents a void. Other 2 loops that are characterized by the cusps are reachable regions of the workspace. Figure4b) shows a zoomed view with plots of the inner branch together with the plots of the f derivatives. It is possible to observe that the derivatives f_r, f_z, and f_w of the cross-section boundary curve meet at the 3 singular points D_1, C_1 and C_2. Their coordinates can be computed as the zeros of the set of the f derivatives. Their coordinates are: D_1=[3.82, 6.62]; C_1=[1.87, 7.12]; C_2=[3.09, 7.16]. The nature of those characteristic points has been checked by Eq.(9). It has been verified that C_1 and C_2 are cusps, since g is equal to zero, and D_1 is a double point, since g is greater than zero. In Fig.5 the workspace boundary of a cuspidal manipulator is shown.

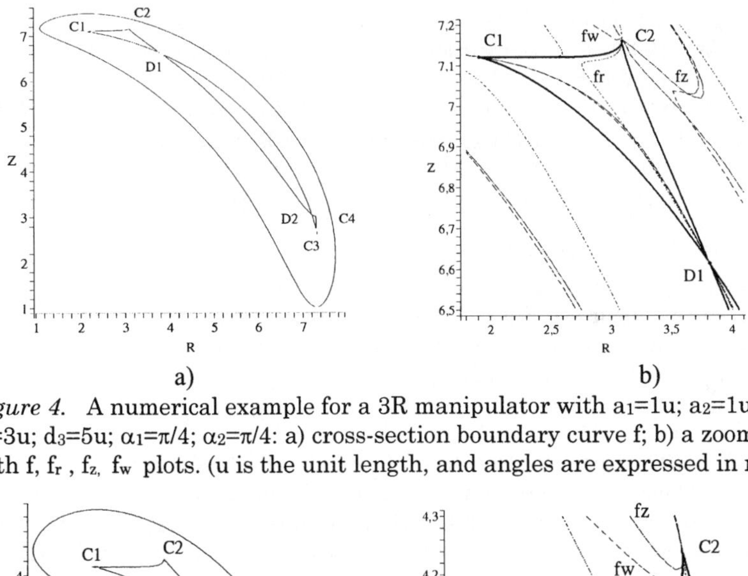

a) b)

Figure 4. A numerical example for a 3R manipulator with a_1=1u; a_2=1u; a_3=1u; d_2=3u; d_3=5u; $\alpha_1=\pi/4$; $\alpha_2=\pi/4$: a) cross-section boundary curve f; b) a zoomed view with f, f_r , f_z, f_w plots. (u is the unit length, and angles are expressed in radians)

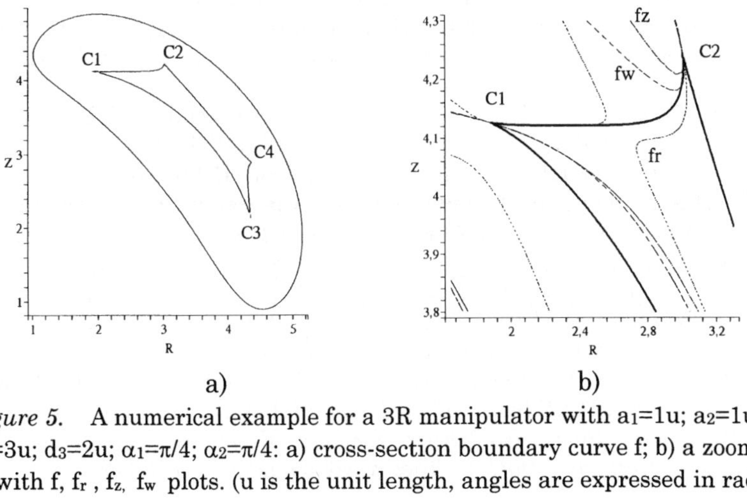

a) b)

Figure 5. A numerical example for a 3R manipulator with a_1=1u; a_2=1u; a_3=1u; d_2=3u; d_3=2u; $\alpha_1=\pi/4$; $\alpha_2=\pi/4$: a) cross-section boundary curve f; b) a zoomed view with f, f_r , f_z, f_w plots. (u is the unit length, angles are expressed in radians)

6. Conclusions

An algebraic formulation has been presented for a Cartesian representation of the boundary workspace of 3R manipulators. The cross-section boundary curve and boundary surface are of 16-th degree and fully cyclic. Geometric singularities of the cross-section boundary curve are identified as delimiting ring void and 4-solution regions. Numerical examples have been presented to outline the practical feasibility of the proposed formulation. The proposed analysis can be useful for design purposes and to deduce an evaluation of the workspace volume by integrating the boundary curve.

References

Angeles J. (1988), Special Loci of the Workspace Spherical Wrists, *Proc. 1st Int. Workshop on Advances in Robotics Kinematics*, Ljubljana, pp. 36-45.

Burdick, J.W. (1995), A Classification of 3R Regional Manipulator Singularities and Geometries, *Mech. and Machine Theory*, vol.30, no.1, pp. 71-89.

Ceccarelli M. (1989), On the Workspace of 3R Robot Arms, *Proc. 5th IFToMM Int. Symp. on Theory and Practice of Mechanism*, Bucharest, Vol. II-1, pp. 37-46.

Ceccarelli M. (1995), A Synthesis Algorithm for Three-Revolute Manipulators by Using an Algebraic Formulation of Workspace Boundary, *ASME Jnl. of Mechanical Design*, vol.117, pp.298-302.

Ceccarelli M. (1996), A Formulation for the Workspace Boundary of General N-Revolute Manipulators, *Mech. and Machine Theory*, vol.31, no.5, pp. 637-646.

Freudenstein F., Primrose E.J.F. (1984), On the Analysis and Synthesis of the Workspace of a Three-Link, Turning-Pair Connected Robot Arm, *ASME Jnl of Mech., Transm. and Autom. in Design*, vol.106, pp. 365-370.

Gibson C.G. (1998), Elementary Geometry of Algebraic Curves, Cambridge University Press.

Hunt, K.H. (1978), Kinematic Geometry, Clarendon Press, Oxford.

Lanni C., Saramago S., Ceccarelli M. (2002), Optimum Design of General 3R Manipulators by Using Traditional and Random Search Optimization Techniques,*Jnl. of the Brazilian Society of Mech. Sciences*, 2002, vol.XXIV, n.4, pp.293-301.

Ottaviano E., Ceccarelli M., Lanni C. (1999), A Characterization of Ring Void in Workspace of Three-Revolute Manipulators, *Proc. 10th World Congress on the Theory of Machines and Mechanisms*, Oulu, vol.3, pp. 1039-1044.

Parenti-Castelli V., Innocenti C. (1988), Position Analysis of Robot Manipulators: Regions and Sub-Regions, *Proc. 1st Int. Conf. on Advances in Robot Kinematics*, Ljubljana, pp.150-158.

Saramago S., Ottaviano E., Ceccarelli M. (2002), A Characterization of the Workspace Boundary of Three-Revolute Manipulators, *Proc. ASME DEC-Mech. and Robotic Conf.*, Montreal, paper DETC2002/MECH-34342.

Smith D.R., Lipkin H. (1993), Higher Order Singularities of Regional Manipulators, *Proc. IEEE Int. Conf. on Rob. and Autom.*, vol.1, pp. 194-199.

Wenger P. (2000), Some Guidelines for the Kinematic Design of New Manipulators, *Mech. and Machine Theory*, vol.35, no.3, pp.437-449.

KINEMATIC ISOTROPY AND DYNAMIC PERFORMANCES IN TRANSLATIONAL PARALLEL MANIPULATORS

Raffaele Di Gregorio
Department of Engineering - University of Ferrara
Via Saragat, 1 – 44100 FERRARA, ITALY
rdigregorio@ing.unife.iti

Vincenzo Parenti-Castelli
DIEM – University of Bologna
Viale Risorgimento, 2 – 40136 BOLOGNA, ITALY
vincenzo.parenticastelli@mail.ing.unibo.it

Abstract The dynamic model of a generic translational parallel manipulator (TPM) is written in a way that highlights the terms influenced by the eventual kinematic isotropy of the manipulator architecture and those that are determinant for obtaining high dynamic performances. The presented model shows that performances are strongly conditioned by the types of motion exhibited by the manipulator leg links for a given platform motion. Finally, the model allows an ideal TPM to be identified which, for the same mobile masses, has the best kinematic and dynamic performances of all TPMs and can be used as a reference to evaluate the performances of any TPMs.

Keywords: Parallel manipulators, kinematic isotropy, dynamic performances

1. Introduction

Translational parallel manipulators (TPMs) are manipulators with three degrees of freedom (dof) where a mobile platform, which can only translate, is connected to the frame (base) through a number of kinematic chains (legs). In the TPMs, the platform translation is the result of the constraints on the platform motion due to the legs.

TPMs have attracted the attention both of the academic and industrial worlds since Clavel proposed the DELTA robot (Clavel, 1988) which is a TPM with high dynamic performances. The success of the DELTA robot led to think that high dynamic performances are a characteristic common to all TPMs. This opinion is partly justified by the fact that a parallel architecture usually allows the actuators to be located on or near to the base, which greatly reduces the mobile masses. Nevertheless, obtaining

J. Lenarčič and C. Galletti (eds.), On Advances in Robot Kinematics, 255-264.

high dynamic performances is mainly related to a good global design of the machine.

In the literature, many TPMs with three topologically equal legs have been proposed (Hervé, 1995; Tsai and Stamper, 1996; Tsai, 1996; Di Gregorio and Parenti-Castelli, 1998; Tsai, 1999; Zhao and Huang, 2000; Di Gregorio, 2001; Carricato and Parenti-Castelli, 2001; Kim and Tsai, 2002; Kong and Gosselin, 2002b). Moreover, type syntheses which identify leg topologies for TPMs have also been presented (Hervé and Sparacino, 1991; Frisoli et al., 2000; Carricato and Parenti-Castelli, 2002; Kong and Gosselin, 2002a).

Parallel manipulators exhibit singularities that are configurations where the relationship between actuated-joint rates and platform twist is no longer one to one (see Gosselin and Angeles, 1990; Zlatanov et al., 1995). When the manipulator is in a singularity, it becomes uncontrollable and, in some of them (direct problem singularities (Gosselin and Angeles, 1990)), it undergoes infinite internal loads. Therefore, all the singularities must be found at the design stage and avoided during operation.

Moreover, from a static point of view, the best load transmission between actuators and platform occurs when the manipulator assumes kinematically isotropic configurations (Yoshikawa, 1985; Gosselin and Angeles, 1991). Manipulators whose configurations are kinematically isotropic in the whole workspace are called globally isotropic.

Because of these reasons, TPMs that are free from singularities and globally isotropic deserve special attention (Carricato and Parenti-Castelli, 2002; Kong and Gosselin, 2002a, 2002b; Kim and Tsai, 2002).

Recently, some indices for measuring the dynamic performances of three-dof manipulators have been presented (Di Gregorio and Parenti-Castelli, 2002). Such indices have been used in (Di Gregorio and Parenti-Castelli, 2003) to compare the dynamic performances of a TPM that is not globally isotropic (the translational 3UPU (Tsai, 1996)), with the ones of a globally isotropic TPM (the Cartesian 3PRRR (Tsai, 2002)). The result of this comparison showed that choosing a globally isotropic architecture does not necessarily lead to improve the dynamic performances of the manipulator.

This paper presents the dynamic model of a TPM in a way that highlights both the terms influenced by the eventual kinematic isotropy of the architecture and the ones that are determinant for obtaining high dynamic performances. Moreover, it shows that the dynamic performances are strongly conditioned by the types of motion exhibited by the legs' links for the same platform motion. Finally, the model identifies an ideal TPM which, for the same mobile masses, has the best kinematic and dynamic performances of all TPMs and can be used as a reference to evaluate the performances of any TPM.

2. General dynamic model of a TPM

Figure 1 shows a generic TPM and the notations. With reference to Fig. 1, n is the number of legs. r_i, i=1,...,n, is the number of links of the i-th leg. Link ij, i=1,...,n and j=1,...,r_i, is the j-th link of the i-th leg. S_b is a Cartesian reference system fixed to the base. S_p is a Cartesian reference system fixed to the platform. O_p is the origin of S_p. $\mathbf{p}\equiv\{x, y, z\}^T$ is the position vector, measured in S_b, which locates O_p.

Figure 2 shows the link ij. With reference to Fig. 2, G_{ij} is the center of mass of the link ij. S_{ij} is a Cartesian reference system that is fixed to the link ij and has the origin in G_{ij}.

Moreover, m_{ij} (m_p) will be the mass of the link ij (the platform). ${}^{ij}\mathbf{I}_{ij}$ will be the inertia tensor of the link ij referred to G_{ij} and measured in S_{ij}, whereas ${}^{b}\mathbf{I}_{ij}$ will be the same tensor measured in S_b. The relationship between ${}^{ij}\mathbf{I}_{ij}$ and ${}^{b}\mathbf{I}_{ij}$ is

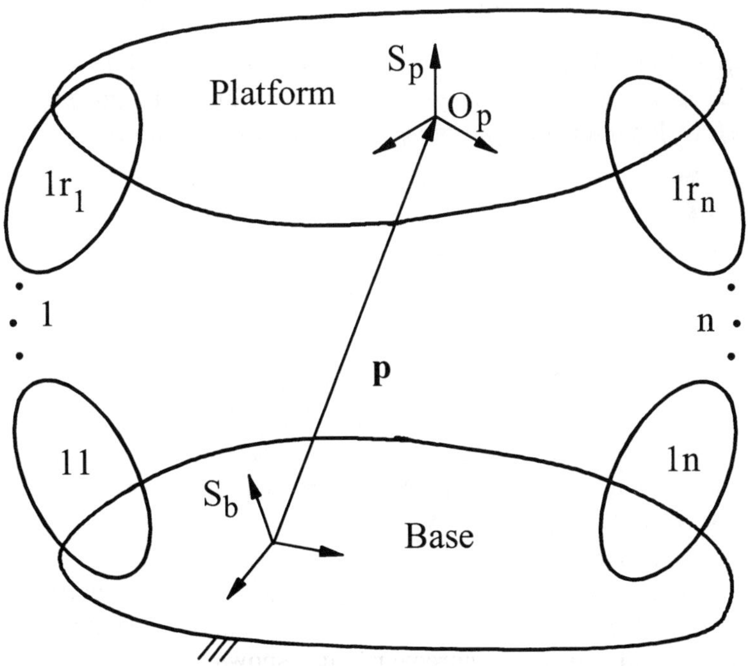

Figure 1. Generic TPM

$$ {}^{b}\mathbf{I}_{ij} = {}^{b}\mathbf{R}_{ij}\, {}^{ij}\mathbf{I}_{ij}\, {}^{b}\mathbf{R}_{ij}^{T} \tag{1} $$

where ${}^{b}\mathbf{R}_{ij}$ is the rotation matrix which transforms vector components measured in S_{ij} into vector components measured in S_b.

Hereafter, x, y and z (the components of $\mathbf{p}$) will be the generalized coordinates of the TPM. $\mathbf{q}\equiv\{q_1, q_2, q_3\}^T$ will be the vector whose components are the actuated-joint variables. $\tau\equiv\{\tau_1, \tau_2, \tau_3\}^T$ will be the vector whose components are the forces and/or torques applied by the actuators in the actuated joints.

With these notations, in the following part of this section, firstly, the expressions of the TPM's kinetic energy and of the generalized forces applied to the TPM will be written; then, by using the Lagrangian formulation of dynamics, the dynamic model of the TPM will be derived.

2.1 Kinetic energy

The kinetic energy of a TPM can be written as follows

$$ E = \frac{1}{2}\dot{\mathbf{p}}^{T}\,\mathbf{N}\,\dot{\mathbf{p}} \tag{2} $$

where $\dot{\mathbf{p}}$ is the velocity of the origin of S_p (i.e. the first time derivative of $\mathbf{p}$) and $\mathbf{N}$ is the generalized inertia matrix, referred to the platform, of the TPM. $\mathbf{N}$ depends only on $\mathbf{p}$ and its explicit expression is

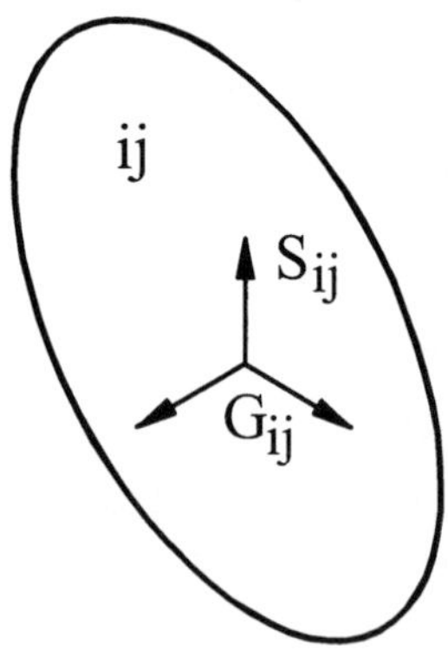

Figure 2. Link ij

$$\mathbf{N} = m_p \mathbf{1} + \sum_{i=1,n} \sum_{j=1,r_i} (m_{ij} \mathbf{T}_{ij}^T \mathbf{T}_{ij} + \mathbf{H}_{ij}^T \, {}^b\mathbf{I}_{ij} \, \mathbf{H}_{ij}) \tag{3}$$

where **1** is the 3×3 identity matrix; whereas $\mathbf{T}_{ij}$ and $\mathbf{H}_{ij}$ are the 3×3 matrices appearing in the relationships

$$\dot{\mathbf{G}}_{ij} = \mathbf{T}_{ij} \, \dot{\mathbf{p}}, \qquad \omega_{ij} = \mathbf{H}_{ij} \, \dot{\mathbf{p}} \tag{4}$$

($\dot{\mathbf{G}}_{ij}$ and ω_{ij} are the velocity of the center of mass and the angular velocity of the link ij respectively, measured in S_b). Relationships (4) can be written in explicit form through the velocity analysis of the TPM. Matrices $\mathbf{T}_{ij}$ and $\mathbf{H}_{ij}$ contain the geometry of the manipulator and are function only of **p**.

2.2 Generalized forces

Since the generalized coordinates of the TPM are the components x, y and z of **p** (see Fig. 1), the vector $\mathbf{Q} \equiv \{Q_x, Q_y, Q_z\}^T$ whose components are the generalized forces applied to the TPM can be expressed as follows

$$\mathbf{Q} = \mathbf{f}_e + \mathbf{M} \, \mathbf{g} + \mathbf{J}^T \, \tau \tag{5}$$

where $\mathbf{f}_e$ is the resultant of the external forces applied to the platform, projected on S_b. **g** is the gravity-acceleration vector projected in S_b. **M** is the following 3×3 matrix

$$\mathbf{M} = m_p \mathbf{1} + \sum_{i=1,n} \sum_{j=1,r_i} m_{ij} \mathbf{T}_{ij}^T \tag{6}$$

Finally, **J** is the 3×3 Jacobian matrix appearing in the relationship

$$\dot{\mathbf{q}} = \mathbf{J} \, \dot{\mathbf{p}} \tag{7}$$

where $\dot{\mathbf{q}}$ is the first time derivative of **q**. Relationship (7) can be written in explicit form through the velocity analysis of the TPM. **J** contains the geometry of the manipulator and is function only of **p**.

2.3 Dynamic model

The Lagrangian formulation of dynamics allows the following dynamic model to be written for the TPM

$$Q_x = \frac{d}{dt}\left(\frac{\partial E}{\partial \dot{x}}\right) - \frac{\partial E}{\partial x} \tag{8.1}$$

$$Q_y = \frac{d}{dt}\left(\frac{\partial E}{\partial \dot{y}}\right) - \frac{\partial E}{\partial y} \tag{8.2}$$

$$Q_z = \frac{d}{dt}\left(\frac{\partial E}{\partial \dot{z}}\right) - \frac{\partial E}{\partial z} \tag{8.3}$$

By using relationship (5) and expanding the derivative at the right-hand sides of Eqs. (8), the dynamic model (8) can be written in matrix form as follows

$$\mathbf{f}_e + \mathbf{M}\,\mathbf{g} + \mathbf{J}^T \tau = \mathbf{N}\ddot{\mathbf{p}} + \mathbf{c}(\mathbf{p},\dot{\mathbf{p}}) \tag{9}$$

where $\ddot{\mathbf{p}}$ is the acceleration of the origin of S_p (i.e. the second time derivative of $\mathbf{p}$) and

$$\mathbf{c}(\mathbf{p},\dot{\mathbf{p}}) = \left(\dot{x}\frac{\partial \mathbf{N}}{\partial x} + \dot{y}\frac{\partial \mathbf{N}}{\partial y} + \dot{z}\frac{\partial \mathbf{N}}{\partial z}\right)\dot{\mathbf{p}} - \frac{1}{2}\left(\dot{\mathbf{p}}^T\frac{\partial \mathbf{N}}{\partial x}\dot{\mathbf{p}},\ \dot{\mathbf{p}}^T\frac{\partial \mathbf{N}}{\partial y}\dot{\mathbf{p}},\ \dot{\mathbf{p}}^T\frac{\partial \mathbf{N}}{\partial z}\dot{\mathbf{p}}\right)^T \tag{10}$$

The term $\mathbf{c}(\mathbf{p},\dot{\mathbf{p}})$ takes into account all the non-linear inertia forces applied to the TPM (i.e. Coriolis' and centrifugal forces). Relationship (10) shows that $\mathbf{c}(\mathbf{p},\dot{\mathbf{p}})$ is roughly proportional to the square of the platform velocity, $\dot{\mathbf{p}}$, and vanishes, in the whole workspace, if the generalized inertia matrix $\mathbf{N}$ does not depend on $\mathbf{p}$ (i.e. is constant).

3. Kinematic isotropy and dynamic performances

In (Gosselin and Angeles, 1991), the concept of kinematically isotropic configuration has been defined. According to this definition, the configuration of a TPM is kinematically isotropic when the condition number of the Jacobian matrix $\mathbf{J}$ (see Eq. (7)) is equal to 1. Since $\mathbf{J}$ is a 3×3 square matrix, this condition is verified when the eigenvalues of $\mathbf{J}$ are all equal, which implies that the column (row) vectors of $\mathbf{J}$ are mutually orthogonal and have the same magnitude (i.e. a constant λ exists so that the matrix $\lambda\mathbf{J}$ is orthogonal). When a manipulator assumes a kinematically isotropic configuration, the loads applied to the platform are distributed at best among the actuators. Moreover, since 1 is the lowest value the condition number of $\mathbf{J}$ can assume, whereas the same condition number become infinite in correspondence of a TPM singularity, the isotropic configurations are the farthest from singularities.

Both these properties of isotropic configurations are searched for during the design stage of a manipulator. A well designed TPM must have the condition number of **J** near to 1 in the whole workspace. The properties of matrix **J** are determined by the architecture of the TPM. Nevertheless, the use of a globally isotropic architecture (i.e. with condition number of **J** equal to 1 in the whole workspace) does not necessarily imply any satisfactory manipulator's dynamic performances which, indeed, are related to the properties of the generalized inertia matrix **N** (Di Gregorio and Parenti-Castelli, 2002).

The explicit expression of **N** (see Eq. (3)) shows that **N** depends on the mass distribution (which is described by the constants m_p, m_{ij} and ${}^{ij}\mathbf{I}_{ij}$, and by the matrix $\mathbf{T}_{ij}$ that, in general, varies with **p**) and on the manipulator architecture (which enters in the matrices $\mathbf{H}_{ij}$ and $\mathbf{T}_{ij}$ that, in general, are both function of **p**). Since the manipulator architecture influences both **J** and **N**, a TPM architecture which is chosen only considering the properties of **J** may happen to have worse dynamic performances than another having **J** with a greater condition number.

From a dynamic point of view, architectures which allow the mass distribution to be improved with suitable choices of material and/or link geometry must be preferred.

For a given value of the mobile masses, the best dynamic performances occur when matrix **N** does not depend on **p** (i.e. it is constant in the whole workspace) and the column (row) vectors of **N** are mutually orthogonal and have the same magnitude (i.e. a constant ν exists so that the matrix $\nu\mathbf{N}$ is orthogonal). Such an **N** matrix has the eigenvalues that are all equal in the whole workspace (i.e. all the manipulator configurations are dynamically isotropic (Di Gregorio and Parenti-Castelli, 2002)). Moreover, it guarantees the cancellation of the non-linear inertia forces (i.e. the term $\mathbf{c}(\mathbf{p}, \dot{\mathbf{p}})$ given by (10)) in the whole workspace.

When the **N** matrix of a TPM satisfies these conditions, by simply orientating the axes of S_b along the direction of the column (row) vectors of **N**, Eqs. (9) (which are the dynamic model of the TPM) become identical to the motion equations of a heavy point whose mass is equal to the eigenvalue with multiplicity 3 of matrix **N**. Moreover, the smaller the eigenvalue, the faster the TPM. For this reason, a TPM that is dynamically isotropic in the whole workspace, and has null non-linear inertia term, will be called point-like TPM.

It is easy to show that a TPM with three legs of type PPP (P stands for prismatic pair), in which the sliding directions of the prismatic pairs are parallel to the coordinate axes of S_b (Fig. 3), is a point-like TPM (point-like 3PPP) if the sums of the masses of the links that translate along the same direction are all equal (i.e., in the case of Fig. 3, if $m_{31}+m_{32}+m_{22}=m_{21}+m_{22}+m_{12}=m_{11}+m_{12}+m_{32}$). Carricato and Parenti-Castelli (2002), showed that the point-like 3PPP is also globally isotropic and free from singularities. Therefore, when considering all the TPMs for a given value of the mobile masses, the point-like 3PPP provides the best kinematic and dynamic performances in the whole workspace. Unfortunately, the fact that the point-like 3PPP is overconstrained and has a great number of passive prismatic pairs discouraged designers from building it for practical applications. For this reason, the point-like 3PPP can be considered an ideal TPM with reference to which the performances of TPMs can be compared for the same mobile masses. Moreover, realizing a commercial point-like 3PPP may be an interesting goal.

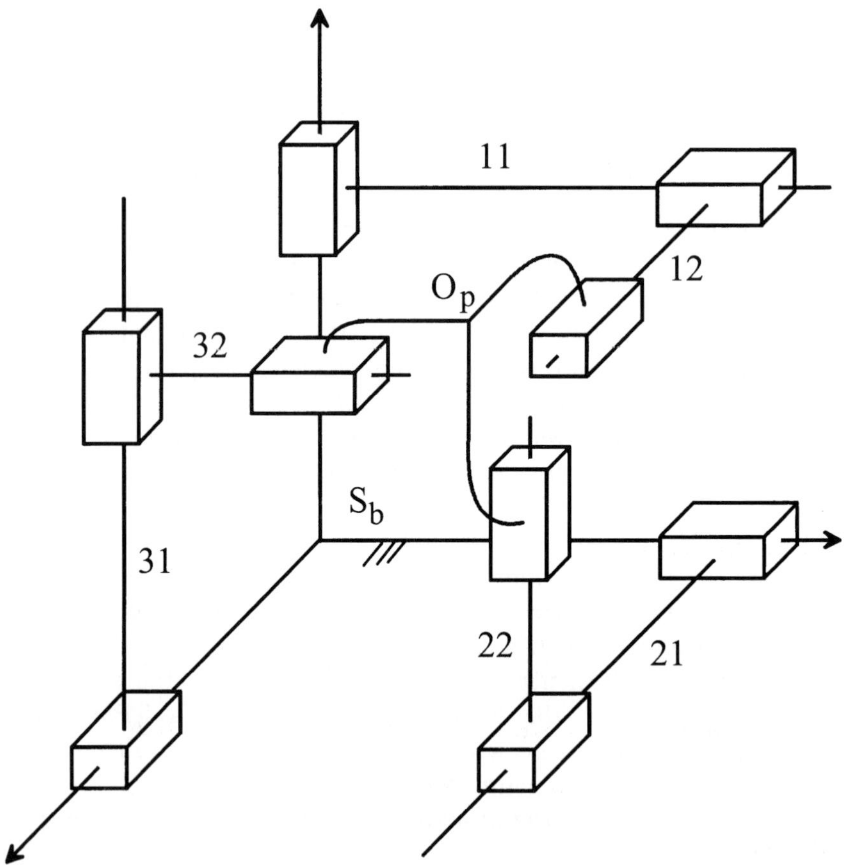

Figure 3. Point-like 3PPP

4. Conclusions

The dynamic model of a general translational parallel manipulator (TPM) has been proposed. By using the model, it has been shown that simply searching kinematic isotropy in the design of a TPM does not usually involve finding a TPM architecture with improved dynamic performances but, sometimes, may leads to worsening the dynamic performances. Finally, by discussing the same model, an ideal TPM that behaves like a heavy point and is kinematically isotropic in the whole workspace, has been identified.

Acknowledgments

This paper has been supported by funds of the Italian MIUR.

References

Carricato, M. and Parenti-Castelli, V. (2001), Position Analysis of a new family of 3-dof parallel manipulator, *Proc. of the 2001 ASME Design Engineering Technical Conferences*, Pittsburgh, USA, Paper No. DETC2001/DAC-21036.

Carricato, M. and Parenti-Castelli, V. (2002), Singularity-free fully-isotropic translational parallel mechanisms, *The Int. J. of Robotics Research*, no. 2, vol. 21, pp. 161-174.

Clavel, R. (1988), DELTA, a fast robot with parallel geometry, *Proc. of the 18th Int. Sym. on Industrial Robot*, Lausanne, Switzerland, pp. 91-100.

Di Gregorio, R. (2001), Kinematics of the translational 3-URC mechanism, *Proc. of the 2001 IEEE/ASME Int. Conf. on Advanced Intelligent Mechatronics*, Como, Italy, pp. 147-152.

Di Gregorio, R., and Parenti-Castelli, V. (1998), A translational 3-dof parallel manipulator, *Advances in Robot Kinematics: Analysis and Control*, J. Lenarcic and M.L. Husty eds., Kluwer Academic Publishers, Dordrecht, pp. 49-58.

Di Gregorio, R., and Parenti-Castelli, V. (2002), Dynamic performance indices for 3-dof parallel manipulators, *Advances in Robot Kinematics: Theory and Applications*, J. Lenarcic and F. Thomas eds., Kluwer Academic Publishers, Netherlands, pp. 11-20.

Di Gregorio, R., and Parenti-Castelli, V. (2003), Comparison of 3-dof parallel manipulators based on new dynamic performance indices, *Proc. of the 11th World Congress on the theory of Machines and Mechanisms*, Tianjin, China.

Frisoli, A., Checcacci, D., Salsedo, F. and Bergamasco, M. (2000), Synthesis by screw algebra of translating in-parallel actuated mechanisms, *Advances in Robot Kinematics*, J. Lenarcic and M.M. Stanisic eds., Kluwer Academic Publishers, Dordrecht, pp. 433-440.

Gosselin, C.M. and Angeles, J. (1990), Singularity analysis of closed-loop kinematic chains, *IEEE Trans. on Robotics and Automation*, no. 3, vol. 6, pp. 281-290.

Gosselin, C.M. and Angeles, J. (1991), A global performance index for the kinematic optimization of robotic manipulators, *ASME J. of Mechanical Design*, no. 2, vol. 113, pp. 220-226.

Hervé, J.M. (1995), Design of parallel manipulators via displacement group, *Proc. of the 9th World Congress on the theory of Machines and Mechanisms*, Milan, Italy, Vol. 3, pp 2079-2082.

Hervé, J.M. and Sparacino, F. (1991), Structural synthesis of parallel robots generating spatial translation, *Proc. of the 5th Int. Conf. on Advanced Robotics*, Pisa, Italy, pp. 808-813.

Kim, H.S. and Tsai, L.W. (2002), Evaluation of a Cartesian parallel manipulator, *Advances in Robot Kinematics: Theory and Applications*, J. Lenarcic and F. Thomas eds., Kluwer Academic Publishers, Netherlands, pp. 21-28.

Kong, X. and Gosselin, C.M. (2002a), Type synthesis of linear translational parallel manipulators, *Advances in Robot Kinematics: Theory and Applications*, J. Lenarcic and F. Thomas eds., Kluwer Academic Publishers, Netherlands, pp. 453-462.

Kong, X. and Gosselin, C.M. (2002b), Kinematics and singularity analysis of a novel type of 3-CRR 3-dof translational parallel manipulator, *The Int. J. of Robotics Research*, no. 9, vol. 21, pp. 791-798.

Tsai, L.W. (1996), Kinematics of a three-dof platform with three extansible limbs, *Recent Advances in Robot Kinematics*, J. Lenarcic and V. Parenti-Castelli eds., Kluwer Academic Publishers, Boston, pp. 401-410.

Tsai, L.W. (1999), The enumeration of a class of three-dof parallel manipulators, *Proc. of the 10th World Congress on the theory of Machines and Mechanisms*, Oulu, Finland, Vol. 3, pp 1121-1126.

Tsai, L.W., and Stamper, R., 1996, A parallel manipulator with only translational degrees of freedom, *Proc. of the 1996 ASME Design Engineering Technical Conferences*, Irvine (CA), USA, Paper No. DETC'96/MECH-1152.

Yoshikawa, T. (1985), Manipulability of robotic mechanisms, , *The Int. J. of Robotics Research*, no. 2, vol. 4, pp. 3-9.

Zhao, T.S. and Huang, Z. (2000), A novel three-dof translational platform mechanism and its kinematics, *Proc. of the 2000 ASME Design Engineering Technical Conferences*, Baltimore, USA, Paper No. DETC2000/MECH-14101.

Zlatanov, D., Fenton, R.G. and Benhabib B. (1995), A unifying framework for classification and interpretation of mechanism singularities, *ASME J. of Mechanical Design*, no. 4, vol. 117, pp. 566-572.

FULLY-ISOTROPIC T3R1-TYPE PARALLEL MANIPULATORS

Grigore Gogu
Laboratory of Researches and Applications in Advanced Mechanics, French Institute of Advanced Mechanics and University Blaise Pascal, Clermont-Ferrand, France
Grigore.Gogu@ifma.fr

Abstract The paper presents a special family of fully-isotropic parallel manipulators with four degrees of mobility. The moving platform of T3R1-type parallel manipulators has three orthogonal translations and one rotation. Only revolute and prismatic joints are used to achieve the four motions that are controlled independently by four actuators situated on the fixed base. It is shown that a one-to-one correspondence exists between the actuated joint space and the operational space of the moving platform. The Jacobian matrix of fully-isotropic T3R1-type parallel manipulators presented in this paper is the identity 4×4 matrix throughout their workspace. A method is proposed for structural synthesis based on the theory of linear transformations. The special family of fully-isotropic T3R1-type parallel manipulators includes 196 solutions. As far as we are aware this paper presents for the first time fully-isotropic parallel manipulators with four degrees of mobility.

Keywords: parallel manipulators, fully-isotropic, four degrees of mobility, structural synthesis

1. Introduction

A fully-isotropic T3R1-type parallel manipulator presented in this paper is composed of a moving platform and a fixed base connected by four complex legs. Only revolute and prismatic joints are used in the kinematic chains (KCs) associated to the legs. Under the action of the total constraints of its four legs the moving platform can only translate along three orthogonal directions and rotate about one of the three orthogonal axes. The four motions can be controlled independently by one rotary and three linear actuators situated on the fixed base. The Jacobian matrix (J) is the identity matrix (I_{4X4}) throughout the entire workspace. As far as we are aware, this paper proposes for the first time fully-isotropic parallel manipulators with four degrees of mobility. Only fully-isotropic parallel robots with three degrees of mobility are known in the literature. Caricato and Parenti-Castelli, 2002, presented for the first time the topological synthesis and classification of fully-isotropic translational parallel mechanisms. They founded their approach on

J. Lenarčič and C. Galletti (eds.), On Advances in Robot Kinematics, 265-272.

constraint and direct singularities investigation. Kim and Tsai, 2002, presented a fully-isotropic Cartesian Parallel Manipulator (3-PRRR), developed in the Department of Mechanical Engineering at University of California. Kong and Gosselin, 2002a, 2002b, presented a fully-isotropic 3-CRR translational parallel manipulator too. The following notations are used for the kinematic pairs: P - prismatic pair, R- revolute pair, C - cylindrical pair.

Several types of parallel robots with four degrees of mobility have been proposed, e.g., in (Koevermans et al., 1975 - quoted by Merlet, 1998, Reboulet et al., 1991, Hesselbach et al., 1998, Tanev, 1998, Wang and Gosselin, 1998, Pierrot and Company, 1999, Company and Pierrot, 1999, Rolland, 1999, Lenarčič, Stanišić and Parenti-Castelli, (2000), Zhao and Huang, 2000, Zlatanov and Gosselin, 2001; Fang and Tsai, 2002, Gao et al. (2002), Huang and Lee, 2003). We note that no one of these robots is fully-isotropic throughout its entire workspace.

The general methods used for structural synthesis of parallel mechanisms can be divided into two approaches: the method based on displacement group theory (Hervé and Sparacino, 1991, Hervé, 1995) and the methods based on screw algebra (Frisoli et al., 2000, Kong and Gosselin, 2001, Fang and Tsai, 2002, Huang and Li 2002, 2003).

The method proposed in this paper for structural synthesis of fully-isotropic T3R1-type parallel manipulator is founded on the theory of linear transformations.

2. Kinematic structure

The general kinematic structure of a T3R1 fully-isotropic robotic manipulator discussed in this paper is presented in Fig. 1 (a). The complex legs are obtained by concatenating the following KCs :

$A_1(1\text{-}\ldots\text{-}n_{A1})$, $A_2(1\text{-}\ldots\text{-}n_{A2})$, $A_3(1\text{-}\ldots\text{-}n_{A3})$, $A_4(1\text{-}\ldots\text{-}n_{A1}\equiv n)$ - elementary open KCs with prismatic and revolute joints serially concatenated,

$B_1 \leftarrow A_1\text{-}A_2$ - single loop mechanism obtained by integrating in the same link ($n_{B1}\equiv n_{A1}\equiv n_{A2}$) the final links of A_1 and A_2 - Fig. 1(b),

$Q_1 \leftarrow B_1\text{-}R_1$ - complex open kinematic chain obtained by serial concatenation of B_1 and a revolute joint (R_1) which axis is parallel with the rotation axis of the moving platform n - Fig. 1(c),

$D_1 \leftarrow Q_1\text{-}A_3$ - complex closed kinematic chain obtained by parallel concatenation of Q_1 and A_3 by integrating in the same link ($n_{D1}\equiv n_{Q1}\equiv n_{A3}$) the final links ($n_{Q1}$ and n_{A3}) of Q_1 and A_3 - Fig. 1(d),

$Q_2 \leftarrow D_1\text{-}R_2$ - complex open kinematic chain obtained by serial concatenation of D_1 and a revolute joint (R_2) which axis is also parallel with the rotation axis of the moving platform - Fig. 1(e),

$A_5 \leftarrow A_3\text{-}R_2$ - elementary open kinematic chain obtained by serial

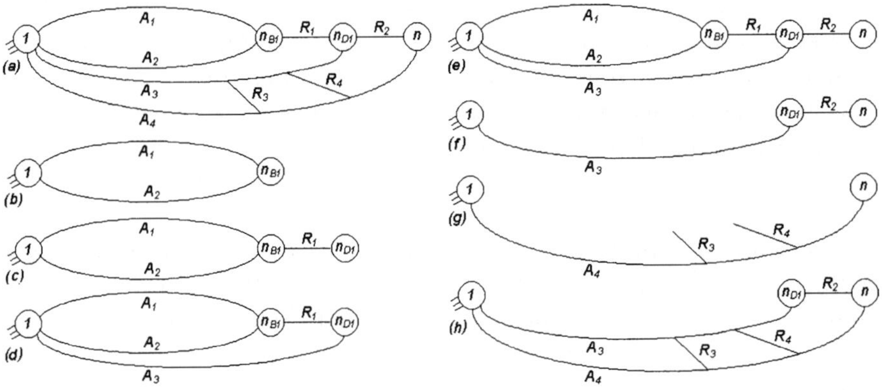

Figure 1. General kinematic structure of T3R1 fully-isotropic parallel robots (a) and its components B_1(b), Q_1(c), D_1(d), Q_2(e), A_5(f), Q_3(g) and D_2(h)

concatenation of A_3 and R_2 and having the moving platform ($n\equiv n_{A5}$) as its final link - Fig. 1(f),

Q_3 - complex open kinematic chain having the moving platform as a final link ($n\equiv n_{Q3}$), obtained by concatenating A_4 and the revolute joints R_3 and R_4 which axes are also parallel with the rotation axis of the moving platform - Fig. 1(g),

$D_2\leftarrow A_5$-Q_3 - complex closed kinematic chain obtained by the concatenation of A_5 and Q_3 by putting together elements of Q_3 and A_5 (e.g., $n\equiv n_{D2}\equiv n_{A5}\equiv n_{Q3}$) - Fig. 1(h),

$D_3\leftarrow Q_2$-Q_3 - complex kinematic chain obtained by parallel concatenation of Q_2 and Q_3 by integrating the final links (n_{Q2} and n_{Q3}) of Q_2 and Q_3 in the moving platform ($n\equiv n_{D3}\equiv n_{Q2}\equiv n_{Q3}$) - Fig. 1(a).

The first joint of each kinematic chain A_1, A_2, A_3 and A_4 is actuated. The manipulator has three linear actuators and one rotary actuator situated on the fixed base. We denote by $\dot{d}_1$, $\dot{d}_2$, $\dot{d}_3$ and $\dot{\varphi}_e$ the velocities in the four actuated joints and by v_1, v_2, v_3 and ω_e ($e=x,y,z$) the velocities of the moving platform. The complex mechanism D_2 integrates a linear ($\dot{d}_3$) and a rotary actuator ($\dot{\varphi}_e$) to achieve a translation and a rotation of the moving platform (v_3 and ω_e). The direction of this translation and the axis of rotation coincide. The other two translations of the moving platform (v_1 and v_2) are achieved by the two linear motors ($\dot{d}_1$ and $\dot{d}_2$) integrated in the single-loop mechanisms B_1.

3. Structural synthesis

The structural synthesis approach proposed focuses on synthesis of the mechanisms B_1 and D_2. The approach is founded on the theory of linear transformations. Let us consider a mechanism L (that can be elementary or complex, open or closed). It achieves a linear transformation $\boldsymbol{F}$ from the finite dimensional vector space U (called joint velocity space) into the finite dimensional vector space W (called external velocity space). That is to each vector $\dot{\boldsymbol{q}}$, in the joint vector space U, the linear transformation $\boldsymbol{F}$ assigns a uniquely determined vector $\boldsymbol{F}(\dot{\boldsymbol{q}})$ in the vector space W

$$\boldsymbol{F}(\dot{\boldsymbol{q}})=[J][\dot{\boldsymbol{q}}]. \tag{1}$$

The vector $\dot{\boldsymbol{q}}$ represents the joint velocity vector and it is composed of the relative velocities in joints. The vector $\boldsymbol{F}(\dot{\boldsymbol{q}})$ in W is the image of the vector $\dot{\boldsymbol{q}}$ in U under the transformation $\boldsymbol{F}$ with the having the Jacobian matrix J. The image of U in W represents the range R_F of $\boldsymbol{F}$. In this paper, the range R_F of $\boldsymbol{F}$ is called operational space with the dimension

$$dim(R_F)=rank(F)=rank[J]. \tag{2}$$

The relative spatiality ($S^L_{a/b}$) between two elements (a and b) in the mechanism L represents the number of relative independent infinitesimal displacements or velocities allowed by the mechanism between the two elements. It is given by the dimension of the vector space $R^L_{a/b}$ of the relative velocities between elements a and b in the mechanism L

$$S^L_{a/b} = dim(R^L_{a/b}). \tag{3}$$

We notice that the range $R^L_{a/b}$ is a sub-space of R^L and implicitly a sub-space of W. Relative spatiality characterises the connectivity of the two elements (a and b) in the mechanism L having the degree of mobility M_L.

Let us consider that $v_1=v_x$, $v_2=v_y$, $v_3=v_z$, and $\omega_e=\omega_z$ $(e=z)$, where x,y,z are 3 orthogonal directions of a reference frame $O_0x_0y_0z_0$. The following conditions are necessary to obtain a T3R1 fully-isotropic parallel manipulator discussed in this paper:

$$v_1=\dot{d}_1=v_x,\ v_2=\dot{d}_2=v_y,\ v_3=\dot{d}_3=v_z,\ \omega_e=\dot{\varphi}_e=\dot{\varphi}_z=\omega_z. \tag{4}$$

A parallel manipulator respecting Eqs. 4 is fully-isotropic. Its Jacobian matrix becomes the identity 4×4 matrix throughout the entire workspace. The condition number and the determinant of the Jacobian matrix being equal to one, the manipulator performs very well with regard to force and motion transmission capabilities.

Eqs. 4 also involve the following fully-isotropic conditions:

1) $M_{B1}=M_{D1}=3$, $S^{B1}_{nB1/1}=S^{D1}_{nD1/1}=3$, $(R^{B1}_{nB1/1})=(R^{D1}_{nD1/1})=(\boldsymbol{v}_x,\boldsymbol{v}_y,\boldsymbol{v}_z)$,

2) $M_{Q1}=M_{Q2}=M_{A5}=M_{D2}=M_{D3}=4$, $S^{Q1}_{nQ1/1}=S^{Q2}_{nQ2/1}=S^{A5}_{nA5/1}=S^{D2}_{nD2/1}=S^{D3}_{nD3/1}=4$, $(R^{Q1}_{nQ1/1})=(R^{Q2}_{nQ2/1})=(R^{A5}_{nA5/1})=(R^{D2}_{nD2/1})=(R^{D3}_{nD3/1})=(\boldsymbol{v}_x,\boldsymbol{v}_y,\boldsymbol{v}_z,\boldsymbol{\omega}_z)$,

3) if $\dot{\boldsymbol{d}}_1=\boldsymbol{0}$ than $S^{B1}_{nB1/1}=S^{D1}_{nD1/1}=2$, $(R^{B1}_{nB1/1})=(R^{D1}_{nD1/1})=(\boldsymbol{v}_y,\boldsymbol{v}_z)$,

$S^{Q1}_{nQ1/1} = S^{Q2}_{nQ2/1} = S^{A5}_{nA5/1} = S^{D2}_{nD2/1} = S^{D3}_{nD3/1} = 3$,

$(R^{Q1}_{nQ1/1}) = (R^{Q2}_{nQ2/1}) = (R^{A5}_{nA5/1}) = (R^{D2}_{nD2/1}) = (R^{D3}_{nD3/1}) = (v_y, v_z, \omega_z)$,

4) if $\dot{d}_2 = 0$ than $S^{B1}_{nB1/1} = S^{D1}_{nD1/1} = 2$, $(R^{B1}_{nB1/1}) = (R^{D1}_{nD1/1}) = (v_x, v_z)$,

$S^{Q1}_{nQ1/1} = S^{Q2}_{nQ2/1} = S^{A5}_{nA5/1} = S^{D2}_{nD2/1} = S^{D3}_{nD3/1} = 3$,

$(R^{Q1}_{nQ1/1}) = (R^{Q2}_{nQ2/1}) = (R^{A5}_{nA5/1}) = (R^{D2}_{nD2/1}) = (R^{D3}_{nD3/1}) = (v_x, v_z, \omega_z)$,

5) if $\dot{d}_3 = 0$ than $S^{B1}_{nB1/1} = S^{D1}_{nD1/1} = 2$, $(R^{B1}_{nB1/1}) = (R^{D1}_{nD1/1}) = (v_x, v_y)$,

$S^{Q1}_{nQ1/1} = S^{Q2}_{nQ2/1} = S^{A5}_{nA5/1} = S^{D2}_{nD2/1} = S^{D3}_{nD3/1} = 3$,

$(R^{Q1}_{nQ1/1}) = (R^{Q2}_{nQ2/1}) = (R^{A5}_{nA5/1}) = (R^{D2}_{nD2/1}) = (R^{D3}_{nD3/1}) = (v_x, v_y, \omega_z)$,

6) if $\dot{\varphi}_e = 0$ than $S^{B1}_{nB1/1} = S^{D1}_{nD1/1} = 3$, $(R^{B1}_{nB1/1}) = (R^{D1}_{nD1/1}) = (v_x, v_y, v_z)$,

$S^{Q1}_{nQ1/1} = S^{Q2}_{nQ2/1} = S^{A5}_{nA5/1} = 4$, $(R^{Q1}_{nQ1/1}) = (R^{Q2}_{nQ2/1}) = (R^{A5}_{nA5/1}) = (v_x, v_y, v_z, \omega_z)$,

$S^{D2}_{nD2/1} = S^{D3}_{nD3/1} = 3$, $(R^{D2}_{nD2/1}) = (R^{D3}_{nD3/1}) = (v_x, v_y, v_z)$.

We recall that by M_L we denoted the degree of mobility of the kinematic chain L, by $S^L_{a/b}$ we denoted the relative spatiality between the elements a and b in the kinematic chain L and by $(R^L_{a/b})$ we denoted the base of the vector space $R^L_{a/b}$ of the relative velocities between elements a and b in the kinematic chain L. The elements a and b represents the extreme links of L; a is the last element (n_L) and b is the fixed base ($1 \equiv 0$). In the previous notations the kinematic chain L could be B_1, D_1, Q_1, Q_2, A_5, D_2 , D_3 and the last element n_L could be n_{B1}, n_{D1}, n_{Q1}, n_{Q2}, n_{A5}, n_{D2} , n_{D3}. We note that bases $(R^{Q1}_{nQ1/1})$, $(R^{Q2}_{nQ2/1})$, $(R^{A5}_{nA5/1})$, $(R^{D2}_{nD2/1})$ and $(R^{D3}_{nD3/1})$ must be unique bases of the corresponding vector spaces.

The elementary open kinematic chains A_1 and A_2 composing single-loop mechanism $B_1 \leftarrow A_1$-A_2 satisfying the fully-isotropic conditions could have the following arrangements of joints:

A_1: $\underline{P_x}$-P_y-P_z, $\underline{P_x}$-P_z-P_y, $\underline{P_x}$-P_y-R_x-R_x, $\underline{P_x}$-P_z-R_x-R_x, $\underline{P_x}$-R_x-R_x-P_y, $\underline{P_x}$-R_x-R_x-P_z, $\underline{P_x}$-R_x-P_{yz}-R_x, $\underline{P_x}$-R_x-R_x-R_x, $\underline{P_x}$-(R_x-R_x-R_x-R_x)-R_x-R_x, $\underline{P_x}$-P_y-(R_x-R_x-R_x-R_x), $\underline{P_x}$-P_z-(R_x-R_x-R_x-R_x), $\underline{P_x}$-(R_x-R_x-R_x-R_x)-P_y, $\underline{P_x}$-(R_x-R_x-R_x-R_x)-P_z, $\underline{P_x}$-(R_x-R_x-R_x-R_x)-(R_x-R_x-R_x-R_x-R_x),

A_2: $\underline{P_y}$-P_z-P_x, $\underline{P_y}$-P_x-P_z, $\underline{P_y}$-P_z-R_y-R_y, $\underline{P_y}$-P_x-R_y-R_y, $\underline{P_y}$-R_y-R_y-P_z, $\underline{P_y}$-R_y-R_y-P_x, $\underline{P_y}$-R_y-P_{zx}-R_y, $\underline{P_y}$-R_y-R_y-R_y, $\underline{P_y}$-(R_y-R_y-R_y-R_y)-R_y-R_y, $\underline{P_y}$-P_z-(R_y-R_y-R_y-R_y), $\underline{P_y}$-P_x-(R_y-R_y-R_y-R_y), $\underline{P_y}$-(R_y-R_y-R_y-R_y)-P_z, $\underline{P_y}$-(R_y-R_y-R_y-R_y)-P_x, $\underline{P_y}$-(R_y-R_y-R_y-R_y)-(R_y-R_y-R_y-R_y-R_y),

The actuated joints are underlined. The notation P_{yz} indicates that the direction of the translation is parallel to the plane yz. The parentheses indicate a planar four-bar parallelogram. We note that 196 combinations of A_1 and A_2 are possible to obtain the single-loop mechanism B_1 satisfying the fully-isotropic conditions. It is not necessary to consider

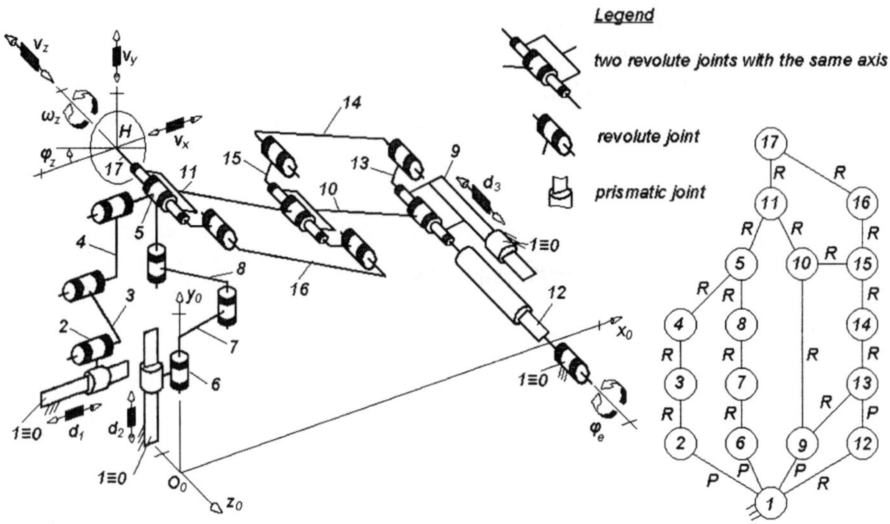

Figure 2. Kinematic structure of a fully-isotropic parallel manipulator with identical structural solutions for A_1 and A_2

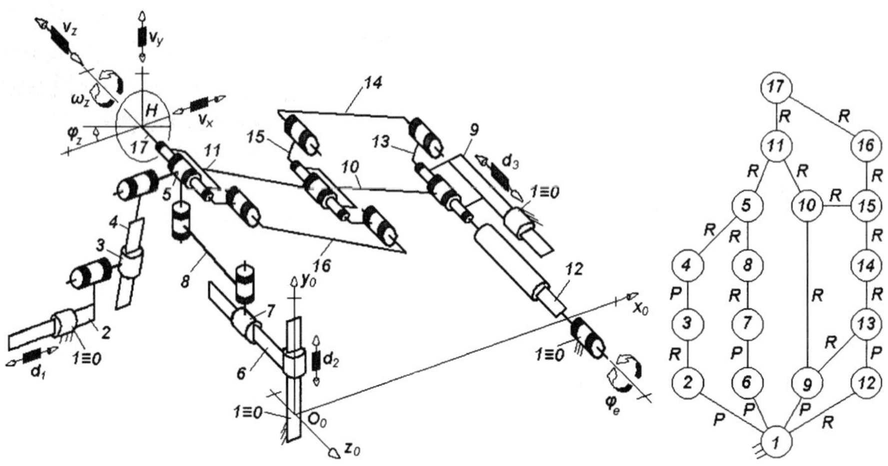

Figure 3. Kinematic structure of a fully-isotropic parallel manipulator with different structural solutions for A_1 and A_2

only the identical structural solutions. We can see that A_1 and A_2 represent orthogonal versions of legs used in translational parallel mechanisms with three degrees of mobility (Hervé and Sparacino, 1991, Frisoli et al., 2000, Carricato and Parenti-Castelli, 2002).

The complex closed kinematic chain D_2 are specially conceived to achieve the translation υ_z and the rotation ω_z of the moving platform by respecting the fully-isotropic conditions. The elementary kinematic chains A_3 and A_4 composing D_2 have the following arrangements of joints: A_3 ($\underline{P}_z$-R_z-R_z) and A_4 ($\underline{R}_z$-P_z-R_z-R_z-R_z-R_z). The complex kinematic chain D_2 is obtained by connecting A_3 and A_4 by the three revolute joints R_2, R_3 and R_4 which axes are parallel with the rotation axis of the moving platform (z axis).

The two examples presented in Figs.2 and 3 have: $A_1(1\equiv0\text{-}2\text{-}3\text{-}4\text{-}5)$, $A_2(1\equiv0\text{-}6\text{-}7\text{-}8\text{-}5)$, $A_3(1\equiv0\text{-}9\text{-}10\text{-}11)$ and $A_4(1\equiv0\text{-}12\text{-}13\text{-}14\text{-}15\text{-}16\text{-}17)$. The moving platform is $n\equiv17$. The example presented in Fig. 2 has identical structural solutions for A_1 ($\underline{P}_x$-R_x-R_x-R_x) and A_2 ($\underline{P}_y$-R_y-R_y-R_y). The example presented in Fig. 3 has other two distinct solutions for A_1 ($\underline{P}_x$-R_x-P_{yz}-R_x) and A_2 ($\underline{P}_y$-P_z-R_y-R_y).

4. Conclusions

An approach has been proposed for structural synthesis of a family of fully-isotropic T3R1-type parallel manipulators. Special legs were conceived to achieve fully-isotropic conditions. As far as we are aware this paper presents for the first time fully-isotropic parallel manipulators with four degrees of mobility and a method for their structural synthesis.

5. Acknowledgement

This work was sustained by CNRS (The French National Council of Scientific Research) in the frame of the project ROBEA-MAX (Complex Architecture Machines) 2002-2003.

References

Carricato, M., and Parenti-Castelli, V. (2002), Singularity-free fully-isotropic translational parallel mechanisms, *The International Journal of Robotics Research*, no. 2, vol. 22, pp. 161-174.

Company, O., and Pierrot, F. (1999), A new 3T-1R parallel robot, *Proceedings of the International Conference on Robotics and Automation*, Tokyo, Japan, pp. 557-562.

Fang, Y., and Tsai, L-W. (2002), Structural synthesis of a class of 4-dof and 5-dof parallel manipulators with identical limb structures, *The International Journal of Robotics Research*, no. 9, vol. 21, pp. 799-810.

Frisoli, A., Checcacci, D., Salsedo, F., and Bergamasco, M. (2000), Synthesis by screw algebra of translating in-parallel actuated mechanisms, Advances in robot kinematics, (J. Lenarčič and M. M. Stanišić, eds), Kluwer Academic Publishers, pp. 433-440.

Gao, F., Li, W., Zhao, X., Jin, Z., and Zhao, H. (2002), Mechanism and Machine Theory, no. 11, vol. 37, pp. 1395-1411.

Hervé, J .M., and Sparacino, F. (1991), Structural synthesis of parallel robots generating spatial translation, *Proceedings of the 5th IEEE International*

Conference on Advanced Robotics, Pisa, Italy, pp. 2079-2082.

Hervé, J. M. (1995), Design of parallel manipulators via the displacement group, *Proceedings of the 9th World Congress on the Theory of Machines and Mechanisms*, Milan, Italy, pp. 2079-2082.

Hesselbach, J., Plitea, N., Frindt, M., and Kusiek, A. (1998), A new parallel mechanism to use for cutting convex glass panels, *Advances in robot kinematics*, (J. Lenarčič and M. L. Husty, eds), Kluwer Academic Publishers, pp. 165-174.

Huang, Z., and Li, Q. C. (2003), Type synthesis of symmetrical lower-mobility paralle mechanisms using the constraint-synthesis method, *The International Journal of Robotics Research*, no. 1, vol. 22, pp. 59-79.

Kong, X., and Gosselin, C. M. (2001), Generation of parallel manipulators with three translational degrees of freedom based on screw theory, *Proceedings of 2001 CCToMM Symposium on Mechanisms, Machines and Mechatronics*, Saint-Hubert, Montreal, Canada.

Kong, X., and Gosselin, C. M. (2002), Type synthesis of linear translational parallel manipulators, *Advances in robot kinematics*, (J. Lenarčič and F. Thomas, eds), Kluwer Academic Publishers, pp. 453-462.

Kong, X., and Gosselin, C. M. (2002), Kinematics and singularity analysis of a novel type of 3-CRR 3-dof translational parallel manipulator, *The International Journal of Robotics Research*, no. 9, vol. 21, pp. 791-798.

Lenarčič, J., Stanišić, M.M., and Parenti-Castelli, V. (2000), A 4-dof parallel mechanism stimulating the movement of the human sternum-clavicle-scapula complex, *Advances in robot kinematics*, (J. Lenarčič and M. M. Stanišić, eds), Kluwer Academic Publishers, pp. 325-332.

Merlet, J.P. (1997), *Les robots parallèls*, 2e édition, Hermès, Paris.

Pierrot, F., and Company, O. (1999), H4: a new family of 4-dof parallel robots, *Proceedings of IEEE/ASME Int. Conf. on Advanced Intelligent Mechatronics*, Atlanta, USA, pp. 508-513.

Reboulet, C. et al. (1991), Rapport d'avancement projet VAP, Rapport de recherche 7743, CNES/DERA, quoted by Merlet, J. P (1998).

Rolland, L. H. (1999), The Manta and the Kanuk: novel 4-dof parallel mechanism for industrial handling. *Proceedings of ASME Dynamic Systems and Control Division, IMECE'99 Conference*, Nashville, USA, pp. 831-844.

Tanev, T.K. Forward displacement analysis of a three legged four-degree-of-freedom parallel manipulator, *Advances in robot kinematics*, (J. Lenarčič and M. L. Husty, eds), Kluwer Academic Publishers, pp. 147-154.

Tsai, L-W., and Kim, H. S. (2002), Evaluation of a Cartesian parallel manipulator, *Advances in robot kinematics*, (J. Lenarčič and F. Thomas, eds), Kluwer Academic Publishers, pp. 21-28.

Wang, J., and Gosselin, C.M. (1998), Kinematic analysis and singularity loci of spatial four-degree-of-freedom parallel manipulators using a vector formulation, *ASME Transactions, Journal of Mechanical Design*, no. 3, vol 123, pp. 375-381.

Zhao, T. S., and Huang, Z. (2000), A novel spatial four-dof parallel mechanism and its position analysis. *Mechanical Science and Technology*, no. 6, vol. 19, pp.927-929.

Zlatanov, D., and Gosselin, C.M. (2001), A new parallel architecture with four degrees of freedom, *Proceedings of the 2nd Workshop on Computational Kinematics*, Seoul, Korea, pp. 57-66.

Analysis of Mechanisms

THE SINGULAR PLANES OF THE ARTICULATED ARM SUBASSEMBLY

C.M. Goehler and M.M. Stanišić
University of Notre Dame
Department of Aerospace and Mechanical Engineering
cgoehler@nd.edu stanisic@nd.edu

Abstract A generic representation of arm-subassembly singularities is possible through the use of singular planes. These singular planes are the loci of instantaneously singular points on the terminal link. The singular planes have been revealed previously, albeit through some relatively sophisticated methods. The purpose of this paper is to redevelop these singular planes for the articulated arm-subassembly using simple kinematic concepts which should be well-known to the general robotics community, entry-level graduate students and upper level undergraduates. The intention is to make the concept of singular planes better understood and more useful as a tool for learning about and understanding the kinematic singularities of industrial manipulators.

Keywords: Singularities, kinematics, Denavit and Hartenberg parameters, articulated arm subassembly

1. Introduction

Singularities are kinematic conditions of robotic manipulators, with both beneficial and negative effects, of which engineers should have a clear understanding. Singularities of industrial manipulators can be generically represented by geometric quantities known as singular planes, Stanišić and Engelberth, 1988, Hayes et al., 2002. If an individual desires to fully understand the development of these singular planes, a graduate course in Screw Theory is in order.

Two singular planes exist for the articulated arm subassembly that can be mathematically shown through use of Screw Theory, Stanišić and Engelberth, 1988. A focused geometric interpretation of how singularities arise was developed so that all robot users could understand these two singular planes without an advanced level of mathematical knowledge, Hayes et al., 2002. However, this method closely resembled Screw Theory.

The motivation of this paper is to present this previously discussed concept in a manner more easily comprehended, by less sophisticated

J. Lenarčič and C. Galletti (eds.), On Advances in Robot Kinematics, 275-282.

methods. This is made possible through use of simple kinematic concepts taught in many undergraduate courses.

The main body of this paper is dedicated to establishing the kinematic equations for the problem and solving for the equations of the singular planes for an articulated arm subassembly.

2. Singularity Analysis

The kinematic equations for the articulated arm subassembly shown in Fig. 1 can be attained using the Denavit and Hartenberg parameters and notation as in Craig, 1989. The frame assignments can be seen in Fig. 1, where the $\hat{Y}_i$ axes follow from the $\hat{X}_i$ and $\hat{Z}_i$ axes and are not shown.

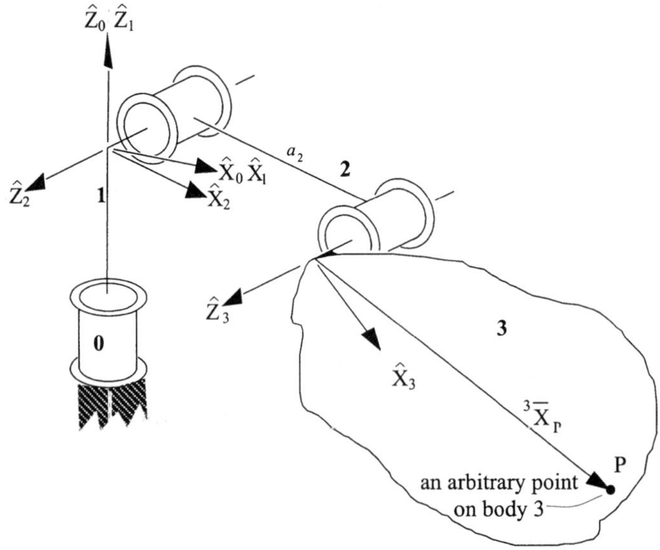

Figure 1. Kinematic Model of an Articulated Arm Subassembly

Table 1. Denavit and Hartenberg parameters.

i	α_{i-1}	a_{i-1}	d_i	θ_i
1	0	0	0	θ_1
2	$\frac{\pi}{2}$	0	0	θ_2
3	0	a_2	0	θ_3

From Tb. 1, the homogeneous transformations can be calculated using Eq. (3.6) in Craig, 1989,

$$ {}^0_1T = \begin{bmatrix} \cos\theta_1 & -\sin\theta_1 & 0 & 0 \\ \sin\theta_1 & \cos\theta_1 & 0 & 0 \\ 0 & 0 & 1 & 0 \\ 0 & 0 & 0 & 1 \end{bmatrix}, \tag{1} $$

$$ {}^1_2T = \begin{bmatrix} \cos\theta_2 & -\sin\theta_2 & 0 & 0 \\ 0 & 0 & -1 & 0 \\ \sin\theta_2 & \cos\theta_2 & 0 & 0 \\ 0 & 0 & 0 & 1 \end{bmatrix}, \tag{2} $$

$$ {}^2_3T = \begin{bmatrix} \cos\theta_3 & -\sin\theta_3 & 0 & a_2 \\ \sin\theta_3 & \cos\theta_3 & 0 & 0 \\ 0 & 0 & 1 & 0 \\ 0 & 0 & 0 & 1 \end{bmatrix}. \tag{3} $$

The homogeneous coordinates of an *arbitrary* point P on body 3, in frame $\{3\}$, are given by,

$$ {}^3\bar{\mathrm{X}}_P = \begin{bmatrix} x \\ y \\ z \\ 1 \end{bmatrix}. \tag{4} $$

The homogeneous coordinates of point P in frame $\{2\}$ are given by,

$$ {}^2\bar{\mathrm{X}}_P = {}^2_3T\,{}^3\bar{\mathrm{X}}_P = \begin{bmatrix} x\cos\theta_3 - y\sin\theta_3 + a_2 \\ x\sin\theta_3 + y\cos\theta_3 \\ z \\ 1 \end{bmatrix}, \tag{5} $$

where ${}^2_3T$ is given in Eq. 3. The homogeneous coordinates of point P in frame $\{1\}$ are given by,

$$ {}^1\bar{\mathrm{X}}_P = {}^1_2T\,{}^2\bar{\mathrm{X}}_P = \begin{bmatrix} x\cos\theta_{23} - y\sin\theta_{23} + a_2\cos\theta_2 \\ -z \\ x\sin\theta_{23} + y\cos\theta_{23} + a_2\sin\theta_2 \\ 1 \end{bmatrix}, \tag{6} $$

where $\theta_{23} = \theta_2 + \theta_3$ and ${}^1_2T$ is given in Eq. 2. The homogeneous coordinates of point P in frame $\{0\}$ are given by,

$$ {}^0\bar{\mathrm{X}}_P = {}^0_1T\,{}^1\bar{\mathrm{X}}_P = \begin{bmatrix} x\mathrm{c}_1\mathrm{c}_{23} - y\mathrm{c}_1\mathrm{s}_{23} + z\mathrm{s}_1 + a_2\mathrm{c}_1\mathrm{c}_2 \\ x\mathrm{s}_1\mathrm{c}_{23} - y\mathrm{s}_1\mathrm{s}_{23} - z\mathrm{c}_1 + a_2\mathrm{s}_1\mathrm{c}_2 \\ x\mathrm{s}_{23} + y\mathrm{c}_{23} + a_2\mathrm{s}_2 \\ 1 \end{bmatrix}, \tag{7} $$

where ${}^0_1T$ is given in Eq. 1. The shorthand notation, $\mathrm{s}_i = \sin\theta_i$ and $\mathrm{c}_i = \cos\theta_i$ is used. Eq. 7 is then differentiated with respect to θ_1, θ_2 and θ_{23} to get the Jacobian, J, shown in Eq. 8,

$$\begin{aligned} J &= \begin{bmatrix} \frac{\partial\, {}^0\bar{\mathrm{X}}_P}{\partial\theta_1} & \frac{\partial\, {}^0\bar{\mathrm{X}}_P}{\partial\theta_2} & \frac{\partial\, {}^0\bar{\mathrm{X}}_P}{\partial\theta_{23}} \end{bmatrix} \\ &= \begin{bmatrix} z\mathrm{c}_1 - \mathrm{s}_1(a_2\mathrm{c}_2 + x\mathrm{c}_{23} - y\mathrm{s}_{23}) & -a_2\mathrm{c}_1\mathrm{s}_2 & -\mathrm{c}_1(x\mathrm{s}_{23} + y\mathrm{c}_{23}) \\ z\mathrm{s}_1 + \mathrm{c}_1(a_2\mathrm{c}_2 + x\mathrm{c}_{23} - y\mathrm{s}_{23}) & -a_2\mathrm{s}_1\mathrm{s}_2 & -\mathrm{s}_1(x\mathrm{s}_{23} + y\mathrm{c}_{23}) \\ 0 & a_2\mathrm{c}_2 & x\mathrm{c}_{23} - y\mathrm{s}_{23} \end{bmatrix}. \end{aligned} \tag{8}$$

Singularities occur when the determinant of this Jacobian vanishes. The determinant of the Jacobian was expanded along the third row,

$$|J| = -a_2\mathrm{c}_2 D_1 + (x\mathrm{c}_{23} - y\mathrm{s}_{23})D_2 = 0, \tag{9}$$

where

$$D_1 = \begin{vmatrix} z\mathrm{c}_1 - \mathrm{s}_1(a_2\mathrm{c}_2 + x\mathrm{c}_{23} - y\mathrm{s}_{23}) & -\mathrm{c}_1(x\mathrm{s}_{23} + y\mathrm{c}_{23}) \\ z\mathrm{s}_1 + \mathrm{c}_1(a_2\mathrm{c}_2 + x\mathrm{c}_{23} - y\mathrm{s}_{23}) & -\mathrm{s}_1(x\mathrm{s}_{23} + y\mathrm{c}_{23}) \end{vmatrix}, \tag{10}$$

and

$$D_2 = \begin{vmatrix} z\mathrm{c}_1 - \mathrm{s}_1(a_2\mathrm{c}_2 + x\mathrm{c}_{23} - y\mathrm{s}_{23}) & -a_2\mathrm{c}_1\mathrm{s}_2 \\ z\mathrm{s}_1 + \mathrm{c}_1(a_2\mathrm{c}_2 + x\mathrm{c}_{23} - y\mathrm{s}_{23}) & -a_2\mathrm{s}_1\mathrm{s}_2 \end{vmatrix}. \tag{11}$$

Expanding Eqs. 10 and 11, substituting into Eq. 9 and grouping like terms yields,

$$|J| = -a_2(a_2\mathrm{c}_2 + x\mathrm{c}_{23} - y\mathrm{s}_{23})[x(\mathrm{c}_2\mathrm{s}_{23} - \mathrm{s}_2\mathrm{c}_{23}) + y(\mathrm{c}_2\mathrm{c}_{23} + \mathrm{s}_2\mathrm{s}_{23})] = 0. \tag{12}$$

As expected, the determinant is independent of θ_1. Taking advantage of double angle trigonometric identities, Eq. 12 reduces to,

$$|J| = -a_2\,(x\mathrm{s}_3 + y\mathrm{c}_3)\,(a_2\mathrm{c}_2 + x\mathrm{c}_{23} - y\mathrm{s}_{23}) = 0. \tag{13}$$

3. The Singular Planes

Eq. 13 contains two non-trivial singularity conditions along with a trivial condition, $a_2 = 0$.

3.1 Case 1

The first non-trivial condition is represented by,

$$a_2\mathrm{c}_2 + x\mathrm{c}_{23} - y\mathrm{s}_{23} = 0, \tag{14}$$

from Eq. 13. This relationship corresponds to setting the x component of Eq. 6 to zero, which means the arbitrary point P lies on the $\hat{\mathrm{Z}}_1$-$\hat{\mathrm{Y}}_1$

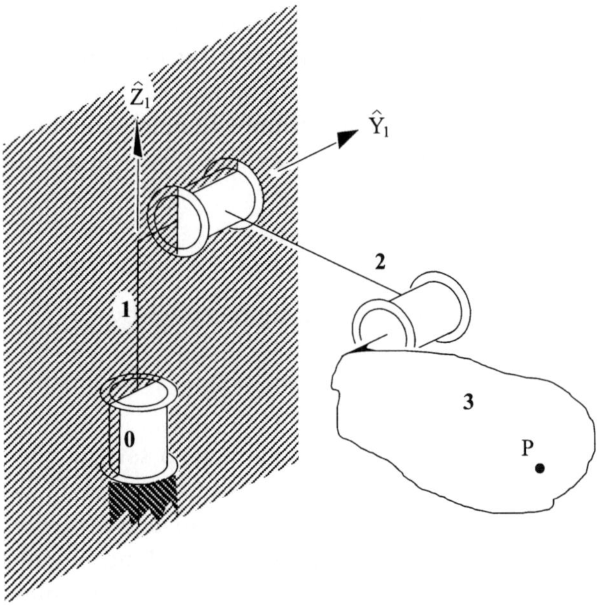

Figure 2. Singular Planes of an Articulated Arm Subassembly

plane of Fig. 1. *The* $\hat{Z}_1$-$\hat{Y}_1$ *plane is a singular plane of the articulated arm subassembly.* This singular plane represents a locus of points on body 3 that are instantaneously singular and is shown in Fig. 2. When the singularity condition in Eq. 14 is substituted into Eq. 8, the singular Jacobian in case 1, J_1, is given by,

$$J_1 = \begin{bmatrix} z\mathrm{c}_1 & -a_2\mathrm{c}_1\mathrm{s}_2 & -\mathrm{c}_1(x\mathrm{s}_{23} + y\mathrm{c}_{23}) \\ z\mathrm{s}_1 & -a_2\mathrm{s}_1\mathrm{s}_2 & -\mathrm{s}_1(x\mathrm{s}_{23} + y\mathrm{c}_{23}) \\ 0 & a_2\mathrm{c}_2 & x\mathrm{c}_{23} - y\mathrm{s}_{23} \end{bmatrix}. \tag{15}$$

In Eq. 15, the first two rows are multiples of the row vector $[z, -a_2s_2, (xs_{23}+yc_{23})]$ so instantaneously the dimension of the row space is two, as is the rank of J_1. In order to see which direction of motion is no longer possible for points in this singular plane, the columns of singular J_1 are written as,

$$J_1 = \tag{16}$$

$$\left[\ z\,{}^0\hat{\mathrm{X}}_1,\quad -a_2(\mathrm{s}_2{}^0\hat{\mathrm{X}}_1 - \mathrm{c}_2{}^0\hat{\mathrm{Z}}_1),\quad -(x\mathrm{s}_{23} + y\mathrm{c}_{23}){}^0\hat{\mathrm{X}}_1 + (x\mathrm{c}_{23} - y\mathrm{s}_{23}){}^0\hat{\mathrm{Z}}_1\ \right],$$

where ${}^0\hat{\mathrm{X}}_1$ and ${}^0\hat{\mathrm{Z}}_1$ come from Eq. 1. We see that the column space of J_1 is instantaneously spanned by ${}^0\hat{\mathrm{X}}_1$ and ${}^0\hat{\mathrm{Z}}_1$ and the velocity of point P has only ${}^0\hat{\mathrm{X}}_1$ and ${}^0\hat{\mathrm{Z}}_1$ components. The point P cannot move in the

${}^{0}\hat{Y}_1$ direction with a finite velocity. This is also the direction in which the arm-subasembly can transmit a theoretically infinite force to the environment through point P (the Principle of Virtual Work) and the subassembly instantaneously has an infinite mechanical advantage. This singular plane could be referred to as the "over-the-shoulder" singular plane.

3.2 Case 2

The second non-trivial condition is represented by,

$$xs_3 + yc_3 = 0, \tag{17}$$

from Eq. 13. This relationship corresponds to setting the y component of Eq. 5 to zero, which means the arbitrary point P lies on the $\hat{Z}_2$-$\hat{X}_2$ plane of Fig. 1. *The* $\hat{Z}_2$ *-*$\hat{X}_2$ *plane is also a singular plane of the articulated arm subassembly.* This singular plane represents a second locus of points on body 3 that are instantaneously singular and is shown in Fig. 3. Expanding the double angles in Eq. 8 and substituting Eq. 17

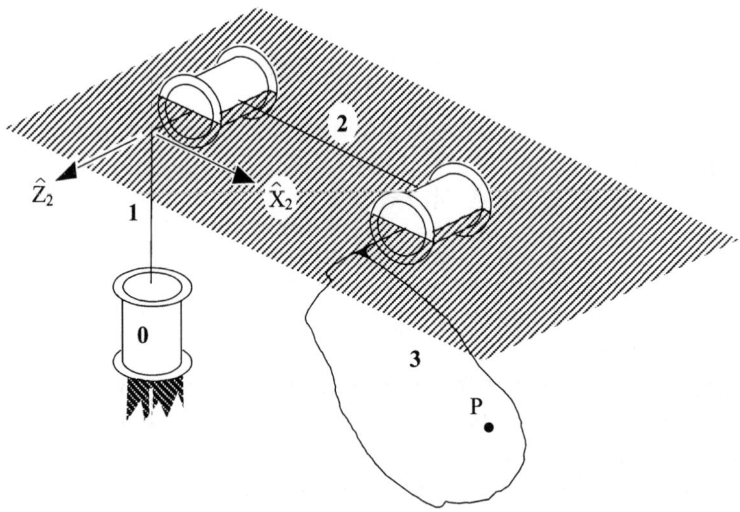

Figure 3. Singular Planes of an Articulated Arm Subassembly

yields the singular Jacobian in case 2, J_2,

$$J_2 = \begin{bmatrix} z\mathrm{c}_1 - \mathrm{s}_1\mathrm{c}_2(a_2 + x\mathrm{c}_3 - y\mathrm{s}_3) & -a_2\mathrm{c}_1\mathrm{s}_2 & -\mathrm{c}_1\mathrm{s}_2(x\mathrm{c}_3 - y\mathrm{s}_3) \\ z\mathrm{s}_1 + \mathrm{c}_1\mathrm{c}_2(a_2 + x\mathrm{c}_3 - y\mathrm{s}_3) & -a_2\mathrm{s}_1\mathrm{s}_2 & -\mathrm{s}_1\mathrm{s}_2(x\mathrm{c}_3 - y\mathrm{s}_3) \\ 0 & a_2\mathrm{c}_2 & \mathrm{c}_2(x\mathrm{c}_3 - y\mathrm{s}_3) \end{bmatrix}. \tag{18}$$

In Eq. 18, the second and third columns are multiples of the column vector $[-c_1s_2, -s_1s_2, c_2]^T$ so instantaneously the dimension of the column space of J_2 is two, as is the rank of J_2. In order to see which direction of motion is no longer possible for points in this singular plane, the columns of singular J_2 are written as,

$$J_2 = \tag{19}$$

$$\left[\ \mathrm{c}_2(z\,{}^0\hat{\mathrm{X}}_2 - (x\mathrm{c}_3 - y\mathrm{s}_3 + a_2)\,{}^0\hat{\mathrm{Z}}_2) - z\mathrm{s}_2\,{}^0\hat{\mathrm{Y}}_2,\quad a_2\,{}^0\hat{\mathrm{Y}}_2,\quad (x\mathrm{c}_3 - y\mathrm{s}_3)\,{}^0\hat{\mathrm{Y}}_2\ \right],$$

where ${}^0\hat{\mathrm{X}}_2$, ${}^0\hat{\mathrm{Y}}_2$ and ${}^0\hat{\mathrm{Z}}_2$ come from the product of Eqs. 1 and 2,

$${}^0_2T = {}^0_1T\ {}^1_2T = \begin{bmatrix} \cos\theta_1\cos\theta_2 & -\cos\theta_1\sin\theta_2 & \sin\theta_1 & 0 \\ \sin\theta_1\cos\theta_2 & -\sin\theta_1\sin\theta_2 & -\cos\theta_1 & 0 \\ \sin\theta_2 & \cos\theta_2 & 0 & 0 \\ 0 & 0 & 0 & 1 \end{bmatrix}. \tag{20}$$

The column space of J_2 is instantaneously spanned by ${}^0\hat{\mathrm{Y}}_2$ and a vector comprised of ${}^0\hat{\mathrm{X}}_2$ and ${}^0\hat{\mathrm{Z}}_2$ components $(z\,{}^0\hat{\mathrm{X}}_2 - (x\mathrm{c}_3 - y\mathrm{s}_3 + a_2)\,{}^0\hat{\mathrm{Z}}_2)$. The vector orthonormal to this basis, $\bar{v}$,

$$\bar{v} = \frac{1}{\sqrt{z^2 + (x\mathrm{c}_3 - y\mathrm{s}_3 + a_2)^2}} \begin{bmatrix} z\mathrm{s}_1 + (x\mathrm{c}_3 - y\mathrm{s}_3 + a_2)\mathrm{c}_1\mathrm{c}_2 \\ -z\mathrm{c}_1 + (x\mathrm{c}_3 - y\mathrm{s}_3 + a_2)\mathrm{s}_1\mathrm{c}_2 \\ (x\mathrm{c}_3 - y\mathrm{s}_3 + a_2)\mathrm{s}_2 \end{bmatrix}, \tag{21}$$

is the unit vector along the position vector, ${}^0\bar{\mathrm{X}}_P$, when point P is located on the $\hat{\mathrm{Z}}_2$ -$\hat{\mathrm{X}}_2$ plane (i.e. when Eq. 17 is substituted into Eq. 7). Point P cannot move in this direction with a finite velocity. This is also the direction in which the arm-subasembly can transmit a theoretically infinite force to the environment through point P and the subassembly instantaneously has an infinite mechanical advantage. This singular plane should be referred to as the workspace boundary singular plane.

4. Conclusion

In this paper we have redeveloped the singular planes of the articulated arm-subassembly using relatively simple methods of kinematic analysis which are well-known to the general kinematics community and kinematics students. No advanced knowledge of Screw Theory or understanding of Screw Theory is required. This is not intended to belittle prior developments which employed Screw Theory. Clearly their developments were more elegant and compact. Our goal however is to try and reach a broader audience which may not be familiar with Screw Theory and to try to increase their awareness and understanding of singular planes as a generic means of describing manipulator singularities. A future paper will continue the procedure developed here to redevelop the singular planes of the remaining geometries of arm-subassemblies and wrist-subassemblies.

References

Craig, J.J. (1989), *Introduction to Robotics: Mechanics and Control, Second Edition*, New York, Addison-Wesley Publishing Company, Inc.

Hayes, M.J.D., Husty, M.L., and Zsombor-Murray, P.J. (2002), Singular Configurations of Wrist-Partitioned 6R Serial Roots: a Geometric Perspective for Users, *Transactions of the CSME*, vol. 26, no. 1, pp. 41–55.

Stanišić, M.M. and Engelberth, J.W. (1988), A Geometric Description of Manipulator Singularities in Terms of Singular Surfaces, *1st Intl. Meeting of Advances in Robot Kinematics*, Ljubljana, Slovenia.

INVERSE KINEMATICS WITH FUZZY REDUNDANCY RESOLUTION FOR A FIRE FIGHTING ROBOT

Bruno Siciliano, Luigi Villani
Dipartimento di Informatica e Sistemistica
Università di Napoli Federico II
{siciliano,lvillani}@unina.it

Abstract The inverse kinematics problem for a fire fighting robot is considered in this paper. The robot is designed for a prompt intervention in road and railway tunnels and is composed by a self-cooling monorail vehicle carrying a fire fighting monitor. The redundant degrees of freedom of the system are used to perform additional control objectives besides the assigned task. A fuzzy technique is adopted to distribute the motion between the monitor and the vehicle while keeping the robot in a security zone. Simulation case studies are developed to demonstrate the effectiveness of the proposed approach.

Keywords: Inverse kinematics, redundancy resolution, fuzzy inference system

1. Introduction

Fire fighting is a very dangerous task, especially in close environments as road or railways tunnels. In fact, the presence of traffic, rails and other obstructions, the high temperature, the intense smoke and heat make a prompt and effective intervention very difficult. Sometimes the fire becomes uncontrollable, like for the 1999 Monte Bianco disaster. On the other hand, the risk of fires in tunnels is very high, due to the possible occurrence of road or rail accidents and to the unavoidable presence of inflammable material.

In such situations robot technology may play a crucial role to guarantee a rapid intervention and continuous water restocking. A possible approach is the ROBOGAT patent (www.robogat.com), consisting in a self-cooling monorail vehicle, mounted on either the tunnel wall or ceiling, carrying a fire fighting monitor. The rail guarantees continuous water supply and the robot may be either teleoperated or autonomous, by using enhanced sensors as thermal and infrared cameras, pyrometers, gas cromotographs.

A typical task for this type of robot may be that of reaching the fire and direct an high pressure water flow at the base of the flames. From a

J. Lenarčič and C. Galletti (eds.), On Advances in Robot Kinematics, 283-292.

kinematical point of view, the system is redundant; hence the redundant degrees of mobility may be exploited to achieve a suitable coordination between the motion of the vehicle and that of the monitor. To this purpose, the task-priority redundancy resolution technique (Nakamura *et al.*, 1897) can be adopted to specify a primary task (e.g., a prescribed trajectory for the water jet) which is fulfilled with higher priority with respect to a secondary task (e.g., maintaining a security distance or a dexterous posture).

In this paper, a redundancy resolution approach is adopted to solve the inverse kinematics problem for a fire fighting robot, based on the weighted pseudo-inverse jacobian and a task-priority strategy integrated with a fuzzy logic algorithm. The approach is inspired to (Antonelli and Chiaverini, 2000) where a fuzzy technique is used to distribute the motion between a manipulator and an underwater vehicle and to handle multiple secondary tasks.

In detail, a fuzzy inference system (FIS) is in charge of distributing the required water-jet motion beteween the vehicle and the monitor, by setting the weights of the weighted pseudo-inverse jacobian. Moreover, the FIS may activate the secondary task to keep the monitor in a dexterous configuration during some phases of the primary task.

Numerical simulations have been developed to show the effectiveness of the proposed approach and to prove its superiority with respect to a fixed weights inverse kinematic scheme.

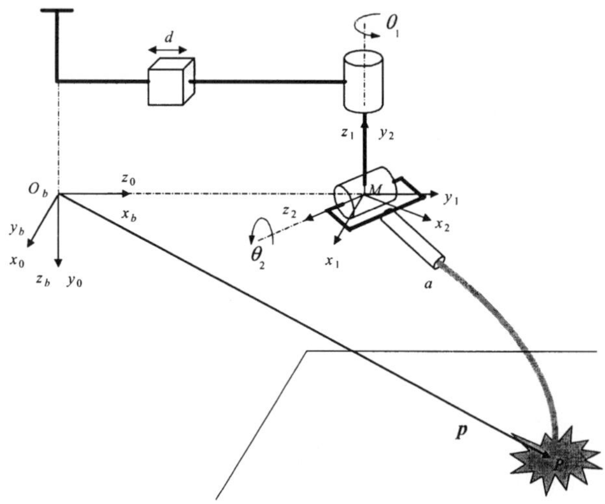

Figure 1. Kinematic model of the fire fighting robot.

2. Direct Kinematics

The kinematic structure of the fire fighting robot (see Fig. 1) can be described by considering a prismatic joint corresponding to the vehicle moving on the monorail (joint variable d), two revolute joints corresponding to the monitor mounted under the vehicle (joint variables θ_1 and θ_2), and the water jet shot by the monitor. Neglecting the influence of the motion of the monitor on the shape of the water jet and considering a quasi-static regime, the water trajectory can be described by a parabolic curve lying on the vertical plane of the monitor, corresponding to the parabolic motion of a particle under gravity (i.e., the so-called projectile motion). It is assumed that the shape of the water jet can be modified also by varying the area a of the nozzle of the monitor.

The relevant frames required to compute the robot kinematics, chosen according to the Denavit-Hartenberg convention (Sciavicco and Siciliano, 2000), are reported in Fig. 1. It can be seen that the axes z_0, z_1 and z_2 intersect at the point M.

The direct kinematics equation characterizes the position vector $\boldsymbol{p} = [x \quad y \quad z]^{\mathrm{T}}$ of a point P of the water jet, referred to the base frame, in terms of the joint position vector $\boldsymbol{q} = [d \quad \theta_1 \quad \theta_2 \quad a]^{\mathrm{T}}$.

The vector $\boldsymbol{p}$ can be expressed as:

$$\widetilde{\boldsymbol{p}} = \widetilde{\boldsymbol{p}}_M + \boldsymbol{T}_2{}^2\widetilde{\boldsymbol{p}} \tag{1}$$

where $\widetilde{\boldsymbol{v}}$ denotes the representation of the vector $\boldsymbol{v}$ in homogeneous coordinates, i.e. $\widetilde{\boldsymbol{v}} = [\boldsymbol{v}^{\mathrm{T}} \quad 1]^{T}$, $\boldsymbol{p}_M$ is the position vector of the point M with respect to the base frame, ${}^2\boldsymbol{p} = [{}^2x \quad {}^2y \quad {}^2z]^{\mathrm{T}}$ is the position vector of the point P with respect to the frame 2, and $\boldsymbol{T}_2$ is the homogeneous transformation matrix of frame 2 with respect to the base frame. The matrix $\boldsymbol{T}_2$ can be easily computed as

$$\boldsymbol{T}_2 = \begin{bmatrix} c_1 & 0 & s_1 & d_1 \\ -s_1 & 0 & c_1 & 0 \\ 0 & -1 & 0 & 0 \\ 0 & 0 & 0 & 1 \end{bmatrix}, \tag{2}$$

where s_1 and c_1 denotes $\sin(\theta_1)$ and $\cos(\theta_1)$ respectively.

The homogeneous representation of vector $\boldsymbol{p}_M$ coincides with the last column of the matrix $\boldsymbol{T}_2$, while the components of the vector ${}^2\boldsymbol{p}$ can be computed on the basis of the projectile equations of motion written in the frame 2 as:

$$ {}^2x = lc_2 + v_0 c_2 t \tag{3}$$

$$ {}^2y = ls_2 + v_0 s_2 t - \frac{1}{2} g t^2 \tag{4}$$

$$^{2}z = 0, \tag{5}$$

where t is the time variable, l is the length of the second link of the monitor, g is the magnitude of the gravity acceleration, and v_0 is the magnitude of the initial velocity of the water flow. This quantity can be expressed in terms of the nozzle's area a as

$$v_0 = \frac{f}{s}, \tag{6}$$

where f is the flow rate, which is assumed to be constant.

The parameter t can be eliminated from Eqs. (3) and (4), and thus ^{2}x can be expressed as a function of ^{2}y or viceversa. Both the options are feasible, however, for the purpose of this work, it is convenient computing ^{2}x as a function of ^{2}y in the form

$$^{2}x = lc_2 + \frac{1}{g}\frac{f^2}{a^2}c_2\left(s_2 + \sqrt{s_2^2 + 2g\frac{a^2}{f^2}(z + ls_2)}\right) = r(\theta_2, a, z), \tag{7}$$

where the equality $^{2}y = -z$ and Eq. (6) have been taken into account.

Hence, from Eq. (1), the position vector $\boldsymbol{p}$ can be computed as

$$\boldsymbol{p}(\boldsymbol{q}, z) = \begin{bmatrix} d + r(\theta_2, a, z)c_1 \\ -r(\theta_2, a, z)s_1 \\ z \end{bmatrix}. \tag{8}$$

Notice that the quantity $r(\theta_2, a, z)$ is the range of the water jet evaluated on a horizontal plane placed at a distance z along the z_b-axis of the base frame.

3. Inverse Kinematics

To solve the inverse kinematics problem, the four joint variables d, θ_1, θ_2 and a must be computed from the three components x, y and z of the vector $\boldsymbol{p}$ by inverting the first two equations in (8). The complexity of these equations and the existence of multiple solutions due to the redundancy of the system, suggest to solve the problem starting from the differential mapping computed from the first two components of (8)

$$\dot{\boldsymbol{p}}_{xy} = \boldsymbol{J}_{xy}(\boldsymbol{q}, z)\dot{\boldsymbol{q}} + \boldsymbol{J}_z(\boldsymbol{q}, z)\dot{z} \tag{9}$$

where $\boldsymbol{p}_{xy} = [x \quad y]^{\mathrm{T}}$. Notice that the Jacobian matrix $\boldsymbol{J}_{xy}(\boldsymbol{q}, z)$ is (2×4), hence the system has two redundant degrees of mobility.

The simplest way to invert the mapping (9) is to use the pseudo-inverse of the Jacobian matrix:

$$\dot{\boldsymbol{q}} = \boldsymbol{J}^{\dagger}(\boldsymbol{q}, z)\dot{\boldsymbol{p}}_{xy} - \boldsymbol{J}^{\dagger}(\boldsymbol{q}, z)\boldsymbol{J}_z(\boldsymbol{q}, z)\dot{z} \tag{10}$$

where

$$\boldsymbol{J}^{\dagger} = \boldsymbol{J}_{xy}^{\mathrm{T}}(\boldsymbol{J}_{xy}\boldsymbol{J}_{xy}^{\mathrm{T}})^{-1}. \tag{11}$$

This solution corresponds to the minimization of the joint velocities in a least-square sense (Sciavicco and Siciliano, 2000).

Because of the different characteristic of the available degrees of mobility, it could be required to modify the velocity distribution with respect to the least-square minimal solution. For example, it would be preferable to perform slow gross motion using the vehicle and fast motion of small amplitude using the monitor. This might be achieved by adopting a weighted pseudo-inverse $\boldsymbol{J}_W^{\dagger}$

$$\boldsymbol{J}_W^{\dagger} = \boldsymbol{W}^{-1}\boldsymbol{J}_{xy}^{\mathrm{T}}(\boldsymbol{J}_{xy}\boldsymbol{W}^{-1}\boldsymbol{J}_{xy}^{\mathrm{T}})^{-1} \tag{12}$$

with the (4×4) matrix $\boldsymbol{W}^{-1} = \mathrm{diag}\{\beta_1, \beta_2, \beta_3, \beta_4\}$ where β_i is a weight factor belonging to the interval $[0, 1]$ such that $\beta = 1$ corresponds to full motion for the i-th degree of mobility and $\beta_i = 0$ corresponds to freeze the corresponding joint.

The redundancy of the system can be further exploited by using a task priority strategy corresponding to a solution to (9) of the form

$$\dot{\boldsymbol{q}} = \boldsymbol{J}_W^{\dagger}(\boldsymbol{q}, z)\dot{\boldsymbol{p}}_{xy} - \boldsymbol{J}_W^{\dagger}(\boldsymbol{q}, z)\boldsymbol{J}_z(\boldsymbol{q}, z)\dot{z} + \left(\boldsymbol{I}_4 - \boldsymbol{J}_W^{\dagger}(\boldsymbol{q}, z)\boldsymbol{J}_{xy}(\boldsymbol{q}, z)\right)\dot{\boldsymbol{q}}_a \tag{13}$$

where $\boldsymbol{I}_4$ is the (4×4) identity matrix, $\dot{\boldsymbol{q}}_a$ is an arbitrary joint velocity vector and the operator $\left(\boldsymbol{I}_4 - \boldsymbol{J}_W^{\dagger}\boldsymbol{J}_{xy}\right)$ projects the joint velocity vector in the null space of the Jacobian matrix. This solution generates an internal motion of the robotic system (secondary task) which does not affect the motion of the water jet end-point P (primary task).

The joint velocity vector $\dot{\boldsymbol{q}}_a$ can be chosen aligned to the gradient of a scalar objective function $W(\boldsymbol{q})$, i.e.:

$$\dot{\boldsymbol{q}}_a = -\alpha k_w \frac{\partial W(\boldsymbol{q})}{\partial \boldsymbol{q}} \tag{14}$$

with $k_w > 0$ and $\alpha \in [0, 1]$, in order to achieve a local minimum for $W(\boldsymbol{q})$ (Liégeois, 1977).

To avoid numerical drift due to discrete-time integration, a Closed Loop Inverse Kinematics (CLIK) algorithm can be adopted (Sciavicco and Siciliano, 2000), which computes $\boldsymbol{q}$ from the integration of the vector:

$$\dot{\boldsymbol{q}} = \boldsymbol{J}_W^{\dagger}(\boldsymbol{q}, z_d)\boldsymbol{v} - \boldsymbol{J}_W^{\dagger}(\boldsymbol{q}, z_d)\boldsymbol{J}_z(\boldsymbol{q}, z_d)\dot{z}_d + \left(\boldsymbol{I}_4 - \boldsymbol{J}_W^{\dagger}(\boldsymbol{q}, z_d)\boldsymbol{J}_{xy}(\boldsymbol{q}, z_d)\right)\dot{\boldsymbol{q}}_a \tag{15}$$

with $\boldsymbol{v} = \dot{\boldsymbol{p}}_{xy,d} + \boldsymbol{K}(\boldsymbol{p}_{xy,d} - \boldsymbol{p}_{xy})$, where $\boldsymbol{K}$ is a (2×2) positive definite matrix gain to be chosen so as to ensure the convergence to zero of the error $\boldsymbol{p}_{xy,d} - \boldsymbol{p}_{xy}$. Notice that in (15) the subscript d denotes the components of the position and velocity vectors that are input to the CLIK algorithm; the position components without the subscript d are those computed from the joint position vector $\boldsymbol{q}$ (the output of the algorithm) via the direct kinematics equation (8).

4. Fuzzy Inverse Kinematics

The CLIK algorithm (15) allows to solve the redundancy of the system by suitably choosing the coefficients β_i of (12) and the coefficient α in (14) for a given function $W(\boldsymbol{q})$ defining the secondary task.

The choice of the above parameters could be made a priori, according to the task requirements. However, setting constant values would mean fix the motion distribution among the degrees of mobility of the system and activate the secondary task during the whole development of the primary task.

In fact, it could be useful to automatically change the values of the parameters β_i and α during the operation. For example, if the task requires to reach a fire that is initially out of the range of the monitor, a satisfactory inverse kinematics solution should generate a joint trajectory that first moves the vehicle close to the fire zone, up to a security distance, and then adjust the water jet to reach the fire while possibly keeping a dexterous posture for the monitor. Hence, β_1 should be set close to one when the robot is far from the fire and close to zero when the robot reaches the security distance; on the other hand, the choice $\alpha = 1$ should be made only when the robot is close to the fire.

A flexible handling of the variables of interest can be achieved by using a fuzzy approach, which is in charge of managing the distribution of the motion between the joints as well as activating the secondary task. This is done by adjusting online the weighting factors β_i and α in (15) according to the Mamdani fuzzy inference system (Driankov *et al.*, 1995).

An example of application of the inverse kinematics algorithm with fuzzy redundancy resolution for the fire fighting robot is described in the case studies.

5. Simulation

The simulation case studies have been performed using MATLAB with the Fuzzy Logic Toolbox. It is assumed that the robot is mounted under the ceiling of a tunnel which is about 6.30 m hight and 10 m large at the base. The length of the second link of the monitor is $l = 0.8\,\text{m}$. The joint limits are $d_m = -100\,\text{m}$ $d_M = -100\,\text{m}$, $\theta_{1m} = -\pi/3\,\text{rad}$, $\theta_{1M} = \pi/3\,\text{rad}$, $\theta_{2m} = -\pi/6\,\text{rad}$, $\theta_{2M} = 0\,\text{rad}$, $a_m = 3.14\,\text{m}^2$, $a_M = 32.0\,\text{m}^2$. The flow rate is $f = 8 \cdot 10^{-3}\,\text{m}^3/\text{s}$. The algorithms have been implemented at a sampling frequency of 1000 Hz, and the gains in (15) have been set to: $\boldsymbol{K} = \text{diag}\{5, 5\}$, $k_w = 200$.

An articulated task has been considered, which includes three phases:

- **Phase 1:** The water jet is required to reach the core of the fire in the position $\boldsymbol{p}_f = [\,30 \quad 1.5 \quad 6.3\,]^{\text{T}}$ m from the initial configuration $\boldsymbol{q}_i = [\,0 \quad 0 \quad -\pi/12 \quad 5.48 \cdot 10^{-4}\,]^{\text{T}}$, corresponding to the initial position $\boldsymbol{p}_i = [\,11.97 \quad 0 \quad 6.3\,]^{\text{T}}$ m. A trajectory is assigned along a segment from $\boldsymbol{p}_i$ to $\boldsymbol{p}_f$, according to a quintic polinomial time law of 5 s duration.

- **Phase 2:** The water jet is required to move alternatively along a segment between the positions $\boldsymbol{p}_f - \boldsymbol{\delta}$ and $\boldsymbol{p}_f + \boldsymbol{\delta}$, with $\boldsymbol{\delta} = [\,1 \quad 1 \quad 0\,]^{\text{T}}$ m, according to a quintic polinomial time law of 0.5 s duration. This phase starts at $t = 5\,\text{s}$ and ends at $t = 9.5\,\text{s}$.

- **Phase 3:** The water jet is required to move backwards from $\boldsymbol{p}_f$ to $\boldsymbol{p}'_f = [\,23.5 \quad 1.5 \quad 0\,]^{\text{T}}$ m according to a quintic polinomial time law of 5.5 s duration.

Notice that in the **Phase 1** the water jet is initially far from the fire and is required to reach the fire; the **Phase 2** corresponds to the operations required to extinguish the fire; the **Phase 3** represents a situation where the fire moves towards the robot.

The secondary task consists in keeping the monitor in a dexterous configuration where all the variables are far from the joint limits, i.e., $\theta_1 = 0\,\text{rad}$, $\theta_2 = -\pi/12\,\text{rad}$ and $a = 5.48 \cdot 10^{-3}\,\text{m}^2$.

Two different case studies are considered, with and without using the fuzzy logic algorithm respectively, based on the weighted pseudo-inverse algorithm with secondary task (15).

In the first case study, the weights in (12) and (14) have been fixed to $\beta_1 = \beta_2 = \beta_3 = \beta_4 = 1$, $\alpha = 1$.

The results are reported in Fig. 2 in terms of the time history of the joint variables and of the norm of the position error $\boldsymbol{p}_d - \boldsymbol{p}$. It can be observed that during the **Phase 1** the pitch variable θ_2 exceeds the upper joint limit $\theta_{2M} = 0$, in spite of the presence of the secondary task. This is due to the fact that the robot, in order to track the trajectory imposed to the end-point of the water jet, tends to raise the second link of the monitor while the vehicle slowly moves toward the fire. During

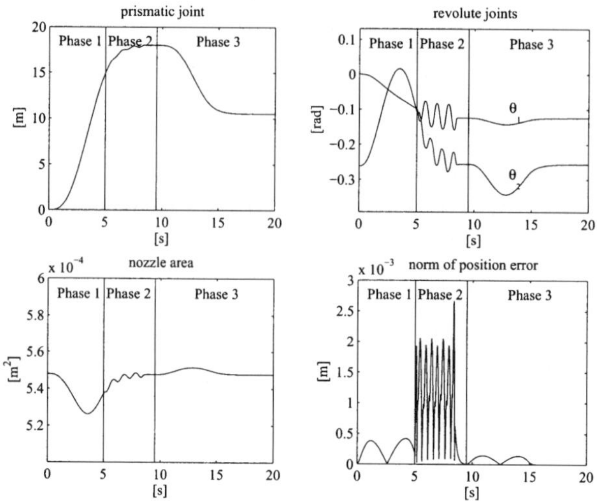

Figure 2. Time histories of the joint positions and of the norm of the position error in the first case study.

the **Phase 2** the vehicle continues to move towards the fire, up to a distance of about 12 m, and keep this distance from the fire also during the **Phase 3**. The position errors remain low along all the motion.

In the second case study, the weights in (12) and (14) are computed via a fuzzy logic approach. To reduce the number of the parameters, they have been chosen as: $\beta_1 = 1 - \beta$, $\beta_2 = \beta_3 = \beta_4 = \beta$.

The Fuzzy Inference System (FIS) receives as input the actual distance of the vehicle from the fire. Hence, a linguistic variable *distance* is considered that can take the values: *distance*={low, medium, high}. The corresponding membership function is reported in Fig. 3. The outputs are given by the linguistic variables β ={vehicle,monitor} and α ={low, high}.

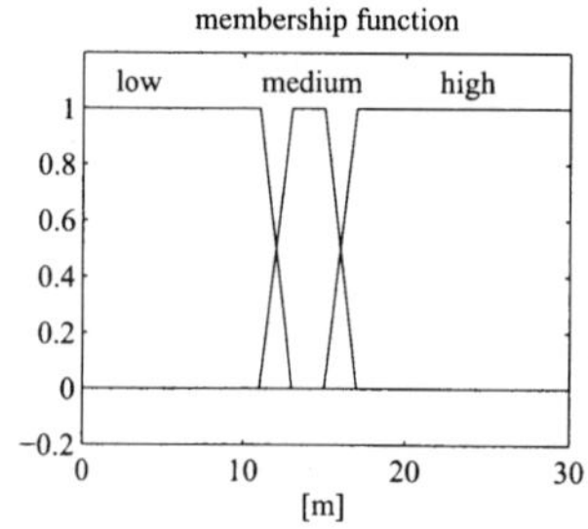

Figure 3. Membership function for the linguistic variable *distance*.

The following fuzzy rules have been used:

1) if (*distance* is high) then (β is vehicle) and (α is low);

2) if (*distance* is medium) then (β is monitor) and (α is high);

3) if (*distance* is low) then (β is vehicle) and (α is high).

On the basis of the first rule, when the robot is far from the fire, the vehicle is asked to approach the fire and the monitor is almost motionless. When the robot reaches a medium distance from the fire (second rule), the monitor is asked to move while keeping a dexterous posture, and the vehicle is left almost motionless. If the fire moves too close to the robot (third rule), the vehicle is asked to move in order to maintain a security distance, possibly keeping the monitor in a dexterous posture.

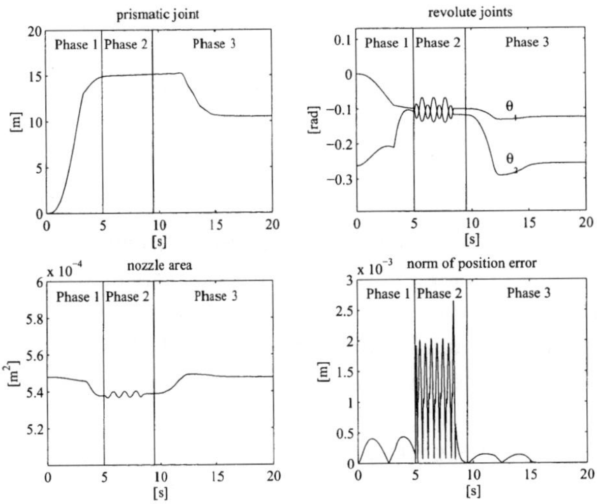

Figure 4. Time histories of the joint positions and of the norm of the position error in the second case study.

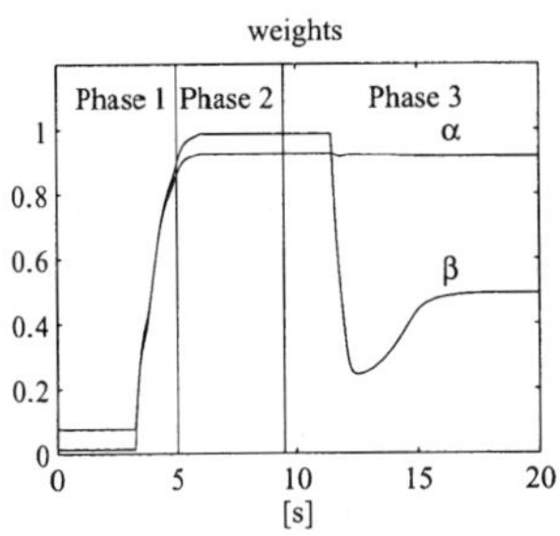

Figure 5. Time histories of the weights α and β in the second case study.

The and–or operations have been calculated by resorting to the *min–max* operations respectively. The defuzzification operation, required to achieve the numerical values of the variables $\beta \in [0,1]$ and $\alpha \in [0,1]$, have been performed using the centroid technique and a normalization.

The results of the simulation are reported in Fig. 4 in terms of the time history of the joint variables and of the norm of the position error $\boldsymbol{p}_d - \boldsymbol{p}$. The corresponding time histories of the variables β and α computed by the fuzzy logic algorithm are reported in Fig. 5. It can be observed that during the **Phase 1** the vehicle moves quickly toward the fire by reaching a distance of about 15 m while all the the joint variables remain far from the joint limits. During the **Phase 2** the vehicle remains motionless while the monitor is in charge to track the prescribed trajectory for the water jet. During the **Phase 3** the robot initially moves the monitor, while the distance from the fire is greater than the security margin; then the vehicle moves back to keep a minimum distance from the fire of about 12 m. Notice that the time history of the position errors is similar to that of the previous case study.

6. Conclusions

An algorithm for redundancy resolution for autonomous tasks of a fire fighting robot has been developed in this paper. The scheme is based on the weighted pseudo-inverse jacobian and a task-priority strategy, combined to a fuzzy technique. Numerical simulations have been developed to demonstrate the effectiveness of the proposed approach.

7. Acknowledgements

The research work reported here was supported by MIUR and ASI.

References

Antonelli G., and Chiaverini S. (2003), Fuzzy redundancy resolution and motion coordination for underwater vehicle-manipulator systems, *IEEE Trans. on Fuzzy Systems*, vol. 11, pp. 109–120.

Driankov D., Hellendoorn H., and Reinfrank M. (1995), *An introduction to fuzzy control*, Sringer-Verlag,Berlin, D.

Liégeois, A. (1977), Automatic supervisory control of the configuration and behavior of multibody mechanisms, *IEEE Trans. on Syst., Man, Cybern.*, vol. SMC-7, pp. 868–871.

Nakamura Y., Hanafusa H., and Yoshikawa T. (1987), Task-priority based redundancy control of robot manipulators, *International Journal of Robotics Research*, vol. 6, no. 2, pp. 3–15.

Sciavicco, L., and Siciliano, B. (2000), *Modelling and Control of Robot Manipulators*, (2nd Ed.), Springer-Verlag, London, UK.

A FORCE BASED APPROACH TO ERROR ANALYSIS OF PARALLEL KINEMATIC MECHANISMS

Andreas Pott and Manfred Hiller
Chair of Mechatronics
University Duisburg-Essen, Germany
{pott, hiller}@imech.de

Abstract Error analysis is an important issue in parallel kinematic synthesis because manufacturing and assembly errors have a significant impact on the mechanical precision of the manipulator. In this paper we propose a method to describe the displacements of the tool center point (TCP) in terms of error amplifications for uncertainties in all kinematic parameters. This is done by using the differential geometric properties of kinetostatic transmission.

The computational effort of this method only depends on the number of rigid bodies and the topological complexity of the mechanism while the number of parameters does not affect computing time. The performance of this method is demonstrated with an example of the six degree-of-freedom parallel kinematic machine *Linapod* that is currently investigated in order to optimize the machine's accuracy.

1. Introduction

The investigation of parallel kinematic machines (PKM) has become a research topic of interest in the past because these structures promise high accuracy and stiffness, good dynamic behavior, and low production costs due to modular design. Investigating the prototypes of PKMs it turned out that these machines could not realize the high requirements yet. The large number of joints results in high compliances, and the complicated kinematic structure can lead to errors at the TCP which are hard to predict. Since the mathematical modelling of PKMs is a challenging task, sophisticated methods are needed to compensate these drawbacks. In order to lower the manufacturing costs, it is essential to analyze the correlation between the tolerances of the mechanical components and the accuracy of the end-effector. Further on it has been proposed to optimize the kinematic parameters in order to reduce PKM's sensitivity with respect to errors in the mechanical components. Thus a method to efficiently estimate the errors of the TCP is needed for the optimization.

J. Lenarčič and C. Galletti (eds.), On Advances in Robot Kinematics, 293-302.

Most approaches to error analysis are based on direct kinematics (Song et al., 1999; Brisan et al., 2002). For a serial mechanical structure the position transmission function can be derived easily with respect to the kinematic parameters. Since in general it is not possible to obtain closed form solutions for the forward kinematics of mechanisms that involve closed kinematic loops, we cannot compute an analytical derivative of the position transmission function. Even if the closed form solution exists, it is complex and difficult to work with. One can use computer algebra to handle the large equations, but still it is hard to interpret the results. To overcome the lack of an analytical position transmission function, the loop closure conditions $\boldsymbol{f}(\boldsymbol{y},\boldsymbol{g}) = \boldsymbol{0}$ can be utilized and differentiation yields

$$\underbrace{\frac{\partial \boldsymbol{f}(\boldsymbol{y},\boldsymbol{g})}{\partial \boldsymbol{y}}}_{\boldsymbol{A}} \delta \boldsymbol{y} + \underbrace{\frac{\partial \boldsymbol{f}(\boldsymbol{y},\boldsymbol{g})}{\partial \boldsymbol{g}}}_{\boldsymbol{B}} \delta \boldsymbol{g} = \boldsymbol{0}, \tag{1}$$

where $\boldsymbol{y} = (\boldsymbol{x}^{\mathrm{T}}, \boldsymbol{\theta}^{\mathrm{T}})^{\mathrm{T}}$ are the degrees-of-freedom (dof) of the end-effector with the position $\boldsymbol{x}$ and the orientation $\boldsymbol{\theta}$ in Bryant angles, and $\boldsymbol{g}$ are the geometrical parameters. Then we directly get the displacements of the TCP $\delta \boldsymbol{y} = -\boldsymbol{A}^{-1}\boldsymbol{B}\,\delta \boldsymbol{g}$ where the matrix $\boldsymbol{A}^{-1}\boldsymbol{B}$ maps the errors in the components to the displacements of the TCP. But this solution has two important drawbacks. The first disadvantage arises from the fact that we have to introduce additional parameters g_i that are 0 in the nominal geometry. This is necessary in order to differentiate with respect to these variables. Although it is theoretically possible to calculate $\boldsymbol{A}^{-1}\boldsymbol{B}$ analytically, the matrix $\boldsymbol{B}$ is large in dimension and every entry of the matrix is an extensive term. For the spatial case $\boldsymbol{A}$ is a 6×6 matrix and we have to invert the matrix numerically since the analytical inverse is useless in practice. Thus, Eq. (1) can hardly be handled. The second problem is that we can only investigate errors that are present in the closure condition. Therefore, errors in the joints are neglected although they significantly affect the accuracy of the mechanism.

Other approaches to error analysis use numerical differentiation of the direct kinematics to calculate the derivatives. This is time consuming, because changes in all parameters must be taken into account, and we have to compute the complex direct kinematics for each parameter.

Section 2 gives a short introduction to the kinetostatic method on which the proposed approach is based. In section 3 virtual error joints are introduced to model the errors in the components and an algorithm is derived that supplies the error amplifications in terms of force transmission. Section 4 presents two examples of the algorithm and its performance. Finally in section 5 some conclusions are drawn.

2. The Kinetostatic Method

The kinetostatic method treats all mechanical components like joints, links, and mass elements as transmission functions that map the kinematic and dynamic state $\boldsymbol{q}, \dot{\boldsymbol{q}}, \ddot{\boldsymbol{q}}, \boldsymbol{Q}$ of one coordinate frame $\mathcal{K}$ to another frame $\mathcal{K}'$ (Kecskeméthy, 1993). The transmission depends on the generalized coordinates $\boldsymbol{s}, \dot{\boldsymbol{s}}, \ddot{\boldsymbol{s}}$. Here the state consists of the pose $\boldsymbol{q}$, the twist $\dot{\boldsymbol{q}}$, the accelerations $\ddot{\boldsymbol{q}}$, and the wrench $\boldsymbol{Q}$. The natural direction of force transmission is the opposite direction of the motion mapping.

It was shown that the constraints for closing kinematic loops can be modelled as transmission functions, too. This results in a modular design that allows calculation of the kinematics of an arbitrary mechanism, including the transmission of velocities, accelerations, and forces. We can represent a single joint as well as a complete mechanism with this transmission schema.

The kinetostatic model can obviously be evaluated for any set of values for the generalized coordinates $\boldsymbol{s}, \dot{\boldsymbol{s}}, \ddot{\boldsymbol{s}}$. For a given pose we can especially apply the different unit twists and accelerations. This method is referred to as kinematic differentials which can be used to set up the equations of motion (Kecskeméthy, 1993) and to calculate the Jacobian $\boldsymbol{J}$ of the end-effector (Hiller and Kecskeméthy, 1989). The complete kinetostatic formalism is implemented in the object oriented programming library M☐BILE that uses the C++ programming language (Kecskeméthy, 1994) and a differential geometric interpretation of the kinetostatics has been given (Kecskeméthy, 1993).

The following approach to error analysis is based on the kinetostatic method and its capability to calculate the kinematics and statics of multibody systems with general topology.

3. Error Amplifications

Let us now consider an arbitrary mechanism with f dof and n articulated joints with the variables $\boldsymbol{q}$. For serial mechanisms we get a closed form solution of the direct kinematics using Denavit-Hartenberg parameters (Denavit and Hartenberg, 1955). Therefore, it is easy to identify the influence of errors in the geometrical parameters $\boldsymbol{g}_i = (\alpha, \theta, d, a)_i^{\mathrm{T}}$ by deriving the position transmission function $\boldsymbol{\varphi}(\boldsymbol{q}, \boldsymbol{g})$ with respect to those parameters $\boldsymbol{g} = (\boldsymbol{g}_1^{\mathrm{T}}, ..., \boldsymbol{g}_n^{\mathrm{T}})^{\mathrm{T}}$ and we get

$$\delta \boldsymbol{e} = \frac{\partial \boldsymbol{\varphi}(\boldsymbol{q}, \boldsymbol{g})}{\partial \boldsymbol{g}} \delta \boldsymbol{g} = \boldsymbol{J}_{\mathrm{err}}\, \delta \boldsymbol{g}, \tag{2}$$

where $\boldsymbol{J}_{\mathrm{err}}$ denotes the Jacobian with respect to the parameters $\boldsymbol{g}$. The coefficients of this Jacobian are referred to as *error amplification co-*

efficients, because they linearly map the geometrical errors $\delta \boldsymbol{g}$ to the displacements of the TCP $\delta \boldsymbol{e}$.

For general mechanisms that involve closed kinematic loops, it is not possible to find an analytic function $\boldsymbol{\varphi}(\boldsymbol{q}, \boldsymbol{g})$ and we cannot directly calculate the Jacobian $\boldsymbol{J}_{\text{err}}$. As stated in section 2, it is possible to calculate the Jacobian with respect to the generalized coordinates $\boldsymbol{q}$ using the kinematic differentials, even if the mechanisms involves closed kinematic loops. Therefore, the basic idea of this approach is to model the errors as generalized coordinates of *virtual error joints.*

3.1 Kinetostatics of Virtual Error Joints

We can describe the geometrical errors by inserting general 6 dof joints between all joints and links of the mechanical system. The variables of the error joints are denoted as $\boldsymbol{g}_i = (\boldsymbol{x}^{\mathrm{T}}, \boldsymbol{\theta}^{\mathrm{T}})^{\mathrm{T}}$, where the vector $\boldsymbol{x}$ refers to the position offset and the vector $\boldsymbol{\theta}$ describes small rotations with Bryant angles. Then the kinetostatic transmission function for the velocities of this joint becomes

$$\begin{pmatrix} \boldsymbol{v}_i \\ \boldsymbol{\omega}_i \end{pmatrix} = \underbrace{\begin{pmatrix} \Delta \boldsymbol{R} & -\tilde{\boldsymbol{x}}\, \Delta \boldsymbol{R} & \boldsymbol{I}_3 & \boldsymbol{0} \\ \boldsymbol{0} & \Delta \boldsymbol{R} & \boldsymbol{0} & \boldsymbol{I}_3 \end{pmatrix}}_{\boldsymbol{J}_{\text{local}}} \begin{pmatrix} \boldsymbol{v}_{i-1} \\ \boldsymbol{\omega}_{i-1} \\ \dot{\boldsymbol{x}} \\ \dot{\boldsymbol{\theta}} \end{pmatrix} \tag{3}$$

where $\boldsymbol{v}_i$ and $\boldsymbol{\omega}_i$ denote the linear and angular velocities of the frame $\mathcal{K}_i$ decomposed in local coordinates, respectively, $\tilde{\boldsymbol{x}}$ is the skew symmetric matrix with $\tilde{\boldsymbol{x}}(\Delta \boldsymbol{R}\, \boldsymbol{\omega}_{i-1}) = \boldsymbol{x} \times (\Delta \boldsymbol{R}\, \boldsymbol{\omega}_{i-1})$, and $\Delta \boldsymbol{R} = \Delta \boldsymbol{R}(\boldsymbol{\theta})$ is the rotation matrix for Bryant angles $\boldsymbol{\theta}$. If we assume ideal transmission, we derive the force transmission from Eq. (3) by transposing the local Jacobian $\boldsymbol{J}_{\text{local}}$ and exchanging the twist $(\boldsymbol{v}, \boldsymbol{\omega})$ for the wrench $(\boldsymbol{f}, \boldsymbol{\tau})$ (Kecskeméthy, 1993). Assuming that we start from the nominal geometry and setting $\boldsymbol{x} = \boldsymbol{\theta} = \boldsymbol{0}$, then

$$\begin{pmatrix} \boldsymbol{f}_{i-1} \\ \boldsymbol{\tau}_{i-1} \\ \boldsymbol{Q}^{(x)} \\ \boldsymbol{Q}^{(\theta)} \end{pmatrix} = \underbrace{\begin{pmatrix} \boldsymbol{I}_3 & \boldsymbol{0} \\ \boldsymbol{0} & \boldsymbol{I}_3 \\ \boldsymbol{I}_3 & \boldsymbol{0} \\ \boldsymbol{0} & \boldsymbol{I}_3 \end{pmatrix}}_{\boldsymbol{J}_{\text{local}}^{\mathrm{T}}} \begin{pmatrix} \boldsymbol{f}_i \\ \boldsymbol{\tau}_i \end{pmatrix} \tag{4}$$

where $\boldsymbol{Q}^{(x)}$ and $\boldsymbol{Q}^{(\theta)}$ denote the generalized forces and torques corresponding to the generalized coordinates $\boldsymbol{x}$ and $\boldsymbol{\theta}$, respectively. It is important to note that for the nominal kinematics (without errors) the internal forces $(\boldsymbol{f}, \boldsymbol{\tau})_{i-1}$ are equal to the generalized forces/torques

$\boldsymbol{Q}^{(x)}, \boldsymbol{Q}^{(\theta)}$ of the joint.

For the modified mechanism with virtual error joints we can calculate the error amplification matrix $\boldsymbol{J}_{\text{err}}$ for the m geometrical parameters containing errors with $m + 1$ evaluations of the direct kinematics. In the first evaluation we use the nominal parameters und then we need m evaluations to set the generalized variables of the virtual error joints to small errors Δg_i, evaluate the kinematic function $\boldsymbol{\varphi}$, and compare the TCP position with the nominal position.

If we use the velocity transmission function instead of the position transmission function, we directly receive the exact linearization for the geometrical errors. In other words we treat the virtual error joint variables $\boldsymbol{g}$ as generalized coordinates and calculate the Jacobian with respect to these coordinates. This can be done with the Jacobian algorithm in Hiller and Kecskeméthy, 1989: first we solve the direct kinematic position problem for the nominal mechanism (without errors). Then we set the jth generalized velocity $\dot{g}_j$ to 1 and all other velocities to 0. The evaluation of the velocity transmission function yields the twist of the TCP which appears in the jth column of the error amplification matrix $\boldsymbol{J}_{\text{err}}$. Then we repeat this procedure for every generalized coordinate in order to obtain to whole Jacobian. Since the kinetostatic formalism is a general approach to kinematics, we can evaluate the velocity transmission function for any topology.

3.2 Force Based Error Analysis

This approach can be further optimized by using the fact that it follows from Eq. (2)

$$\boldsymbol{Q} = \boldsymbol{J}_{\text{err}}^{\text{T}} \boldsymbol{w}_{\text{TCP}}, \tag{5}$$

if we assume ideal transmission, which means that power is neither produced nor consumed. Here $\boldsymbol{w}_{\text{TCP}}$ denotes the wrench at the TCP and $\boldsymbol{Q}$ are the generalized forces/torques corresponding to the errors $\boldsymbol{g}$. The use of the force transmission is similar to a method for clearance analysis (Parenti-Castelli and Venanzi, 2002). As shown above, the generalized forces $\boldsymbol{Q}$ are identical to the internal forces. Hence, we can remove the added virtual error joints, because it is sufficient to consider the internal forces. Consequentially, we receive a simple algorithm to compute the error amplifications for all joints and links with one evaluation of the direct position kinematics and six evaluations of the force transmission:

1 Solve the direct kinematics for the desired posture of the mechanism.

2 Compute the internal forces/torques for all rigid bodies and joints

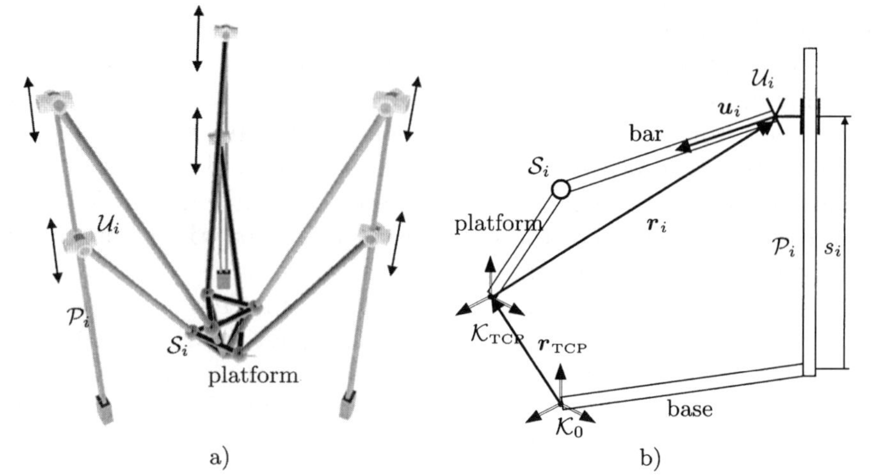

Figure 1. a) 6 dof. Parallel Kinematic Maschine *Linapod* with fixed length legs. b) Geometry of one leg of the PKM *Linapod*. $\mathcal{P}$,$\mathcal{U}$,$\mathcal{S}$ denote prismatic, universal and spherical joints, respectively.

of the mechanism by applying successively the six unity forces and torques to the TCP. The internal forces/torques (in local coordinates) are the coefficients of the error amplification matrix $\boldsymbol{J}_{\text{err}}$.

Note that this algorithm works efficiently with the kinetostatic method, but can be applied to any method that supplies the internal forces.

The Jacobian $\boldsymbol{J}$ depends on the *operation point* on the TCP, thus the effect of errors changes with the considered point. This problem is related to the formulation of a measure for the dexterity of a robot. Due to space limitation we will not go into detail, but it must be stated that it is possible to find an optimal operation point with respect to the errors of the kinematic parameters. This can be done by minimizing the condition number of the error amplification matrix following the approach for the Jacobian of the TCP (Ranjbaran et al., 1996).

4. Examples

In this section we give an examples of a 6 dof fully parallel manipulator where no closed form solution for direct kinematics is known. A performance comparison is presented.

4.1 6 Dof Parallel Robot

The 6 dof parallel kinematic machine tool *Linapod* (Wurst, 1998) was developed at the Institute for Control Engineering of Machine Tools and Manufacturing Units at the University Stuttgart (Germany) and is depicted in Figure 1a. Six rigid links connect the mobile platform to the fixed frame with spherical/universal joints. The lengths of the upper and lower links are $l_u = 1.7$m and $l_l = 1.25$m, respectively. The pivot points on the frame are actuated with linear motors, moving all parallelly to the z-axis. The radii of the the upper and lower platform pivot points and the base frame are $r_u = 0.2$m, $r_l = 0.22$m and $r_b = 0.886$m, respectively. The distance between the two plains of the platform is $\Delta h = 0.2$m and the triangles of the pivot points have a mutual rotational offset of 70°. We define the home position in the center of the workspace and assume small errors in the geometry of every leg. Applying the algorithm from section 3, we can calculate the error amplification matrix $\boldsymbol{J}_l$ for errors in the length of the bars. Therefore, we have to solve the direct kinematic problem to get the position $\boldsymbol{r}_i$ and the direction $\boldsymbol{u}_i$ of the six legs with respect to the TCP (Figure 1b). Then we calculate the internal forces in the bars from force equilibrium conditions when applying the unit forces/torques to the TCP, and we get the matrix equation

$$\underbrace{\begin{pmatrix} \boldsymbol{u}_1 & \boldsymbol{u}_2 & \boldsymbol{u}_3 & \boldsymbol{u}_4 & \boldsymbol{u}_5 & \boldsymbol{u}_6 \\ \boldsymbol{p}_1 & \boldsymbol{p}_2 & \boldsymbol{p}_3 & \boldsymbol{p}_4 & \boldsymbol{p}_5 & \boldsymbol{p}_6 \end{pmatrix}}_{\boldsymbol{A}} \underbrace{(\boldsymbol{f}_1\ \boldsymbol{f}_2\ \boldsymbol{f}_3\ \boldsymbol{f}_4\ \boldsymbol{f}_5\ \boldsymbol{f}_6)}_{\boldsymbol{F}} = \boldsymbol{I}_6, \tag{6}$$

where $\boldsymbol{p}_i = \boldsymbol{u}_i \times \boldsymbol{r}_i$, and the vectors $\boldsymbol{f}_i$ are the internal forces in the direction of each leg, respectively, that result from the unit wrenches applied to the TCP. Here we find $\boldsymbol{J}_l^{\mathrm{T}} = \boldsymbol{F} = \boldsymbol{A}^{-1}$. With the parameters of the *Linapod* $\boldsymbol{J}_l$ becomes at its home position

$$\boldsymbol{J}_l^{\mathrm{T}} = \begin{pmatrix} 0.058 & 0.558 & -0.616 & -0.009 & 0.567 & -0.557 \\ -0.678 & 0.389 & 0.288 & -0.649 & 0.316 & 0.333 \\ -0.153 & -0.153 & -0.153 & -0.230 & -0.230 & -0.230 \\ -0.905 & -1.219 & 2.125 & 0.103 & 2.411 & -2.515 \\ 1.931 & -1.749 & -0.181 & -2.844 & 1.511 & 1.332 \\ -2.231 & -2.231 & -2.231 & 2.020 & 2.020 & 2.020 \end{pmatrix}. \tag{7}$$

Now we assume that the length error for all bars is $\Delta\boldsymbol{e} = k_e(1,1,1,1,1,1)^{\mathrm{T}}$ and for $k_e = 10\,\mu$m we get a total position error $|\Delta r_{\mathrm{TCP}}| = 11.528\,\mu$m. We calculate the exact solution with the direct kinematics to $|\Delta r_{\mathrm{TCP}}| = 11.528\,\mu$m as well. In Figure 2 the relative error of the approximation is compared with the exact solution. It turns out that the approximation is almost accurate up to a geometrical error of about $k_e = 1$ mm. For $k_e = 100$ mm the relative error is about 1% which is still enough for most applications.

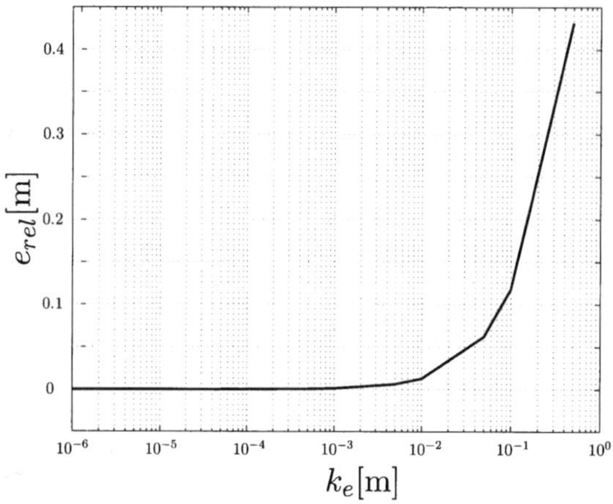

Figure 2. Difference between discrete error calculation (exact) and linearization.

Table 1. Performance of different algorithms on an AMD Athlon 1 GHz for the error analysis of all 252 parameters for *Linapod*. Relative times compared to force based Jacobian.

	all parameters		*optimized for* Linapod	
Algorithm	*time [ms]*	*relative time*	*time [ms]*	*retative time*
numerical differentiation	163.92	68.87	12.31	5.17
velocity based Jacobian	52.13	21.90	4.33	1.82
force based Jacobian	2.38	1.00	2.38	1.00

4.2 Performance Comparison

In this section we compare the computational effort of different algorithms to calculate the error amplification matrix $\boldsymbol{J}_{\text{err}}$. The total cost of an algorithm depends on the number of kinematic evaluations, while the administrative overhead can be neglected.

For the numerical differentiation approach we need one evaluation to solve the nominal direct kinematics and one evaluation of the position forward kinematics for each geometrical parameter. The total numerical effort basically depends on the number of geometrical parameters which are considered to have errors.

The velocity based method to calculate the Jacobian needs one evaluation of the velocity forward kinematics for each parameter. The force based approach always needs six evaluation of force transmission independently of the number of geometrical errors that are taken into account. In Table 1 the computational time from MOBILE (Kecskeméthy, 1994) is listed. We can see that the numerical differentiation approach needs more time than the velocity based method although both need the same number of direct kinematic evaluations. The reason for the difference is that the numerical routines for the velocity transmission are about 68% faster than the position transmission. With the force based method we can save 98.5% computational time and receive the exact error amplification matrix instead of the approximation from the numerical differentiation approach.

We can use the fact that some parameters do not affect the precision and that the mechanism is symmetric to optimize the calculations by reducing the parameter set. But still the force based method outperforms the other approaches (Table 1). The numerical effort of the force based method does not depend on the number of parameters, but only on the total number of elements and the topological complexity of the mechanism. In particular the number of closed kinematic loops and the number of coupled constraints affect the evaluation time of the proposed method. Note that the other methods suffer from a higher system complexity, too.

5. Conclusions

The proposed force based method delivers insight in the impacts of geometrical errors in serial and parallel mechanisms and provides an efficient way to calculate the error amplification coefficients. As shown in section 3, the error amplifications can be calculated from the internal forces in the mechanism, when we apply unity wrenches to the TCP. The example in section 4 demonstrate that the algorithm is suitable for parallel mechanisms. The algorithm is capable to find the parameters that affect the precision as well as to quantify the exact influence of these parameters on the accuracy of the end-effector. For the typical dimension of errors that arise from manufacturing and assembly the proposed method is exact. The examples show that the force based method needs less time to rate all parameters than the other approaches need to calculate the error amplification matrix for a preselected set of parameters. The performance of the algorithm does not depend on the number of parameters. Another useful advantage of the proposed method is that the model of the PKM needs not to be modified in order to apply the algorithm.

We use this method for PKM synthesis in order to find geometrical parameters for the *Linapod* that minimize the influence of manufacturing and assembly errors on the machine accuracy. Further on we utilize this algorithm to analyze the errors within the universal joints and to select the parameters that have to be taken into account for a calibration.

Acknowledgment

This work is supported by the German Research Council (*Deutsche Forschungsgemeinschaft*) under HI370/19-2.

The authors would like to thank our partners of the Institute for Control Engineering of Machine Tools and Manufacturing Units at the University Stuttgart (Germany) for their support with respect to the *Linapod* project.

References

Brisan, C., Franitza, D., and Hiller, M. (2002). Modelling and Analysis of Errors for Parallel Robots. In *Proceedings of the Kolloquium of SFB 562.*

Denavit, J. and Hartenberg, R. (1955). A kinematic notation for lower pair mechanisms based on matrices. *Journal of Applied Mechanics*, 77.

Hiller, M. and Kecskeméthy, A. (1989). Equations of motion of complex multibody systems using kinematical differentials. *Transactions of the Canadian Society of Mechanical Engineering*, 13(4):113–121.

Kecskeméthy, A. (1993). *Objektorientierte Modellierung der Dynamik von Mehrkörpersystemen mit Hilfe von Übertragungselementen.* Fortschrittberichte VDI, Reihe 20, Nr. 88. VDI Verlag, Düsseldorf.

Kecskeméthy, A. (1994). *M∞BILE– Users guide and reference manual.* Fachgebiet Mechatronik, University of Duisburg-Essen.

Parenti-Castelli, V. and Venanzi, S. (2002). A new Deterministic Method for Clearance Influence Analysis in Spatial Mechanisms. In *Proceedings of ASME International Mechanical Engineering Congress & Exposition*, New Orleans, Louisiana.

Ranjbaran, F., Angeles, J., and Kecskeméthy, A. (1996). On the kinematic conditioning of robotic manipulators. In *IEEE International Conference on Robotics and Automation*, pages 3167–3172.

Song, J., Mou, J.-I., and King, C. (1999). Error modeling and compensation for parallel kinematic machines. In *Parallel Kinematic Maschines*, pages 172–187. Springer-Verlag.

Wurst, K.-H. (1998). Linapod – Machine Tools as Parallel Link System in a Modular Design. In *Proceedings of the 1st European-American Forum on Parallel Kinematic Machines*, Milan, Italy.

ANALYSIS AND OPTIMIZATION OF A PLANAR TENDON-DRIVEN PARALLEL MANIPULATOR

JA Snyman and AM Hay
Multidisciplinary Design Optimization Group (MDOG)
Department of Mechanical and Aeronautical Engineering
University of Pretoria
South Africa
jan.snyman@eng.up.ac.za

Abstract In this paper new methodologies for determining the tension distribution, and optimal configurations of tendon-driven planar parallel manipulators are presented. Tendon-driven parallel manipulators are characterized by the use of cables in place of the linear actuators used in most parallel manipulators. The numerical methodologies developed here have potential for easy application to more complex spatial cases.

Keywords: Tendon-driven manipulator, optimization, tension distribution

1. Introduction

Tendon-driven parallel manipulators (TDPMs) represent a relatively recent technology, characterized by the use of cables in place of the linear actuators generally used in parallel manipulators. Possible applications of this technology include overhead cranes for materials handling (Dagalakis et al., 1989, Bostelman et al., 1994, Verhoeven et al., 1998) and as fast-moving micro-manipulators (Verhoeven et al., 1998).

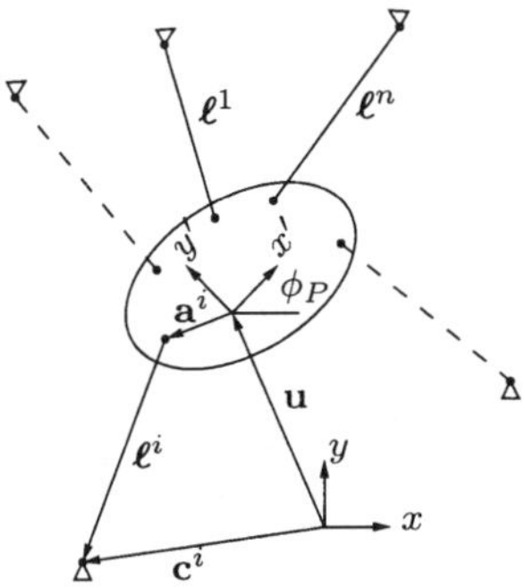

Figure 1. Schematic representation of the planar tendon-driven manipulator

J. Lenarčič and C. Galletti (eds.), On Advances in Robot Kinematics, 303-312.

Three separate topics are examined in this paper, namely the determination of cable forces for overconstrained tendon-driven parallel manipulators, workspace determination of TDPMs and dimensional synthesis of TDPMs for a large dextrous workspace. Numerical methodologies for addressing these topics are proposed. As far as the current state-of-the-art relating to these topics is concerned, Verhoeven and Hiller, 2002 present a method for determining the cable tension distribution in overconstrained TDPMs. This method is limited though to the homogenous case, where no external forces are applied to the manipulator end-effector. Lafourcade et al., 2002 determine cable tensions based on the minimum norm solution. In terms of workspace calculation, Verhoeven and Hiller, 2000 have proposed a method for determining planar TDPM workspaces considering external forces, but not torques applied to the moving platform. Other authors, for example, Fattah and Agrawal, 2002 have used a discretization approach to determining and optimizing manipulator workspaces.

2. The tendon-driven parallel manipulator

As shown schematically in Fig. 1, the tendon-driven parallel manipulator considered here consists of a moving platform connected to a fixed frame by means of n cables, with associated displacement vectors $\boldsymbol{\ell}^i = [\ell_x^i, \ell_y^i]^\top$, $i = 1, 2, \ldots, n$. The lengths of the cables, denoted l_i, $i = 1, 2, \ldots, n$ where $l_i = \ell^i = \|\boldsymbol{\ell}^i\|$, can be varied by winches attached to the fixed frame. Here $\|\cdot\|$ denotes the Euclidean norm. A coordinate frame $x' - y'$ is attached to the moving platform. The position of the origin P of the platform coordinate frame is $\mathbf{u} = [u_1, u_2]^\top = [x_P, y_P]^\top$, and the platform frame is inclined at an angle $w = \phi_P$ to the global $x - y$ coordinate frame. The cables are attached to the fixed frame at $\mathbf{c}^i$, $i = 1, 2, \ldots, n$ and to the moving platform at points $\bar{\mathbf{a}}^i, i = 1, 2, \ldots, n$, defined relative to the platform coordinate frame $x' - y'$. These vectors may of course be transformed to the global coordinate frame by means of a transformation matrix $\mathbf{T}(\phi_P)$ such that the global attachment vectors $\mathbf{a}^i$ are given by $\mathbf{a}^i = \mathbf{T}(\phi_P)\bar{\mathbf{a}}^i$. The forces in each of the cables are described by vectors $\mathbf{f}^{Ci} = [f_x^{Ci}, f_y^{Ci}]^\top$, $i = 1, 2, \ldots, n$. The corresponding magnitudes of the tensions in these cables are denoted $t_i, i = 1, 2, \ldots, n$ where $t_i = f^{Ci} = \|\mathbf{f}^{Ci}\|$. External forces and torques acting on the platform are $\mathbf{f}^P = [f_x^P, f_y^P, \tau^P]^\top$.

Force equilibrium implies that for any fixed position of the moving platform, specified by $\mathbf{u}$ and w, the following equations must hold:

$$\textstyle\sum_{i=1}^n \mathbf{f}^{Ci} + [f_x^P, f_y^P]^\top = \mathbf{0} \text{ and } \sum_{i=1}^n \mathbf{a}^i \times \mathbf{f}^{Ci} + \tau^P = 0 \quad (1)$$

Noting that the tensions $\mathbf{f}^{Ci}$ act parallel to their corresponding cable vectors $\boldsymbol{\ell}^i$, it follows that $\mathbf{f}^{Ci} = \frac{\boldsymbol{\ell}^i}{l_i} t_i$. Eqs. (1) can thus be rewritten as

$$\sum_{i=1}^{n} \frac{\boldsymbol{\ell}^i}{l_i} t_i + [f_x^P, f_y^P]^\top = \mathbf{0} \text{ and } \sum_{i=1}^{n} \mathbf{a}^i \times \frac{\boldsymbol{\ell}^i}{l_i} t_i + \tau^P = 0 \tag{2}$$

Writing (2) in matrix form gives

$$\begin{bmatrix} f_x^P \\ f_y^P \\ \tau^P \end{bmatrix} = - \left[\begin{array}{ccc|c|c} \frac{\ell_x^1}{l_1} & \frac{\ell_x^2}{l_2} & \frac{\ell_x^3}{l_3} & \cdots & \frac{\ell_x^n}{l_n} \\ \frac{\ell_y^1}{l_1} & \frac{\ell_y^2}{l_2} & \frac{\ell_y^3}{l_3} & \cdots & \frac{\ell_y^n}{l_n} \\ \frac{(\mathbf{a}^1 \times \boldsymbol{\ell}^1)_z}{l_1} & \frac{(\mathbf{a}^2 \times \boldsymbol{\ell}^2)_z}{l_2} & \frac{(\mathbf{a}^3 \times \boldsymbol{\ell}^3)_z}{l_3} & \cdots & \frac{(\mathbf{a}^n \times \boldsymbol{\ell}^n)_z}{l_n} \end{array} \right] \left[\begin{array}{c} t_1 \\ t_2 \\ t_3 \\ \hline \vdots \\ t_n \end{array} \right] \tag{3}$$

$$\text{or } \mathbf{f}^P = -\mathbf{S}\mathbf{t}$$

where $\mathbf{S}$ is called the *structure matrix* of the manipulator (Verhoeven et al., 1998) and a subscript z denotes the z-component of a vector.

3. Calculation of the cable tensions

It is evident for the planar case, as pointed out by Verhoeven and Hiller, 2002, that system (3) is overconstrained and thus has many solutions if $n > 3$. For $n = 3$ there are 3 equations and 3 unknowns, and thus if the equations are linearly independent, there will be a unique solution simply given by

$$\mathbf{t} = -\mathbf{S}^{-1}\mathbf{f}^P \tag{4}$$

For $n > 3$ there are many solutions, assuming $\mathbf{S}\mathbf{S}^\top$ is invertible. The general solution to (3) in this case is of the form $\mathbf{t} = \mathbf{t}^{\text{mn}} + \mathbf{t}^{\text{nul}}$ where $\mathbf{t}^{\text{mn}}$ is the minimum norm solution of (3) and $\mathbf{t}^{\text{nul}}$ is a vector belonging to the nullspace $\mathcal{N}(\mathbf{S})$ of $\mathbf{S}$. The minimum norm solution $\mathbf{t}^{\text{mn}}$ is determined by means of the Moore-Penrose inverse, defined as $\mathbf{S}^+ = \mathbf{S}^\top(\mathbf{S}\mathbf{S}^\top)^{-1}$, and is given by (Fattah and Agrawal, 2002):

$$\mathbf{t}^{\text{mn}} = -\mathbf{S}^+\mathbf{f}^P = -\mathbf{S}^\top(\mathbf{S}\mathbf{S}^\top)^{-1}\mathbf{f}^P \tag{5}$$

The new approach proposed here is aimed at determining the minimum or maximum possible constrained ℓ_1-norm, $\|\mathbf{t}\|_1 = \sum_{i=1}^{n} |t_i|$, of the cable tension vector $\mathbf{t}$, and is posed as a linear programming problem. Using the partitioning indicated in system (3) to define

$$\mathbf{f}^P = -\left[\ \mathbf{A} \mid \mathbf{B}\ \right] \left[\ \mathbf{t}^{A\top} \mid \mathbf{t}^{B\top}\ \right]^\top \tag{6}$$

where $\mathbf{A}$ is 3×3 and $\mathbf{B}$ is $3 \times (n-3)$, system (3) can be rewritten as:

$$\mathbf{t}^A(\mathbf{t}^B) = -\mathbf{A}^{-1}\mathbf{f}^P - \mathbf{A}^{-1}\mathbf{B}\mathbf{t}^B \tag{7}$$

which gives the values of the dependent tensions $\mathbf{t}^A \in \Re^3$ as a function of the independent tensions $\mathbf{t}^B \in \Re^{n-3}$. For a given position $\mathbf{u} = [x_P, y_P]^\top$ and orientation $w = \phi_P$ of the platform, and for identical prescribed lower and upper bounds, $t^{\min}$ and $t^{\max}$ on the tension magnitudes t_i, $i = 1, 2, \ldots, n$, the constraints on the three dependent cable tensions $\mathbf{t}^A$ are used to calculate the feasible region for the independent cable tensions $\mathbf{t}^B \in \Re^{n-3}$. For each dependent cable, minimum and maximum tension limits dictate that the following inequalities hold:

$$\begin{aligned} t^{\min} - t_i^A(\mathbf{t}^B) &\leq 0, \;\; i = 1, 2, 3 \\ \text{and } t_i^A(\mathbf{t}^B) - t^{\max} &\leq 0, \;\; i = 1, 2, 3 \end{aligned} \tag{8}$$

Defining $\mathbf{M} = -\mathbf{A}^{-1}\mathbf{f}^P$ and $\mathbf{N} = -\mathbf{A}^{-1}\mathbf{B}$, substituting (7) into (8), and including the inequalities limiting the tensions of the independent cables $\mathbf{t}^B$, the feasible region in the independent cable tension space $\mathbf{t}^B \in \Re^{n-3}$ is bounded by the inequality constraints of the following linear optimization problem:

$$\begin{aligned} &\min\nolimits_{\mathbf{t}^B} \|\mathbf{t}\|_1 = \min\nolimits_{\mathbf{t}^B} \left(\textstyle\sum_{p=1}^{n} t_p^A(\mathbf{t}^B) + \sum_{q=1}^{n-3} t_q^B \right) \\ &\text{such that } \textstyle\sum_{j=1}^{n-3} N_{ij} t_j^B \geq t^{\min} - M_i, \;\; i = 1, 2, 3; \\ &\qquad \textstyle\sum_{j=1}^{n-3} N_{ij} t_j^B \leq t^{\max} - M_i, \;\; i = 1, 2, 3 \\ &\text{and } 0 \leq t^{\min} \leq t_k^B \leq t^{\max}, \;\; k = 1, 2, \ldots, (n-3) \end{aligned} \tag{9}$$

The minimum or maximum (where a maximization is performed instead) allowable ℓ_1-norm of the cable tensions may be determined by solving optimization problem (9). For $n \geq 5$ this can be done efficiently using linear programming methods. For the case where $n = 4$, the inequalities contained in optimization problem (9) reduce to

$$\begin{aligned} N_i t^B &\geq (t^{\min} - M_i), \;\; i = 1, 2, 3 \\ N_i t^B &\leq (t^{\max} - M_i), \;\; i = 1, 2, 3 \\ 0 \leq t^{\min} &\leq t^B \leq t^{\max} \end{aligned} \tag{10}$$

In this case, the existence of a feasible solution, and the minimum $t^{B\min}$ and maximum $t^{B\max}$ allowable independent cable tensions, may easily and efficiently be determined analytically by examining the extreme values of t^B defined by (10). If $t^{B\min} > t^{B\max}$, then no feasible solution exists for the cable tensions. If feasible, then each extreme value of t^B will correspond to either a minimum or maximum constrained ℓ_1-norm solution, whichever case to be determined by comparison. Of course, once a feasible t^B is known, the dependent cable tensions $\mathbf{t}^A$

may be calculated using (7). In terms of control of such a manipulator, any cable tensions $\mathbf{t}$ corresponding to independent cable tension in the range $[t^{B\min}, t^{B\max}]$ may be used.

4. Workspace determination

4.1 Constant orientation workspace

The workspace for TDPMs is primarily determined by the allowable forces in the cables (Verhoeven et al., 1998). The tension in each of the cables should lie between a pretension and maximum tension:

$$t^{\min} \leq t_i \leq t^{\max}, \ i = 1, 2, \ldots, n \tag{11}$$

Here it is also required that each cable has a minimum allowable length $l^{\min}$:

$$l_i \geq l^{\min}, \ i = 1, 2, \ldots, n \tag{12}$$

The constant orientation workspace $W^C[w^{\text{fix}}]$ of the TDPM for a fixed platform orientation $w^{\text{fix}} = \phi_P^{\text{fix}}$, and under a given prescribed load, (which may contain a torsional component), is defined here as

$$\begin{aligned} W^C[w^{\text{fix}}] \ = \ & \Big\{\mathbf{u} \in \Re^2 : \mathbf{\Phi}(\mathbf{u}, \mathbf{v}, w^{\text{fix}}) = \mathbf{0}; \\ & t^{\min} \leq t_i \leq t^{\max} \text{ and } l_i \geq l^{\min}, \ i = 1, 2, \ldots, n\Big\} \end{aligned} \tag{13}$$

where $\mathbf{\Phi}(\mathbf{u}, \mathbf{v}, w^{\text{fix}}) = \mathbf{0}$ are the kinematic constraint equations defining the geometrical relationships between the input and output variables. The cable tensions $\mathbf{t}$ are calculated using the constrained ℓ_1-norm approach proposed in the previous section, and the cable lengths $\mathbf{v}$ can be determined by solving the inverse kinematics of the manipulator. Other conditions such as platform stiffness and proximity to singularities can also be easily included in the workspace definition by the addition of further inequality constraints. In this paper, for the the 3-cable manipulator, a lower limit is placed on the determinant of the kinematic Jacobian, to ensure that the manipulator does not approach a singular position:

$$\det(\mathbf{J}) \geq D^{\min} \tag{14}$$

The workspace for a given manipulator configuration may be determined by means of a discretization approach, by successively determining whether constraints (11), (12) and (14) are satisfied at each of the node points of the discretized output space of the manipulator. Another, more efficient and accurate approach is the chord optimization method for workspace determination (Snyman and Hay, 2000) which has successfully been applied to the determination of TDPM workspaces by Hay, 2003.

4.2 Dextrous workspace

The dextrous workspace $W^D[\phi^{\min}, \phi^{\max}]$ of the planar TDPM is defined as:

$$\begin{aligned} W^D[\phi^{\min}, \phi^{\max}] \;=\; & \{\mathbf{u} \in \Re^2 : \boldsymbol{\Phi}(\mathbf{u}, \mathbf{v}, w) = \mathbf{0}; \\ & t^{\min} \leq t_i \leq t^{\max},\ i = 1, 2, \ldots, n; \text{ and} \\ & l_i \geq l^{\min}\ i = 1, 2, \ldots, n \text{ for all } w \in [\phi^{\min}, \phi^{\max}]\} \end{aligned} \tag{15}$$

The dextrous workspace can be thought of as the intersection of all constant orientation workspaces in the range $[\phi^{\min}, \phi^{\max}]$. Du Plessis and Snyman, 2001 suggest a numerical approach for determining parallel manipulator dextrous workspaces. Firstly constant orientation workspaces are determined for a finite number m^{sl} of regularly spaced ϕ_P values in the range $[\phi^{\min}, \phi^{\max}]$. The intersection of these constant orientation workspaces then yields the dextrous workspace of the manipulator. In many cases the dextrous workspace may be efficiently and accurately described by simply computing the intersection of the two extreme constant orientation workspaces corresponding to $\phi^{\min}$ and $\phi^{\max}$. Du Plessis and Snyman point out however that it is necessary in this case to also check the validity of this efficient approach by ensuring that the constant orientation workspace at the intermediate central value of $(\phi^{\min} + \phi^{\max})/2$ fully contains the dextrous workspace computed using the extreme values. In the case where the above-mentioned approach is invalid, the dextrous workspace may be progressively more accurately approximated by increasing m^{sl}, the number of regularly spaced values of ϕ_P between $\phi^{\min}$ and $\phi^{\max}$. The discretization algorithm outlined in the previous section can be easily modified to compute the dextrous workspace, by testing for compliance for a *range* of values of ϕ_P, instead of just one fixed value, as is the case for the constant orientation workspace.

5. Dextrous workspace maximization

5.1 Optimization formulation

The manipulator workspace is highly dependent on the manipulator design, load on the end-effector, and end effector orientation. With this in mind, the problem addressed here is to design the manipulator so that it yields the greatest dextrous workspace for a given load on the platform. It is assumed that the cable frame attachment points $\mathbf{c}^i$ can be arbitrarily positioned anywhere along a fixed square frame (see Figs. 2 and 3). The angle between the global x-axis and the vector $\mathbf{c}^i$, which connects the global origin at the center of the frame with the associated attachment point on the square, is denoted β^i. The design vector

Table 1. TDPM moving platform cable attachment points

	3-cable	4-cable
$\bar{\mathbf{a}}^1$	$\frac{1}{\sqrt{2}}[0.1, 0.1]^\top$	$\frac{1}{\sqrt{2}}[0.1, 0.1]^\top$
$\bar{\mathbf{a}}^2$	$\frac{1}{\sqrt{2}}[-0.1, 0.1]^\top$	$\frac{1}{\sqrt{2}}[-0.1, 0.1]^\top$
$\bar{\mathbf{a}}^3$	$[0.0, -0.1]^\top$	$[-0.1, -0.1]^\top$
$\bar{\mathbf{a}}^4$	-	$[0.1, -0.1]^\top$

for the optimization problem is thus $\mathbf{d} = [\beta^1, \beta^2, \ldots, \beta^n]^\top$. The only constraints imposed on the optimization are that a lower limit $\beta^{\min}$ is placed on the angular separations of the frame cable attachment points. The workspace maximization problem is simply

$$\begin{aligned} &\max_{\mathbf{d}} W^D[\phi^{\min}, \phi^{\max}] \\ \text{such that } &\left|\beta^{i+1} - \beta^i\right| \geq \beta^{\min},\ i = 1, 2, \ldots, n-1 \\ \text{and } &\left|\beta^n - \beta^1\right| \geq \beta^{\min} \end{aligned} \tag{16}$$

5.2 Numerical results

Optimization formulation (16) is used in this section to determine optimum manipulator designs with respect to maximal dextrous workspace for the 3 and 4-cable manipulators. For the specific planar TDPM used to illustrate the methodology, the square frame dimensions are 2×2 in arbitrary units. The global origin $x - y$ is positioned at the center of the frame. Two different configurations, a 3-cable and a 4-cable manipulator will be considered. For each of these configurations, the cable attachment points on the moving platform are given in Tb. 1.

For each manipulator type, optimal manipulator configurations were determined for a static load of $\mathbf{f}^P = [5, -10, 1]^\top$, in arbitrary units, applied at the origin of the moving $x' - y'$ frame. The optimization problem (16) was solved using the Dynamic-Q optimization algorithm (Snyman and Hay, 2002). This optimization algorithm is suitable for problems where some numerical noise is present in the optimization problem. This is indeed the case here, since the discretization method is used for determining the manipulator dextrous workspaces. Parameters used for the Dynamic-Q method are a move limit of $\rho = 0.1$, a finite difference interval of $\Gamma = 0.05$ for determining the gradients for the 3-cable manipulator, and $\Gamma = 0.1$ for the 4-cable manipulator. Central finite differences were used in calculating these gradients. For the 3-cable manipulator the $[-0.1, 0.1]$ dextrous workspace was computed using a discretization resolution of $m = 100$ points in each direction, and m^{sl} of 3. The range

$[\phi^{\min}, \phi^{\max}]$ is given in radians. For the 4-cable manipulator, which is capable of reaching a larger workspace, the $[-0.2, 0.2]$ dextrous workspace was computed using a resolution $m = 50$ points, and m^{sl} of 21. Note that the number of intermediate values of ϕ_P used for the 3- and 4-cable manipulator differs significantly. This is because just three values of ϕ_P can be used to calculate the 3-cable manipulator dextrous workspaces accurately and efficiently. For the 4-cable manipulator, however, m^{sl} must be increased in order to obtain an accurate description of the dextrous workspace. For both the 3- and 4-cable manipulator and each load case, four feasible random starting designs were chosen for the numerical optimization embodied in equation (16). The values of the various maximum and minimum limits required in inequalities (11), (12), (14) and (16), as used for all numerical examples in this paper, are $t^{\min} = 5$, $t^{\max} = 100$, $l^{\min} = 0.1$, $\beta^{\min} = \pi/18$ and $D^{\min} = 10^{-3}$.

Results obtained for the 3-cable manipulator are given in Tb. 2. This table gives the number of gradient evaluations N^g of the Dynamic-Q algorithm required to find the solution, the randomly chosen starting design $\mathbf{d}^0$ and area A_d^0 of the associated dextrous workspace, the optimized design $\mathbf{d}^*$ and area A_d^* of the optimized dextrous workspace. Fig. 2 shows results of the dimensional synthesis for representative optimization Run R1. Fig. 2(a) shows the starting design and associated workspace, and Fig. 2(b) the optimal design and workspace. Optimization results for the 4-cable manipulator are reported in Tb. 3. Fig. 3 once more shows representative results from the synthesis. For the 4-cable case, local maxima are found during the design optimization. See for example results for Run 3.

6. Conclusions

The new constrained ℓ_1-norm approach for determining cable tensions in overconstrained TDPMs is reliable, and indeed critical for the accurate and correct determination of TDPM workspaces. Two methodologies for determining workspaces of planar TDPMs may be used. The discretization method, for which results are specifically given here, is robust and reliable, but has a high computational requirement in comparison to the accuracy of the workspace determination. The chord method, previously reported by Snyman and Hay, 2000, is accurate, reliable and more efficient, but may require some user interaction in selecting the initial point for starting the workspace computation. In practice a combination of the two methods (Hay, 2003) proves to be the most reliable and accurate, first using the discretization method at a low resolution to obtain a rough estimate of the workspace, and then reverting to the chord method to obtain an accurate and efficient mapping of the workspace boundary. The dimensional synthesis yields TDPM designs with maximal dextrous workspaces for given static loads on the platform. The high dependance of TDPM workspaces on their design is illustrated, demonstrating the importance of dimensional synthesis of such manipulators. Clearly, the numerical methodologies developed here, have potential for easy application to more complex spatial cases.

References

Bostelman, R., Albus, J., Dagalakis, N., Jacoff, A. and Gross, A. (1994), Applications of the NIST Robocrane, *Robotics and Manufacturing*, vol. 5, pp. 403–410.

Dagalakis, N.G. and Albus, J.S. and Wang, B.L. and Unger, J. and Lee, J.D. (1989), Stiffness study of a parallel link robot crane for shipbuilding applications, *ASME Journal of Offshore Mechanics and Arctic Engineering*, vol. 111, no. 3, pp. 183–193.

Du Plessis, L.J. and Snyman, J.A. (2001), A numerical method for the determination of dextrous workspaces of Gough-Stewart platforms, *International Journal for Numerical Methods in Engineering*, vol. 52, pp. 345–369.

Fattah, A. and Agrawal, S.K. (2002), Design of cable-suspended planar parallel robots for an optimal workspace, *Proceedings of the Workshop on Fundamental Issues and Future Research Directions for Parallel Mechanisms and Manipulators*, Quebec City, Quebec, pp. 195–202.

Hay, A.M. (2003), *Optimal dimensional synthesis of planar parallel manipulators with respect to workspaces*, Ph.D. Thesis, Department of Mechanical and Aeronautical Engineering, University of Pretoria.

Lafourcade, P., Llibre, M. and Reboulet, C. (2002), Design of a parallel wire-driven manipulator for wind tunnels, *Proceedings of the Workshop on Fundamental Issues and Future Research Directions for Parallel Mechanisms and Manipulators*, Quebec City, Quebec, pp. 187–194.

Snyman, J.A. and Hay, A.M. (2000), The chord method for the determination of non-convex workspaces of planar parallel platforms, In J. Lenarčič and M.M. Stanišić, editors, *Advances in Robot Kinematics*, Piran, Slovenia, Kluwer Academic Publishers, pp. 285–294.

Snyman, J.A. and Hay, A.M. (2002), The Dynamic-Q optimization method: An alternative to SQP?, *Computers and Mathematics with Applications*, vol. 44, pp. 1589–1598.

Verhoeven, R., Hiller, M. and Tadokoro, S. (1998), Workspace of tendon-driven Stewart platforms: Basics, classification, details on the planar 2-dof class, *Proceedings of the 4th International Conference on Motion and Vibration Control*, ETH Zurich, Switzerland.

Verhoeven, R. and Hiller, M. (2000), Estimating the controllable workspace of tendon-based Stewart platforms, In J. Lenarčič and M.M. Stanišić, editors, *Advances in Robot Kinematics*, Piran, Slovenia, Kluwer Academic Publishers, pp. 277–284.

Verhoeven, R. and Hiller, M. (2002), Tension distribution in tendon-based Stewart platforms, In J. Lenarčič and F. Thomas, editors, *Advances in Robot Kinematics*, Caldes de Malavella, Spain, Kluwer Academic Publishers, pp. 117–124.

Table 2. 3-cable TDPM optimized designs

Run	N^g	$\mathbf{d}^0$	A_d^0	$\mathbf{d}^*$	A_d^*
1	83	$[0.7983, 1.281, 3.373]^\top$	0.0500	$[0.6543, 2.357, 4.252]^\top$	1.024
3	5	$[0.6097, 2.319, 4.197]^\top$	0.8364	$[0.6550, 2.371, 4.215]^\top$	1.019
3	48	$[0.3040, 1.220, 5.040]^\top$	0.0028	$[0.6358, 2.366, 4.263]^\top$	1.019
4	45	$[0.5922, 1.718, 4.865]^\top$	0.1736	$[0.6500, 2.369, 4.233]^\top$	1.021

Table 3. 4-cable TDPM optimized designs

Run	N^g	$\mathbf{d}^0$	A_d^0	$\mathbf{d}^*$	A_d^*
1	55	$[1.347, 1.622, 2.036, 5.416]^\top$	0.0048	$[0.7572, 2.922, 3.097, 5.449]^\top$	1.3008
2	17	$[0.4202, 2.962, 3.531, 6.158]^\top$	0.1968	$[0.7636, 2.929, 3.104, 5.439]^\top$	1.2944
3	37	$[0.6810, 0.9393, 3.676, 5.416]^\top$	0.0016	$[1.452, 1.626, 3.843, 5.406]^\top$	0.8304
4	23	$[0.1010, 2.683, 4.338, 4.407]^\top$	0.2032	$[0.8095, 2.330, 4.444, 4.619]^\top$	1.1920

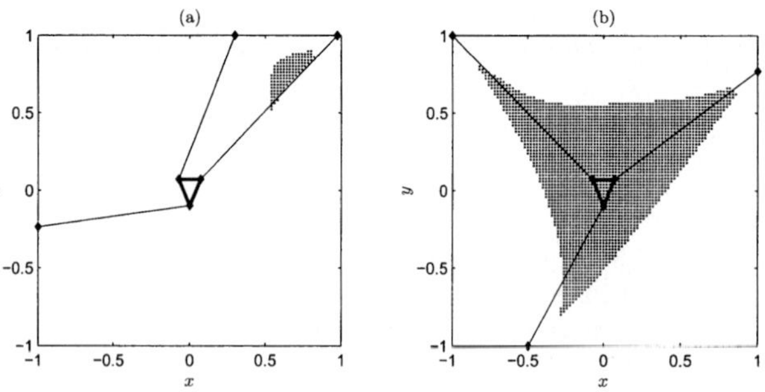

Figure 2. 3-cable TDPM (a) starting and (b) optimized design (Run 1)

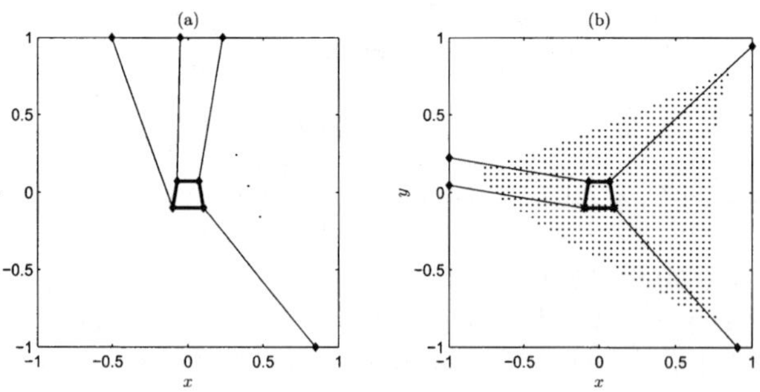

Figure 3. 4-cable TDPM (a) starting and (b) optimized design (Run 1)

CLOSED-FORM EQUILIBRIUM ANALYSIS OF A PLANAR TENSEGRITY STRUCTURE

Jahan Bayat
University of Florida
Department of Mechanical and Aerospace Engineering
jahan_bayat@hotmail.com

Carl D. Crane III
University of Florida
Department of Mechanical and Aerospace Engineering
ccrane@ufl.edu

Abstract This paper presents a closed-form analysis of a two-spring planar tensegrity mechanism to determine all possible equilibrium configurations for the device when no external forces or moments are applied. The equilibrium position is determined by identifying the configurations at which the potential energy stored in the two springs is a minimum. A 28th degree polynomial expressed in terms of the length of one of the springs is developed where this polynomial identifies the cases where the change in potential energy with respect to a change in the spring length is zero. A numerical example is presented.

Keywords: Tensegrity, compliance, equilibrium.

1. Introduction

The word *tensegrity* is a combination of the words *tension* and *integrity* (Edmondson, 1987 and Fuller, 1975). Tensegrity structures are spatial structures formed by a combination of rigid elements in compression (struts) and connecting elements that are in tension (ties). No pair of struts touch and the end of each strut is connected to three non-coplanar ties (Yin et al, 2002). The entire configuration stands by itself and maintains its form solely because of the internal arrangement of the struts and ties (Tobie, 1976).

The development of tensegrity structures is relatively new and the works related have only existed for approximately twenty five years. Kenner, 1976, established the relation between the rotation of the top and bottom ties. Tobie, 1976, presented procedures for the generation of tensile structures by physical and graphical means. Yin, 2002, obtained Kenner's results using energy considerations and found the equilibrium

J. Lenarčič and C. Galletti (eds.), On Advances in Robot Kinematics, 313-320.

position for unloaded tensegrity prisms. Stern, 1999, developed generic design equations to find the lengths of the struts and elastic ties needed to create a desired geometry for a symmetric case. Knight, 2000, addressed the problem of stability of tensegrity structures for the design of deployable antennae.

2. Problem Statement

Figure 1a shows a tensegrity prism with a three sided base and top. Figure 1b shows a four sided parallel prism where the side ties are parallel and Figure 1c shows a four sided tensegrity prisms. Duffy et al, 2000, based on the works of Tobie, 1976, and Kenner, 1976, states that an n sided tensegrity prism can be obtained from the parallel prism by rotating the top of the parallel prism by an angle α given by

$$\alpha = \frac{\pi}{2} - \frac{\pi}{n} . \qquad (1)$$

This paper focuses on the analysis of a two-strut tensegrity system. In this case n=2 and α=0 which means that the parallel and tensegrity prisms are the same. Figure 2 shows a two strut tensegrity prism.

The objective of this effort is to determine, in closed-form, all possible equilibrium deployed positions of a planar tensegrity system wherein two of the ties are compliant. Figure 3 shows the system which is comprised of two struts (compression members a_{12} and a_{34}), two non-compliant ties (tension members a_{41} and a_{23}), and two elastic tensile members (springs), one connected between points 1 and 3 and one between points 2 and 4. It should be noted in Figure 2 that strut a_{34} passes through a slit cut in strut a_{12} and as such the two struts do not intersect or collide.

The problem statement can be explicitly written as:

given:	a_{12}, a_{34}	lengths of struts,
	a_{23}, a_{41}	lengths of non-compliant ties,
	k_1, L_{01}	spring constant and free length of compliant tie between points 4 and 2,
	k_2, L_{02}	spring constant and free length of compliant tie between points 3 and 1.
find:	L_1	length of spring 1 at equilibrium position,
	L_2	length of spring 2 at equilibrium position corresponding to length of spring 1, i.e. L_1.

It should be noted that the problem statement could be formulated in a variety of ways, i.e. a different variable (such as the relative angle between strut a_{34} and tie a_{41}) could have been selected as the generalized parameter for this problem. Attempts at using this alternate approach did not successfully yield a closed form solution for the equilibrium positions.

3. Development of Geometric and Potential Energy Constraint Equations

Figure 3 shows the nomenclature that is used. L_1 and L_2 are the extended lengths of the compliant ties between points 4 and 2 and points 3 and 1. Figure 4 shows the triangle formed by sides a_{34}, a_{23}, and L_1. A cosine law for this triangle can be written as

$$\frac{L_1^2}{2} + \frac{a_{34}^2}{2} + L_1 a_{34} \cos\theta_4' = \frac{a_{23}^2}{2} \quad . \tag{1}$$

Solving for $\cos\theta_4'$ yields

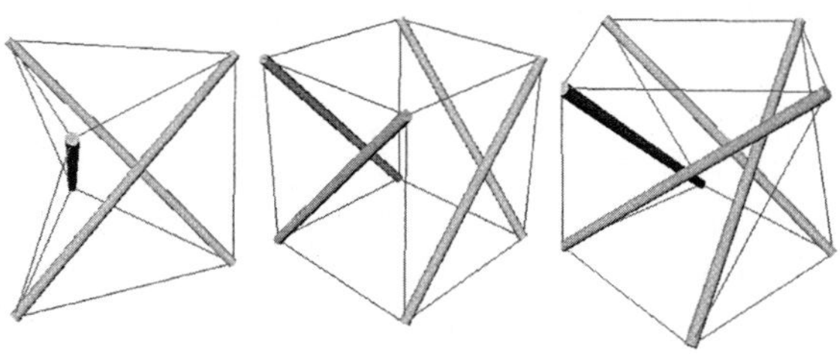

(a) (b) (c)

Figure 1. Tensegrity Structures

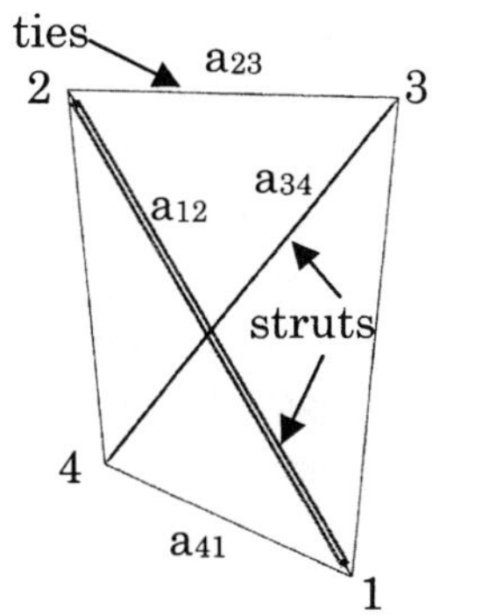

Figure 2. Planar Tensegrity Structure

$$\cos\theta_4' = \frac{a_{23}^2 - L_1^2 - a_{34}^2}{2 L_1 a_{34}} \quad . \tag{2}$$

Figure 5 shows the triangle formed by sides a_{41}, a_{12}, and L_1. A cosine law for this triangle can be written as

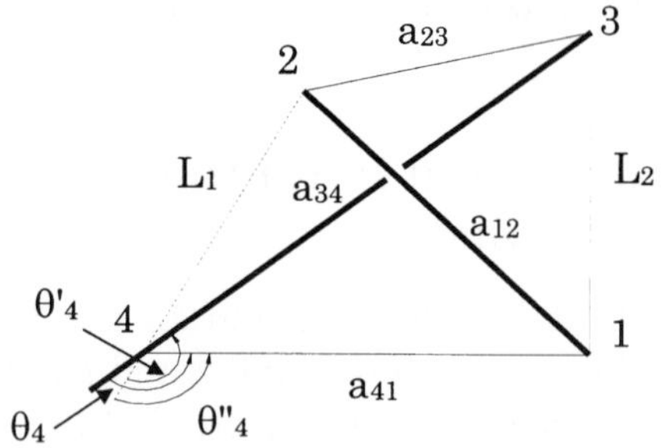

Figure 3. Planar Tensegrity Structure

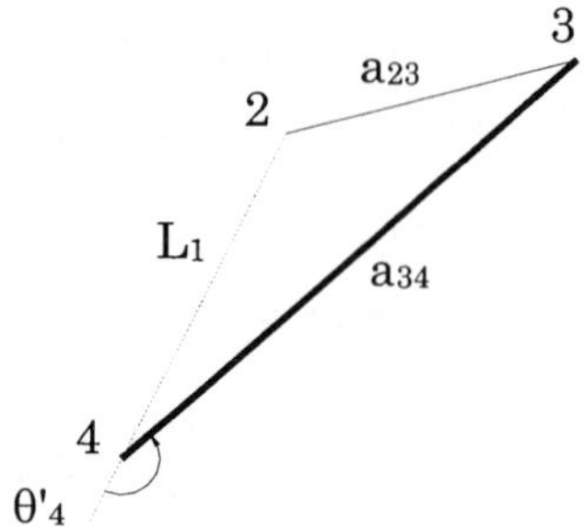

Figure 4. Triangle 4-3-2

$$\frac{L_1^2}{2}+\frac{a_{41}^2}{2}+L_1 a_{41}\cos\theta_4''=\frac{a_{12}^2}{2} \quad . \tag{3}$$

Solving for $\cos\theta_4''$ yields

$$\cos\theta_4''=\frac{a_{12}^2-L_1^2-a_{41}^2}{2L_1 a_{41}} \quad . \tag{4}$$

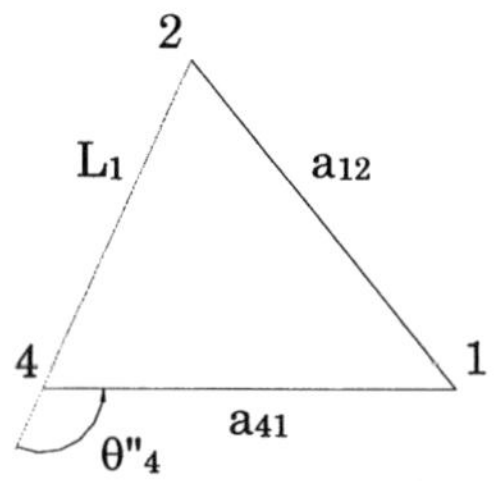

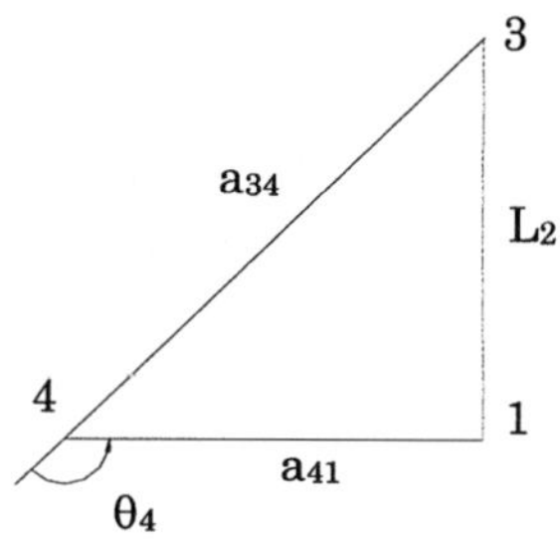

Figure 5. Triangle 4-1-2 *Figure 6.* Triangle 4-1-3

Figure 6 shows the triangle formed by sides a_{41}, a_{34}, and L_2. A cosine law for this triangle can be written as

$$\frac{a_{34}^2}{2}+\frac{a_{41}^2}{2}+a_{34}a_{41}\cos\theta_4=\frac{L_2^2}{2} \quad . \tag{5}$$

Solving for $\cos\theta_4$ yields

$$\cos\theta_4=\frac{L_2^2-a_{34}^2-a_{41}^2}{2a_{34}a_{41}} \quad . \tag{6}$$

From Figure 3 it is apparent that

$$\theta_4+\theta_4'=\pi+\theta_4'' \tag{7}$$

Equating the cosine of the left and right sides of (7) yields

$$\cos(\theta_4 + \theta_4') = \cos(\pi + \theta_4'') \tag{8}$$

and expanding this equation yields

$$\cos\theta_4 \cos\theta_4' - \sin\theta_4 \sin\theta_4' = -\cos\theta_4'' . \tag{9}$$

Rearranging (9) yields

$$\cos\theta_4 \cos\theta_4' + \cos\theta_4'' = \sin\theta_4 \sin\theta_4' . \tag{10}$$

Squaring both sides of (10) gives

$$(\cos\theta_4)^2 (\cos\theta_4')^2 + 2 \cos\theta_4 \cos\theta_4' \cos\theta_4'' + (\cos\theta_4'')^2 = (\sin\theta_4)^2 (\sin\theta_4')^2 . \tag{11}$$

Substituting for $(\sin\theta_4)^2$ and $(\sin\theta_4')^2$ in terms of $\cos\theta_4$ and $\cos\theta_4'$ gives

$$(\cos\theta_4)^2 (\cos\theta_4')^2 + 2 \cos\theta_4 \cos\theta_4' \cos\theta_4'' + (\cos\theta_4'')^2 = (1-\cos^2\theta_4)(1-\cos^2\theta_4') . \tag{12}$$

Equations (2), (4), and (6) are substituted into (12) to yield a single equation in the parameters L_1 and L_2 which can be written as

$$A L_2^4 + B L_2^2 + C = 0 \tag{13}$$

where

$$A = L_1^2, \quad B = L_1^4 + B_2 L_1^2 + B_0, \quad C = C_2 L_1^2 + C_0 \tag{14}$$

and where

$$\begin{aligned} B_2 &= -(a_{23}^2 + a_{34}^2 + a_{41}^2 + a_{12}^2), \\ B_0 &= (a_{12} - a_{41})(a_{12} + a_{41})(a_{23} - a_{34})(a_{23} + a_{34}), \\ C_2 &= (a_{34} - a_{41})(a_{34} + a_{41})(a_{23} - a_{12})(a_{23} + a_{12}), \\ C_0 &= (a_{41}a_{23} + a_{34}a_{12})(a_{41}a_{23} - a_{34}a_{12})(a_{41}^2 + a_{23}^2 - a_{12}^2 - a_{34}^2). \end{aligned} \tag{15}$$

The potential energy of the system can be evaluated as

$$U = \frac{1}{2} k_1 (L_1 - L_{01})^2 + \frac{1}{2} k_2 (L_2 - L_{02})^2 . \tag{16}$$

At equilibrium, the potential energy will be a minimum. This condition can be determined as the configuration of the mechanism whereby the derivative of the potential energy taken with respect to the length L_1 equals zero, i.e.

$$\frac{dU}{dL_1} = k_1 (L_1 - L_{01}) + k_2 (L_2 - L_{02}) \frac{dL_2}{dL_1} = 0 . \tag{17}$$

The derivative dL_2/dL_1 can be determined via implicit differentiation from equation (13) as

$$\frac{dL_2}{dL_1} = \frac{-L_1 [L_2^2 (L_2^2 + 2L_1^2 - a_{23}^2 - a_{41}^2 - a_{34}^2 - a_{12}^2) + (a_{12}^2 - a_{23}^2)(a_{41}^2 - a_{34}^2)]}{L_2 [L_1^2 (L_1^2 + 2L_2^2 - a_{23}^2 - a_{41}^2 - a_{34}^2 - a_{12}^2) + (a_{12}^2 - a_{41}^2)(a_{23}^2 - a_{34}^2))]} . \tag{18}$$

Substituting (18) into (17) and regrouping gives

$$D L_2^5 + E L_2^4 + F L_2^3 + G L_2^2 + H L_2 + J = 0 \tag{19}$$

where

$$D = D_1 L_1 , E = E_1 L_1 , F = F_3 L_1^3 + F_2 L_1^2 + F_1 L_1, G = G_3 L_1^3 + G_1 L_1,$$
$$H = H_5 L_1^5 + H_4 L_1^4 + H_3 L_1^3 + H_2 L_1^2 + H_1 L_1 + H_0, J = J_1 L_1 \tag{20}$$

and where,

$$\begin{aligned}
&D_1 = k_2,\ \ E_1 = -k_2 L_{02}, \\
&F_3 = 2 (k_2 - k_1), F_2 = 2 k_1 L_{01}, F_1 = - k_2 (a_{12}^2 + a_{23}^2 + a_{34}^2 + a_{41}^2), \\
&G_3 = -2 k_2 L_{02}, G_1 = k_2 L_{02} (a_{12}^2 + a_{23}^2 + a_{34}^2 + a_{41}^2), \\
&H_5 = - k_1, H_4 = k_1 L_{01}, H_3 = k_1 (a_{12}^2 + a_{23}^2 + a_{34}^2 + a_{41}^2), \\
&H_2 = - k_1 L_{01} (a_{12}^2 + a_{23}^2 + a_{34}^2 + a_{41}^2), \\
&H_1 = - k_1 (a_{34}^2 - a_{23}^2) (a_{41}^2 - a_{12}^2) + k_2 (a_{34}^2 - a_{41}^2) (a_{23}^2 - a_{12}^2), \\
&H_0 = k_1 L_{01} (a_{34}^2 - a_{23}^2) (a_{41}^2 - a_{12}^2), \\
&J_1 = k_2 L_{02} (a_{34}^2 - a_{41}^2) (a_{12}^2 - a_{23}^2) .
\end{aligned} \tag{21}$$

4. Solution of Geometry and Energy Equations

Equations (19) and (13) represent two equations in the two unknowns L_1 and L_2. These equations can be solved by using Sylvester's variable elimination procedure by multiplying equation (13) by L_2, L_2^2, L_2^3, and L_2^4 and equation (19) by L_2, L_2^2, L_2^3 to yield a total of nine equations that can be written in matrix form as

$$\begin{bmatrix} 0 & 0 & 0 & D & E & F & G & H & J \\ 0 & 0 & 0 & 0 & A & 0 & B & 0 & C \\ 0 & 0 & 0 & A & 0 & B & 0 & C & 0 \\ 0 & 0 & D & E & F & G & H & J & 0 \\ 0 & 0 & A & 0 & B & 0 & C & 0 & 0 \\ 0 & D & E & F & G & H & J & 0 & 0 \\ 0 & A & 0 & B & 0 & C & 0 & 0 & 0 \\ D & E & F & G & H & J & 0 & 0 & 0 \\ A & 0 & B & 0 & C & 0 & 0 & 0 & 0 \end{bmatrix} \begin{bmatrix} L_2^8 \\ L_2^7 \\ L_2^6 \\ L_2^5 \\ L_2^4 \\ L_2^3 \\ L_2^2 \\ L_2 \\ 1 \end{bmatrix} = \begin{bmatrix} 0 \\ 0 \\ 0 \\ 0 \\ 0 \\ 0 \\ 0 \\ 0 \\ 0 \end{bmatrix}. \tag{22}$$

A solution to this set of equations can only occur if the determinant of the 9×9 coefficient matrix is equal to zero. Expansion of this determinant yields a 30th degree polynomial in the variable L_1. When the determinant was expanded symbolically, it was seen that the two lowest order coefficients were identically zero. Thus the polynomial can be divided throughout by L_1^2 to yield a 28th degree polynomial. The coefficients of the 28th degree polynomial were obtained symbolically in terms of the given quantities, but are not presented here due to their length and complexity.

Values for L_2 that correspond to each value of L_1 can be determined by first solving (13) for four possible values of L_2. Only one of these four values also satisfies equation (19).

5. Numerical Example

The following parameters were selected to show the results of a numerical example:

strut lengths: $a_{12} = 3$ in. $a_{34} = 3.5$ in.
non-compliant tie lengths: $a_{41} = 4$ in. $a_{23} = 2$ in.
spring 1 free length & spring constant: $L_{01} = 0.5$ in. $k_1 = 4$ lbf/in.
spring 2 free length & spring constant: $L_{02} = 1$ in. $k_2 = 2.5$ lbf/in.

Eight real and twenty complex roots were obtained for L_1. The real values for L_1 and the corresponding values of L_2 are shown in Table 1.

Table 1. Eight Real Solutions

Case	1	2	3	4	5	6	7	8
L1, in.	-5.485	-5.322	-1.741	-1.576	1.628	1.863	5.129	5.476
L2, in.	2.333	-2.901	-1.495	1.870	1.709	-1.354	-3.288	2.394

The values of L_1 and L_2 listed in Table 1 satisfy the geometric constraints defined by equation (13) and the energy condition defined by equation (19). Each of these eight cases was analyzed to determine whether it represented a minimum or maximum potential energy condition and cases 3, 4, 5, and 6 were found to be minimum states. A free body analysis of struts a_{12} and a_{34} was performed to show that these bodies were indeed in equilibrium.

Efforts were undertaken to obtain a numerical example that would yield more than eight real roots. One example of each type of Grashof and non-Grashof mechanism was analyzed, yet for all these cases eight real roots were found. The complex values of L_1 were analyzed and corresponding complex values of L_2 were determined. In every case it was possible to obtain complex pairs of L_1 and L_2 that satisfied the geometric constraint equation, (13), and the derivative of potential energy equation (19). Thus no extraneous roots were introduced during the variable elimination procedure.

6. Conclusions

This paper presents a technique to obtain all possible equilibrium positions of a planar tensegrity system that incorporates two compliant members. The approach of satisfying geometric constraints while simultaneously finding positions where the derivative of the total potential energy with respect to the generalized coordinate equaled zero resulted in a 28th degree polynomial in a single variable. Although the resulting polynomial was of higher degree than anticipated, an analysis of the real and complex solutions indicates that no extraneous solutions were introduced during the variable elimination procedure.

7. Acknowledgements

The authors would like to gratefully acknowledge the support of the Air Force Office of Scientific Research, Grant Number F49620-00-1-0021 and of the Department of Energy, grant number DE-FG04-86NE37967.

References

Duffy, J., Rooney, J., Knight, B., and Crane, C., (2000), A Review of a Familly of Self-Deploying Tensegrity Structures with Elastic Ties, *The Shock and Vibration Digest*, Vol. 32, No. 2, March 2000, pp. 100-106.

Edmondson, A., (1987) *A Fuller Explanation: The Synergetic Geometry of R. Buckminster Fuller*, Birkhauser, Boston.

Fuller, R., (1975), *Synergetics: The Geometry of Thinking*, MacMillan Publishing Co., Inc., New York, 1975.

Kenner, H., (1976), *Geodesic Math and How to Use It*, University of California Press, Berkeley and Los Angeles, CA.

Knight, B.F., (2000), Deployable Antenna Kinematics using Tensegrity Structure Design, *Ph.D. thesis*, University of Florida, Gainesville, FL.

Stern, I.P., (1999), Development of Design Equations for Self-Deployable N-Strut Tensegrity Systems, *M.S. thesis*, University of Florida, Gainesville, FL.

Tobie, R.S., (1976), A Report on an Inquiry into The Existence, Formation and Representation of Tensile Structures, *Master of Industrial Design thesis*, Pratt Institute, New York.

Yin, J., Duffy, J., and Crane, C., (2002), An Analysis for the Design of Self-Deployable Tensegrity and Reinforced Tensegrity Prisms with Elastic Ties, *International Journal of Robotics and Automation*, Special Issue on Compliance and Compliant Mechanisms, Volume 17, Issue 1.

SINGULARITIES AND MOBILITY OF A CLASS OF 4-DOF MECHANISMS

Dimiter Zlatanov[1], Matteo Zoppi[2] and Clément M. Gosselin[3]
[1] *Universität Innsbruck, Technikerstr. 13, A-6020, Innsbruck, Austria*
[2] *Università di Genova, via Opera Pia 15A, 16145, Genova, Italia*
[3] *Laboratoire de robotique, Université Laval, Québec, Canada, G1K 7P4*
[zlatanov, gosselin]@gmc.ulaval.ca, zoppi@dimec.unige.it

Abstract The paper discusses mobility and singularities of a class of 4-dof parallel wrists with four legs. The singular configurations are classified in detail and their geometric interpretation is discussed. The velocity kinematics and the Jacobian operator are formulated via a screw-system approach. A fully parameterized package of Maple tools was developed and used to visualize singularities and their consequences.

Keywords: singularities, 4dof mechanisms

1. Introduction

The paper studies the mobility and singularities of a class of parallel mechanisms (PMs), Figure 1. When first introduced (Zlatanov and Gosselin, 2001) these were the only known fully-parallel 4-dof chains. Later, other 4-dof PMs have been presented (Huang and Li, 2002).

Four 5R legs, labelled $A, \ldots, D$, connect base and platform. Consider leg L, $L \in \{A, B, C, D\}$. The joint screws (and their axes), $\boldsymbol{\xi}_1^L, \ldots, \boldsymbol{\xi}_5^L$, as well as their direction vectors, $\mathbf{k}_1^L, \ldots, \mathbf{k}_5^L$, are numbered from the

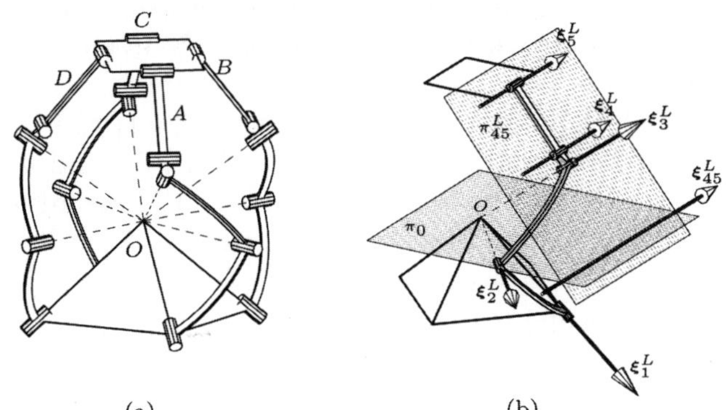

Figure 1. The mechanism architecture with 4-dof.

J. Lenarčič and C. Galletti (eds.), On Advances in Robot Kinematics, 321-328.

base. The first three axes intersect in O, the fixed *rotation centre* of the PM, common to all legs. The fourth and fifth are parallel. The fifth axes, $\xi_5^A, \ldots, \xi_5^D$, are parallel to a common *platform plane*, π_h (fixed in the platform), but are not all parallel to each other. Then, the platform has four dof: it rotates arbitrarily around O and translates orthogonally to π_h. The extrusion (heave), h, of the platform is the distance between π_h and its parallel 0-*extrusion plane*, π_0, through O.

Singularities are classified using six types, as in Zlatanov et al., 1995: redundant input/output (RI/RO), impossible output/input (IO/II), redundant passive motion (RPM), increased instantaneous mobility (IIM).

A fully parametric mathematical model of the mechanism kinematics has been developed in Maple to support the analysis and illustrate and verify the obtained conditions for singularity. The created library of graphical and computational tools simplifies the analysis and provides a sophisticated and effective graphic output. All presented 3D figures have been automatically generated by Maple and are exact. The images can be modified immediately for different values of the geometric parameters.

2. Analysis of the platform constraint

We use the rotating reference frame $O\mathbf{ijk}$ with a fixed origin and vectors constant in the platform, $\mathbf{k} \perp \pi_0$. For any frame, we associate a standard basis, $\{\boldsymbol{\varrho}_x, \boldsymbol{\varrho}_y, \boldsymbol{\varrho}_z, \boldsymbol{\tau}_x, \boldsymbol{\tau}_y, \boldsymbol{\tau}_z\}$, of the space of twists, $\mathcal{S}$. The basis elements are the three rotations and three translations about the coordinate axes. The dual wrench basis is $\{\boldsymbol{\varphi}_x, \boldsymbol{\varphi}_y, \boldsymbol{\varphi}_z, \boldsymbol{\mu}_x, \boldsymbol{\mu}_y, \boldsymbol{\mu}_z\}$.

For leg L, the twist systems $\mathcal{T}_L \subset \mathcal{P}_L$ are spanned by all joint screws and by the passive (non-actuated) joint screws, respectively. The reciprocal system $\mathcal{W}_L = \mathcal{T}_L^\perp$ consists of the platform wrenches the leg can resist (transmit to the base) when all its joints are free to move. The wrenches resisted when the actuated joint is locked form $\mathcal{V}_L = \mathcal{P}_L^\perp$.

For any nonsingular leg posture (with linearly independent leg joint screws) $\operatorname{Span}(\boldsymbol{\varphi}_0^L) = \mathcal{W}_L \subsetneq \mathcal{V}_L$, $\dim \mathcal{V}_L = 2$, where $\boldsymbol{\varphi}_0^L$ is the force at O with direction $\mathbf{k}_5^L$ ($\mathbf{k}_4^L$). In any leg singularity, $\operatorname{Span}(\boldsymbol{\varphi}_0^L) \subsetneq \mathcal{W}_L \subset \mathcal{V}_L$.

The spaces of *structural constraints*, $\mathcal{W} = \sum_L \mathcal{W}_L$, and *actuated constraints*, $\mathcal{V} = \sum_L \mathcal{V}_L$, represent the wrenches the PM can resist, when the actuated joints are free and blocked, respectively. If no leg is singular, $\mathcal{W} = \operatorname{Span}(\boldsymbol{\varphi}_0^A, \ldots, \boldsymbol{\varphi}_0^D) = \operatorname{Span}(\boldsymbol{\varphi}_x, \boldsymbol{\varphi}_y)$. (Since $\mathbf{k}_5^A, \ldots \mathbf{k}_5^D$, are coplanar and not all parallel, the four $\boldsymbol{\varphi}_0^L$ span the planar pencil of forces, $\mathcal{W}_0 = \operatorname{Span}(\boldsymbol{\varphi}_x, \boldsymbol{\varphi}_y)$, at O in π_o. The PM cannot have a constraint singularity (Zlatanov et al., 2002), where $\dim \mathcal{W} > 2$.)

With no leg singular, the platform freedoms, $\mathcal{T} = \cap_L \mathcal{T}_L = \mathcal{W}^\perp$, are given by $\mathcal{M} = \mathcal{W}_0^\perp = \operatorname{Span}(\boldsymbol{\varrho}_x, \boldsymbol{\varrho}_y, \boldsymbol{\varrho}_z, \boldsymbol{\tau}_z)$ ($\mathcal{M}^* = \operatorname{Span}(\boldsymbol{\varphi}_z, \boldsymbol{\mu}_x, \boldsymbol{\mu}_y, \boldsymbol{\mu}_z)$

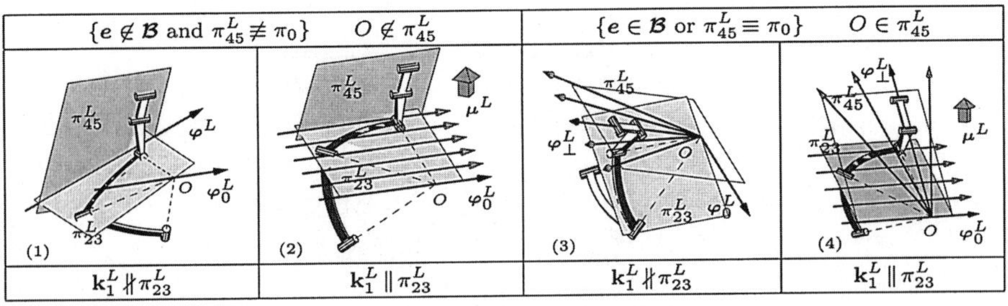

Figure 2. Wrench spaces and leg geometry for each posture class.

is the dual). In IO-type singularities $\mathcal{T}$ is smaller. A leg singularity leads to a loss of a platform freedom (IO) only if $\mathcal{W}_L - \mathcal{W}_0 \neq \varnothing$ ($\mathcal{W}_L \not\subset \mathcal{W}_0$).

Depending on the relative location of the passive joints in a leg, we distinguish four basic classes of leg postures, i.e. subsets, $\mathcal{C}_i^L$, $i = 1, \ldots, 4$, of the feasible leg configuration space, $\mathcal{C}^L$. To describe geometrically a leg posture, $\boldsymbol{q}^L$, we refer to two planes: π_{23}^L through $\boldsymbol{\xi}_2^L$ and $\boldsymbol{\xi}_3^L$; and π_{45}^L defined by $\boldsymbol{\xi}_4^L$ and $\boldsymbol{\xi}_5^L$. The link angle $\alpha_{i\,i+1}^L$, $i < 3$, is between $\mathbf{k}_i^L$ and $\mathbf{k}_{i+1}^L$, while α_{34}^L is between $\mathbf{k}_3^L$ and the plane defined by $\boldsymbol{\xi}_4^L$ and O.

For each class, $\mathcal{C}_i^L$, a subclass, $\mathcal{C}_{i1}^L$, is defined by the additional condition $\mathbf{k}_1^L \parallel \pi_{23}^L$ (i.e., when the first three joint axes are coplanar).

Class 1. $\mathcal{C}_1^L = \{\boldsymbol{q}^L \mid \mathbf{k}_4^L \nparallel \pi_{23}^L$ and $O \notin \pi_{45}^L\}$ (Figure 2–1). $\mathcal{V}_L = \mathrm{Span}\,(\boldsymbol{\varphi}_0^L, \boldsymbol{\varphi}^L)$, where $\boldsymbol{\varphi}^L$ is the force along $\pi_{23}^L \cap \pi_{45}^L$. The force $\boldsymbol{\varphi}^L$ is not through O and not on π_0, therefore the intersection $\mathcal{V}_L \cap \mathcal{W}_0 = \mathrm{Span}\,(\boldsymbol{\varphi}_0^L)$, and $\mathcal{V}_L$ is a cylindroid. Since $O \notin \pi_{45}^L$, then $\pi_{45}^L \not\equiv \pi_0$. If $\boldsymbol{q}^L \in \mathcal{C}_{11}^L$ ($\mathbf{k}_1^L \parallel \pi_{23}^L$), $\mathcal{W}_L = \mathcal{V}_L$ and $\mathcal{W}_L \not\subset \mathcal{W}_0$, else $\mathcal{W}_L = \mathrm{Span}\,(\boldsymbol{\varphi}_0^L)$.

Class 2. $\mathcal{C}_2^L = \{\boldsymbol{q}^L \mid \mathbf{k}_4^L \parallel \pi_{23}^L$ and $O \notin \pi_{45}^L\}$ (Figure 2–2). $\mathcal{V}_L = \mathrm{Span}\,(\boldsymbol{\varphi}_0^L, \boldsymbol{\mu}^L)$ is the 2-system of the pure forces parallel to $\mathbf{k}_4^L$ and lying on π_{23}^L ($\boldsymbol{\mu}^L$ is a pure moment orthogonal to π_{23}^L). We note that $\mathcal{V}_L \cap \mathcal{W}_0 = \mathrm{Span}\,(\boldsymbol{\varphi}_0^L)$. Since $O \notin \pi_{45}^L$, $\pi_{45}^L \not\equiv \pi_0$. $\mathcal{W}_L = \mathcal{V}_L$ and $\mathcal{W}_L \not\subset \mathcal{W}_0$ iff $\boldsymbol{q}^L \in \mathcal{C}_{21}^L$ ($\mathbf{k}_1^L \parallel \pi_{23}^L$), else $\mathcal{W}_L = \mathrm{Span}\,(\boldsymbol{\varphi}_0^L)$.

Class 3. $\mathcal{C}_3^L = \{\boldsymbol{q}^L \mid \mathbf{k}_4^L \nparallel \pi_{23}^L$ and $O \in \pi_{45}^L\}$ (Figure 2–3). $\mathcal{V}_L = \mathrm{Span}\,(\boldsymbol{\varphi}_0^L, \boldsymbol{\varphi}_\perp^L)$ is the planar pencil of forces centred at O and lying in π_{45}^L. $\boldsymbol{\varphi}_\perp^L$ is a pure force through O orthogonal to $\mathbf{k}_4^L$ and lying on π_{45}^L. If $\pi_{45}^L \equiv \pi_0$, then $\mathcal{V}_L = \mathcal{W}_0$; else, $\mathcal{V}_L \cap \mathcal{W}_0 = \mathrm{Span}\,(\boldsymbol{\varphi}_0^L)$. $\mathcal{W}_L = \mathcal{V}_L$ everywhere in $\mathcal{C}_3^L$. Iff $\pi_{45}^L \not\equiv \pi_0$, $\mathcal{W}_L \not\subset \mathcal{W}_0$.

Class 4. $\mathcal{C}_4^L = \{\boldsymbol{q}^L \mid \mathbf{k}_4^L \parallel \pi_{23}^L$ and $O \in \pi_{45}^L\}$ (Figure 2–4). $\mathcal{V}_L = \mathrm{Span}\,(\boldsymbol{\varphi}_0^L, \boldsymbol{\mu}^L, \boldsymbol{\varphi}_\perp^L)$ is a 3-system spanned by a planar pencil in π_{45}^L and a moment perpendicular to π_{23}^L. (If $\alpha_{34}^L \neq 0$, then $\pi_{23}^L \not\equiv \pi_{45}^L$ and $\mathcal{V}_L$ has screws of all pitches. Else, if $\alpha_{34}^L = 0$, then $\pi_{23}^L \equiv \pi_{45}^L$ and $\mathcal{V}_L$ is spanned by the pure forces in a plane.) If $\pi_{45}^L \equiv \pi_0$, then $\mathcal{W}_0 \subset \mathcal{V}_L$; else $\mathcal{V}_L \cap \mathcal{W}_0 = \mathrm{Span}\,(\boldsymbol{\varphi}_0^L)$. $\mathcal{W}_L = \mathcal{V}_L$ if $\boldsymbol{q}^L \in \mathcal{C}_{41}^L$ ($\mathbf{k}_1^L \parallel \pi_{23}^L$),

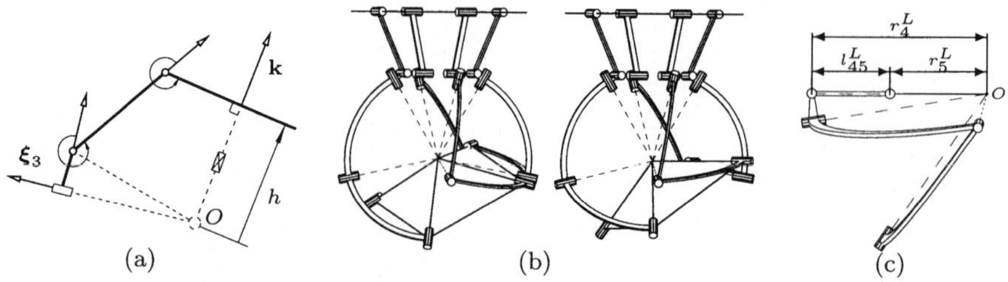

Figure 3. (a) The heave section of leg L. (b) Fixed platform and moving base at maximum heave. (c) A leg allowing $\pi_{45}^L \equiv \pi_0$.

else $\mathcal{W}_L = \text{Span}\,(\varphi_0^L, \varphi_\perp^L)$. In $\mathcal{C}_4^L$, we have extra constraints, $\mathcal{W}_L \not\subset \mathcal{W}_0$, unless both $\pi_{45}^L \equiv \pi_0$ and $\mathbf{k}_1^L \nparallel \pi_{23}^L$.

Note that a leg singularity always leads to extra structural constraints (and an IO singularity), $\mathcal{W}_L - \mathcal{W}_0 \neq \varnothing$, except if $\pi_{45}^L \equiv \pi_0$ and $\boldsymbol{q}^L \notin \mathcal{C}_{41}^L$.

3. The workspace and its boundary

Let $\boldsymbol{\mathcal{X}} \subset SO(3) \times \mathbb{R}$ be the 4-dimensional workspace of the mechanism.

The *extrusion boundary*, $\boldsymbol{\mathcal{B}} \subsetneq \partial\boldsymbol{\mathcal{X}}$, consists of all feasible platform poses, $\boldsymbol{e} = (\mathbf{R}, h)$, where the heave, h, achieves an extremal value for some L. The PRRR planar loop in Figure 3(a) represents the heave section of the PM along one of the leg chains. Extremal extension occurs when $O \in \pi_{45}$, so a platform pose is in $\boldsymbol{\mathcal{B}}$ iff for at least one leg $\pi_{45}^L \not\equiv \pi_0$ and $O \in \pi_{45}^L$ ($\boldsymbol{q}^L \in \mathcal{C}_3^L \cup \mathcal{C}_4^L$). Figure 3(b) shows a mechanism in different configurations, with extremal heave and different platform orientations. (A PM may not reach extremal values of h for some platform rotation. In fact, if α_{12}^L and α_{23}^L are sufficiently small, $\boldsymbol{\mathcal{B}}$ can be empty.)

The remainder of the boundary, $\partial\boldsymbol{\mathcal{X}} - \boldsymbol{\mathcal{B}}$, consists of mechanism configurations where, at least in one leg $\mathbf{k}_1^L \parallel \pi_{23}^L$, i.e., $\boldsymbol{q}^L \in \mathcal{C}_{11}^L \cup \mathcal{C}_{21}^L$.

It can be proven rigorously that every configuration with $\boldsymbol{e} \in \partial\boldsymbol{\mathcal{X}}$ is an IO-type singularity. Conversely, an IO-type singularity generally implies $\boldsymbol{e} \in \partial\boldsymbol{\mathcal{X}}$, the only exception being configurations where $\boldsymbol{\xi}_1^L \equiv \boldsymbol{\xi}_3^L$. Thus, only if $\alpha_{12}^L = \alpha_{23}^L$ can there be IO singularities for $\boldsymbol{e} \in \text{int}\,(\boldsymbol{\mathcal{X}})$.

4. The equivalent mechanism and $\pi_{45}^L \equiv \pi_0$

If all $\pi_{45}^L \not\equiv \pi_0$, the PM is instantaneously equivalent to a simplified mechanism (Figure 1b). The fourth and fifth joints of each leg are replaced by a single revolute joint, $\boldsymbol{\xi}_{45}^L$, at the intersection of π_{45}^L and π_0.

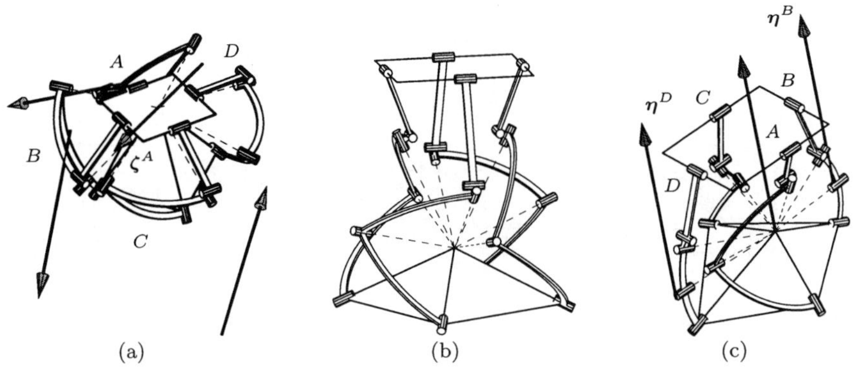

Figure 4. (a) An (IIM, RI, RO)-singularity with $\pi_{45}^D \equiv \pi_0$; $\boldsymbol{\zeta}^D$ is through O on π_0. (b) An (RPM, IIM, IO, II)-singularity with two $\mathcal{C}_4$ legs. (c) An (RPM, RI, IIM, IO)-singularity with $\boldsymbol{e} \in \boldsymbol{\mathcal{B}}$, $\boldsymbol{q}^A \in \boldsymbol{\mathcal{C}}_4^A$, $\boldsymbol{q}^C \in \boldsymbol{\mathcal{C}}_1^C$, $\boldsymbol{q}^L \in \boldsymbol{\mathcal{C}}_{11}^L$, $L = B, D$.

(If $\pi_{45}^L \parallel \pi_0$, the new joint is prismatic.) The new joint velocity is the sum of the velocities it replaces.

When $\pi_{45}^L \equiv \pi_0$, there is no equivalence. This occurs when $O \in \pi_{45}^L$ and the distance, r_5^L, from $\boldsymbol{\xi}_5^L$ to Oz equals $r_4^L \pm l_{45}^L$, where r_4^L is between Oz and $\boldsymbol{\xi}_4^L$, and l_{45}^L is the length of the last leg link (Figure3c).

Suppose $\pi_{45}^L \equiv \pi_0$, while the other legs are nonsingular. An IO occurs only if $\boldsymbol{q}^L \in \boldsymbol{\mathcal{C}}_{41}^L$. Then, there is also RPM and RI. The 5 joints of leg L span a 3-system and the platform has 3 dof. Therefore, the PM has 5 dof (like a closed loop with 8 joints spanning a 3 system) and IIM is present. The configuration also belongs to both or neither of types II and RO depending on the other 3 leg chains. (Indeed, replace leg L with a passive nonsingular leg with the same 3-dof. Then this equivalent PM can have only a singularity of class (RO, II).) Therefore, the configuration is either (RI, RPM, IO, IIM) or with all six types.

Otherwise, if $\boldsymbol{q}^L \notin \boldsymbol{\mathcal{C}}_{41}^L$, although leg L is singular, the platform is constrained only by $\mathcal{W}_0$ and $\dim \mathcal{T} = 4$ (no IO). Either $\boldsymbol{q}^L \in \boldsymbol{\mathcal{C}}_4^L - \boldsymbol{\mathcal{C}}_{41}^L$ and we have an RPM singularity, or $\boldsymbol{q}^L \in \boldsymbol{\mathcal{C}}_3^L$ and there is RI. In either case, since there is no IO, there must be an IIM-type singularity. (This follows from the interdependence rules of the singularity types, Zlatanov et al., 1995.) Moreover, if $\boldsymbol{q}^L \in \boldsymbol{\mathcal{C}}_3^L$ and there is no RPM, an RO-type singularity must be present as well. Thus we have at least (RPM, IIM) or (RI, RO, IIM) and each can be augmented with (RO, II) if the other three legs cooperate. Figure 4(a) shows an example of a singularity of class (IIM, RI, RO). Thus, $\pi_{45}^L \equiv \pi_0$ allows the six singularity classes *with* IIM and *without* IO, impossible for "usual" PMs (Zlatanov et al., 1994). Below, we generally assume that no $\pi_{45}^L \equiv \pi_0$.

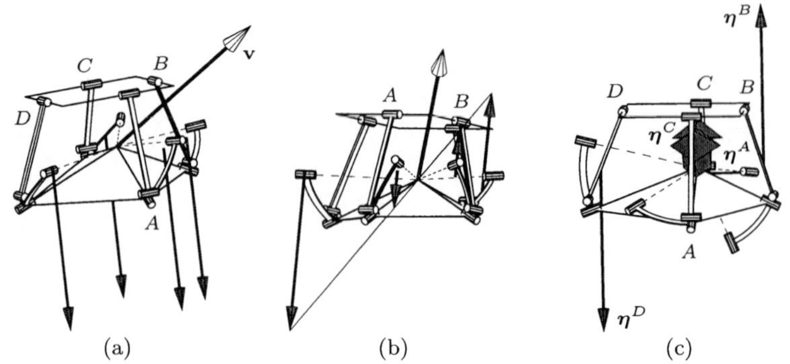

Figure 5. IIM-type singularity with $\boldsymbol{e} \in \operatorname{int}(\boldsymbol{\mathcal{X}})$ and $\boldsymbol{\xi}_1^L \equiv \boldsymbol{\xi}_3^L, \forall L$. In (b) the wrenches $\boldsymbol{\eta}^L$ of (a) are drawn with the lengths of their pitches, to show the linear dependence. In (c) the moment components of all $\boldsymbol{\eta}^L$ are coplanar.

5. Leg Singularities

5.1 RPM- and IIM-type singularities

An RPM-type singularity occurs iff a leg chain has a linear dependence of the passive joints screws, $\dim \mathcal{P}_L < 4$, i.e., $\dim \mathcal{V}_L > 2$. According to Section 2, this is equivalent to $\boldsymbol{q}^L \in \boldsymbol{\mathcal{C}}_4^L$ for some L.

Each $\boldsymbol{\mathcal{C}}_4$ leg has one passive dof, so the redundant passive freedoms of the mechanism are as many as the $\boldsymbol{\mathcal{C}}_4$ legs. Figure 4b represents a configuration with two RPM freedoms.

Provided that all $\pi_{45}^L \not\equiv \pi_0$, a necessary condition for $O \in \pi_{45}^L$ is $\boldsymbol{e} \in \boldsymbol{\mathcal{B}}$. Therefore, under this hypothesis, all RPM-type singular configurations have an end-effector in $\boldsymbol{\mathcal{B}}$ and the workspace interior $\operatorname{int}(\boldsymbol{\mathcal{X}})$ is free of them. (As we saw, if the geometry of some leg is such that $\pi_{45}^L \equiv \pi_0$, it is possible to have an RPM-type singularity and $\boldsymbol{e} \in \operatorname{int}(\boldsymbol{\mathcal{X}})$.)

A method for the study of IIM-type singularities in certain parallel mechanisms is given in Zlatanov et al., 1994. The approach can be applied to the instantaneously equivalent PM. A necessary and sufficient condition for IIM is the existence of four linearly dependant wrenches, $\boldsymbol{\eta}^A, \ldots, \boldsymbol{\eta}^D$, not all zero, such that $\boldsymbol{\eta}^L \in \mathcal{W}_L \cap \mathcal{M}^*, \forall L$.

If $\pi_{45}^L \not\equiv \pi_0$ and $\alpha_{12}^L \neq \alpha_{23}^L$ for all L, the mechanism configuration has an end-effector in $\partial \boldsymbol{\mathcal{X}}$ and $\operatorname{int}(\boldsymbol{\mathcal{X}})$ is free from IIM-type singularity. Examples of IIM-type configurations are presented in Figures 4–5.

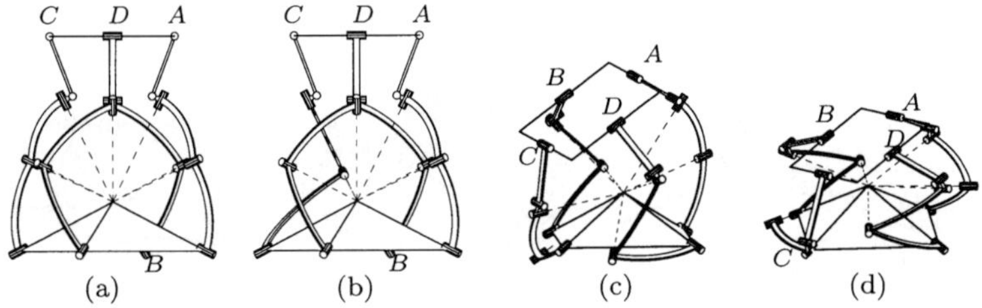

Figure 6. Singularities with IO and RI. (a) $\boldsymbol{e} \in \boldsymbol{\mathcal{B}}$, $\boldsymbol{q}^L \in \boldsymbol{\mathcal{C}}_3^L$, $L = A, C$. (b)$\boldsymbol{q}^C \in \boldsymbol{\mathcal{C}}_4^C$ also RPM-type singular.(c)$\boldsymbol{e} \in \partial\boldsymbol{\mathcal{X}} - \boldsymbol{\mathcal{B}}$, $\boldsymbol{q}^A \in \boldsymbol{\mathcal{C}}_{11}^A$, $\boldsymbol{q}^L \in \boldsymbol{\mathcal{C}}_1^L$, $L = B, C, D$. (d)$\boldsymbol{e} \in \mathrm{int}\,(\boldsymbol{\mathcal{X}})$, $\boldsymbol{\xi}_1^C \equiv \boldsymbol{\xi}_3^C$, $\boldsymbol{q}^L \in \boldsymbol{\mathcal{C}}_1^L$, $L = A, B, D$.

5.2 Singularities on and off $\boldsymbol{\mathcal{B}}$

For configurations with $\boldsymbol{e} \in \boldsymbol{\mathcal{B}}$, either $\boldsymbol{q}^L \in \boldsymbol{\mathcal{C}}_3^L$ or $\boldsymbol{q}^L \in \boldsymbol{\mathcal{C}}_4^L$ ($\pi_{45}^L \not\equiv \pi_0$) for some L. All these are IO-type singularities with respect to extrusion. If $\boldsymbol{q}^L \in \boldsymbol{\mathcal{C}}_4^L$ an RPM-type singularity occurs. Else, if $\boldsymbol{q}^L \in \boldsymbol{\mathcal{C}}_3^L$, there is an RI-type singularity. If $\boldsymbol{q}^L \in \boldsymbol{\mathcal{C}}_{41}^L$, both RI- and RPM-type are present.

When $\boldsymbol{e} \in \partial\boldsymbol{\mathcal{X}} - \boldsymbol{\mathcal{B}}$, $\boldsymbol{q}^L \in \boldsymbol{\mathcal{C}}_{11}^L \cup \boldsymbol{\mathcal{C}}_{21}^L$ for all legs. Assuming $\pi_{45}^L \not\equiv \pi_0$, such a configuration is an (RI, IO) singularity. Iff $\alpha_{12}^L = \alpha_{23}^L$, there are leg singularities with $\boldsymbol{e} \in \mathrm{int}\,(\boldsymbol{\mathcal{X}})$. They are (at least) in the IO and RI types, but never in RPM. Figure 6 shows singularities of class (RI, IO).

6. Singular configurations with nonsingular legs

The analysis in the previous sections shows that if no leg is singular, the only possible singularity class is (RO, II), Figure 7. When $\pi_{45}^L \not\equiv \pi_0$ and $\alpha_{12}^L \neq \alpha_{23}^L$, "no singular legs" is equivalent to $\boldsymbol{e} \in \mathrm{int}\,(\boldsymbol{\mathcal{X}})$.

By using the equivalent PM with 4R legs we obtain the 4×4 Jacobian:

$$\mathbf{Z} = \begin{bmatrix} -\frac{1}{r_{45}^A}\mathbf{n}_{23}^A \cdot \mathbf{k}_4^A & | & \mathbf{n}_{23}^A \\ -\frac{1}{r_{45}^B}\mathbf{n}_{23}^B \cdot \mathbf{k}_4^B & | & \mathbf{n}_{23}^B \\ -\frac{1}{r_{45}^C}\mathbf{n}_{23}^C \cdot \mathbf{k}_4^C & | & \mathbf{n}_{23}^C \\ -\frac{1}{r_{45}^D}\mathbf{n}_{23}^D \cdot \mathbf{k}_4^D & | & \mathbf{n}_{23}^D \end{bmatrix} \tag{1}$$

The rows of $\mathbf{Z}$ represent wrenches in $\mathcal{M}^*$. These can be transformed to an equivalent 4-element spanning system with three (or four) pure moments. The condition for singularity is obtained as a 3×3 determinant, and in case of singularity, the RO-motion screw is derived symbolically.

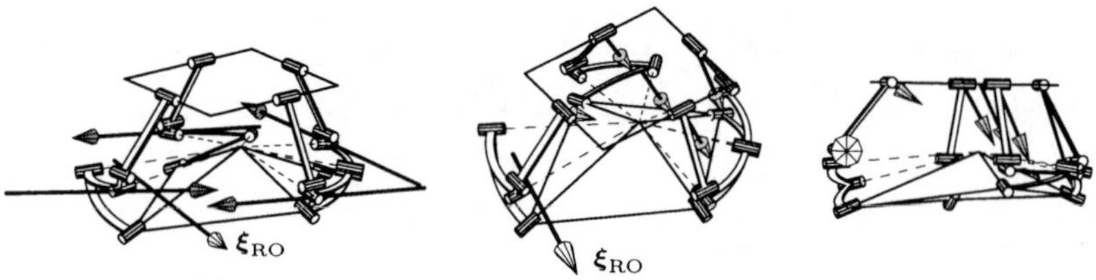

Figure 7. (RO, II) singularities. (a) The RO motion , $\boldsymbol{\xi}_{\mathrm{RO}}$ (pitch −0.00104) and the forces φ^L are shown. (b-c) $\boldsymbol{\xi}_{\mathrm{RO}}$ and the velocities at the platform points are shown.

7. Conclusions

The paper studies the possible singularity types of a family of 4-dof parallel wrists. Under certain hypotheses, the RPM-, RI-, IO- and IIM-type singularities are restricted to the workspace boundary. The typical interior singularities of type (RO, II) are identified by checking a system of 4 wrenches. Under the most general assumptions, these PMs allow all 21 singularity classes, including those 6 not possible for classic PMs.

The method of analysis is general. A fully parametric 3D virtual mock-up of the mechanism, based on a strong mathematic model, has been implemented in Maple via a library of computational and graphical routines created to analyze the architecture, visualize the mechanism behaviour and verify the results of the singularity analysis. Since most such routines are general, they can be reused for different mechanisms. The software is available from the authors.

References

Huang, Z., Li, Q.C., "On the Type Synthesis of Lower-Mobility Parallel Manipulators". Proc. of the workshop Fundamental Issues and Future Research Directions for Parallel Mechanisms and Manipulators, Québec City, Québec, Canada, 2002.

Zlatanov, D., Benhabib, B., Fenton, R.G., "Velocity and singularity analysis of Hybrid Chain Manipulators". Proc. of the 23rd Biennial Mechaism Conference, DE-Vol.72, pp.467-476, 1994.

Zlatanov, D., Benhabib, B.,Fenton, R.G., "A unifying framework for classification and interpretation of mechanism singularities". Trans. of the ASME Journal of Mechanical Design, Vol. 117, pp. 566-572, December 1995.

Zlatanov, D., Bonev, I., Gosselin, C., "Constraint Singularities of Parallel Mechanisms". Proc. of the 2002 IEEE International Conference on Robotics and Automation, ICRA 2002, May 11-15, Washington, DC, USA, pp.496-502, 2002.

Zlatanov, D., Gosselin, C., "A family of new parallel architectures with four degrees of freedom". In F.C. Park C.C. and Iurascu, editor, Computational Kinematics, pp.57-66, Mai 2001

KINEMATICS AND DYNAMICS OF A 6-RUS HUNT-TYPE PARALLEL MANIPULATOR BY USING NATURAL COORDINATES

J. J. Gil, I. Zabalza, J. Ros, J. M. Pintor and J. M. Jiménez
Public University of Navarra, Mechanical Engineering Dep.
Campus Arrosadia, 31006 Pamplona, Spain.
javier.gil@unavarra.es

Abstract In this paper the kinematics, (position, velocities and accelerations) and the dynamics of a 6-RUS parallel manipulator by using the natural coordinates are analysed. One numerical example of kinematics and inverse dynamics is presented.

Keywords: Kinematics, Dynamics, Multibody, Robotics, Parallel Manipulator.

1. Introduction

Many authors have published different methods on the kinematics and dynamics of different types of parallel manipulators. References on the subject can be found in Merlet 2000 and Urls:

-http://www.sop.inria.fr/coprin/equipe/merlet/merlet.html

-http://www.parallemic.org/

The manipulators with rotating actuators compose a special group within the parallel manipulators. Researches on this type of manipulators can be find references Pierrot et al. 1990, Benea 1996, and Zanganeh et al. 1997.

The dynamic equations can be based on different types of generalized coordinates, but most of the authors select Cartesian translations and Euler rotations because they can be easily related to the manipulator degrees of freedom. In this article, the kinematic and dynamic analysis of a 6-RUS parallel manipulator using natural coordinate is presented.

In this case, the natural coordinates are the Cartesian coordinates of the kinematic pairs between the links that compose the manipulator. The advantage of this method resides in the structure of the constraint equations derivatives and the simplicity of the mass matrices.

2. Manipulator topology and notation

The 6-RKS parallel manipulator was introduced by Hunt 1983 and is composed of two triangular platforms one of them fixed to the ground. On

J. Lenarčič and C. Galletti (eds.), On Advances in Robot Kinematics, 329-335.

- A_{pq}: length of the edge of the moving platform that links the p-th and q-th rods.

In order to obtain a constant mass matrix for the moving platform, its displacements and rotations must be referred to the position of four non coplanar points. To this end, an auxiliary point (200) is used. It is the vertex of a pyramid whose base is the triangle formed by the vertices of the moving platform.

- A_{2pq}: length of the lateral edge of the pyramid corresponding to p-th and q-th rods.

3. Kinematics

The kinematic analysis comprises the study of position, velocities and accelerations of the manipulator.

3.1 Position

To model the 6-RUS parallel manipulator, a set of *natural coordinates,* García de Jalón et al. 1994, is used. This set of coordinates defines the position of all the elements of the mechanical system with respect to an inertial reference frame. In this case, the generalized coordinates are Cartesian coordinates of different key points as described in section 2. An inertial reference frame is considered on the fixed platform, as shown in Fig. 2, taking the Z-axis from the ground towards the moving platform. Giving the lengths R_i, L_i and A_{ij}, the set of constraint equations that define the kinematics of the robot can be written as follows:

- One constant length condition for each actuator:

$$(x_{11}-x_{01})^2+(y_{11}-y_{01})^2+(z_{11}-z_{01})^2-R_1^2=0 \qquad \text{(1 to 6)}$$

- One perpendicular condition for each actuator:

$$(x_{11}-x_{01})(x_{02}-x_{01})+(y_{11}-y_{01})(y_{02}-y_{01})+(z_{11}-z_{01})(z_{02}-z_{01})=0 \qquad \text{(7 to 12)}$$

- One constant length condition for each rod:

$$(x_{161}-x_{11})^2+(y_{161}-y_{11})^2+(z_{161}-z_{11})^2-L_1^2=0 \qquad \text{(13 to 18)}$$

- One constant length condition for each edge of the moving platform:

$$(x_{161}-x_{123})^2+(y_{161}-y_{123})^2+(z_{161}-z_{123})^2-A_{12}^2=0 \qquad \text{(19 to 21)}$$

- One constant length condition for each lateral edge of the pyramid:

$$(x_{200}-x_{161})^2+(y_{200}-y_{161})^2+(z_{200}-z_{161})^2-A_{261}^2=0 \qquad \text{(21 to 24)}$$

In this manipulator is advantageous to introduce 18 linear equations of cranks tips position, of the type:

$$(x_{11}-x_{01})-R_1\cos\theta_1=0 \qquad \text{(25 to 42)}$$

Equations 1 to 42 define a system of non-linear algebraic equations that must be satisfied for all positions. Grouping the coordinates in a vector $\mathbf{q}$, the equations 1 to 42 are expressed as:

$$\mathbf{\Phi}(\mathbf{q}) = \mathbf{0} \tag{43}$$

System (43) is non-linear in the position unknowns. Depending on the number of known positions in vector $\mathbf{q}$, some equations will be directly fulfilled, resulting in a compatible system in the rest of unknowns.

This equation system can be solved using the Newton-Raphson method. The iterative procedure can be written as:

$$\mathbf{\Phi}(\mathbf{q}_i) = \mathbf{\Phi}_q(\mathbf{q}_i)(\mathbf{q}_i - \mathbf{q}_{i+1}) \tag{44}$$

Where $\mathbf{\Phi}_q$ is the jacobian matrix of constraint conditions with respect to the coordinates vector $\mathbf{q}$.

3.2 Velocities

Once the manipulator position is determined, the velocities analysis can be made. The equations system (43) respect to the time is derived obtaining the following system.

$$\mathbf{\Phi}_q\dot{\mathbf{q}} + \mathbf{\Phi}_t = \mathbf{0} \tag{45}$$

Where $\dot{\mathbf{q}}$ is the vector of velocities and $\mathbf{\Phi}_t$ is the explicit derivative of the constraint conditions with respect to the time.

System (45) is linear in the velocity unknowns. Depending on the number of known velocities in vector $\dot{\mathbf{q}}$, some equations will be directly fulfilled, resulting in a compatible linear system in the rest of unknowns.

3.3 Accelerations

Once the manipulator velocities are determined, the accelerations analysis can be made. The equations system (45) with respect to the time is derived obtaining the following system.

$$\mathbf{\Phi}_q\ddot{\mathbf{q}} + \dot{\mathbf{\Phi}}_q\dot{\mathbf{q}} + \dot{\mathbf{\Phi}}_t = \mathbf{0} \tag{46}$$

Where $\dot{\mathbf{\Phi}}_q$ is the derivative of the jacobian matrix $\mathbf{\Phi}_q$ with respect to the time, $\ddot{\mathbf{q}}$ is the accelerations vector and $\dot{\mathbf{\Phi}}_t$ is the second explicit derivative of the constraint conditions with respect to the time.

System (46) is linear in the acceleration unknowns. Depending on the number of known accelerations in vector $\ddot{\mathbf{q}}$, some equations will be directly fulfilled, resulting in a compatible linear system in the rest of unknowns.

4. Dynamics

Applying the method proposed by García de Jalón et al. 1994, the following equations system leads the manipulator dynamics:

$$\mathbf{M}\ddot{\mathbf{q}} + \boldsymbol{\Phi}_{\mathbf{q}}^{T}\boldsymbol{\lambda} = \mathbf{Q} \tag{47}$$

where $\mathbf{M}$ is the mass matrix referred to the vector of natural coordinates $\mathbf{q}$, $\boldsymbol{\lambda}$ is the Lagrange multipliers vector and $\mathbf{Q}$ is the vector of applied forces.

In this case, $\mathbf{M}$ contains the dynamic information of the actuator cranks, the rods and the moving platform. The mass matrices of cranks and rods can be referred to two points under certain assumptions and therefore, their mass matrices are of dimension 6x6. The moving platform is represented by four points so as to obtain a 12x12 constant mass matrix.

Taking into a count that all cranks are fixed at one of their ends and the sharing points at the different joint positions, the resulting global mass matrix is of dimension 30x30.

In the more general dynamic analysis, the equations (28) and (29) must be fulfilled at the same time, resulting in:

$$\begin{bmatrix} \mathbf{M} & \boldsymbol{\Phi}_{\mathbf{q}}^{T} \\ \boldsymbol{\Phi}_{\mathbf{q}} & \mathbf{0} \end{bmatrix} \begin{bmatrix} \ddot{\mathbf{q}} \\ \boldsymbol{\lambda} \end{bmatrix} = \begin{bmatrix} \mathbf{Q} \\ -\dot{\boldsymbol{\Phi}}_{\mathbf{q}}\dot{\mathbf{q}} - \dot{\boldsymbol{\Phi}}_{t} \end{bmatrix} \tag{48}$$

System (48) is linear in $\ddot{\mathbf{q}}$ and $\boldsymbol{\lambda}$, but could be reordered for different sets of unknowns. In this case, the unknown $\boldsymbol{\lambda}$'s are directly related to the compression forces in the cranks and rods. Actuator torques can be applied as generalized forces in the corresponding redundant coordinate (see equations 25-42).

In order to solve the inverse dynamic problem, a compatible solution for $\ddot{\mathbf{q}}$ is proposed and system (47) could then be used to determine the unknown actuator torques.

5. Numerical example

In this numerical example, the inverse dynamics problem is solved to determine the actuator torques for a given manipulator kinematics.

The inertia properties of the moving platform are represented in this case by an 80 kg. cylinder of 0.5 m. of diameter and 1.2 m. high which is located on the barycentre of the moving platform.

Fig. 3 shows the topological data of the 6-RUS parallel manipulator. The axes of the actuators mounted on the fix platform form an equilateral triangle of a meter of edge. The actuators are symmetrically mounted with respect to the triangle symmetry axes, being 0,1 m. the distance between the points of the cranks on the axes of the actuators of a same edge of the triangle, that is, the distances between points 01 to 02, 03 to 04 and 05 to 06. The cranks length is 0,1 m. for all actuators. The rods lengths are also all equal to 0,6 m. and the moving platform is an equilateral triangle of 0,5 m. of edge.

The position problem requires six coordinates to be set. In this case, an angle of 30° measured from the interior of the fixed platform is selected for all six crank actuators. The velocity problem is solved by forcing the angular velocity of the actuators to 10, -5, -5, 5, -5 and 10 rad/s. respectively.

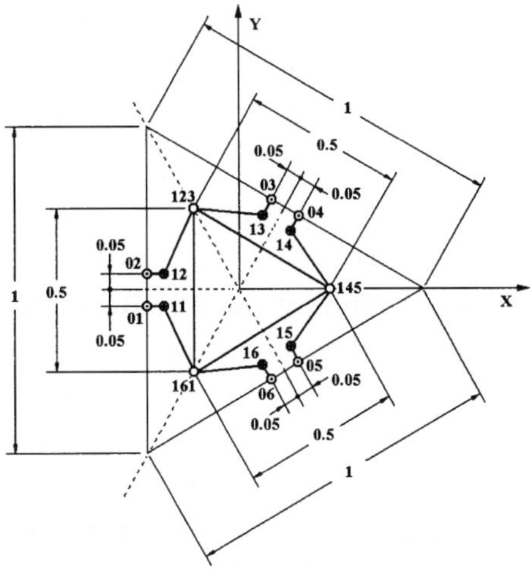

Figure 3. Example of 6-RUS parallel manipulator

5.1 Position

Under the previously mentioned conditions, and using the coordinate system described in Fig. 3 with the Z axis towards the moving platform, the position of all actuators are respectively:

$$
\begin{aligned}
01 &= (-0.288675134, -0.05, \quad 0)\\
02 &= (-0.288675134, \ 0.05, \quad 0)\\
03 &= (0.101036297, \quad 0.275, \ 0)\\
04 &= (0.187638837, \quad 0.225, \ 0)\\
05 &= (0.187638837, \ -0.225, \ 0)\\
06 &= (0.101036297, \ -0.275, \ 0)
\end{aligned}
$$

Knowing the actuator angles, all crank tip coordinates can be determined by solving the reduced equation system formed by equations (25) to (42).

The position solutions for points 11 to 16 are respectively:

11 = (-0.202072594, -0.05, 0.05)
12 = (-0.202072594, 0.05, 0.05)
13 = (0.057735027, 0.20, 0.05)
14 = (0.144337567, 0.15, 0.05)
15 = (0.144337567, -0.15, 0.05)
16 = (0.057735027, -0.20, 0.05)

Applying the method exposed in section 3.1 the vertices of the moving platform and the vertex of the pyramid (in this case a tetrahedron) are determined, being the following coordinates:

123 = (-0.144337568, 0.25, 0.612731434)
145 = (0.2886751346, 0, 0.612731434)
161 = (-0.144337568,- 0.25, 0.612731434)
200 = (0, 0, 1.020979724)

5.2 Velocities

Knowing the angular velocities of the actuators, the components of the cranks tips velocities directly determined as:

v11 = (-0.5, 0, 0.866025)
v12 = (0.25, 0, -0.433013)
v13 = (-0.125, -0.216506, -0.433013)
v14 = (0.125, 0.216506, 0.433013)
v15 = (-0.125, 0.216506, -0.433013)
v16 = (0.25, -0.433012, 0.866025)

Applying the method exposed in section 3.2, the velocities of the moving platform vertices and the point 200 are determined as follows:

v123 = (0.441165, -0.764120, -0.181050)
v145 = (0, -1.528241, 0)
v161 = (-0.441165, -0.764120, 0.588414)
v200 = (0,192033 -0.390562, 0.135788)

5.3 Accelerations

Considering that the angular velocities of the actuators are constant, the components of the cranks tips accelerations are determined directly:

A11 = (-8.660254, 0, -5)
A12 = (-2.165064, 0, -1.25)
A13 = (1.082532, 1.875, -1.25)
A14 = (1.082532, 1.875, -1.25)

A15 = (1.082532,	-1.875,	-1.25)
A16 = (4.330128,	-7.5,	-5)

Applying the method exposed in section 3.3, the accelerations of the moving platform vertices and the point 200 are determined, being the following components:

A 123 = (0.610334,	-1.057130,	-2.374374)
A 145 = (-0.652916,	0,	-7.075207)
A 161 = (0.972277,	1.684033,	-6.570443)
A 200 = (4.981187	-3.894854,	-6.397191)

5.4 Dynamics

Applying the method exposed in section 4 for the present inverse dynamics case, the following torques to the motors of 1 to 6 are obtained:

$$T = (110.82, 91.35, 0, 31.51, -133.37, -145.32)$$

The traction-compression force on each rod can be deduced directly from Lagrange multipliers as $|F_i| = 2\lambda_i L_i$ resulting:

$$F = (-1450.24, -1195.46, -1.18, -412.37, 1745.42, 1901.84)$$

6. Conclusions

In this paper kinematics, (position, velocities and accelerations) and dynamics of a 6-RUS parallel manipulator by using natural coordinates have been analysed and one numerical example of kinematics and inverse dynamics has been presented.

References

Benea, R., (1996), *Contribution à l'ètude des robots pleinement parallèles de type 6R-RR-S*, Doctoral Thesis, Université de Savoie, Annecy, France.

García de Jalón, J., Bayo, E., (1994), *Kinematic and dynamic simulation of multibody system,* New York, Springer-Verlag.

Hunt, K.H., (1983), Structural kinematics of in-parallel-actuated robot-arms, *J. of Mech. Trans and Auto. in Design*, no. 4, vol. 105, pp.705-712.

Merlet, J.P., (2000), *Parallel Robots*, Dordrecht, Kluwer Academic Publishers.

Pierrot, F., Uchiyama, M., Dauchez, P., Fournier, A. (1990), A new design of a 6-DOF parallel robot, *Journal of Robotics and Mechatronics*, no. 4, vol. 2, pp.92-99.

Zanganeh, K.E., Sinatra, R., Angeles, J., (1997), Kinematics and dynamics of a six-degree-of-freedom parallel manipulator with revolute legs, *Robotica*, vol. 15, pp.385-394.

Design of Mechanisms

K. Al-Widyan and J. Angeles
The robust design of Schönflies-motion generators

K. Wohlhart
Polyhedral Zig-Zag linkages

H. Pottman, M. Hofer and B. Ravani
Variational motion design

M. Karouia and J.M. Hervé
New parallel wrists: Special limbs with motion dependency

T.G. Ghuneim, J. Angeles and S.P. Bai
The design of a novel pitch-roll wrist with spherical cam-roller pairs

S.-H. Lee, T.A. Lasky and S.A. Velinsky
Design of a hybrid manipulator for painting highway signage

D. Sallé and P. Bidaud
An optimal kinematic design for a highly dextrous MIS instrument

G. Schreiber, A. Albu-Schäffer, M. Fischer, F. Schoeppe and G. Hirzinger
Achievable Cartesian stiffness with passively compliant, uncoupled joints

THE ROBUST DESIGN OF SCHÖNFLIES-MOTION GENERATORS

K. Al-Widyan and J. Angeles
Department of Mechanical Engineering & Centre for Intelligent Machines
McGill University, 817 Sherbrooke St. West, Montreal, Canada H3A 2K6
widyan@cim.mcgill.ca, angeles@cim.mcgill.ca

Abstract The subject of this paper is the robust design of a four-degree-of-freedom manipulator of the SCARA type, i.e., a Schönflies-motion generator. The optimally robust design is secured by means of mechanical isotropy, which thus leads to robustness in the main aspects of the performance, namely, kinetostatic, elastostatic, and elastodynamic. It is shown that a mechanically isotropic design is possible for a parallel manipulator.

Keywords: Mechanical performance, mechanical isotropy, SCARA, Schönflies motions, parallel robots, robustness

1. Introduction

During the design stage of robotic systems, the designer is confronted with the lack of a specific task, which is the case in the design of conventional, single-degree-of-freedom machines e.g., as pertaining to four-bar linkages for motion generation (Bottema and Roth, 1990). A class of tasks, for example, can consist of motion in a plane, albeit at different levels, such as in pick-and-place operations in the assembly of electronic PC boards. Along with the definition of the class of tasks, design specifications include items such as maximum reach, positioning-precision, and load-carrying capacity. Moreover, when in operation, the various sensors on which the system relies produce signals that are unavoidably corrupted with noise. The outcome is a host of measurement errors, which just add to the machining and assembly errors that are inherent to any manufactured mechanical system.

Apparently, task uncertainty and noise in the design and operation of robotic mechanical systems pose a major challenge to the designer of multi-degree-of-freedom (multi-dof) robotic mechanical systems. As a means to cope with the said uncertainties, we adopt the concept of robust design, as propounded by Taguchi and his school (Taguchi, 1987; Taguchi, 1993).

J. Lenarčič and C. Galletti (eds.), On Advances in Robot Kinematics, 339-350.

Robustness means insensitivity to changes in the design-environment conditions (Wu and Wu, 2000). An effective means of achieving robustness, in connection with robotic mechanical systems, is by rendering the mechanical subsystems in question *mechanically isotropic* (Al-Widyan and Angeles, 2001), which leads to optimum behaviour under three operation conditions: *kinetostatic*; *elastostatic*; and *elastodynamic*. In fact, kinetostatic isotropy guarantees positioning robustness to errors due to: joint-encoder noisy readouts; actuator noisy response; and manufacturing and assembly inaccuracies. Elastostatic isotropy, in turn, guarantees robustness against static random loads stemming from the environment, while the outcome of elastodynamic isotropy is that a) the structure exhibits one single natural frequency; and b) all vectors of the six-dimensional motion space of the moving platform are associated with natural modes of the system. The manipulator, thus, behaves as a single-dof system as far as linear vibrations arising from external excitations is concerned.

The use of kinetostatic isotropy as a design criterion in multi-dof systems is not new. We can cite in this regard the work of Salisbury and Craig, 1982, who minimized the condition number of the Jacobian matrix of two-link mechanical fingers in an attempt to optimize their positioning accuracy. However, attempts to optimize the elastostatic and the elastodynamic behaviours of multi-dof systems, along the same lines, are rather scant, but we can cite a recent work by Portman and Sandor, 2001, whereby the authors address all these issues in their design of a micropositioning device.

The theoretical foundations for a model-based robust design of robotic mechanical systems in general was laid down in (Al-Widyan and Angeles, 2001). With the aid of the principles and methodology proposed therein, we aim here at drawing guidelines as to how to design the systems of interest to be accurate and robust against changes in the operation conditions.

The focus of this paper is mechanical systems capable of producing *Schönflies displacements*, namely, *Schönflies-Motion generators* (SMG). Indeed, *Schönflies displacements* are a subgroup of the rigid-body displacement group. The eponym is due to the German mathematician-mineralogist Arthur Moritz Schönflies (1853–1928), who found that this set of displacements has the algebraic structure of a group, as Hervé, 1978, 1999, pointed out.

The first mechanical realization of the Schönflies displacement subgroup of the six-dof displacement group was achieved by means of a SCARA system (Furuya et al., 1983). An alternative Schönflies-Motion Generator, as proposed by Angeles et al., 2000, is built upon two similar units, the proximal and the distal pan-tilt mechanisms. Each has one actuated revolute axis and one kinematic pair of the class that Hervé, 1999, calls Π-pairs, namely, the parallelogram linkage of each mechanism. In fact, Π-pairs were concurrently proposed, independently, by Wohlwhart, 1992, and Hervé and Sparacino, 1992.

In connection with the serial architecture of SMG, a kinetostatically isotropic design is possible for this type of manipulator (Al-Widyan and Angeles, 2002). However, elastostatic and elstodynamic isotropies are not possible because of the serial type of this architecture.

In pursuing an optimally robust design of a SMG, while taking into account the impossibility of devising a mechanically isotropic serial manipulator, we attempt now a parallel architecture for the same. This architecture was proposed by Angeles and Morozov, 2003.

In this paper, we focus on the geometric dimensioning and inertial parameters of a parallel architecture of the SMG, aimed at endowing the manipulator with mechanical isotropy.

2. A Parallel Architecture

The system at hand, shown in Fig. 1, is composed of two identical four-degree-of-freedom serial manipulators in a parallel array, sharing one common base and one common end-effector. Each manipulator is a Schönflies-motion generator, as reported in (Angeles et al., 2000), and hence, each produces a subgroup of the displacement group of rigid-body motions. The intersection of the two displacement subgroups, known as the *product* of the two subgroups in group-theoretical language (Hervé, 1999; Angeles, 2004), is the same Schönflies subgroup. Each manipulator comprises two cascaded modules producing, each, a set of two-degree-of-freedom displacements: one rotation about a vertical line $\mathcal{V}_i$ and one translation along a circle lying in a plane parallel to $\mathcal{V}_i$, for $i = 1, 2$. While lines $\mathcal{V}_i$ of the two modules are parallel, the planes of the two circles make a dihedral angle of variable magnitude. Each module comprises two kinematic pairs: one revolute of axis parallel to $\mathcal{V}_i$ and one parallelogram four-bar linkage, i.e., a Π joint, of plane parallel to $\mathcal{V}_i$.

Of the four kinematic pairs of the two modules, only two need be actuated. While any combination of two axes can be actuated, the most convenient ones are those of the module closer to the base, thereby allowing for motors fixed to the base, which leads to a lighter device. The two motors transmit a pan and a tilt motion to the lower parallelogram of each leg by means of a gear train, an instance of which is the differential mechanism described in (Angeles et al., 2000). This mode of actuation thus allows for four identical motors.

Figure 1. A parallel Schönflies-motion generator

3. Design For Robustness

In this section, we make use of the special structure of the parallel SMG at hand to find a set of conditions under which SMG can be rendered mechanically robust by means of isotropy.

The methodology we will adopt here for guaranteeing robustness is consistent with the novel approach proposed in (Alwidyan and Angeles, 2004) for the robust design of engineering systems and elements thereof. Under this approach, we first render the Jacobian, the stiffness, and the frequency matrices isotropic—all their singular values identical—and then minimize the resulting common singular value. What isotropy brings about is a design whose performance sensitivity against parameters on which the designer has no control is a minimum. Additionally, under isotropy, the designer can dispense with a priori knowledge of the statistical distributions of the parameters over which the designer has no control.

3.1 The Kinetostatic Design

The kinematic relations of parallel manipulators usually involve two Jacobian matrices, that are commonly referred to as the forward and the inverse Jacobians, depending on the role they play in the kinetostatic transformation between the joint and the Cartesian variables.

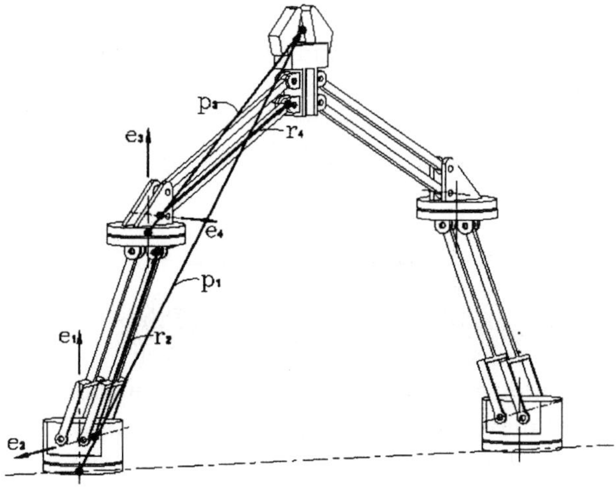

Figure 2. kinematic chain of a parallel Schönflies-motion generator

The Jacobian Matrices. The kinematic structure of the system under design comprises two serial subchains, its legs, that connect a fixed base to a common moving platform, thereby forming a closed kinematic chain, as shown in Fig. 2. The motion of the system is provided, in turn, by actuating some joints in the serial subchains while leaving the other joints unactuated. A system of kinematic relations, free of unactuated joint rates, is obtained, that can be cast in the form

$$\mathbf{A}\mathbf{t} \quad = \quad \mathbf{B}\dot{\boldsymbol{\theta}} \tag{1}$$

where $\mathbf{t}$ is the four-dimensional twist vector of the moving platform, and $\dot{\boldsymbol{\theta}}$ is the four-dimensional vector of actuated joint rates, namely,

$$\mathbf{t} = [\ \dot{\phi} \quad \dot{\mathbf{p}}^T\ ]^T, \quad \dot{\boldsymbol{\theta}} = [\ \dot{\boldsymbol{\theta}}_I^T \quad \dot{\boldsymbol{\theta}}_{II}^T\ ]^T \tag{2}$$

with $\dot{\boldsymbol{\theta}}_I$ and $\dot{\boldsymbol{\theta}}_{II}$, defined for legs I and II, respectively, as

$$\dot{\boldsymbol{\theta}}_J = [\ \dot{\theta}_{J1} \quad \dot{\theta}_{J2} \quad \dot{\theta}_{J3} \quad \dot{\theta}_{J4}\ ]^T, \quad J = I, II \tag{3}$$

Moreover, $\mathbf{A}$ is the 4×4 *forward-kinematics Jacobian matrix*, and $\mathbf{B}$ is the 4×4 *inverse Jacobian matrix* of the manipulator. Lack of space prevents us from including the derivation of the two foregoing Jacobians, which take the forms

$$\mathbf{A} = \begin{bmatrix} \mathbf{A}_I \\ \mathbf{A}_{II} \end{bmatrix}, \qquad \mathbf{B} = \mathrm{diag}(\mathbf{B}_I, \mathbf{B}_{II}) \tag{4}$$

whose blocks are defined below, for $J = I, II$:

$$\mathbf{A}_J = \begin{bmatrix} -\mathbf{e}_{J4}^T(\mathbf{e}_1 \times \mathbf{p}_3)_J & \mathbf{e}_{J4}^T \\ 0 & \mathbf{r}_{J4}^T \end{bmatrix}, \quad \mathbf{B}_J = \begin{bmatrix} \mathbf{e}_{J4}^T(\mathbf{e}_1 \times \mathbf{p}_{13})_J & \mathbf{e}_{J4}^T(\mathbf{e}_2 \times \mathbf{r}_2)_J \\ \mathbf{r}_{J4}^T(\mathbf{e}_1 \times \mathbf{p}_{13})_J & \mathbf{r}_{J4}^T(\mathbf{e}_2 \times \mathbf{r}_2)_J \end{bmatrix}$$

The Isotropy Conditions. Now we find the conditions for rendering the Jacobian matrices of the system at hand isotropic.

The isotropy conditions for parallel manipulators can be stated in a fashion similar to serial manipulators (Angeles, 2002). It is recalled that a design is isotropic if both **A** and **B** are isotropic, i.e., if positive scalars α and β exist such that

$$\mathbf{A}^T\mathbf{A} = \alpha^2\mathbf{1}_{4\times 4} \quad \text{and} \quad \mathbf{B}^T\mathbf{B} = \beta^2\mathbf{1}_{4\times 4} \tag{5}$$

where $\mathbf{1}_{4\times 4}$ is the 4×4 identity matrix. Moreover, α and β are the identical singular values of **A** and **B**, respectively.

The joint rates are transformed into both translation and rotation, which leads to different units in the Jacobian entries. Hence, two characteristic lengths, L_A and L_B, are needed to dimensionally homogenize the vector spaces of the end-effector twists (Angeles, 2002). Thus, we redefine the two Jacobians of eq.(4) as

$$\overline{\mathbf{A}} \equiv \begin{bmatrix} -\mathbf{e}_{I4}^T(\mathbf{e}_1 \times \mathbf{p}_3)_I/L_A & \mathbf{e}_{I4}^T \\ 0 & \mathbf{r}_{I4}^T/L_A \\ -\mathbf{e}_{II4}^T(\mathbf{e}_1 \times \mathbf{p}_3)_{II}/L_A & \mathbf{e}_{II4}^T \\ 0 & \mathbf{r}_{II4}^T/L_A \end{bmatrix} \tag{6}$$

and, for $J = I, II$,

$$\overline{\mathbf{B}} \equiv \text{diag}(\overline{\mathbf{B}}_I, \overline{\mathbf{B}}_{II}), \ \overline{\mathbf{B}}_J \equiv \begin{bmatrix} \mathbf{e}_{J4}^T(\mathbf{e}_1 \times \mathbf{p}_{13})_J & \mathbf{e}_{J4}^T(\mathbf{e}_2 \times \mathbf{r}_2)_J \\ \mathbf{r}_{J4}^T(\mathbf{e}_1 \times \mathbf{p}_{13})_J/L_B & \mathbf{r}_{J4}^T(\mathbf{e}_2 \times \mathbf{r}_2)_J/L_B \end{bmatrix} \tag{7}$$

Substitution of $\overline{\mathbf{A}}$ of eq.(6) into eq.(5), and simplification of the expression thus resulting leads to the conditions for an isotropic design, namely,

$$\overline{\mathbf{A}}^T\overline{\mathbf{A}} \equiv \begin{bmatrix} h & \mathbf{v}^T \\ \mathbf{v} & \mathbf{N} \end{bmatrix} = \alpha^2 \begin{bmatrix} 1 & \mathbf{0}^T \\ \mathbf{0} & \mathbf{1}_{3\times 3} \end{bmatrix} \tag{8}$$

with

$$h = (\|(\mathbf{e}_1 \times \mathbf{p}_3)_I\|^2 + \|(\mathbf{e}_1 \times \mathbf{p}_3)_{II}\|^2)/L_A^2 \tag{9a}$$

$$\mathbf{v} = -[\mathbf{e}_{I4}^T(\mathbf{e}_1 \times \mathbf{p}_3)_I\mathbf{e}_{I4} + \mathbf{e}_{I4}^T(\mathbf{e}_1 \times \mathbf{p}_3)_{II}\mathbf{e}_{II4}]/L_A \tag{9b}$$

$$\mathbf{N} = \mathbf{e}_{I4}\mathbf{e}_{I4}^T + \left(\frac{\mathbf{r}_{I4}}{L_A}\right)\left(\frac{\mathbf{r}_{I4}^T}{L_A}\right) + \mathbf{e}_{II4}\mathbf{e}_{II4}^T + \left(\frac{\mathbf{r}_{II4}}{L_A}\right)\left(\frac{\mathbf{r}_{II4}^T}{L_A}\right) \tag{9c}$$

Observing that $\mathbf{e}_{I4} = \mathbf{e}_{II4}$, eqs.(8) and (9c) leads to

$$2\,\mathbf{e}_{I4}\mathbf{e}_{I4}^T + \left(\frac{\mathbf{r}_{I4}}{L_A}\right)\left(\frac{\mathbf{r}_{I4}^T}{L_A}\right) + \left(\frac{\mathbf{r}_{II4}}{L_A}\right)\left(\frac{\mathbf{r}_{II4}^T}{L_A}\right) = \alpha^2\mathbf{1}_{3\times 3} \tag{10}$$

By rendering the triplet $\{\sqrt{2}\mathbf{e}_{I4}, \mathbf{r}_{I4}/L_A, \mathbf{r}_{II4}/L_A\}$ orthogonal and of equal Euclidean norm, eq.(10) can be satisfied, thereby obtaining the conditions

$$\alpha = \sqrt{2}, \quad \|\mathbf{r}_{I4}\| = \|\mathbf{r}_{II4}\| = l_2, \quad L_A = \frac{\sqrt{2}}{2} l_2, \quad \mathbf{r}_{I4}^T \mathbf{r}_{II4} = 0 \tag{11}$$

Moreover, eqs.(8) and (9b) lead, in turn, to

$$\mathbf{e}_{I4}\mathbf{e}_{I4}^T[(\mathbf{e}_1 \times \mathbf{p}_3)_I + (\mathbf{e}_1 \times \mathbf{p}_3)_{II}] = \mathbf{0} \Longrightarrow \mathbf{e}_{I4} \cdot [\mathbf{e}_1 \times (\mathbf{p}_{I3} + \mathbf{p}_{II3})] = 0 \tag{12}$$

which implies that the projections of $\mathbf{p}_{I3}$ and $\mathbf{p}_{II3}$ onto the horizontal plane must be equal. Moreover, combining eqs.(9a), (11), and (12), we obtain

$$\|(\mathbf{e} \times \mathbf{p})_{J1}\|^2 = \frac{1}{2} L_A^2 \alpha^2 = \frac{1}{2} l_2^2$$

As a consequence of the above discussion, for matrix $\overline{\mathbf{A}}$ to be isotropic, the angles between the two beams of the upper parallelepipeds and the horizontal plane should be 45°, i.e., $\theta_{J4} = 45°$, as shown in Fig. 3.

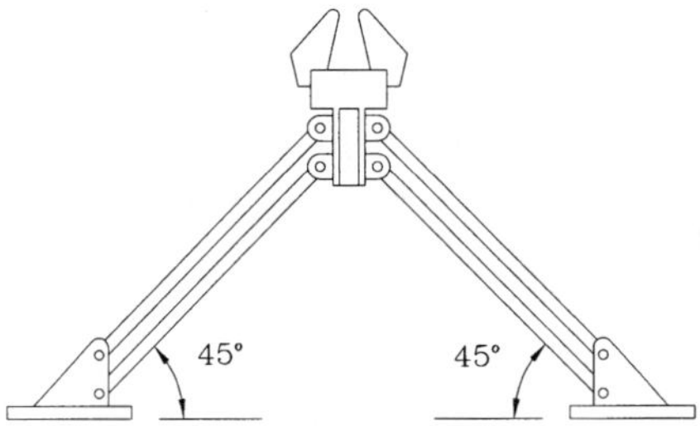

Figure 3: An isotropic condition for $\overline{\mathbf{A}}$

Matrix $\overline{\mathbf{B}}$ of eq.(4) is block-diagonal; to render $\overline{\mathbf{B}}$ isotropic, the product $\overline{\mathbf{B}}^T\overline{\mathbf{B}}$ should be proportional to the 4×4 identity matrix, i.e.,

$$\overline{\mathbf{B}}_J^T \overline{\mathbf{B}}_J = \begin{bmatrix} S_J & T_J \\ T_J & V_J \end{bmatrix} = \beta^2 \mathbf{1}_{2\times 2}, \quad J = I, II \tag{13}$$

with

$$S_J = \{\mathbf{e}_{J4}^T(\mathbf{e}_1 \times \mathbf{p}_{13})_J\}^2 + \{\mathbf{r}_{J4}^T(\mathbf{e}_1 \times \mathbf{p}_{13})_J\}^2 / L_B^2$$
$$T_J = \mathbf{e}_{J4}^T(\mathbf{e}_2 \times \mathbf{r}_2)_J(\mathbf{e}_1 \times \mathbf{p}_{13})_J^T \mathbf{e}_4 + \mathbf{r}_{J4}^T(\mathbf{e}_2 \times \mathbf{r}_2)_J(\mathbf{e}_1 \times \mathbf{p}_{13})_J^T \mathbf{r}_{J4}/L_B^2$$
$$V_J = \{\mathbf{e}_{J4}^T(\mathbf{e}_2 \times \mathbf{r}_2)_J\}^2 + \{\mathbf{r}_{J4}^T(\mathbf{e}_2 \times \mathbf{r}_2)_J\}^2 / L_B^2$$

For isotropy, we must have, for $J = I, II$,

$$\mathbf{e}_{J4}^T(\mathbf{e}_2 \times \mathbf{r}_2)_J(\mathbf{e}_1 \times \mathbf{p}_{13})_J^T \mathbf{e}_4 + \mathbf{r}_{J4}^T(\mathbf{e}_2 \times \mathbf{r}_2)_J(\mathbf{e}_1 \times \mathbf{p}_{13})_J^T \mathbf{r}_{J4}/L_B^2 = 0 \tag{14a}$$

$$[\mathbf{e}_{J4}^T(\mathbf{e}_1 \times \mathbf{p}_{13})_J]^2 + [\mathbf{r}_{J4}^T(\mathbf{e}_1 \times \mathbf{p}_{13})_J]^2 / L_B^2 = [\mathbf{e}_{J4}^T(\mathbf{e}_2 \times \mathbf{r}_2)_J]^2 + [\mathbf{r}_{J4}^T(\mathbf{e}_2 \times \mathbf{r}_2)_J]^2 / L_B^2 \tag{14b}$$

The isotropy conditions stated in eqs.(14a & b) can be verified in a variety of ways. A simple one is shown in Fig. 4a, for which we chose with: $L_A = \sqrt{2}l_2/2$, $L_B = 0.8093l_2$, and

$$\overline{\mathbf{A}} = \begin{bmatrix} 1 & 0 & -1 & 0 \\ 0 & 1 & 0 & 1 \\ 1 & 0 & 1 & 0 \\ 0 & -1 & 0 & 1 \end{bmatrix}, \quad \overline{\mathbf{B}} = (0.4226l_1)\begin{bmatrix} 1 & 0 & 0 & 0 \\ 0 & -1 & 0 & 0 \\ 0 & 0 & 1 & 0 \\ 0 & 0 & 0 & -1 \end{bmatrix} \tag{15}$$

An alternative isotropic posture is depicted in Fig. 4b, in which

$$\overline{\mathbf{A}} = \begin{bmatrix} 1 & -0.908 & -0.419 & 0 \\ 0 & 0.419 & -0.908 & 1 \\ 1 & 0.908 & 0.419 & 0 \\ 0 & -0.419 & 0.908 & 1 \end{bmatrix}, \quad L_A = \frac{\sqrt{2}}{2}l_2$$

and

$$\overline{\mathbf{B}} = \begin{bmatrix} 0.143l_1 & 0.853l_1 & 0 & 0 \\ 0.853l_1 & -0.143l_1 & 0 & 0 \\ 0 & 0 & 0.143l_1 & 0.853l_1 \\ 0 & 0 & 0.853l_1 & -0.143l_1 \end{bmatrix}, \quad L_B = 0.257l_2$$

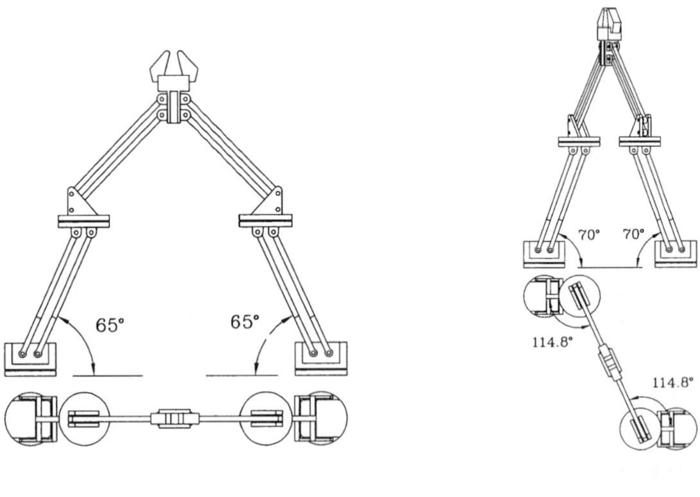

(a) (b)

Figure 4. Isotropic postures; (a) $\theta_{J1} = \theta_{J3} = 0°$, $\theta_{II1} = 180°$, $\theta_{J2} = 65°$, $\theta_{J4} = 45°$ (b) $\theta_{J2} = 70°$, $\theta_{J3} = -65.2°$

3.2 The Elastostatic Design

The elastostatic performance of the manipulator of interest, when subject to "small" environment forces and torques while the actuators are locked at one given posture, is a key design criterion. Under the above conditions, the system becomes a linearly elastic structure, which can be made least sensitive to perturbations from the environment by means of robustness. For brevity, we shall assume that only the joints of the system are flexible, its link being rigid. Therefore, the stiffness matrix $\mathbf{K}$, expressed in the joint space, turns out to be a 4×4 diagonal matrix of the form

$$\mathbf{K} = \mathrm{diag}(k_1, k_2, k_3, k_4) \tag{16}$$

where k_i, for $i = 1, \ldots, 4$, denotes the torsional stiffness of the ith actuated joint.

We can safely assume that the actuated joints are identical, and hence, have identical stiffness k, that is, $k_1 = k_2 = k_3 = k_4 = k$; under these conditions, the stiffness matrix becomes a multiple of the 4×4 identity matrix, which is therefore isotropic. Consequently, the elastostatically isotropic design is possible for this manipulator in the joint space.

3.3 Elastodyamic Design

Under the assumption that only the inertia of the rigid moving platform is significant—a plausible assumption, since parallel manipulators are used first and foremost because of their large load-carrying capacity—it can be shown that its mass matrix in the space of elastic joint deflections, at the isotropic posture displayed in Fig 4a, takes the form

$$\mathbf{M} = (.04465ml_1^2)\begin{bmatrix} \gamma^2+1 & 0 & \gamma^2-1 & 0 \\ 0 & 2 & 0 & 0 \\ \gamma^2-1 & 0 & \gamma^2+1 & 0 \\ 0 & 0 & 0 & 2 \end{bmatrix}, \quad \gamma = \frac{\rho}{L_A}$$

with ρ indicating the radius of gyration of the moving platform. Moreover, the corresponding stiffness matrix is a multiple of the 4×4 identity matrix, that is

$$\mathbf{K} = k\mathbf{1}_{4\times4}$$

Correspondingly, the *frequency matrix* $\mathbf{\Omega}^2$, introduced in (Angeles, 1992), takes the form

$$\mathbf{\Omega}^2 \equiv \sqrt{\mathbf{M}^{-1}}\mathbf{K}\sqrt{\mathbf{M}^{-1}} = 22.3956\left(\frac{k}{m\gamma}\right)\begin{bmatrix} \gamma+1 & 0 & \gamma-1 & 0 \\ 0 & 2 & 0 & 0 \\ \gamma-1 & 0 & \gamma+1 & 0 \\ 0 & 0 & 0 & 2 \end{bmatrix}$$

whence it is apparent that, to render $\boldsymbol{\Omega}^2$ isotropic, we should choose

$$\gamma = 1 \Longrightarrow \rho = L_A = \frac{\sqrt{2}}{2} l_2$$

which implies that elastodynamic isotropy requires that the radius of gyration of the platform be equal to the characteristic length of matrix $\mathbf{A}$, i.e., $\sqrt{2}l_2/2$.

Under the foregoing condition, then,

$$\boldsymbol{\Omega}^2 = \omega^2 \mathbf{1}_{3\times 3}, \quad \omega = 4.7324\sqrt{\frac{k}{m}}$$

ω thus being the quadruple natural frequency of the manipulator. Thus, a manipulator of the kind at hand can be designed with elastodynamic isotropy. In order to render the manipulator minimally sensitive to environment perturbations, then, ω should be chosen as large as budgetary and physical constraints permit.

4. Conclusions

The subject of the paper is the mechanical design of a four-degree-of-freedom manipulator of the SCARA type, i.e., a Schönflies-motion generator, with a parallel architecture. The concept of mechanical isotropy was used here as a design criterion to produce a robust design. With the aid of this concept, we showed how to determine the pertinent geometric, elastic and inertial parameters involved for a robust performance.

We derived a set of conditions that render the Schönflies-Motion Generator mechanically isotropic. Upon imposing these conditions, we showed that kinetostatic, elastostatic and elastodynamic isotropies are possible for this class of manipulators. Moreover, we found general geometric conditions that lead to a family of kinetostatically isotropic designs.

The isotropic posture is an optimum configuration at which a manipulator is maximally robust in terms of its kinetostatic performance. To improve the performance during task execution, the task should be planned within a region of the workspace close to the isotropic posture.

5. Acknowledgments

The work reported here is supported by NSERC (Canada's Natural Sciences and Engineering Research Council) under Strategic Project 215729-98.

References

Al-Widyan, K. and Angeles, J., "Uncertainty, Isotropy, and Optimality in the Design of Robotic Mechanical Systems," *Proc. International Conference on Multidisciplinary Design in Engineering*, Montreal, Available on CD ROM, SCGM-CMI, 2001.

Al-Widyan, K. and Angeles, J., "The Kinetostatic Design of a Schönfiles-Motion Generator," *Advances in Robot Kinematics: Theory and Applications*, (Lenarčič, J. and Thomas, F.), Kluwer Academic Publishers, Dordrecht, pp. 311–318, 2002.

Al-Widyan, K. and Angeles, J., *A Model-Based Formulation of Robust Design*, Technical Report TR-CIM, Department of Mechanical Engineering and Centre for Intelligent Machines, McGill University, 2004.

Angeles, J., "On the Application of Mohr's Circle to 2-DOF Vibration Analysis: A Tutorial," *J. Sound and Vibration*, Vol. 154, pp 556–567, 1992.

Angeles, J., *Fundamentals of Robotic Mechanical Systems*, Second Edition, Spring-Verlage, New York, 2002.

Angeles, J, "The Qualitative Synthesis of Parallel manipulators," to appear in *ASME Journal of Mechanical Design*, 2004.

Angeles, J., Morozov, A., "A Device for the Production of Schöflies Motions," Patent filed to the PCT on November 6, 2003.

Angeles, J., Morozov, A., and Navarro, O., "A Novel Manipulator Architecture for the Production of SCARA Motions," *Proc. IEEE Int. Conf. Robotics & Automation,* San Francisco, pp. 2370–2375, 2000.

Bottema, O. and Roth, B., *Theoretical Kinematics*, Dover Publications, Mineola, NY, 1990.

Furuya, N., Soma, K., Chin, E., and Makino H., "Research and development of Selective Compliance Assembly Robot Arm. II. Hardware and Software of SCARA Controller," *J. Japan Society of Precision Engineering/Seimitsu Kogaku Kaishi,* Vol. 49, No. 7, pp. 835–841, 1983.

Hervé, J.M., "Analyse structurelle des mécanismes par groupes de déplacements," *Mechanism and Machine Theory*, Vol. 13, pp. 437–450, 1978.

Hervé, J.M., "The Lie group of rigid body displacements, a fundamental tool for mechanism design," *Mechanism and Machine Theory*, Vol. 34, No. 7, pp. 719–730, 1999.

Hervé, J. and Sparacino, F., "Star, a New Concept in Robotics," *Proc. 3rd Int. Workshop on Advances in Robot Kinematics*, pp. 104–111, Ferrara, 1992.

Portman, V. and Sandor, B., "Rigid 6-DOF Parallel Platform for Precision 3-D Micromanipulation," *The International Journal of Machine Tool & Manufacture,* Vol. 41, No. 3, pp. 1229–1220, 2001.

Salisbury, J. and Craig J., "Articulated hands: force control and kinematics issues," *The Int. J. Robotics Res.*, Vol. 1, No. 1, pp. 4–17, 1982.

Taguchi, G., *System of Experimental Design, Vols.I & II*, jointly published by Kraus International Publications, New York and American Supplier Institute, Inc., Dearborn, MI, 1987.

Taguchi, G., *Taguchi on Robust Technology Development: Bringing Quality Engineering Upstream*, ASME Press, New York, 1993.

Wohlhart, K. , "Displacement Analysis of the General Spatial Parallelogram-manipulator," *Proc. 3rd Int. Workshop on Advances in Robot Kinematics*, Ferrara, pp. 176–183, 1992.

Wu, Y. and Wu, A., *Taguchi Methods for Robust Design* , ASME Press, New York, 2000.

POLYHEDRAL ZIG-ZAG LINKAGES

K. Wohlhart
Graz University of Technology
A-8010 Graz, Austria
wohlhart@mechanik.tu-graz.ac.at

Abstract In this paper new overconstrained mechanisms are described which are synthesized by inserting equal basic mechanisms known as Nuremberg scissors into the edges of a Platonian polyhedron and interconnecting them at their ends by different kinds of articulations. The basic mechanism is a planar mechanism and can be inserted into an edge of a Platonian polyhedron in two different ways: either its plane becomes identical with the plane given by the edge line and the center of the polyhedron or orthogonal to this plane. For each of these two insertions two different kinds of articulation (located at the apexes of the polyhedron) are proposed in the paper. Four different kinds of new polyhedral linkages are thus obtained whose syntheses are decribed in detail. A corkscrew on the market under the commercial name Zig-Zag is based on Nuremberg scissors. As the common parts in the newly designed linkages are Nuremberg scissors we adopt this name for the whole family of new polyhedral linkages which are, though highly overconstrained, movable with one degree of freedom.

Keywords: Spatial kinematics, overconstrained mechanisms, polyhedral linkages

1. Introduction

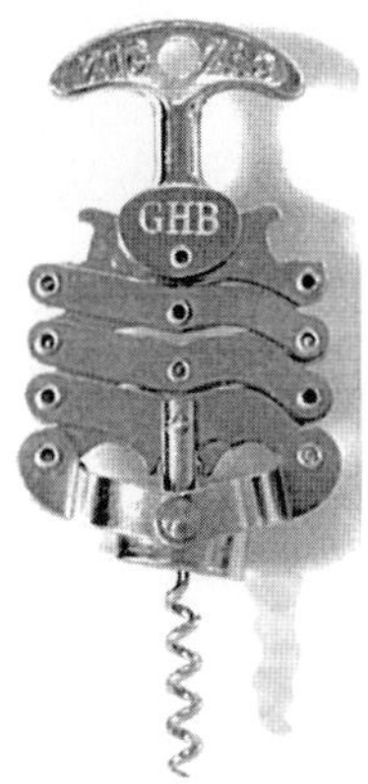

The search for the inventor of the "Nuremberg scissors" seems to be futile. It might well be that he is even identical with that ancient (Nuremberg?) cutler who, by joining two knives first obtained a pair of scissors and driven by his success proceeded to invent the multiple scissors which became known as Nuremberg scissors. This latter invention subsequently found quite a variety of applications. In the extensive museum for toys and games in Nuremberg diverse funny applications in toys are shown, the more serious applications range from medieval warfare scissor ladders (to surmount fortress walls) up to modern hoisting apparatuses and diverse lifting platforms in mechanical engineering. An example of a modern application of this very ancient mechanism is represented by the robot arm of Surveyor 3,which, in 1967, was used to dig out the very first sample on an extraterrestrial body. A com-

J. Lenarčič and C. Galletti (eds.), On Advances in Robot Kinematics, 351-360.

mon example of an application of the Nuremberg Scissors is the corkscrew called "Zig-Zag" (English patent 1884), which will lend us its name for all the new polyhedral linkages which we are going to describe.

To "mobilize" a polyhedron several procedures have been proposed. By inserting appropriate link groups into the faces of a regular polyhedron and connecting them by gussets highly overconstrained polyhedral linkages have been derived (Wohlhart, K., 1999, 2001, 2001). Agraval, S.K., et al., 2001, showed that expanding structures with a single degree of freedom can be obtained from irregular polyhedra by telescoping the edges, i.e. by inserting P-joints into the edges of the polyhedron and mutually fixing the P-joints meeting at an apex. Recently another method of mobilizing an irregular polyhedron has been proposed, where only rotary joints are used (Wohlhart, K., 2004).

In the present paper new regular linkages are synthesized by placing equal Nuremberg scissors onto the edges of a Platonian polyhedron and connecting their ends at each corner by some articulations. Though overconstrained the linkages obtained are, movable with one degree of freedom; they disobey the topological Gruebler Kutzbach condition in the form in which it is commonly stated.

2. Nuremberg Scissors

The Nuremberg scissors (Fig.1) is a planar mechanism consisting of b identical bodies held together by $j = 3b/2 - 2$ rotary joints thus forming

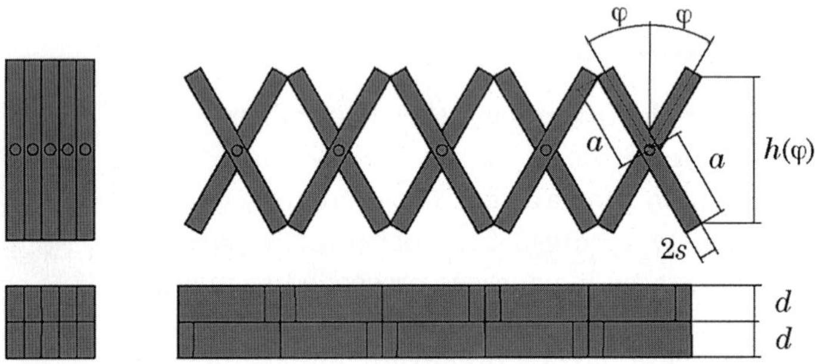

Figure1. The Nuremberg Scissors

$l = b/2 - 1$ fundamental loops. The topological structure formula of Grübler Kutzbach gives for its mobility $F = \Sigma f_i - 3l = 1$. The three system parameters a, s, d determine the form of the Nuremberg scissors together with the position angle φ.

3 Synthesis of a Tetrahedral Zig-Zag Linkage

We start with the Tetrahedral Zig-Zag Linkage and demonstrate a first kind of articulation between the inserted Nuremberg scissors, which we shall call the "Rotor Shield Articulation". It consists of two polygonal bodies (shields) connected by a rotary joint onto which via gussets three

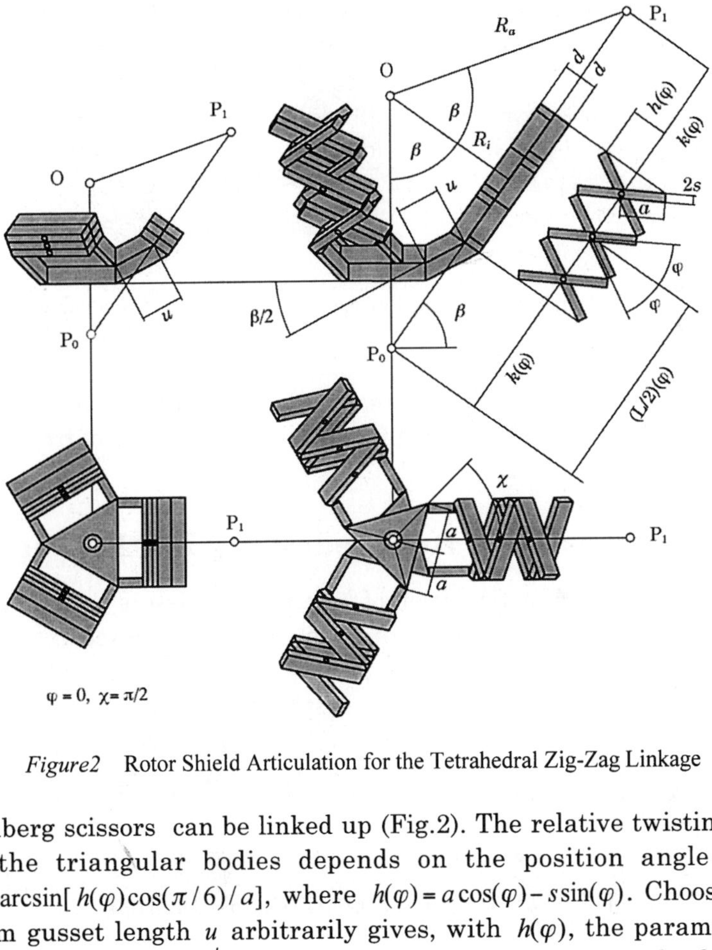

Figure2 Rotor Shield Articulation for the Tetrahedral Zig-Zag Linkage

Nuremberg scissors can be linked up (Fig.2). The relative twisting angle χ of the triangular bodies depends on the position angle φ by: $\chi(\varphi) = \arcsin[\,h(\varphi)\cos(\pi/6)/a]$, where $h(\varphi) = a\cos(\varphi) - s\sin(\varphi)$. Choosing the medium gusset length u arbitrarily gives, with $h(\varphi)$, the parameter set $\{a, s, d\}$ and $\beta = \arccos[-\sqrt{2/3}] - \pi/2 = 54.73^\circ$ for the edge length of the tetrahedron:

$$L(\varphi) = \frac{2}{\cos\beta}\left[\frac{a}{\cos(\pi/6)}\cos\chi(\varphi) + u\cos\left(\frac{\beta}{2}\right)\right] + 6(a\sin\varphi + s\cos\varphi).$$

Containing three Sarrus linkages (Sarrus, P. T., 1853) the mechanism shown in Fig.3 is overconstrained but movable with one degree of freedom. With 6 Nuremberg scissors and 4 Rotor Shield Articulations the

Tetrahedral Zig-Zag Linkage can be built up. Its maximum expansion is approximatively: $\varepsilon = L(\varphi = \pi/2)/L(\varphi = 0)$. Fig. 3 shows the synthesized linkage in its smallest extension: $\varphi = 0$ (left side) and in a medium

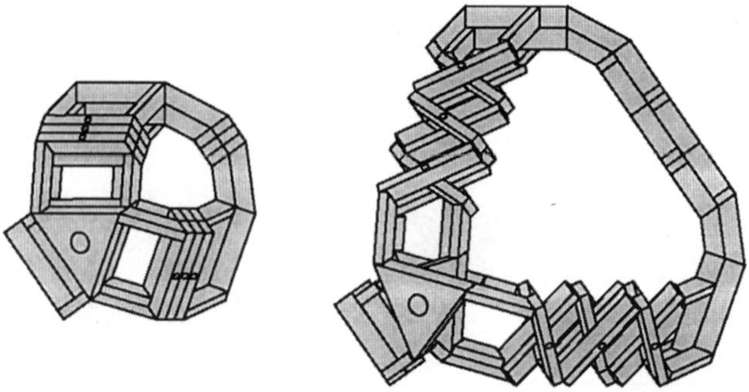

Figure 3 Tetrahedral Zig-Zag Linkage with Rotor Shield Articulations

$\varphi = \pi/6$ (right side) For the greatest $\varphi = \pi/2$ extension the Tetrahedral Zig-Zag Linkage is shaky with many infinitesimal degrees of freedom.

4 Synthesis of a Hexahedral Zig-Zag Linkage

The second kind of articulation of Nuremberg scissors on a polyhedron will be demonstrated on the hexahedron. It shall be called the "Parallel Plate Articulation"(Fig.4). It consists of two polygonal plates onto which triangular scissors are attached. Its system parameters are a, b, β. With the angle $\beta = \arccos(-1/\sqrt{3}) - \pi/2 = 35{,}26^{\circ}$ and $h(\varphi) = a\cos\varphi - s\sin\varphi$ we find for $v(\varphi) = \sqrt{\{a^2(1+[\tan(\beta/2)]^2) - h(\varphi)^2\}}$ and therewith for the edge length of the hexahedron;

$$L(\varphi) = \frac{2}{\cos\beta}[b + v(\varphi)(1+\cos\beta)] + 6(a\sin\varphi + s\cos\varphi).$$

At an apex of a hexahedron three edges meet. Therefore, three Nuremberg scissors are to be attached to the Parallel Plate Articulation at anapex of a hexahedron. The Parallel Plate Articulation represents a triple Sarrus linkage (Sarrus, P. T., 1853) and from this it follows that the articulation together with the attached Nuremberg Scissors is mobile with one degree of freedom. With 8 Parallel Plate Articulations and 12 Nuremberg scissors the Hexahedral Zig-Zag Linkage can be built up which evidently is also mobile with one degree of freedom. (Fig.5) shows the Hexahedral Zig-Zag Linkage in its smallest extension: $\varphi = 0$ (left side) and in a medium extension $\varphi = \pi / 6$ (right side).

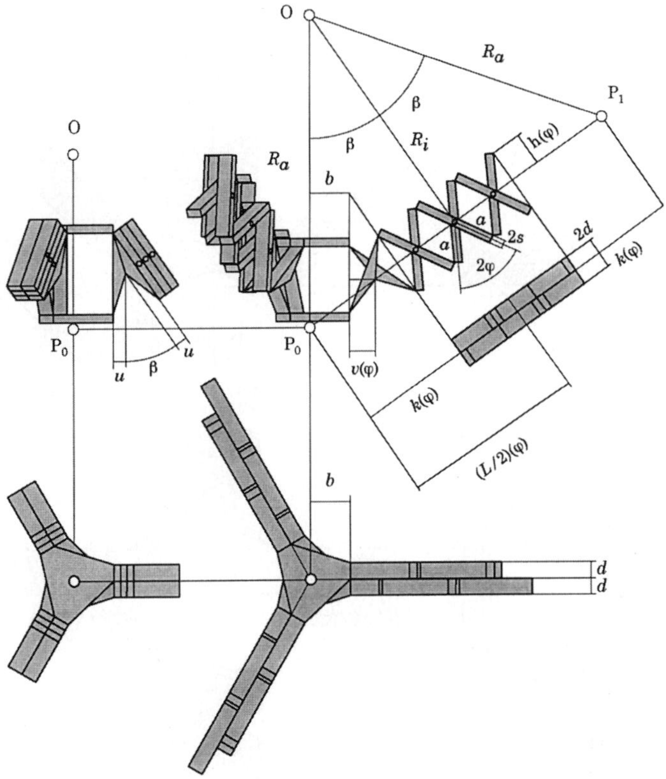

Figure4 Parallel Plate Articulations for the Hexahedral Zig-Zag Linkage

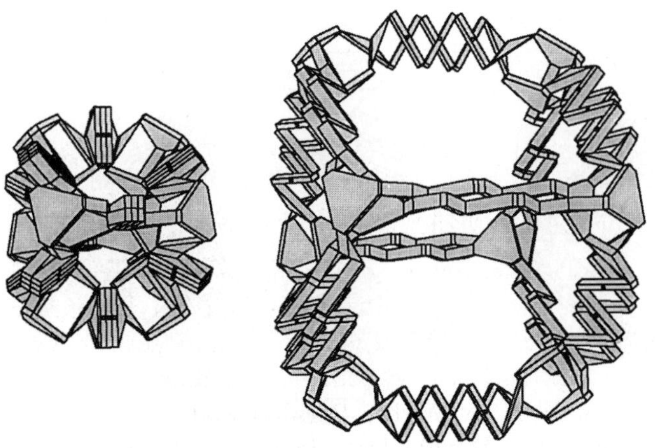

Figure5 Hexahedral Zig-Zag Linkage with Parallel Plate Articulations

5 Synthesis of a Octahedral Zig-Zag Linkage

The third kind of articulation of the Nuremberg scissors will be called the "Flat Ring Articulation" which will be demonstrated at the octa-

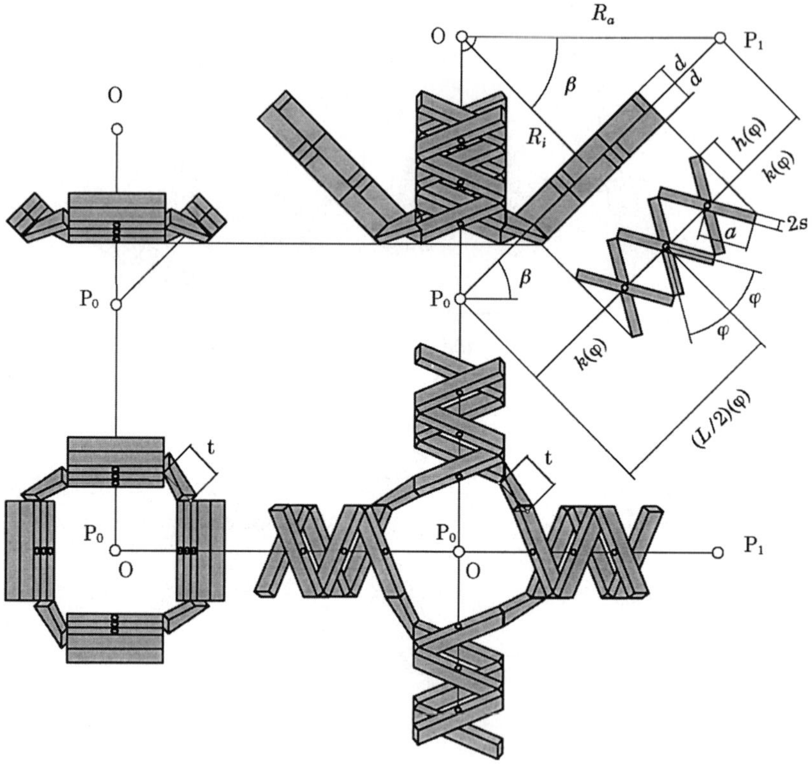

Figure 6 Flat Ring Articulation for the Octahedral Zig-Zag Linkage

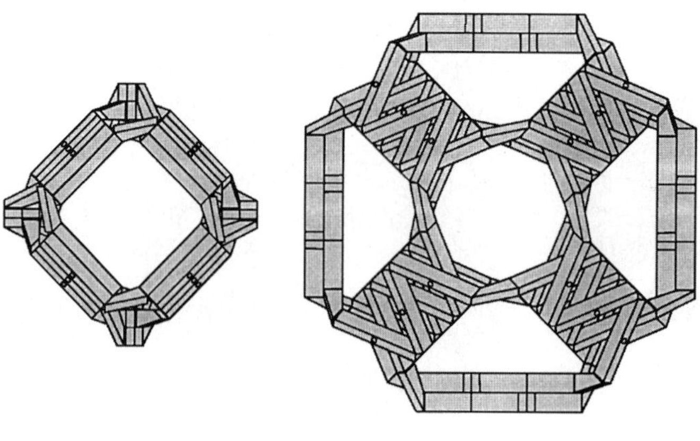

Figure 7 The Octahedral Zig-Zag Linkage with Flat Ring Articulations

hedron. It consists of as many gussets as there are edges meeting at the polyhedral apex (four gussets in the octahedron case). The gussets surround the apex in the form of a ring (Fig.6). The orientation of the gussets is independent of the position angle φ. The parameter set $\{a,s,d,t\}$ together with $\beta=\pi/4$ determines the length of the octahedron edge:

$$L(\varphi)=2[t+\sqrt{2}(a\cos\varphi-s\sin\varphi)]+6(a\sin\varphi+s\cos\varphi).$$

Unlike the mechanisms in Fig. 2 and Fig.4 the mechanism shown in Fig.6 is not overconstrained but mobile with six degrees of freedom. This cannot however, be proved by carelessly applying the structure formula $F=\Sigma f_i-6l$ because this mechanism is not purely spatial, as it contains planar part mechanisms. Nevertheless $F=\Sigma f_i-6l$ can be applied if the planar Nuremberg scissors are taken as "nonrigid bodies" and the number of fundamental loops is determined by disregarding the loops of the planar part mechanisms and the sum of the degrees of freedom due to the defomations of the nonrigid bodies is finally added. In our case then we have $l=1$, $\Sigma f_i=4\times2+4$ and get: $F=\Sigma f_i-6l=6$. The Octahedral Zig-Zag Linkage built up with Flat Ring Articulations is certainly mobile with one degree of freedom, but it could be that it is not overconstrained and is movable with more then one degree. Therefore we determine F with $l=13$ (disregarding the loops in the Nuremberg Scissors), $j=6\times8$ (disregarding the joints in the Nuremberg Scissors) and $\Sigma f_i=j+12=60$ and find: $F=\Sigma f_i-6l=60-6\times13=-18$, which proves overconstraintness.

6 Synthesis of a Dodecahedral Zig-Zag Linkage

The forth kind of articulation of the Nuremberg Scissors will be called the "Coronal Articulation" and demonstrated on the dodecahedron. Fig.8 shows that the three Nuremberg scissors, positioned "upright" on the dodecahedral edges, are connected by three gussets with nonintersecting rotary joint axes whose orientation now depends on the angle φ. With the set of parameters $\{a,s,d,t,\beta=20.91^\circ\}$, $h(\varphi)=a\cos\varphi-s\sin\varphi$ and the components $t_1=k\cos\beta+h\sin\beta+(k\cos\beta-h\sin\beta)/2$, $t_2=(k\cos\beta-h\sin\beta)\sqrt{3}/2$ and $t_3=2h\cos\beta$ of the gusset vector $\vec{t}$. the distance $k(\varphi)$ can be found from the condition $t_1^2+t_2^2+t_3^2=t^2$ and therewith the variable length of the dodecahedral edges is given by:

$$L(\varphi)=2\sqrt{-5h(\varphi)^2+2t^2-3h(\varphi)^2\cos2\beta}/(\cos\beta\sqrt{6})+6(a\sin\varphi+s\cos\varphi).$$

For the mechanism in Fig.8 we find $F=6+3-6\times1=3$ and for the Dodecahedral Zig-Zag Linkage (Fig.9): $F=\Sigma f_i-6l=(20\times6+30)-6\times31=-46$.

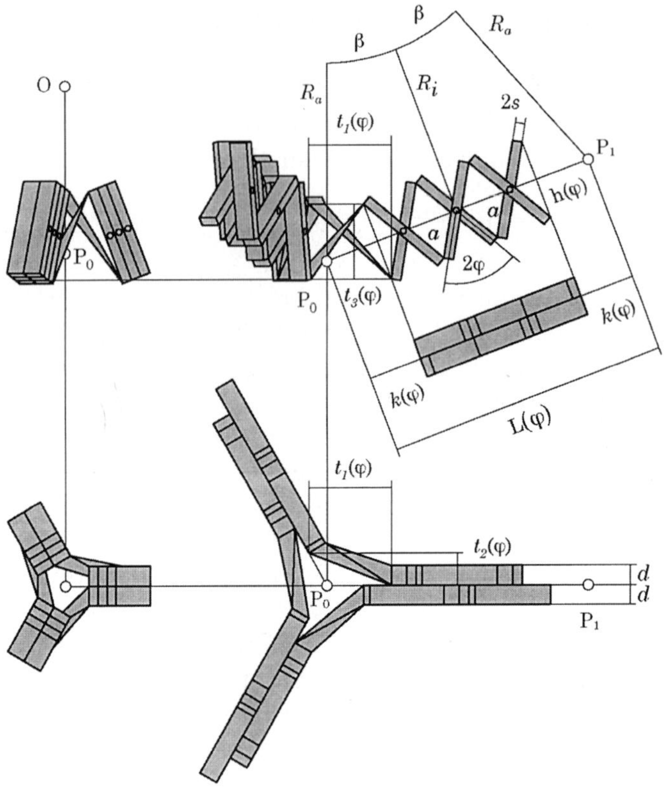

Figure 8 Coronal Articulations the Dodecahedral Zig-Zag Linkage

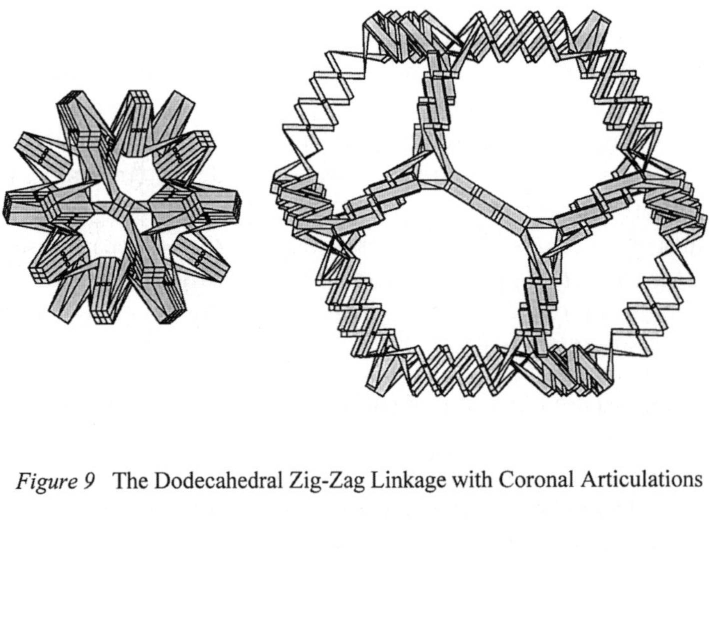

Figure 9 The Dodecahedral Zig-Zag Linkage with Coronal Articulations

7 Synthesis of a Icosahedral Zig-Zag Linkage

For thIcosahedral Zig-Zag Linkage we use the Flat Ring Articulation. With the $\{a,s,d,t,\beta = \arccos[(-3+\sqrt{5})/\sqrt{(20-8\sqrt{5})} - \pi/2 = 31,71^{o}\}$ and

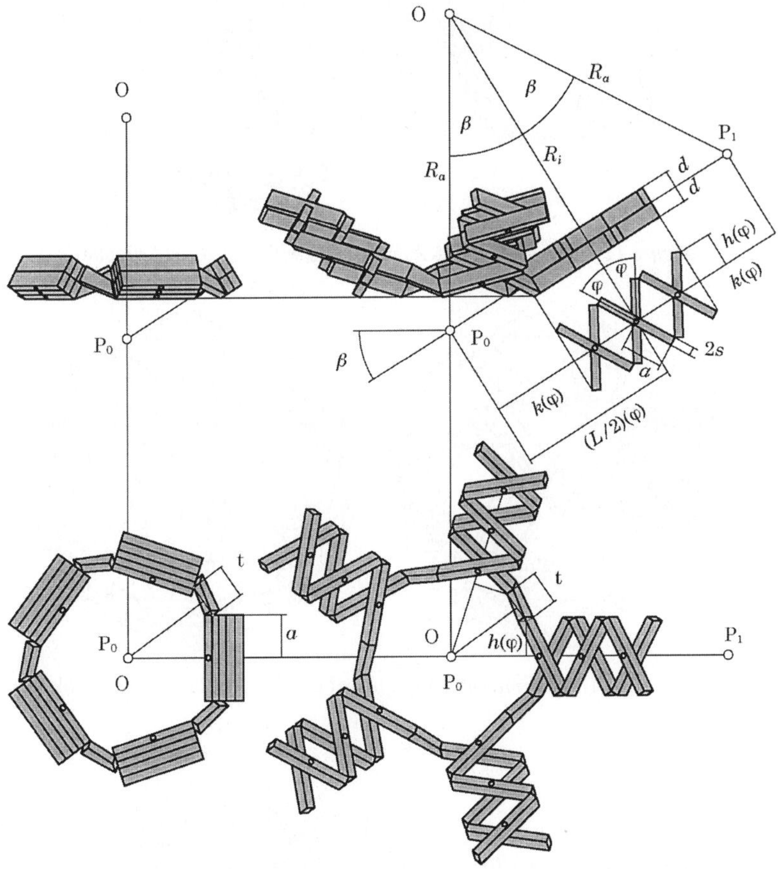

Figure 10 Flat Ring Articulation for the Icosahedral Zig-Zag Linkage

$h(\varphi) = a\cos\varphi - s\sin\varphi$ the edge length of the icosahedron can be calculated:

$$L(\varphi) = 2[(t/2) + h(\varphi)\cos(\pi/5)]/[\cos\beta\sin(\pi/5)] + 6(a\sin\varphi + s\cos\varphi).$$

The mobility of the mechanism shown in Fig.10 (Flat Ring Articulation and 5 Nuremberg scissors) is found by $F = \Sigma f_i - 6l = (10+5) - 6\times 1 = 9$.
For the Ikosahedral Zig-Zag Linkage (Figs. 11,12) we obtain with $\Sigma f_i = j + 30 = 12\times 10 + 30$ and $l = 31$ from $F = \Sigma f_i - 6l = -36$, which indicates overconstraintness. This linkage is movable with one degree of freedom which follows from the fact that for any value of the position angle φ the 30 Nuremberg scissors can be built up with the $12\times 5 = 60$ gussets.

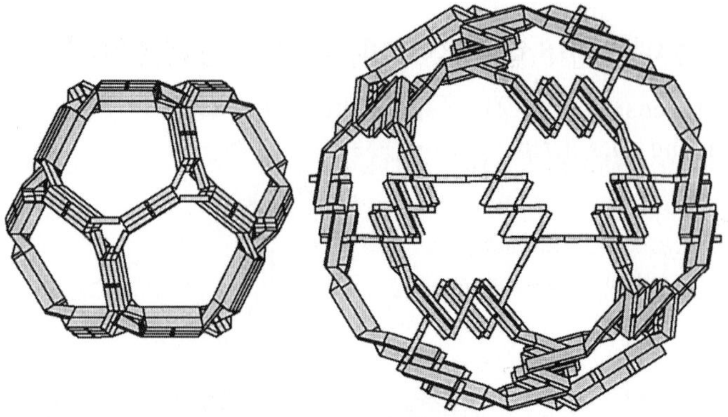

Figure 11 Icosahedral Zig-Zag Linkage with Flat Ring Articulations

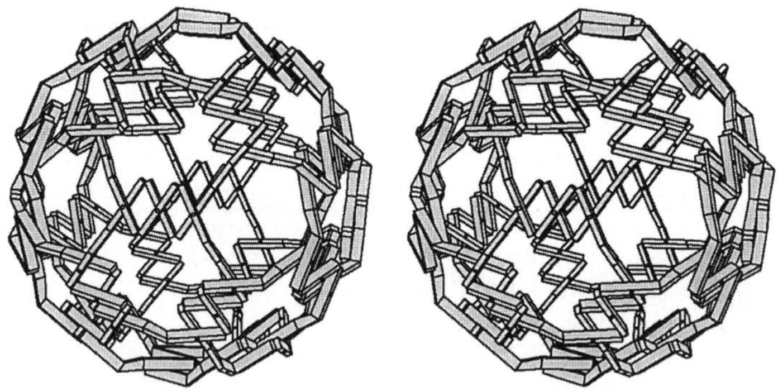

Figure 12 Three-dimensional Picture of the Icosahedral Zig-Zag Linkage

References

Agraval, S.K., Yim, M.,Suh, W. (2001), Polyhedral Single Degree of Freedom Expanding Structures, *Proceedings of IEEE, Seoul,Korea.*

Sarrus, P. T., (1853), Note sur la transformation des mouvements rectilignes alternatifs en mouvements circulaires; et réciproquement, *Comptes Rendus Hebdomadaires des Séances de l' Academie des Sciences,* Vol. 36, pp. 1036-1038, Paris

Wohlhart, K (1999), Deformable Cages, *Proceedings of the 10th World Congress on the Theory of Machines and Mechanisms,* Oulo, Finland,.

Wohlhart, K. (2001), Regular Polyhedral Linkages, *Proceedings of the 2nd Workshop on Computational Kinematics, Seoul,Korea.*

Wohlhart, K. (2001), New Regular Polyhedral Linkages, *Proceedings of the 8th IFTOMM Symposion on the Theory of Machines and Mechanism, Bucharest, Romania.*

Wohlhart, K. (2004), Irregular Polyhedral Linkages, *Proceedings of the 11th World Congress on the Theory of Machines and Mechanisms, Tianjin, China.*

VARIATIONAL MOTION DESIGN

Helmut Pottmann and Michael Hofer
Geometric Modeling and Industrial Geometry Group
Vienna University of Technology, A-1040 Wien, Austria
pottmann,hofer@geometrie.tuwien.ac.at

Bahram Ravani
Department of Mechanical and Aeronautical Engineering
University of California, Davis, CA 95616, USA
bravani@ucdavis.edu

Abstract We discuss the following motion design problem: Given are N positions $\Sigma(t_i)$ of a moving body $\Sigma^0 \subset \mathbb{R}^3$ at time instances t_i. Compute a smooth rigid body motion $\Sigma(t)$ which interpolates or approximates the given key or control positions $\Sigma(t_i)$. Moreover, the motion shall minimize a certain energy which is expressed with help of the energies of trajectories of points on the moving body. Based on very recent results on splines in manifolds we present a characterization of motions which are analogous to known energy minimizing spline curves, such as C^2 cubic splines or splines in tension.

Keywords: Motion design, motion planning, spline motion, splines in manifolds, variational design, screw theory

1. Introduction

Smooth interpolation or approximation of given positions of a moving body is a fundamental task in computer animation and robot motion planning (Latombe, 2001). There exists a large body of literature dealing with this subject, which is rooted in kinematics and Computer Aided Geometric Design (for a recent survey, see Jüttler and Wagner, 2002).

From a more abstract point of view, the problem is a curve interpolation or approximation problem in a manifold. In the present case, this manifold is formed by the elements of the group SE(3) of Euclidean congruence transformations. Depending on the application, various embeddings in a higher dimensional space are appropriate.

Recent work (Belta and Kumar, 2002; Hofer et al., 2004; Wallner, 2004) uses an embedding as 6-dimensional manifold M^6 in a 12-dimensional affine space, which arises as kinematic image space of affine maps in $\mathbb{R}^3$. With an appropriate Euclidean metric in $\mathbb{R}^{12}$, the orthogonal

J. Lenarčič and C. Galletti (eds.), On Advances in Robot Kinematics, 361-370.

projection onto M^6 can be computed explicitly. Now smooth motions arise via the projection of smooth curves $\mathbf{c}(t) \subset \mathbb{R}^{12}$ onto M^6. If $\mathbf{c}(t)$ minimizes an energy functional, its projection $\bar{\mathbf{c}} \subset M^6$ is in general not a minimizer. Therefore, this simple projection principle is not suitable for variational motion design. For contributions to the latter topic, see Barr et al., 1992; Jüttler and Wagner, 2002; Park and Ravani, 1997; Ramamoorthi and Barr, 1997.

In continuation of work started in Hofer et al., 2004, we discuss here variational motion design as a variational curve design problem on $M^6 \subset \mathbb{R}^{12}$. Very recent general results on energy minimizing splines in manifolds (Pottmann and Hofer, 2004) are used to characterize energy minimizing motions. The energy is formulated with help of the energies of trajectories of a finite or infinite set of points (feature points) on the moving body. We find characterizations which are analogous to elegant known results in classical spline theory. For example, the L^2 norm of the second derivative as an energy leads in the unrestricted classical curve interpolation case to cubic C^2 splines; for motion interpolation, we obtain C^2 motions for which – at each time instant – the force system, defined by the 4-th derivative vectors attached to the feature points, is in balance. Geodesic motions in the present setting appear as free motions in the sense of mechanics (Arnol'd, 1989).

The focus of the present paper lies on the description of the theoretical framework and the derivation of the main characterization of energy minimizing motions. We address the computation, which is a special case of the computation of splines in manifolds, only briefly, and finally present illustrative examples.

2. The Group of Euclidean Motions Embedded in the Affine Group

Consider a rigid body moving in Euclidean three-space E^3. We use Cartesian coordinates and denote points of the moving system Σ^0 by $\mathbf{x}^0, \mathbf{y}^0, \ldots$, and points of the fixed system by $\mathbf{x}, \mathbf{y}$, and so on.

A *one-parameter motion* Σ^0/Σ is a smooth family of Euclidean congruence transformations depending on a parameter t which can be thought of as time. A point $\mathbf{x}^0$ of Σ^0 is, at time t, mapped to the point

$$\mathbf{x}(t) = A(t) \cdot \mathbf{x}^0 + \mathbf{a}_0(t) \tag{1}$$

of Σ, where $A(t) \in SO(3)$ and $\mathbf{a}_0(t) \in \mathbb{R}^3$.

If we do not impose any restriction to the matrix A in Eq. (1), we get, for each t, an *affine map*.

In the following, we will use a kinematic mapping that views affine maps as points in 12-dimensional affine space. For that, consider the affine map $\mathbf{x} = \alpha(\mathbf{x}^0) = \mathbf{a}_0 + A \cdot \mathbf{x}^0$. Let us denote the three column vectors of A as $\mathbf{a}_1, \mathbf{a}_2, \mathbf{a}_3$. They describe the images of the basis vectors of Σ^0 in Σ. Of course, we have $\mathbf{x} = \mathbf{a}_0 + x_1^0\mathbf{a}_1 + x_2^0\mathbf{a}_2 + x_3^0\mathbf{a}_3$. Now we associate with the affine map α a *point in 12-dimensional affine space* $\mathbb{R}^{12}$, represented by the vector $\mathsf{A} = (\mathbf{a}_0, \ldots, \mathbf{a}_3)$.

The images of Euclidean congruence transformations (rigid body motions) $\alpha \in SE(3)$ form a 6-dimensional manifold $M^6 \subset \mathbb{R}^{12}$. Its 6 equations are given by the orthogonality conditions of A, i.e., $\mathbf{a}_i \cdot \mathbf{a}_j = \delta_{ij},\ i = 1, 2, 3$.

It will be necessary to introduce a meaningful *metric* in $\mathbb{R}^{12}$. Following Hofer et al., 2004, this is done with help of a collection X of points $\mathbf{x}_1^0, \mathbf{x}_2^0, \ldots, \mathbf{x}_N^0$ in the moving system (body), which shall be called *feature points* henceforth. The squared distance between two affine maps α and β is now defined as sum of squared distances of feature point positions after application of α and β, respectively,

$$d^2(\alpha, \beta) = \|\mathsf{A} - \mathsf{B}\|^2 := \sum_i [\alpha(\mathbf{x}_i^0) - \beta(\mathbf{x}_i^0)]^2. \tag{2}$$

With $\mathsf{A} = (\mathbf{a}_0, \ldots, \mathbf{a}_3)$, $\mathsf{B} = (\mathbf{b}_0, \ldots, \mathbf{b}_3)$, $\mathsf{C} := \mathsf{A} - \mathsf{B} = (\mathbf{c}_0, \ldots, \mathbf{c}_3)$, and $\mathbf{x}_i^0 = (x_{i,1}^0, x_{i,2}^0, x_{i,3}^0)$ the distance becomes

$$\|\mathsf{A} - \mathsf{B}\|^2 = \|\mathsf{C}\|^2 = \sum_i [\mathbf{c}_0 + x_{i,1}^0\mathbf{c}_1 + x_{i,2}^0\mathbf{c}_2 + x_{i,3}^0\mathbf{c}_3]^2 = \mathsf{C}^T \cdot M \cdot \mathsf{C}. \tag{3}$$

This expression with help of a positive definite symmetric matrix M immediately reveals the following result (Hofer et al., 2004): *The metric (2) in the space of affine maps only depends on the barycenter* $\mathbf{s}_x = (1/N)\sum_i \mathbf{x}_i^0$ *and on the inertia tensor*

$$J := \sum_i \mathbf{x}_i^0 \cdot \mathbf{x}_i^{0T} \tag{4}$$

of the set of feature points $\mathbf{x}_i^0$ *in the moving system.* $\mathbb{R}^{12}$ *equipped with this metric is a Euclidean space* E^{12}.

We also see that we need not use unit point masses at a discrete number of feature points. We could instead work with another positive measure on a domain of interest D (the moving body) in Σ^0, e.g. the Lebesgue measure times the characteristic function χ_d of D. Of course, we then replace summation in (2) by integration.

By a well-known result from mechanics, we can replace the points $\mathbf{x}_1^0, \ldots, \mathbf{x}_N^0$ by the six special points

$$\mathbf{s}_x \pm \sqrt{\frac{\lambda_j}{2}}\mathbf{e}^j, \qquad j = 1, 2, 3, \tag{5}$$

without changing the barycenter and the inertia tensor of X. There, $\lambda_1, \lambda_2, \lambda_3$ and $\mathbf{e}^1, \mathbf{e}^2, \mathbf{e}^3$ are the eigenvalues and corresponding unit eigenvectors of the matrix J. Let us choose the barycenter as origin in the moving system and the eigenvectors of J as coordinate axes. Then the six points have coordinates $(\pm f_1, 0, 0), (0, \pm f_2, 0), (0, 0, \pm f_3)$, and the norm in $\mathbb{R}^{12}$ becomes

$$\|\mathsf{C}\|^2 = 6\mathbf{c}_0^2 + 2\sum_{i=1}^{3} f_i^2 \mathbf{c}_i^2. \tag{6}$$

For the following considerations it will be important to know about the tangent spaces at points $\mathsf{A} \subset M^6$ and to characterize orthogonality to these tangent spaces.

A tangent vector at an arbitrary point $\mathsf{A} \in E^{12}$ can be interpreted in E^3 as a velocity vector field of an affine motion. In particular, a tangent vector of M^6 belongs to a velocity vector field of a Euclidean motion, which is of the form $\mathbf{v}(\mathbf{x}) = \overline{\mathbf{c}} + \mathbf{c} \times \mathbf{x}$. The coordinate representation of this tangent vector in E^{12}, attached to $\mathsf{A} = (\mathbf{a}_0, \ldots, \mathbf{a}_3) \in M^6$, reads

$$\mathsf{T} = (\overline{\mathbf{c}} + \mathbf{c} \times \mathbf{a}_0, \mathbf{c} \times \mathbf{a}_1, \ldots, \mathbf{c} \times \mathbf{a}_3). \tag{7}$$

We would like to express orthogonality of an arbitrary vector $\mathsf{D} = (\mathbf{d}_0, \mathbf{d}_1, \ldots, \mathbf{d}_3) \in \mathbb{R}^{12}$ to the tangent space at A. If we align origin and axes of the coordinate system in Σ^0 with center and axes of the inertia ellipsoid, the inner product is expressed in view of (6) as

$$\langle \mathsf{D}, \mathsf{T} \rangle = 6\mathbf{d}_0 \cdot (\overline{\mathbf{c}} + \mathbf{c} \times \mathbf{a}_0) + 2\sum_{i=1}^{3} f_i^2 \mathbf{d}_i \cdot (\mathbf{c} \times \mathbf{a}_i).$$

This equation may also be written as

$$\langle \mathsf{D}, \mathsf{T} \rangle = 6\mathbf{d}_0 \cdot \overline{\mathbf{c}} + 6\mathbf{c} \cdot (\mathbf{a}_0 \times \mathbf{d}_0) + 2\mathbf{c}\sum_{i=1}^{3} f_i^2 (\mathbf{a}_i \times \mathbf{d}_i). \tag{8}$$

Let us attach to the six feature point positions $\mathbf{a}_0 \pm f_i\mathbf{a}_i$, $i = 1, 2, 3$, the vectors of the linear vector field determined by $\mathsf{D} \in \mathbb{R}^{12}$. These vectors are $\mathbf{d}_0 \pm f_i\mathbf{d}_i$, and shall be interpreted as forces (for such concepts from statics, see e.g. Pottmann and Wallner, 2001, pp. 191–194). The moments of these forces are

$$(\mathbf{a}_0 \pm f_i \mathbf{a}_i) \times (\mathbf{d}_0 \pm f_i \mathbf{d}_i).$$

The sums of force vectors and moment vectors is the screw

$$(\mathbf{s}_d, \overline{\mathbf{s}}_d) = (6\mathbf{d}_0, 6\mathbf{a}_0 \times \mathbf{d}_0 + 2\sum_{i=1}^{3} f_i^2(\mathbf{a}_i \times \mathbf{d}_i)). \tag{9}$$

The screw resulting from the action of an instantaneous affine motion (linear vector field) on these points is the same whether we use the 6 special feature points or the original set of feature points (or the mass distribution). We call this the *screw or force system* S_d *induced by the linear vector field* D.

The inner product (8) between T and D is now expressed as

$$\langle \mathsf{D}, \mathsf{T} \rangle = \overline{\mathbf{c}} \cdot \mathbf{s}_d + \mathbf{c} \cdot \overline{\mathbf{s}}_d. \tag{10}$$

This is the *virtual work done by the force system* S_d *on the body which moves instantaneously with the velocity field determined by* T. Orthogonality between D and all T requires $(\mathbf{s}_d, \overline{\mathbf{s}}_d) = 0$, i.e., balance of the induced force system S_d. Thus, we have proved the following result.

Theorem 1 *A vector* $\mathsf{D} \in \mathbb{R}^{12}$*, attached to a point* $\mathsf{A} \in M^6$*, is orthogonal to* $M^6 \subset E^{12}$ *iff the force system* S_d *induced by* D *is in balance.*

3. Variational Motion Design as Variational Curve Design on $M^6 \subset E^{12}$

A Euclidean one-parameter motion corresponds to a curve on the manifold $M^6 \subset E^{12}$. The design of such curves via projection onto M^6 is a principle of transfer from curve design to motion design (Hofer et al., 2004). Such an approach, however, does not preserve minimum energy properties. Thus, we are now dealing with the problem of constructing energy minimizing interpolating spline curves on M^6. In this way, we are computing energy minimizing Euclidean motions which interpolate given positions.

Consider an m-dimensional manifold Φ, embedded in Euclidean $\mathbb{R}^n$, $m < n$. Moreover, a sequence of points $\mathbf{p}_i \in \Phi$, $i = 1, \dots, N$ and real numbers $t_1 < \cdots < t_N$ are given. We are seeking interpolating splines in the manifold.

Let us recall the situation, where we are not confined to a manifold: Among all curves $\mathbf{x}(t) \subset \mathbb{R}^n$, whose first and second derivative satisfy

$\dot{\mathbf{x}} \in AC(I), \ddot{\mathbf{x}} \in L^2(I)$ on $I = [t_1, t_N]$, and which interpolate the given data, $\mathbf{x}(t_i) = \mathbf{p}_i$, the unique minimizer of

$$E_2(\mathbf{x}) = \int_{t_1}^{t_N} \ddot{\mathbf{x}}^2 dt, \tag{11}$$

is the interpolating C^2 *cubic spline* $\mathbf{c}(t)$.

Pottmann and Hofer, 2004 extended this well-known result to the case where the admissible curves $\mathbf{x}(u)$ are restricted to the given manifold Φ. We are considering the restriction to Φ as a constraint, rather than formulating the problem in terms of the intrinsic geometry of the manifold. The solution can be characterized as follows (Pottmann and Hofer, 2004):

Theorem 2 *Consider real numbers $t_1 < \ldots < t_N$ and points $\mathbf{p}_1, \ldots, \mathbf{p}_N$ on an m-dimensional C^4 manifold Φ in Euclidean $\mathbb{R}^n$. Then among all C^1 curves $\mathbf{x} : [t_1, t_N] \to \Phi \subset \mathbb{R}^n$, whose restrictions to the intervals $[t_i, t_{i+1}], i = 1, \ldots, N-1$ are C^4 and which interpolate the given points, $\mathbf{x}(t_i) = \mathbf{p}_i$, $i = 1, \ldots, N$, a curve $\mathbf{c}$ which minimizes the functional E_2 of Eq. (11) is C^2 and possesses segments $\mathbf{c}|[t_i, t_{i+1}]$, whose fourth derivative vectors are orthogonal to Φ. Moreover, at the end points $\mathbf{p}_1 = \mathbf{c}(t_1)$ and $\mathbf{p}_N = \mathbf{c}(t_N)$ of the solution curve, the second derivative vector is orthogonal to Φ.*

We are viewing t_i as given time instances, at which given positions, i.e., points $\mathsf{P}_i \in M^6$, have to be interpolated. The energy (11) of a curve $\mathsf{X}(t) \subset E^{12}$ uses the norm induced by (2) and thus it may also be interpreted as sum of the corresponding energies (L^2 norms of the second derivatives) of the feature point trajectories.

According to Theorem 2, the solution curve $\mathsf{C}(t) \subset M^6$ has 4-th derivative vectors $\mathsf{C}^{(4)}(t)$, which are orthogonal to M^6. The force system $\mathsf{S}_4(t)$ induced by $\mathsf{C}^{(4)}(t)$ consists of the fourth derivative vectors of the feature point trajectories at a given instant t. Using Theorems 1 and 2, we obtain the following result.

Theorem 3 *Consider N input positions, corresponding time instances t_i and differentiability assumptions as in Theorem 2. Then, an interpolating motion minimizing the sum of energies (L^2 norms of the second derivatives) of the feature point trajectories is characterized as follows. The motion is C^2, has at each time instant $t \neq t_i$ a balanced 4-th derivative force system $\mathsf{S}_4(t)$, and at the end positions balanced force systems $\mathsf{S}_2(t_1), \mathsf{S}_2(t_N)$ of second derivatives. In particular, the trajectory of the barycenter of the feature points is an interpolating cubic C^2 spline.*

The result on the trajectory of the barycenter follows from the vanishing of the force components $\mathbf{s}_k$ of the involved k-th derivative systems $\mathsf{S}_k = (\mathbf{s}_k, \overline{\mathbf{s}}_k)$.

We see that these motions somehow balance the deviations of the point trajectories from C^2 cubic splines; there the 4-th derivatives would vanish everywhere. Note that a motion with only C^2 cubic spline trajectories must be translational (the image curve lies in a 3-dimensional affine subspace contained in M^6 and is a cubic spline itself).

It is also quite natural that the moving body via its mass distribution (barycenter and inertia tensor) enters the variational formulation and the interpretation of the solution. That the present approach is natural from the viewpoint of mechanics and kinematics, is also nicely seen if we replace the energy in (11) by the L^2 norm of the first derivative,

$$E_1(\mathbf{x}) = \int_{t_1}^{t_N} \dot{\mathbf{x}}^2 dt. \tag{12}$$

Moreover, we just prescribe the two end positions. Then, one finds with a known counterpart of Theorem 2 (see e.g. Pottmann and Hofer, 2004) a curve $\mathsf{C}(t) \subset M^6$, whose second derivative vectors $\ddot{\mathsf{C}}$ are orthogonal to M^6. From this we conclude immediately that *the minimizers of E_1 are geodesics on M^6 in a scaled arc length parameterization, i.e., $\|\dot{\mathsf{C}}\| = const$.* This proves the following theorem.

Theorem 4 *Motions which join two given positions and arise from minimization of (12) correspond to geodesics on M^6, parameterized by a constant multiple of arc length. At any time instant, such a geodesic motion possesses a balanced force system $\mathsf{S}_2(t)$ of second derivatives. The trajectory of the barycenter of the feature point set on the moving body is a straight line traced with constant speed. These motions are free motions of a body in the sense of mechanics.*

Note that a helical motion is a geodesic motion between two positions $\Sigma(t_1)$ and $\Sigma(t_2)$, if the positions of the inertia ellipsoids share a common axis, and if this is the axis of the helical motion joining the two positions. For more results on free motions, we refer to Arnol'd, 1989.

By minimization of a combination $E_t := E_2 + wE_1$ with a constant positive factor w one obtains the counterparts of *splines in tension* for motions. These are characterized as in Theorem 3, but instead of a balanced 4-th derivative force system, the linearly combined force system $\mathsf{S}_4 - w\mathsf{S}_2$ is in balance. A proof follows from results in (Pottmann and Hofer, 2004). Various other spline types, interpolating or approximating, can be transferred to motion design within the present setting. This will be developed in the future.

Note that the functionals we are considering are also optimizing the parameterization. This is useful for motion design where the time t as parameter plays an important role for applications.

Since even for geodesics we do not have simple results on uniqueness, the more involved case of splines will hardly allow us a characterization of situations with a unique solution. For a proof of the existence of the solution, we refer to (Bohl, 1999; Wallner, 2003).

4. Computational Approach and Examples

Although the presented spline motions possess nice geometric characterizations, the problem is still nonlinear and does not admit an explicit solution as in the unrestricted curve design case. Thus, we have to use numerical algorithms based on a discretization. Geometrically this means that we replace the curve $\mathsf{C}(t) \subset M^6$ by a sufficiently dense polygon P_c with vertices $\mathsf{C}_i \in M^6$. These vertices represent positions of the motion at discrete time instances.

Discretizing a quadratic functional, such as E_1, E_2, E_t, results in a quadratic function in the coordinates of the vertices C_i of the approximating polygon P. Since the C_i's are restricted to M^6, we end up with the constrained minimization of a quadratic function in a rather high dimensional space. From a geometric viewpoint, we have to construct the closest point $\mathbf{p}^*$ on some manifold S to a given point $\mathbf{p}$. Here, $\mathbf{p}$ represents the unrestricted (discretized) spline curve in E^{12}. For the solution of this problem, we developed a geometrically motivated optimization algorithm of the quasi-Newton type (Hofer and Pottmann, 2004). It has been used for the computation of the examples presented in Figs. 1,2.

Another computational approach directly uses the characterization of the motions in a discretized way. For example, a motion minimizing E^2 has a balanced 4th central difference screw at any instance in the time discretization. This results in a system of nonlinear equations, which can be solved with a Newton iteration. In view of the bad global behaviour of a pure Newton algorithm, we used the quasi-Newton approach mentioned above.

5. Conclusions and Future Research

We have used a kinematic mapping of SE(3) to points of a six dimensional manifold M^6 embedded in 12-dimensional Euclidean space E^{12}. The latter space corresponds to affine maps, and the metric therein is defined naturally with help of feature points (a mass distribution) of the moving body. Variational motion design is thus transferred to variational curve design on M^6. Based on recent results on splines in manifolds,

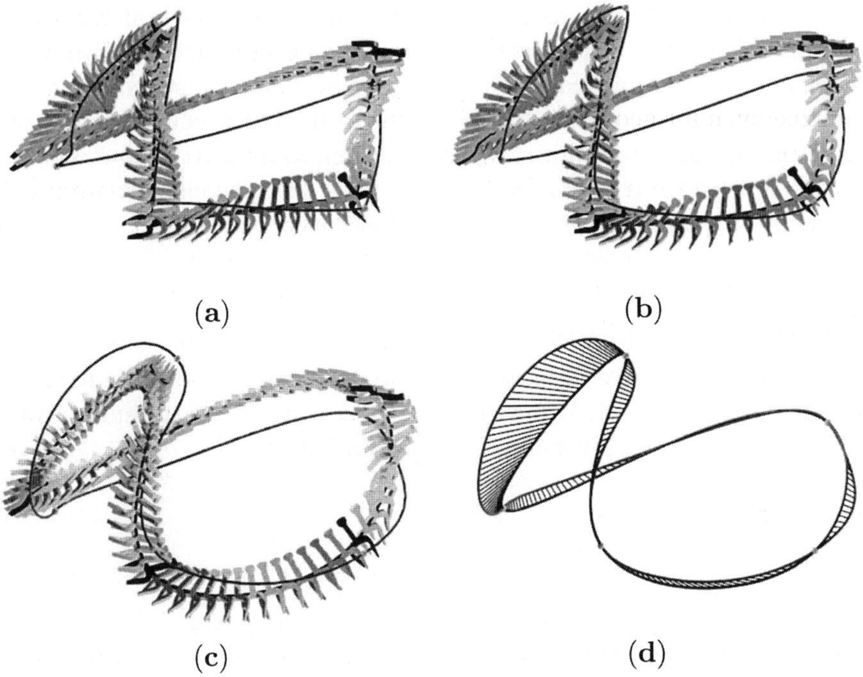

(a) (b) (c) (d)

Figure 1. Cyclic motion of a robot gripper interpolating 5 positions (dark), and the paths of the barycenter and the feature point $(f_1, 0, 0)$: Motion minimizing (**a**) E_1, (**b**) E_t with $w = 0.05$, (**c**) E_2; (**d**) Path of feature point compared to cubic spline.

Figure 2. Open motion interpolating 8 positions and minimizing E_2.

we could characterize and compute the solutions of several variational motion design problems that generalize known variational curve design schemes. The present contribution is not aiming at a special representation of the solution, e.g. in NURBS form, which could be obtained after the design by an approximation scheme (see Jüttler and Wagner, 2002). Future research will be devoted to a more efficient computation based on a quadratically convergent Newton-type iteration, to motion smoothing, and to obstacle avoidance in motion design.

Acknowledgements

This research has been supported by the Austrian Science Fund (FWF) under grant P16002-N05, and in part by California Department of Transportation through University of California-Davis.

References

Arnol'd, V.I. (1989), *Mathematical methods of classical mechanics.* 2nd ed., Springer, New York.

Barr, A.H., Currin, B., Gabriel, S., and Hughes, J.F. (1992), Smooth interpolation of orientations with angular velocity constraints using quaternions, *Computer Graphics* 26 (SIGGRAPH '92), pp. 313–320.

Belta, C., and Kumar, V. (2002), An SVD-projection method for interpolation on SE(3), *IEEE Trans. on Robotics and Automation* 18, 334–345.

Bohl, H. (1999), *Kurven minimaler Energie auf getrimmten Flächen*, PhD thesis, Stuttgart.

Hofer, M., and Pottmann, H. (2004), Energy-minimizing splines in manifolds, submitted for publication.

Hofer, M., Pottmann, H., and Ravani, B. (2004), From curve design algorithms to the design of rigid body motions, *The Visual Computer*, to appear.

Jüttler, B., and Wagner, M. (2002), Kinematics and Animation, In: G. Farin, J. Hoschek, M.S. Kim, eds., *Handbook of Computer Aided Geometric Design*, Elsevier, pp. 723–748.

Latombe, J.-C. (2001), *Robot Motion Planning*, 6th printing, Kluwer.

Park, F.C., and Ravani, B. (1997), Smooth invariant interpolation of rotations, *ACM Transactions on Graphics* 16, pp. 277–295.

Pottmann, H., and Hofer, M., (2004), A variational approach to spline curves on surfaces, submitted for publication.

Pottmann, H., and Wallner, J. (2001), *Computational Line Geometry*, Springer-Verlag, Berlin/Heidelberg/New York.

Ramamoorthi, R., and Barr, A., (1997), Fast construction of accurate quaternion splines, *Computer Graphics (Proceedings of SIGGRAPH 1997)*, pp. 287–292.

Wallner, J. (2003), Energy minimizing curves on surfaces, Preprint, Institute of Geometry, Vienna Univ. of Technology, December 2003.

Wallner, J. (2004), Gliding spline motions and applications, *Computer Aided Geom. Design*, vol. 21, pp. 3–21.

NEW PARALLEL WRISTS: SPECIAL LIMBS WITH MOTION DEPENDENCY

Mourad Karouia
Medical Robotics – Institut de Cardiologie Groupe Hospitalier Pitié-Salpêtrière
47-83, Bd de l'Hôpital 75013 Paris, France
karouia@yahoo.fr

Jacques M. Hervé
Ecole Central Paris
Grande Voie des Vignes 92295 Chatenay-Malabry, France
jherve@ecp.fr

Abstract The paper addresses the structural type synthesis of novel non-overconstrained 3-dof spherical parallel mechanisms. A non-overconstrained spherical parallel mechanism has three limbs; each limb generates a 5-dimensional manifold of the displacement Lie group, which has to contain a Lie subgroup of spherical displacements. In each of the limbs that are considered in this article, two distinct kinematic sub-chains produce two motion subsets that are dependent. Limbs having this special property are described and an enumeration of the deducted mechanisms is disclosed.

Keywords: Spherical parallel mechanisms, non-overconstrained, group theory.

1. Introduction

A spherical parallel mechanism (SPM) is made of three limbs that connect a moving platform (end-effector) to a fixed base, Merlet, 1997. The end-effector has 3-dof rotation around a fixed point with respect to the base. This point denoted O is called the SPM center. Overconstrained SPM of the 3-*RRR* type were studied by Asada and Cro Granito, 1985, and by Gosselin and Angeles, 1994. Karouia and Hervé, 2000, have shown that non-overconstrained SPMs have to implement three 5-dof limbs. Each limb generates a 5-dimensional manifold of displacements, which has to contain the 3-dimensional Lie subgroup $\{S(\mathrm{O})\}$ of spherical displacements around the SPM centre. Karouia and Hervé., 2000, 2002a and 2002b, used a synthesis method, which is based on the algebraic properties of the displacement set and especially on its algebraic structure of a Lie group. They enumerated most of the structural types of SPM limbs. Neglecting the trivial cases of limbs with inactive (or idle) pairs that have a minor interest, truly novel limbs were classified in

J. Lenarčič and C. Galletti (eds.), On Advances in Robot Kinematics, 371-380.

three main families, namely limbs including a generator of planar gliding motion, limbs with a spherical motion generator and limbs with cylindrical motion generators. A comprehensive study of these families is the essential topic of the doctoral dissertation of Karouia, 2003. The capability of the structural type 3-*UPU* , *U* standing for universal joint, for the generation of spherical motion was early discovered by Leguay-Durand, 1998. Karouia and Hervé, 2000, described the 3-*UPU* wrist as a special case of 3-$R^O\underline{G}\,R^O$ where R^O designates a revolute pair whose axis passes through the center O and $\underline{G}$ any 3-dof sub-chain that is equivalent to a planar pair *G*. By the screw theory, Kong and Gosselin, 2002, confirmed most of these results. By a vector explanation, Di Gregorio proposed parallel wrists with inactive pairs, 2001, and detected singularities of some of the previous SPM, 2002.

In the sequel, additional structural types of SPM that implement generators of dependent displacement subgroups are described.

2. Theoretical background

The displacement (or rigid body motion) set {*D*} is endowed with the algebraic structure of a 6-dimensional Lie group, the closed product of the group being the product of displacements. In any Cartesian frame of reference, a matrix product can represent the displacement product and the corresponding Lie group of matrices is usually denoted SE(3), which stands for special Euclidean group acting on $\Re^3$, Selig, 2000.

In a given kinematic chain, the subset of allowed displacements of a rigid body with respect to another body is called "kinematic bond" between the two bodies. Generally a kinematic bond is a n-dimensional manifold included in {*D*}; $0 \leq n \leq 6$; n is usually called degree of freedom (dof) of the bond. A kinematic bond contains the identical displacement *E*, because else the mechanism can't be constructed. A kinematic bond may be a Lie subgroup of {*D*} and then the corresponding kinematic chain is a generator of the subgroup.

The set of spherical rotations about a given point N, {*S*(N)}, is a 3-dimensional Lie subgroup of {*D*}. In any Cartesian frame of reference having origin at N, the elements of {*S*(N)} are represented by orthogonal matrices belonging to the matrix group usually denoted SO(3), where SO(3) stands for "special orthogonal matrices of dimension 3x3". The proper Lie subgroups of {*S*(N)} are the sets of rotations around any axes passing through the center N and they are denoted {*R*(N,**w**)}, the unit vector **w** characterizing the direction of the axis and (N,**w**) being a frame of reference of the axis. (N, **u**, **v**, **w**) being a Cartesian frame of reference, {*R*(N,**w**)} can be represented in the frame (N, **u**, **v**) by the matrix group SO(2), the Lie group of special orthogonal matrices acting on $\Re^2$.

Using intrinsic geometrical entities instead of frame-dependent matrices, one can readily show that a sequence of two revolute pairs of axes (N, **u**) and (N, **v**) generates a kinematic bond, which is the product $\{R(\mathrm{N},\mathbf{u})\}.\{R(\mathrm{N},\mathbf{v})\}$. By definition, in any group, the product of subsets is the set of products.

Because of the product closure in the subgroup $\{S(\mathrm{N})\}$

$$\{R(\mathrm{N},\mathbf{u})\} \subset \{S(\mathrm{N})\} \text{ and } \{R(\mathrm{N},\mathbf{v})\} \subset \{S(\mathrm{N})\} \Rightarrow \{R(\mathrm{N},\mathbf{u})\}.\{R(\mathrm{N},\mathbf{v})\} \subseteq \{S(\mathrm{N})\}$$

and also

$$\{R(\mathrm{N},\mathbf{u})\}.\{R(\mathrm{N},\mathbf{v})\} \subseteq \{S(\mathrm{N})\} \text{ and } \{R(\mathrm{N},\mathbf{w})\} \subset \{S(\mathrm{N})\}$$
$$\Rightarrow \{R(\mathrm{N},\mathbf{u})\}.\{R(\mathrm{N},\mathbf{v})\}.\{R(\mathrm{N},\mathbf{w})\} \subseteq \{S(\mathrm{N})\}$$

Provided **u**, **v**, **w** are linearly independent vectors, the product $\{R(\mathrm{N},\mathbf{u})\}.\{R(\mathrm{N},\mathbf{v})\}.\{R(\mathrm{N},\mathbf{w})\}$ is a 3-dimensional manifold. A difference between this manifold and $\{S(\mathrm{N})\}$ can only originate from the amplitude of the parameter variations. Dealing only with motion types the following equality between finite displacement subsets can be written: $\{R(\mathrm{N},\mathbf{u})\}.\{R(\mathrm{N},\mathbf{v})\}.\{R(\mathrm{N},\mathbf{w})\} = \{S(\mathrm{N})\}$

This set equality expresses the equivalence between a serial arrangement $R^N\, R^N\, R^N$ of 3 revolute pairs R with axes converging at N and a spherical pair S of center N.

One can readily verify that any non-overconstrained 3-dof parallel mechanism has to be made of three 5-dof limbs connecting a fixed base to a moving platform. If the limbs are serial arrangements of lower 1-dof kinematic pairs, then the kinematic bond generated by each limb is the product of the five 1-dof subgroups of $\{D\}$ that are generated by the five lower pairs. The set of permitted displacements (or kinematic bond) of the end-effector with respect to the base is the intersection of the limb bonds. In a parallel wrist of center N, any limb bond has to include a subgroup of spherical rotations $\{S(\mathrm{N})\}$.

A limb bond, which obeys the previous conditions is expressed by any factor permutation in the product $\{R(\mathrm{N},\mathbf{u})\}.\{D_1\}.\{R(\mathrm{N},\mathbf{k})\}.\{D_2\}.\{R(\mathrm{N},\mathbf{w})\}$ where $\{D_1\}$ and $\{D_2\}$ are two arbitrarily chosen 1-dof subgroups. In other words, the limb is a sequence of five 1-dof lower pairs, which includes three revolute pairs with converging axes. The spherical motion is achieved by having $\{D_1\} = \{D_2\} = \{E\}$, improper subgroup made only by the identical transform E. In other words, in a possible limb, two pairs remains inactive (or idle) whereas the three revolute pairs with converging axes are active.

However by an adequate choice of $\{D_1\}$ and $\{D_2\}$ and a factor permutation in the product $\{R(\mathrm{N},\mathbf{u})\}.\{D_1\}.\{R(\mathrm{N},\mathbf{k})\}.\{D_2\}.\{R(\mathrm{N},\mathbf{w})\}$, a subset of factors can be a manifold in $\{D\}$, which can be equated to another subset of factors because of the product closure in any subgroup. In their previous articles, Karouia and Hervé, 2000, 2002a, 2002b, the authors focused on the cases of manifolds that are Lie subgroups of $\{D\}$.

Adequate manifolds that are products of dependent subgroups can also be found and more novel limbs can be derived. Two subgroups are dependent if and only if the intersection of these subgroups is a proper subgroup, i.e. not equal to $\{E\}$. There are two cases of subgroup dependency. In a first case, a subgroup is included in the other subgroup and then the intersection is one of the two intersecting subgroups. In a second case the intersection is a proper subgroup strictly included in each of the intersecting subgroups. Refering to the Hervé list of subgroup products with the second dependency, 1978, two 3-dimensional manifolds that are equal to the product of two dependent 2-dimensional subgroups can be identified and employed in our context because they include a subgroup of rotation like $\{R(\mathrm{O},\mathbf{k})\}$.

3. Motion dependency, elimination of passive mobility

In the product $\{C(\mathrm{A},\mathbf{u})\}.\{C(\mathrm{B},\mathbf{u})\}$, $\mathrm{B} \notin$ line $(\mathrm{A},\mathbf{u})$ of two subgroups of cylindrical displacements with parallel axes, the bond dependency is explained by the intersection $\{C(\mathrm{A},\mathbf{u})\}\cap\{C(\mathrm{B},\mathbf{u})\} = \{T_1(\mathbf{u})\}$, which is the 1-dimensional subgroup of linear translation parallel to $\mathbf{u}$. A kinematic chain denoted $C/C/$, which generates such a bond is made of a serial setting of two cylindrical pairs C with parallel axes. The body that intermediates the distal bodies undergoes passive 1-dof translation. The passive mobility can be eliminated in several ways. For example we can write $\{C(\mathrm{A},\mathbf{u})\}.\{C(\mathrm{B},\mathbf{u})\} = [\{H(\mathrm{A},\mathbf{u},\ \mathrm{p}).\ \{T_1(\mathbf{u})\}].[\{T_1(\mathbf{u})\}.\{H(\mathrm{B},\mathbf{u},\ \mathrm{q})\}]$ where H indicates a subgroup of helical displacements, and using $\{T_1(\mathbf{u})\}.\{T_1(\mathbf{u})\} = \{T_1(\mathbf{u})\}$ that arises from the product closure in the subgroup $\{T_1(\mathbf{u})\}$, we obtain $\{C(\mathrm{A},\mathbf{u})\}.\{C(\mathrm{B},\mathbf{u})\} = \{H(\mathrm{A},\mathbf{u},\ \mathrm{p})\}.[\{T_1(\mathbf{u})\}.\{H(\mathrm{B},\mathbf{u},\ \mathrm{q})\}]$ which is also equal to $\{H(\mathrm{A},\mathbf{u},\ \mathrm{p})\}.\{C(\mathrm{B},\mathbf{u})\}$.

That way, the equivalency of the two kinematic chains of the upper part of figure 1 can be proven. The other equivalencies that can be use for the synthesis of SPM with no idle pair are shown below. They are symbolized by $H^{O}/C/$, $H^{O}/(RH)/$, $H^{O}/(RH)/$, $H^{O}/(RH)/$ where ^{O} means that the axis passes through O and ( ) indicates the equivalency with a C pair, / / the axis parallelism.

The product $\{C(\mathrm{N},\mathbf{u})\}.\{T_2(\mathrm{Pl})\}$ where $\{T_2(\mathrm{Pl})\}$ designates the subgroup of planar translation parallel to Pl is a 3 dimensional manifold if $\mathbf{u}$ is parallel to Pl. As a matter of fact, then the intersection $\{C(\mathrm{N},\mathbf{u})\}\cap\{T_2(\mathrm{Pl}) = \{T_1(\mathbf{u})\}$ represents a passive 1-dof motion. The passive motion can readily be eliminated considering the product closure in the subgroups $\{C(\mathrm{N},\mathbf{u})\}$ and $\{T_2(\mathrm{Pl})\}$: $\{C(\mathrm{N},\mathbf{u})\} = \{H(\mathrm{N},\mathbf{u}, \mathrm{p})\}.\{T_1(\mathbf{u})\}$, $\forall \mathrm{p} \in \Re$, and $\{T_2(\mathrm{Pl})\} = \{T_1(\mathbf{u})\}.\{T_1(\mathbf{v})\}$, $\mathbf{v} \neq \mathbf{u}$ and $\mathbf{v}//\mathrm{Pl}$, and therefore

$$\{C(\mathrm{N},\mathbf{u})\}.\{T_2(\mathrm{Pl})\} = \{H(\mathrm{N},\mathbf{u}, \mathrm{p})\}.\{T_1(\mathbf{u})\}.\{T_1(\mathbf{u})\}.\{T_1(\mathbf{v})\}$$
$$= \{H(\mathrm{N},\mathbf{u}, \mathrm{p})\}.\{T_1(\mathbf{u})\}.\{T_1(\mathbf{v})\} = \{H(\mathrm{N},\mathbf{u}, \mathrm{p})\}.\{T_2(\mathrm{Pl})\}$$

This last set equality is illustrated by the figure 2. The equivalent kinematic pair sequences are denoted $C\!\!\not\,\underline{PP}$ and $H\!\!\not\,\underline{PP}$ where the underline indicates a planar system and $\not\,$ an axis parallel to the plane. In the type synthesis of new SPM with no inactive pair nor passive motion, only the sequence $H^{O}\!\!\not\,\underline{PP}$ will be relevant.

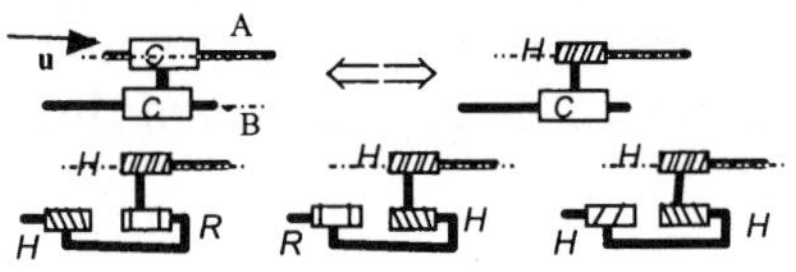

Figure 1. Equivalencies of C/C/

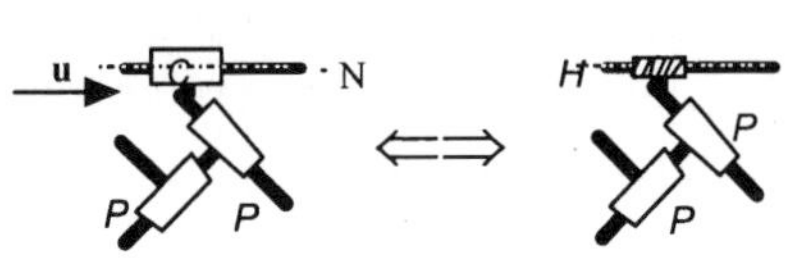

Figure 2. Equivalency of $C\!\!\not\,\underline{PP}$

4. Family of the limb type $R^{O}R^{O}C^{O}/C/$

We demonstrated that the dof of a subchain $C/C/$ is 3; hence a limb of type $R^{O}R^{O}C^{O}/C/$ produces a 5-dof kinematic bond. A limb $R^{O}R^{O}C^{O}/C$ generates a 5-dof motion set that contains the 3-dof motion set generated by $R^{O}R^{O}R^{O}$, which is made of spherical rotations around O. Consequently we can construct SPM with three limbs of the type $R^{O}R^{O}C^{O}/C/$ or analogous types, namely $R^{O}R^{O}C/C^{O}/$, $R^{O}C^{O}/C/R^{O}$. Replacing C^{O}/C by its kinematic equivalencies with no passive mobility (see figure 1), we obtained the SPM limbs of figure 3.

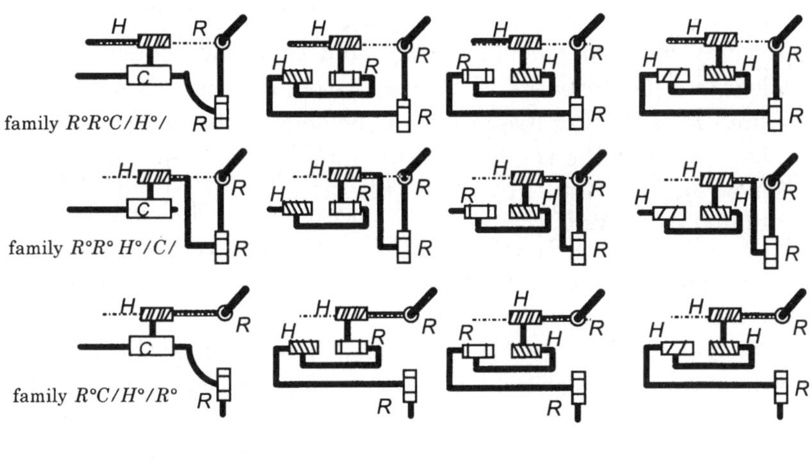

Figure 3. Limbs of the general family $R^{O}R^{O}C^{O}/C/$

5. Family of the limb type $R^{O}R^{O}C^{O}\rfloor/\underline{PP}$

In a limb of type $R^{O}R^{O}C^{O}\rfloor/\underline{PP}$ two R pair axes and a C pair axis are converging at the fixed center O. As previously demonstrated $C^{O}\rfloor/\underline{PP}$ produces 3-dof motion between the distal bodies, hence the limb $R^{O}R^{O}C^{O}\rfloor/\underline{PP}$, generates a 5-dof displacement set, which contains the 3-dof set of spherical rotations around O. The SPM limbs of the family $R^{O}R^{O}C^{O}\rfloor/\underline{PP}$ are described by the combinations : $R^{O}R^{O}\underline{PP}\,C^{O}\rfloor$, $R^{O}R^{O}C^{O}\rfloor/\underline{PP}$, $R^{O}C^{O}\rfloor/\underline{PP}\,R^{O}$. The elimination of the passive mobility in the limb bond (see figure 2), is readily obtained replacing C^{O} by H^{O}. The first three sketches of the figure 4 show these structural types and the following two drawings are special geometrical arrangements of the new limb architectures.

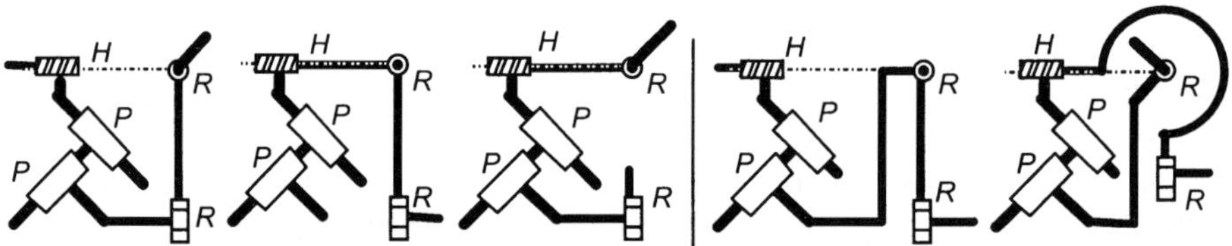

$R^{\circ}R^{\circ}\underline{PP}\,\underline{H^{\circ}}/$ $R^{\circ}R^{\circ}\,\underline{H^{\circ}}/\underline{PP}$ $R^{\circ}\,\underline{H^{\circ}}/\underline{PP}\,R^{\circ}$ [$R^{\circ}R^{\circ}\underline{PP}\,\underline{H^{\circ}}/$] [$R^{\circ}\,\underline{H^{\circ}}/\underline{PP}\,R^{\circ}$]

Figure 4. Family of the limb type $R^{\circ}R^{\circ}\,\underline{PP}\,C^{\circ}\rfloor$

5.1 Family $R^{O}\underline{G}\,H^{O}\rfloor$, family $H^{O}\rfloor/\underline{G}\,H^{O}\rfloor$ and subfamilies

5.1.1 Family $R^{O}\underline{G}\,H^{O}\rfloor$

In a special geometrical arrangement of a limb $R^{O}R^{O}\underline{PP}\,H^{O}\rfloor$, the second pair R^{O} may be perpendicular to $\underline{PP}$, and then the sequence $R^{O}\underline{PP}$ is equivalent to a planar pair G. That way we obtain a new family $R^{O}\underline{G}\,H^{O}\rfloor$. In the limb $R^{O}\underline{G}\,H^{O}\rfloor$ the fixed center O belongs to the H pair axis, which has to be parallel to the G plane. All the limbs of the family $R^{O}\underline{G}\,H^{O}\rfloor$ that are derived substituting G by other generators of planar gliding motion, are exhibited in figure 5.

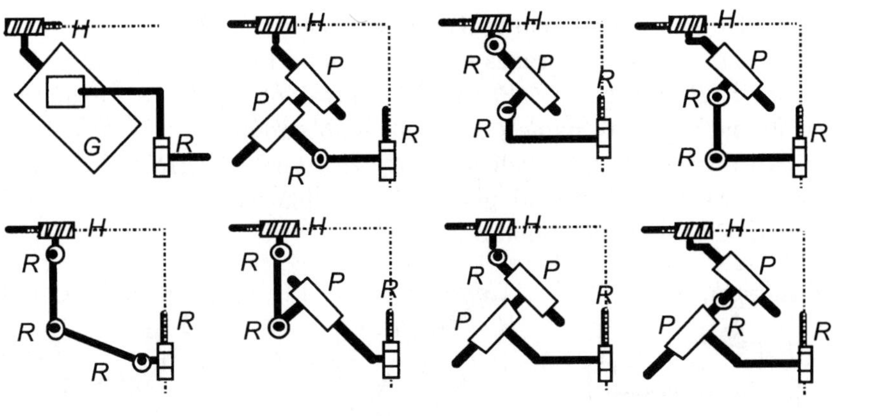

Figure 5. Subfamily $R^{o}\underline{G}\,H^{o}\rfloor$

5.1.2 Family $H^{o}\underline{G}\,H^{o}\rfloor$

In a limb $H^{o}\rfloor\underline{RPP}R^{o}$, the adjacent pairs $\underline{PR}^{o}$ may become equivalent to a $C^{o}\rfloor$pair (fig.6(a) and 6(b)). The $C^{o}\rfloor$pair can be replaced by $(\underline{P}H^{o}\rfloor)$. The resulting limb (figure 6(c)) has the type $H^{o}\rfloor\underline{RP(P}H^{o}\rfloor)$ which is equivalent to $H^{o}\rfloor\underline{G}\,H^{o}\rfloor$ (figure 6(d)). Indeed $\underline{RPP}$ generates planar motion.

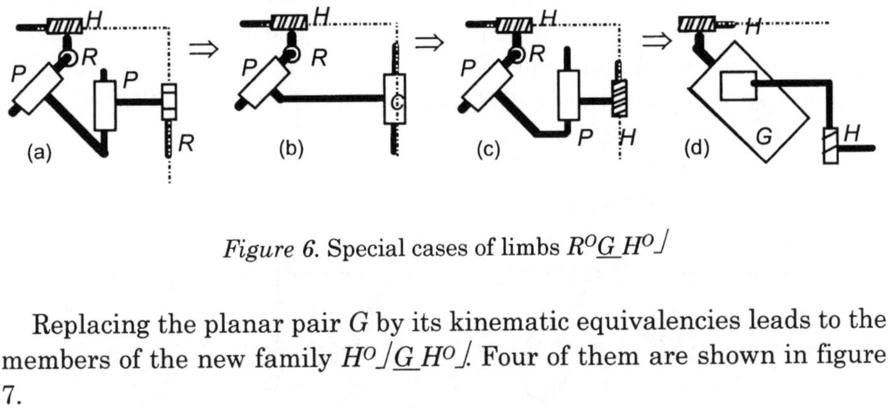

Figure 6. Special cases of limbs $R^{o}\underline{G}\,H^{o}\rfloor$

Replacing the planar pair G by its kinematic equivalencies leads to the members of the new family $H^{o}\rfloor\underline{G}\,H^{o}\rfloor$. Four of them are shown in figure 7.

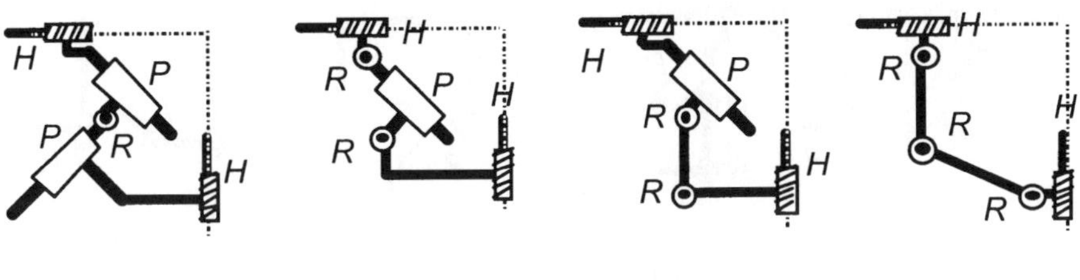

Figure 7. 4 examples of the subfamily $H^{o}\rfloor G\,H^{o}\rfloor$

5.1.3 Subfamilies with cylindrical motion

In special geometrical cases of the previous limbs of type $R^{O}\underline{G}\,H^{O}\rfloor$ a sequence $R^{O}\underline{P}$ may be $(R^{O}\rfloor\underline{P})$ that is equivalent to $C^{O}\rfloor$. An example is given in the figure 6, namely it is the limb subfamily $C^{O}\rfloor\underline{PR}\,H^{O}\rfloor$. We have two analogous subfamilies $C^{O}\rfloor\underline{RP}\,H^{O}\rfloor$ and $C^{O}\rfloor\underline{RR}\,H^{O}\rfloor$ (figure 8). Furthermore the cylindrical pair can be replaced by its equivalencies thus leading to more combinations.

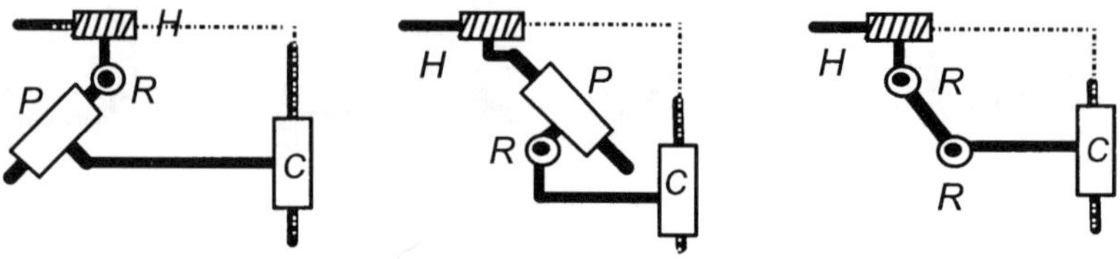

Figure 8. Subfamilies with cylindrical motion

The complementary subfamily $C^{O}\rfloor\underline{R}\,C^{O}\rfloor$ is already described by Karouia and Hervé, 2002b.

5.2 Family $R^{O}H^{O}\rfloor\underline{G}$

In a limb $R^{O}H^{O}\rfloor\underline{PP}\,R^{O}$, the second R^{O} pair may be perpendicular to the plane of $\underline{PP}$. Then the sequence $\underline{PP}\,R^{O}$ becomes $\underline{PPR^{O}}$, which is equivalent to a planar pair G. We obtain a new family $R^{O}H^{O}\rfloor\underline{G}$ employing the equivalencies of G. In a limb of the family $R^{O}H^{O}\rfloor\underline{G}$ the H pair axis passes through the fixed point O and is parallel to the plane of G. The limbs of the family $R^{O}H^{O}\rfloor\underline{G}$, are deducted by replacing G by equivalent generators of planar gliding motion (figure 9).

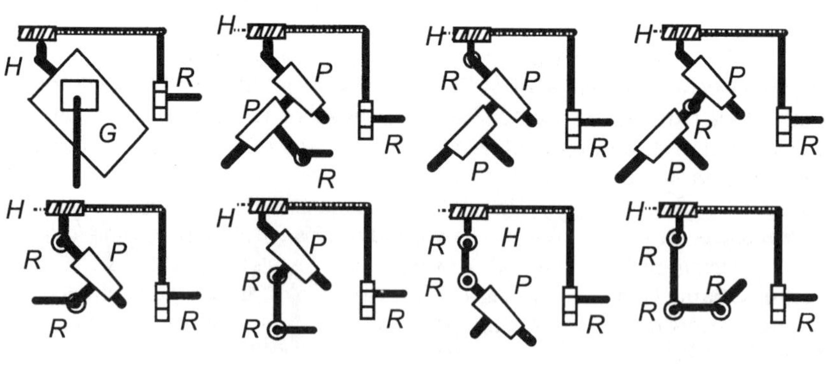

Figure 9. Family $R^{O}H^{O}\rfloor\underline{G}$

5.3 Family $R^{o}\underline{C}^{o}\lrcorner\underline{P}\,H^{o}\lrcorner$ and family $\underline{C^{o}P}\,H^{o}\lrcorner R^{o}$

In a limb of the family $R^{o}R^{o}\underline{PP}\,H^{o}\lrcorner$, the sequence $R^{o}\underline{P}$ may become ($R^{o}\underline{P}$) equivalent to $\underline{C}^{o}\lrcorner$ and, we obtain the subfamily $R^{o}\underline{C}^{o}\lrcorner\underline{P}\,H^{o}\lrcorner$ (figure 10 left) and its special subfamily $R^{o}H^{o}\lrcorner\underline{PP}\,H^{o}\lrcorner$ (figure 10 right) derived from $R^{o}(H^{o}\lrcorner\underline{P})\underline{P}\,H^{o}\lrcorner$ (figure 10 midle).

By the same way, a limb of the type $R^{o}\underline{PP}H^{o}\lrcorner R^{o}$, may be ($R^{o}\underline{P}$) $\underline{P}H^{o}\lrcorner R^{o}$, which leads to the subfamily $\underline{C}^{o}\lrcorner\underline{P}\,H^{o}\lrcorner R^{o}$ (figure 11).

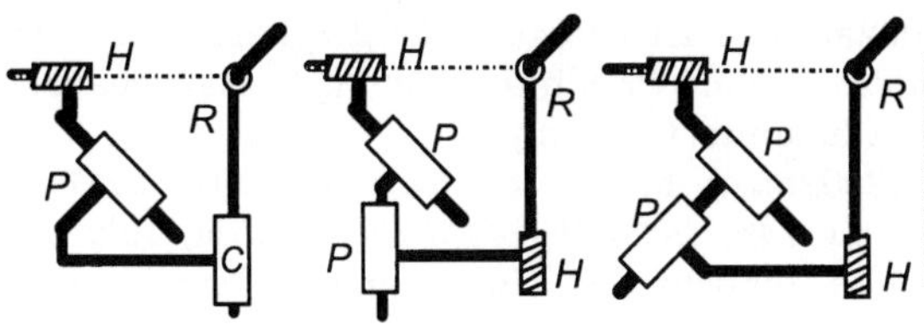

Figure 10. Subfamily $R^{o}\underline{C}^{o}\lrcorner\underline{P}\,H^{o}\lrcorner$

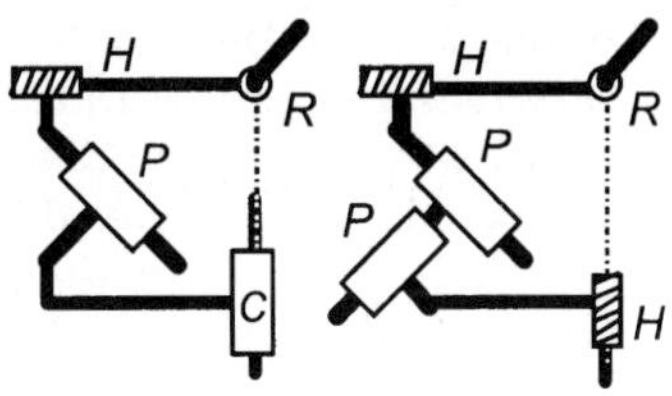

Figure 11. Subfamily $\underline{C}^{o}\lrcorner\underline{P}\,H^{o}\lrcorner R^{o}$

6. Conclusions

The paper describes new special SPM limbs with no inactive pair nor passive motion. In the new special 5-dof limbs, two generators of two dependent displacement Lie subgroups can be identified and the passive motion that results of the dependency is eliminated. Furthermore particular geometrical arrangements yield several subfamilies of the limb main families.

References

Hervé, J.M. (1978), Analyse structurelle des mécanismes par groupe des déplacements, *Mechanism and Machine Theory*, vol.13, pp.437-450.

Asada, H. and Cro Granito, J.A. (1985), Kinematic and static characterization of wrist joints and their optimal design, *Proceedings IEEE Int. Conf. On Robotics and Automation*, St Louis, pp. 244-250.

Gosselin, C. and Angeles, J. (1989), The optimum kinematic design of a spherical three-degree-of-freedom parallel manipulator, *ASME Journal of Mechanism, Transmission and Automation in Design*, vol.111, no.2, pp. 202-207.

Gosselin, C. and Hamel, J.F. (1994), The agile eye : a high performance three-degree-of-freedom camera-orienting device, *IEEE Int. conference on Robotics and Automation,* San Diego, pp. 781-787.

Merlet, J.-P. (1997), *Les robots parallèles*, Hermes, Paris.

Leguay-Durand, S. (1998), Conception et optimisation de mécanismes parallèles à mobilité restreinte, *Doctoral thesis*, June 23, Ecole Nationale Supérieure d'Aéronautique et de l'Espace, Toulouse, France.

Hervé, J.M. (1999), The Lie group of rigid body displacements, a fundamental tool for mechanism design, *Mechanism and Machine Theory*, vol. 34, pp. 719-730.

J.M., Selig (2000), *Geometrical foundations of robotics*, World Scientific, Singapore.

Karouia, M. and Hervé, J.M. (2000), A Three-DOF Tripod For Generating Spherical Rotation, *Advances in Robot Kinematics*, (ed. J. Lenarcic and M.L. Husty), Kluwer Academic Publishers, pp. 395-402.

Di Gregorio; R. (2001), A new parallel wrist employing only revolute pairs, the 3-RUU wrist, *Robotica*, vol.19, no. 3, pp. 305-309.

Karouia, M., and Hervé, J.M. (2002a), An orientational 3-dof parallel mechanism, *Proceedings Parallel Kinematics Seminar PKS'2002*, (Ed. H. R. Neugebauer), April 23-25, 2002, pp. 139-150, Chemnitz, Germany.

Karouia, M., and Hervé, J.M. (2002b), A family of novel orientational 3-dof parallel robots, *Proceedings of the 14th CISM-IFToMM Symposium*, July 1-4, Udine, Italy, *RoManSy 14* (Ed. G. Bianchi, J.C. Guinot and C. Rzymkowski), Springer Vienna, pp. 359-368.

Hervé, J.M. and Karouia, M. (2002c), The novel 3-RUU wrist with no idle pair, *Proceedings of the Workshop on Fundamental Issues and Future Research Directions for Parallel Mechanisms and Manipulators*. (ed. Clément M. Gosselin et Imme Ebert-Uphoff), Oct. 3-4, Québec, Canada, pp. 284-286.

Kong, X.-W. and Gosselin, C.M. (2002), Type synthesis of 3-dof spherical parallel manipulators based on screw theory, *Proceedings of ASME-DETC'02*, Montréal, Canada, 29 sept.-2 Oct.

Di Gregorio, R. (2002), The 3-RRS wrist: a new, very simple and not overconstrained spherical manipulator, *Proceedings of ASME-DETC'02*, Montréal, Canada, 29 sept.-2 Oct.

Karouia, M. (2003), Conception structurale de mécanismes parallèles sphériques, *Doctoral thesis*, Sept.25, Ecole Centrale Paris, France.

THE DESIGN OF A NOVEL PITCH-ROLL WRIST WITH SPHERICAL CAM-ROLLER PAIRS

T.G. Ghuneim, J. Angeles and S.P. Bai
Mechanical Engineering Department &
Centre for Intelligent Machines
McGill University, Canada
{angeles, mspbai}@cim.mcgill.ca

Abstract The design of an innovative robotic wrist with spherical cam-roller pairs is reported here. The geometry of the wrist is determined based on cam-roller mechanisms, which are back-driveable. In this mechanism, two roller carrying discs (RCDs) drive a conjugate-cam subassembly to produce roll and pitch motions. Kinematically equivalent to a bevel-gear differential mechanism, the designed mechanism offers low friction, low backlash, and high stiffness.

Keywords: Robotic Wrist, Cam Synthesis, Differential Mechanism, Pitch-Roll Wrist

1. Introduction

Robotic wrists are commonly used in manipulators for applications that require a large dexterous workspace [Howard, 2002; Tsai, 1988; Yang and Rauchfuss, 2001]. As a robotic wrist is to be attached to a moving arm, it is desirable to design such wrists with light weight, low backlash and high stiffness. Some industrial wrists, like the Cincinnati Milacron T^3 and the Bendix wrist, were designed with bevel-gear trains in a differential array [Tsai, 1988]. For bevel-gear wrists, the kinematics is simple and the manufacturing cost is relatively low if gears with straight teeth are used. However, such gear trains function under rolling and sliding, the latter bringing about noise and vibration. To alleviate these drawbacks, spiral teeth can be used, but at such a cost that, to our knowledge, no robot manufacturer has adopted spiral bevel gears for wrist design. Moreover, unavoidable backlash introduces unmodeled dynamics, which mars robot performance. Attempts have been made to design wrists with different approaches for a better performance. Wiitala and Stanisic proposed a wrist design based on a symmetrically actuated spherical eight-bar linkage [Wiitala and Stanisic, 2000], which can eliminate the singularity within its hemisphere workspace. A three degree-of-freedom (three-dof) parallel spherical wrist, called "Agile Wrist", was

J. Lenarčič and C. Galletti (eds.), On Advances in Robot Kinematics, 381-388.

developed for high accuracy and high precision [Bidault et al., 2001]. The Agile Wrist isotropic design is based on Gosselin's "Agile Eye" [Gosselin and Hamel, 1994], whose link shapes were revised for both higher load-carrying capacity and lower likelihood of link-interference. Other wrist designs can also be found in [Howard, 2002; Yang and Rauchfuss, 2001].

In this paper, a wrist for pitch-roll motion with spherical cam-rollers is introduced. Cam mechanisms outperform gears in some aspects, for cams offer lower friction, lower backlash, and higher stiffness. In addition, since cam mechanisms can be housed in a relative small space, this feature makes them more attractive to robotic applications. In our design, a cam mechanism with conic rollers is adopted to generate two-degree-of-freedom motions: pitch and roll. Two roller-carrying discs (RCDs) transmit the rotations of the two motors. A conjugate cam subassembly carrying the gripper is driven by the RCDs to produce pitch-roll motion. Kinematically equivalent to a bevel-gear differential mechanism, our mechanism offers an attractive alternative.

2. Principles of Cam Design

A cam mechanism comprises a minimum of three elements: frame, cam and follower. The frame is the base on which the moving parts are mounted; the cam and the follower are separately coupled to the frame via lower pairs, while the cam and follower are coupled via a higher kinematic pair. A spherical cam mechanism with an added conic roller placed between cam and follower is shown in Fig. 1. In this figure, all four elements move with relative pure rolling, their instant axes of rotation intersecting at one common point, the center of the sphere; moreover, the roller is coupled to the follower via a revolute pair. The higher kinematic pair is that between the cam and the roller. The addition of the roller serves other purposes besides eliminating any sliding friction (which may or may not be present): it allows for a smaller pressure angle and for a simpler geometry of the follower.

The principles used here for cam-mechanism synthesis are those pertaining to indexing cam mechanisms (ICM), which are used to transmit intermittent motion [Angeles and González-Palacios, 2002]. An ICM essentially consists of one cam and multiple rollers connected to one follower. In practice, coaxial conjugate cams are used to drive corresponding sets of rollers mounted on two opposite sides of the roller-carrier. As each roller comes into contact with one cam, a temporary single cam mechanism is formed; as contact is lost on one follower side, a roller

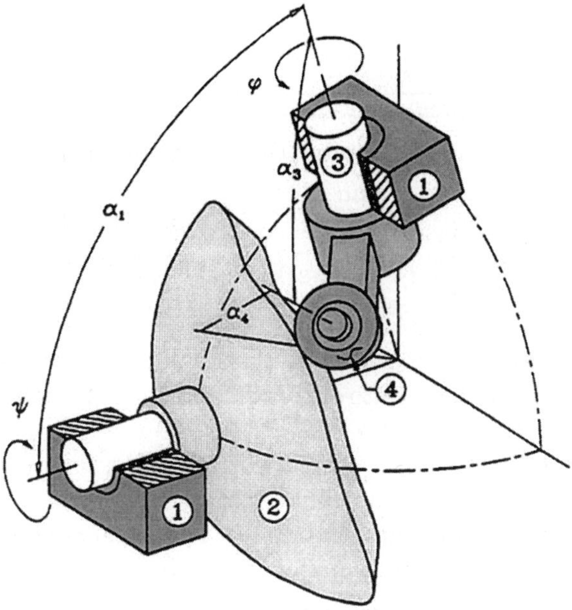

Figure 1. Four-link spherical cam mechanism

from the other side begins to engage the second cam, thus guaranteeing positive action throughout the whole cycle.

3. Cam Profile Generation

In order for the mechanism to transmit motion continuously as a gear mechanism would, the relation between the cam output ψ and the follower input ϕ is constrained to be linear and behave as:

$$\phi = \frac{\pi}{N}(N - 1 + \frac{\psi}{\pi}), \quad N = \text{number of rollers} \tag{1}$$

such that the follower makes one turn for every N turns of the cam, thereby producing a speed reduction of $N : 1$.

Based on the principles stated in Section 2, the Cartesian coordinates of a point on the sphere of radius l traces a spherical curve that determines the cam profile. For perpendicular axes of cam and follower, α_1, illustrated in Fig. 1, is 90°, the coordinates of a point on the abovementioned spherical curve being [González-Palacios and Angeles, 1993]

$$\begin{aligned} x &= l(-k_1\sin\psi + k_2\cos\psi) \\ y &= l(-k_1\cos\psi + k_2\sin\psi) \\ z &= l(\cos\theta_2\cos(\theta_3 - \alpha_4) - \sin\theta_2\sin(\theta_3 - \alpha_4)\cos\delta) \end{aligned} \tag{2}$$

where

$$\theta_2 = \pi - \arctan(1/N) \tag{3}$$

$$\theta_3 = \arctan\left(\frac{\sqrt{(\sin\theta_2\cos\psi\sin\alpha_3 + cos\alpha_3\cos\theta_2)^2 + (\sin\alpha_3\sin\phi)^2}}{\cos\alpha_3\cos\theta_2 - \cos\psi\sin\alpha_3\cos\theta_2}\right) \tag{4}$$

$$\delta = \arctan\left(\frac{\sin\alpha_3\sin\phi}{\sin\alpha_3\sin\theta_2\cos\phi + \cos\alpha_3\cos\theta_2}\right) \tag{5}$$

$$k_1 = \sin\theta_2\cos(\theta_3 - \alpha_4) + \cos\theta_2\sin(\theta_3 - \alpha_4)\cos\delta \tag{6}$$

$$k_2 = -\sin(\theta_3 - \alpha_4)\sin\delta \tag{7}$$

In all equations, α_3 is the angle between follower axis and roller axis, while α_4 is the angle between roller axis and its outer surface, which is in contact with the cam.

Upon varying l in the $l_{min} \leq l \leq l_{max}$ range, the forgoing coordinates generate a conical surface, the contact surface between cam and roller. A cam profile generated with the above equations is shown in Fig. 2, which is illustrated on a unit sphere.

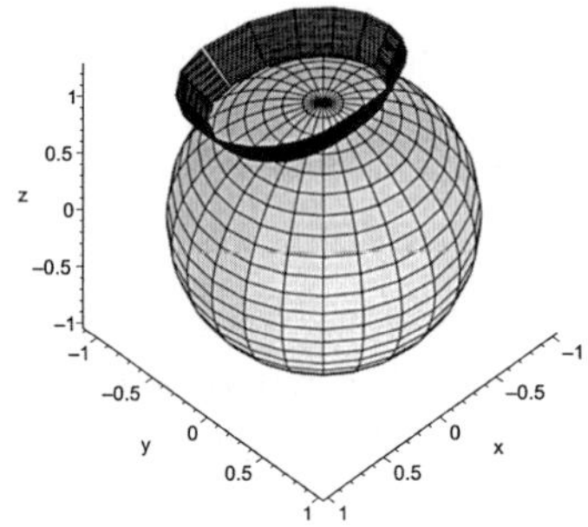

Figure 2. 3D view of a cam profile displayed on a unit sphere

4. Mechanism Design

4.1 Design Parameters

Once the design parameters have been chosen, cam profile data are imported into Pro/Engineer for further development and integration into the design of the mechanism. In choosing the parameters, the negative action (the follower drives the cam) pressure angle [Angeles and López-Cajún, 1991; González-Palacios and Angeles, 1993] has to be minimized with the aim of generating a feasible mechanism, while allowing enough space for the three cam shafts, in order to have the mechanism statically and dynamically balanced.

The parameters used in our design are

$$N = 3, \ \ \alpha_3 = 126.9^\circ, \ \ \alpha_4 = -22^\circ \tag{8}$$

While the cam works under negative action, the pressure angle depends on the velocity direction of the cam contact point. For the cam with the parameters given in Eq. (8), the pressure angle for negative action is in the range $56^\circ \sim 71.6^\circ$, as plotted in Fig. 3.

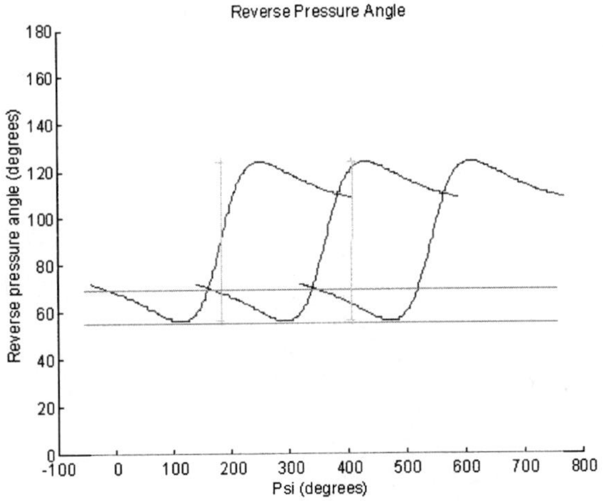

Figure 3. Pressure angle for negative action

4.2 Virtual Prototype

The pitch-roll mechanism under design is illustrated in Fig. 4. The two input shafts connected to motors (not shown in the figure) drive independently each of the identical roller-carrying discs like the one shown in Fig. 5, which produce two outputs, the pitch and the roll motions of the gripper. Three cam-shaft subassemblies, shown in Fig. 6, are used in order to balance the mechanism, each subassembly being a pair of conjugate spherical cams.

In designing the mechanism, care was taken to prevent interference of parts and ease of assembly, as well as minimizing machining work. Most of the parts can be manufactured using general-purpose CNC machine tools. The cams are the most complex parts and require five-axis CNC machine tools, thus making them the most expensive parts.

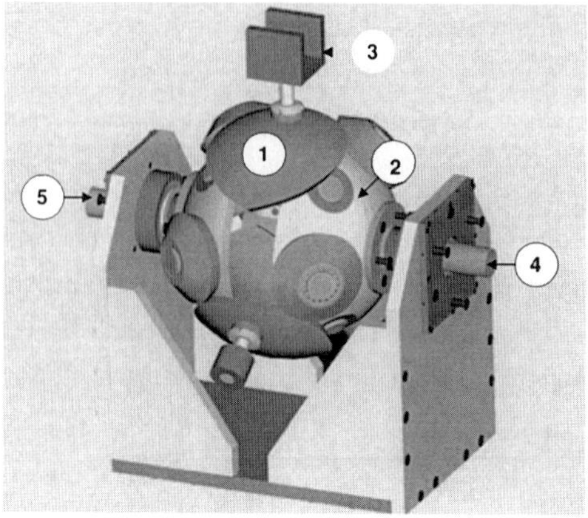

Figure 4. Pro/E model of pitch-roll mechanism: 1. cam; 2. RCD; 3. gripper; 4&5. input shafts connected to two motors

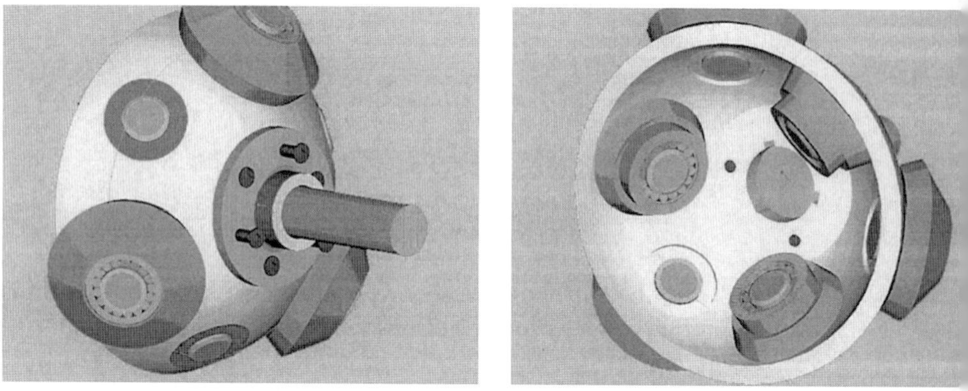

(a) (b)

Figure 5. A roller-carrying disc: (a) front view; (b) rear view

5. Conclusions

In this paper, the design of a novel robotic pitch-roll wrist that is based on spherical cam mechanisms was reported. The wrist has two inputs, the rotation of the roller carrying-discs, and two outputs, the pitch and the roll motions of the gripper. The wrist has all the advantages of a cam mechanism: high stiffness; low backlash; and low friction.

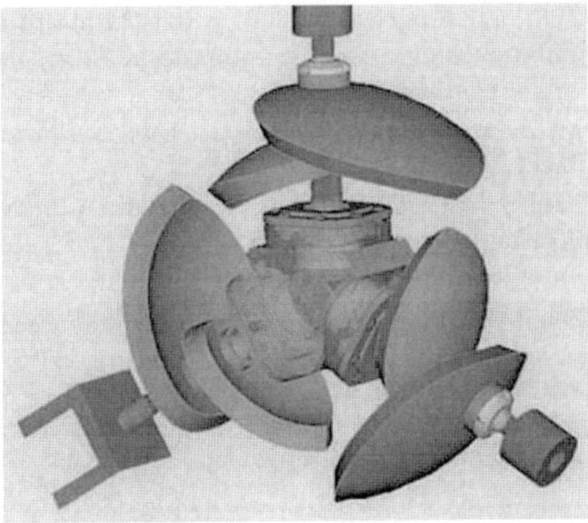

Figure 6. Cam-shaft assembly.

It is noted that the wrist was designed with pitch-roll motion only. The wrist can be used for applications where two-degree-of-freedom rotations are required. For those requiring three-dof, one more rotation can be added to the housing of our wrist.

Our design is not, however, problem-free. The complex geometry of the cams makes them expensive. Tolerances must be the tightest possible; otherwise, interference or backlash will occur. Even so, backlash is likely to occur upon changing roller engagement from one cam to another. As a solution, the mounting of the cam shafts should be specially designed such that the shaft is allowed to undergo submillimetric adjustments to compensate for manufacturing and assembly errors. Such a solution has been applied to planar cam mechanisms, with partial success. The dynamic balancing of the conjugate cams, which are inevitably of different size, is another issue. Further R&D work is currently underway addressing this and other issues.

6. Acknowledgements

The research work reported here was made possible by Grant No. 246488-01 of Canada's Natural Sciences and Engineering Research Council.

References

Angeles, J. and González-Palacios, M. A. (2002), Transmission device, United States Patent, 6,382,038.

Angeles, J. and López-Cajún, C. S.(1991), *Optimization of Cam Mechanisms*, Kluwer Academic, Boston.

Bidault, F., Teng, C. P., and Angeles, J. (2001), Structural optimization of a spherical parallel manipulator using a two-level approach, *Proc. of ASME 2001 Design Engineering Technical Conferences.*

González-Palacios, M. A. and Angeles, J., (1993), *Cam Synthesis*, Kluwer Academic Publishers, Dordrecht.

Gosselin, C. and Hamel, J.-F. (1994), The Agile Eye: A high-performance three-degree-of-freedom camera-orienting device, *Proc. of IEEE Int. Conf. on Robotics and Automation*, pp. 781–786.

Howard, R. D. (2002), Design of a robotic wrist and tool-exchange mechanism for satellite servicing, *Proc. of* 36th *Aerospace Mechanism Symposium*, NASA Glenn Research Center.

Tsai, L.-W. (1988), The kinematics of spatial robotic bevel-gear trains, *IEEE Journal of Robotics and Automation*, 4(2):150–156.

Wiitala, J. M. and Stanisic, M. M. (2000), Design of an overconstrained and dextrous spherical wrist, *ASME Journal of Mechanical Design*, 122:347–353.

Yang, D. and Rauchfuss, J. (2001), A new zero-dimension robot wrist: Design and accessibility analysis, *The International Journal of Robotics Research*, 20(2):163–173.

DESIGN OF A HYBRID MANIPULATOR FOR PAINTING HIGHWAY SIGNAGE

S.-H. Lee, T. A. Lasky, and S. A. Velinsky
Dept. of Mechanical & Aeronautical Engineering
University of California - Davis
shlee@ahmct.ucdavis.edu, talasky@ucdavis.edu, savelinsky@ucdavis.edu

Abstract This paper presents the design of a parallel/serial manipulator aimed at painting signage on highway pavement. The manipulator consists of a planar 2-link parallel mechanism coupled to a translational mechanism. Detailed kinematics are employed to understand the resultant workspace, feasible paths, and manipulability. Based on the analysis, optimal design parameters are selected to allow for the painting of symbols within a full highway lane width.

Keywords: Parallel manipulator, highway maintenance, redundancy resolution

1. Introduction

Parallel manipulators, e.g. Stewart Platform (Stewart 1965), exhibit advantages of high rigidity, accuracy, and load-carrying capacity relative to serial manipulators. Applications include high-precision machining tools (Chae *et al.* 1998), mining machines (Nakashima *et al.* 1993), and motion simulators (Mimura and Funahashi 1995).

In the last two decades, significant research has focused on improving the performance of parallel manipulators. Delta (Pierrot *et al.* 1997) and Hexaglide (Honegger *et al.* 1997) mechanisms provide rapid platform motion by locating heavy actuating components nearer the fixed base, while the hybrid manipulator (Waldron *et al.* 1989) combines a 3-DOF serial mechanism with a 3-DOF parallel one. Stoughton and Arai (1993) provide improvement over the conventional Stewart Platform by modifying the actuator arrangement and structure to optimize a dexterity measure over a subregion. Arai, *et al.* (1990) showed that the conventional Stewart Platform does not have an optimal configuration and suggested a new configuration based on the condition number.

The current research presents a parallel manipulator combined with a sliding mechanism, targeted for painting letters and symbols on the roadway. This hybrid manipulator provides a large dexterous workspace combined with high rigidity, accuracy, and load capacity, and is essentially a planar mechanism consisting of two linear actuators and one sliding mechanism. Painting a letter on the roadway requires a

J. Lenarčič and C. Galletti (eds.), On Advances in Robot Kinematics, 389-398.

rectangular aperture to fit the outline of the letter. With this application-specific focus, one can introduce a *practical workspace* which covers the required tasks, and design the manipulator structure and parameters accordingly (Gosselin and Angeles 1991). The parallel manipulator combined with a sliding mechanism introduces the needed longitudinal freedom to cope with the parallel manipulator's umbrella-shaped workspace to obtain the desired rectangular aperture.

2. Workspace of the parallel manipulator

A painting robot can write a letter or symbol on a roadway without a stencil (Jung and Ravani 1998). The symbol or combination of symbols (e.g. "STOP") on the roadway may typically have an aperture of 4 m width and 2.5 m height. While there are many possibilities, a viable parallel manipulator geometry is shown in Figure 1. Here, $\mathbf{a}_1$ and $\mathbf{a}_2$ denote vectors from the coordinate base to the actuators' bases, $\mathbf{s}_1$ and $\mathbf{s}_2$ are the unit direction vectors of each leg with lengths l_1 and l_2 respectively, and vector $\mathbf{p}$ is to the end-effector of the painting robot.

The parallel manipulator design parameters include minimum leg or actuator length $l_{\min}$, maximum actuator length $l_{\max}$, and separation distance a between actuator installation points on the vehicle.

Based on the manipulator geometry and the painting application of Figure 1, the umbrella-shaped workspace of the parallel manipulator should be optimized to guarantee the desired workspace. The parallel manipulator shown in Figure 1 has an excellent workspace in the x-direction but not in the y-direction. It may be easy to provide the desired workspace width of 4 m, but it is difficult to provide the desired height of 2.5 m. As the painting robot need not reach outside of the letter aperture, the most important workspace is a rectangular workspace—this *practical workspace* satisfying the task requirements will drive the system design. Parameters are first selected to provide workspace width of 4 m (highway lane width). However, a fixed parallel manipulator design will not provide full and optimal coverage of this practical workspace, motivating the hybrid mechanism discussed below.

Using the key design parameters, $l_{\max}$, $l_{\min}$, and a, the workspace aspect ratio, can be adjusted for a given fixed x-direction workspace. The y-axis workspace height H_1 corresponding to a fixed workspace half-width w_x is given by

$$H_1 = \sqrt{l_{\max}^2 - (w_x + a)^2} - \sqrt{l_{\min}^2 - (w_x - a)^2}, \text{ for } a \ge w_x \text{, and} \tag{1a}$$

$$H_1 = \sqrt{l_{\max}^2 - (w_x + a)^2} - l_{\min} \quad \text{for } a < w_x . \tag{1b}$$

For the current case, w_x is fixed at 2.0 m.

3. Sliding mechanism and redundancy resolution

For fixed a, H_1 increases as l_{max} increases. However, increasing l_{max} is not desirable, as the wasted workspace area is simultaneously increased. On the other hand, while decreasing a can increase H_1, this approach has negative implications for manipulability, as noted later. Thus, a sliding mechanism is introduced to improve the y-direction workspace, adding a redundant degree of freedom. Figure 2 shows a combination workspace of the parallel manipulator with the added sliding mechanism. Introducing slide displacement $\mathbf{p}_s$ and a displacement $\mathbf{p}_p$ by the parallel manipulator now relative to the sliding base, the position $\mathbf{p}$ of the end-effector in the fixed frame x - y is given by

$$\mathbf{p} = \mathbf{p}_p + \mathbf{p}_s . \tag{2}$$

Since the moving frame x_0 - y_0 does not rotate, the relation is a simple translation. There are three region types in the workspace: Parallel #1 (P1), Parallel #2 (P2), and Sliding & Parallel (SP). The hatched areas, P1 and P2, can be performed purely by the parallel manipulator. However, the hatched area, P2, cannot cover the remaining practical workspace, as it is restricted to lie inside the umbrella workspace. The left and right upper corners, outside of the umbrella workspace, will be provided by the parallel manipulator and the sliding mechanism simultaneously.

For desired trajectories completely contained in the P2 region, no redundancy resolution is required. If the desired trajectory crosses the boundary between P2 and SP, then a critical problem exists at the boundary of the umbrella-shaped workspace. At this boundary point, q_2, the end-effector meets a kinematical boundary of the parallel manipulator, and it cannot go beyond the boundary, in either the x - or y -direction. While the y -direction can be provided by the sliding mechanism, the x -direction cannot. The x -direction must be provided by the parallel manipulator. A trajectory outside of the umbrella shaped workspace (i.e. in SP) should be covered by the parallel manipulator and the sliding mechanism simultaneously, including redundancy resolution.

Consider a point-to-point motion starting from the lower-right point of the workspace q_1 and ending at the upper-left point of the workspace q_3. Such a trajectory must pass through all region types. Two general path types can provide this point-to-point motion. In the first type (not preferred), the end-effector proceeds to a kinematical boundary point q_2 using the parallel manipulator alone, and traverses the kinematical boundary to a point directly below the desired endpoint, and then the sliding mechanism extends to translate the end-effector to q_3. In the second (preferred) path type, the end-effector proceeds directly from q_1 to q_3 using the parallel and sliding mechanisms simultaneously. The x -

distance of the trajectory is provided by the parallel manipulator, and the parallel manipulator shares the y-distance with the sliding mechanism. The contributions to any y displacement Δy can be expressed as

$$y = \Delta y + y_{\text{offset}} = y_P + y_S + y_{\text{offset}}, \text{ and} \tag{3a}$$

$$y_P = \frac{(H^n - \Delta y^{n-1} H_2)}{H^n} \Delta y, \quad y_S = \frac{H_2}{H^n} \Delta y^n, \quad H = H_1 + H_2, \tag{3b}$$

where y_P is the contribution due to the parallel manipulator, y_S is the contribution of the sliding mechanism, y_{offset} is the displacement from the base to the beginning of the practical workspace with the slide fully retracted, and H_1 and H_2 are the total displacements performed by the parallel manipulator and sliding mechanism, respectively, to achieve a full y displacement H. Integer n is the combination order—its effect will be considered herein.

For the first path type, at boundary q_2 one actuator length reaches its limit and is saturated, and the other actuator length begins decreasing according to the kinematic constraint, and the sliding mechanism starts to move. At the switching point the joint velocities are discontinuous, which can lead to control problems. If trajectory tracking is desired, rather than just point-to-point motion, this path type will not suffice. For the second path, all joint velocities are continuous and trajectory tracking can be supported.

4. Manipulability of the hybrid manipulator

Manipulability is computed based on the Jacobian matrix, which relates joint rates and end-effector velocity. The end-effector vector $\mathbf{p} = \{p_x \quad p_y\}^T$ is rewritten as

$$\mathbf{p} = \mathbf{p}_p + \mathbf{p}_s = \mathbf{a}_i + l_i \mathbf{s}_i + \mathbf{p}_s, \tag{4}$$

where $\mathbf{a}_i$ denotes the vector pointing to the ith prismatic actuator's base position based on the moving frame at the base of the parallel manipulator, and l_i and $\mathbf{s}_i$ denote the length and unit direction vector of the ith actuator, respectively. The vector $\mathbf{p}_s$ denotes the displacement provided by the sliding mechanism. Equation (4) can be differentiated with respect to time as

$$\dot{\mathbf{p}} = l_i (\boldsymbol{\omega}_i \times \mathbf{s}_i) + \dot{l}_i \mathbf{s}_i + \dot{\mathbf{p}}_s, \tag{5}$$

where since $\mathbf{a}_i$ is a constant kinematic parameter, its time derivative becomes zero. Equation (5) includes a passive component $\boldsymbol{\omega}_i$. In order to eliminate the passive component, taking the inner product of both sides of equation (5) with $\mathbf{s}_i$ yields

$$\mathbf{s}_i^T \dot{\mathbf{p}} = l_i \mathbf{s}_i^T (\boldsymbol{\omega}_i \times \mathbf{s}_i) + \dot{l}_i \mathbf{s}_i^T \mathbf{s}_i + \mathbf{s}_i^T \dot{\mathbf{p}}_s = \dot{l}_i + \mathbf{s}_i^T \dot{\mathbf{p}}_s . \tag{6}$$

Since the sliding vector $\mathbf{p}_s$ has only a y-direction component, its time derivative is $\{0 \quad 1\}^T \dot{d}$, introducing scalar sliding velocity $\dot{d}$. Finally, the Jacobian matrix $\mathbf{J}$ is written as

$$\dot{\mathbf{p}} = \begin{bmatrix} \mathbf{s}_1^T \\ \mathbf{s}_2^T \end{bmatrix}^{-1} \begin{Bmatrix} \dot{l}_1 \\ \dot{l}_2 \end{Bmatrix} + \begin{Bmatrix} 0 \\ 1 \end{Bmatrix} \dot{d} = \mathbf{J} \begin{Bmatrix} \dot{l}_1 \\ \dot{l}_2 \\ \dot{d} \end{Bmatrix}, \quad \mathbf{J} = \begin{bmatrix} \begin{bmatrix} \mathbf{s}_1^T \\ \mathbf{s}_2^T \end{bmatrix}^{-1} & \begin{Bmatrix} 0 \\ 1 \end{Bmatrix} \end{bmatrix}. \tag{7}$$

Note that for the second path type, the Jacobian matrix is always as provided in equation (7), while for the first path type, J_{22} becomes zero when the end-effector proceeds from q_2 to q_3.

Manipulability can be considered for each trajectory planning operation. Three principal manipulabilities are typically used in the literature (Murray *et al.* 1994): the minimum singular value of the Jacobian matrix, $\sigma_{\min}(\mathbf{J})$, the inverse of the Condition Number of the Jacobian matrix, $c = \sigma_{\min}(\mathbf{J}) / \sigma_{\max}(\mathbf{J})$, and the determinant of the Jacobian matrix, $\det(\mathbf{J}) = \Pi \sigma_i$. The minimum singular value and condition number of a Jacobian matrix explicitly denote the least ratio and isotropic condition for each axis of velocity transmission, respectively. The determinant of the Jacobian denotes an area of the ellipsoid which is constructed based on each axis of singular value of the Jacobian. In practice, the third manipulability definition, determinant of a Jacobian matrix, is often unreliable. For example, in the case of a large condition number (near 1.0), a large determinant-based manipulability implies good velocity transmission performance with respect to every direction. However, an end-effector near a singular position can still have a large manipulability, if the maximum singular value is very large.

The Jacobian-based manipulability is configuration-dependent, and the test trajectory from lower-right to upper-left of the workspace is used to compute manipulability for both path types—independent and n^{th}-order combination. For the first path, all three manipulabilities show an expected sharp increase at the boundary point. However, those for the second path are continuous. In addition, minimum singular value and condition number manipulability for the second path are slightly better.

Figure 3 provides the sliding mechanism motion profile for several combination ratios of sliding for integer $n \in [1,5]$ of equation (3). Case $n = 1$ is the linear case, and as n increases, the distance profile shared by the sliding mechanism approaches that of the independent motion case. Here, independent motion mode implies sequential movement by the parallel and sliding mechanisms. In this sense, the linear and independent motion modes represent the extreme cases.

Figure 4 provides the manipulability for varying order combination ratio. As expected, for increasing n, the manipulability of the nonlinear combined motion mode begin to resemble that of the independent motion mode. However, while manipulability for the independent mode are discontinuous at the boundary point, those of the combined mode are smooth. In addition, while the condition number manipulability of the combined mode appears to approach a limit as the order of the combination ratio equation (3b) increases, it is decreasing (similarly for the singular value). However, higher order combination ratios may be less desirable based on an increased computational load.

5. Prototype design

For the current application, l_{max} and l_{min} are assumed as 5 m and 2.5 m, respectively, i.e. 100% extension is assumed. The separation a is determined below based on manipulability. With the resulting parallel-only umbrella workspace, one can then determine the required distance H_2 to be provided by the sliding mechanism to achieve the full practical workspace. Since the sliding mechanism is independent of the parallel manipulator it can be designed independently.

For the same test trajectory used above, Figure 5 shows several manipulability ellipsoids along the test trajectory with respect to distance a. As the distance a decreases, the eccentricity of the ellipsoids increases. This means that the increased determinant-based manipulability may be misleading and that while beyond the scope of this paper, force transmission aspects must be considered. Table 1 provides selected design parameters suitable for the given workspace based on the condition number manipulability. Figure 6 shows the final workspace and actuator arrangement.

6. Conclusions

This work has developed a hybrid manipulator painting robot, motivated by application to roadway sign painting. Here, the end-effector does not have to point outside of the symbol aperture, motivating use of the *practical workspace* concept.

For the hybrid manipulator, manipulability analysis considering least singular value, condition number, and determinant of a Jacobian matrix was performed. It was shown that optimization based on condition number manipulability was more appropriate for this type of mechanism, and appropriate parameters were selected leading to design of a specific hybrid manipulator for the painting application.

References

Arai, T., Cleary, K., Nakamura, T., Adachi, H., and Homma, K. (1990), "Design, analysis and construction of a prototype parallel link manipulator," *Proceedings of IROS*, pp. 205-212.

Chae, S., Chung, C., Yang, H., Choi, Y., and Lee, S. (1998), "Adaptive control of a 6 D.O.F. Stewart platform based machine tool," *4th Intl. Conference on Motion and Vibration Control*, Zurich, Switzerland, pp. 461-466.

Gosselin, C., and Angeles, J. (1991), "A global performance index for the kinematic optimization of robotic manipulators." *ASME Journal of Mechanical Design*, no. 3, vol. 113, pp. 220-226.

Honegger, M., Codourey, A., and Burdet, E. (1997), "Adaptive control of the Hexaglide, a 6-DOF parallel manipulator," *IEEE International Conference on Robotics and Automation*, pp. 543-548.

Jung, S., and Ravani, B. (1998), "Precise position control of stenciling robot manipulator using neural networks," *IEEE International Conference on Systems, Man, and Cybernetics*, San Diego, CA, pp. 3554-3559.

Mimura, N., and Funahashi, Y. (1995), "A new analytical system applying 6 DOF parallel link manipulator for evaluating motion sensation," *IEEE International Conference on Robotics and Automation*, pp. 227-233.

Murray, R.M., Li, Z., and Sastry, S.S. (1994), *A mathematical introduction to robotic manipulation*. Boca Raton, FL, CRC Press.

Nakashima, K., Kataoka, M., Nagata, O., Wada, T., and Arai, T. (1993), "Development of the parallel manipulator," in *Robotics, mechatronics and manufacturing systems*, Elsevier Science, pp. 27-32.

Pierrot, F., Fournier, A., and Dauchez, P. (1997), "Towards a fully parallel 6-DOF robot for high speed applications," *IEEE International Conference on Robotics and Automation*, pp. 1288-1293.

Stewart, D. (1965), "A platform with six degrees of freedom." *Proceedings of the Institute of Mechanical Engineering*, no. 15, vol. 180, pp. 371-386.

Stoughton, R.S., and Arai, T. (1993), "A modified Stewart platform manipulator with improved dexterity." *IEEE Transactions on Robotics and Automation*, no. 2, vol. 9, pp. 166-173.

Waldron, K.J., Raghavan, M., and Roth, B. (1989), "Kinematics of a hybrid series-parallel manipulation system." *ASME Journal of Dynamic Systems, Measurement, and Control*, no. 2, vol. 111, pp. 211-221.

Table 1. Design parameters

Parameter	Specification
Leg length l_{max} , l_{min}	5 m, 2.5 m
H_1 (for parallel)	1.128 m
H_2 (for sliding)	1.372 m
Separation distance a	1.9 m

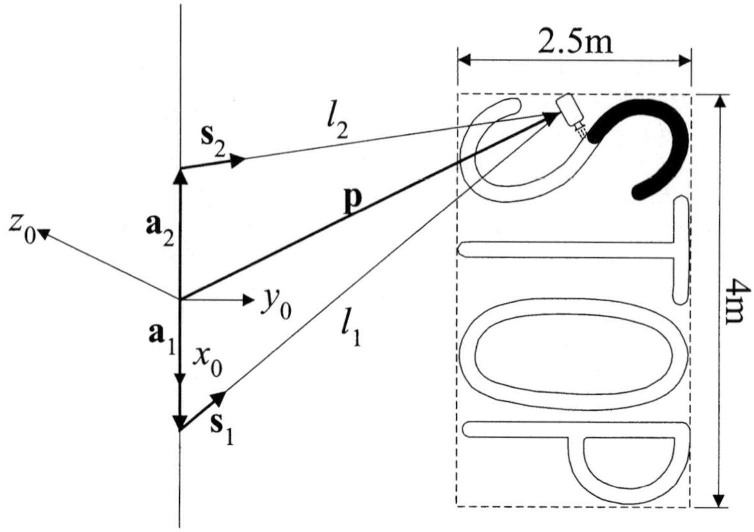

Figure 1. A parallel manipulator for painting roadway symbols

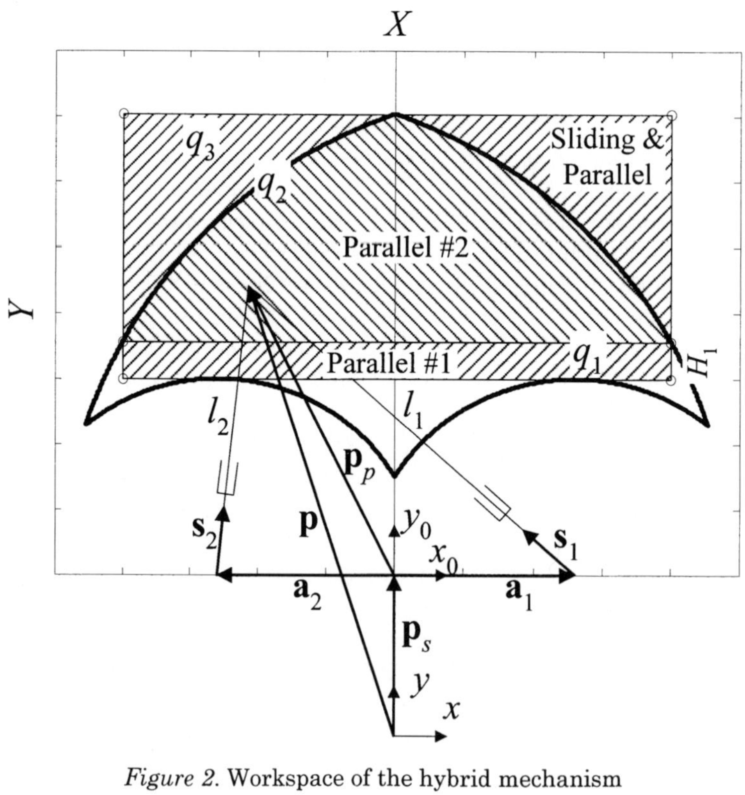

Figure 2. Workspace of the hybrid mechanism

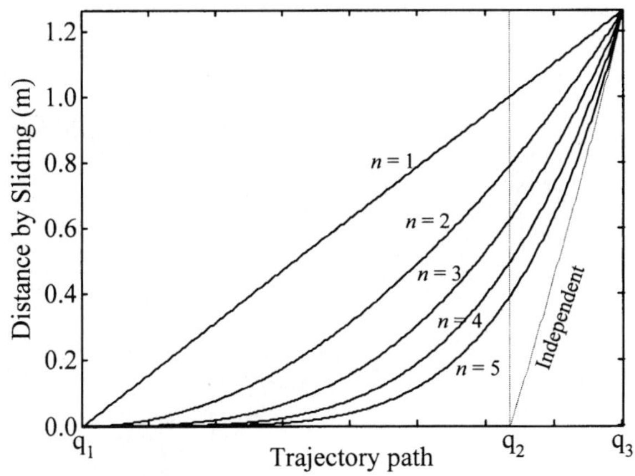

Figure 3. Distance shared by sliding over path w.r.t. combination order n

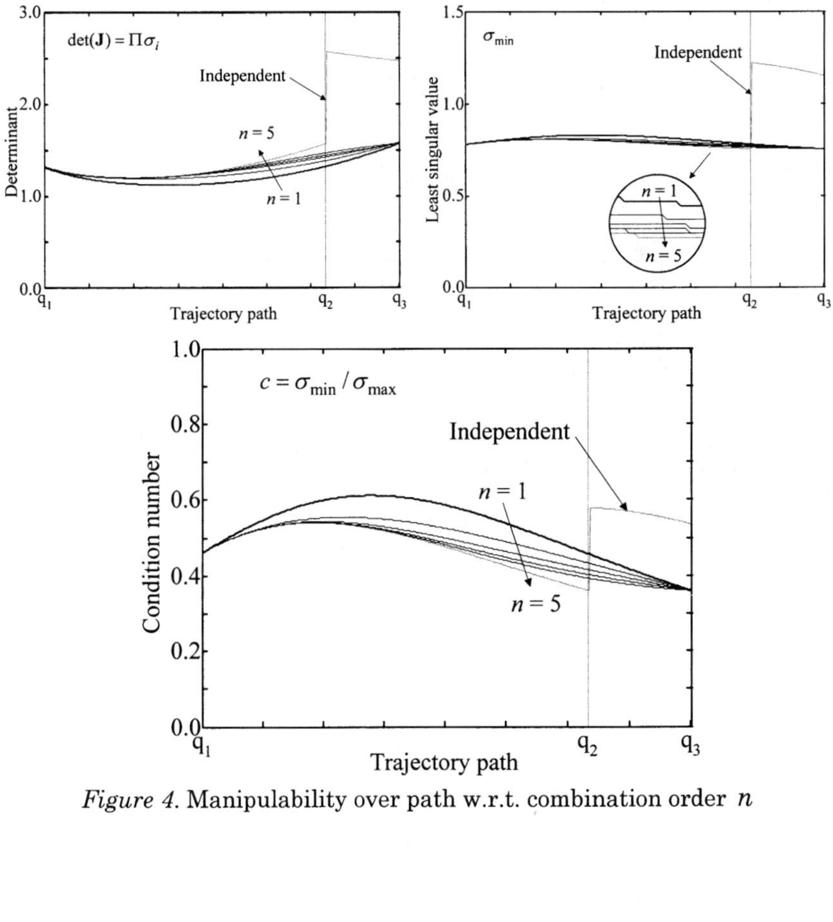

Figure 4. Manipulability over path w.r.t. combination order n

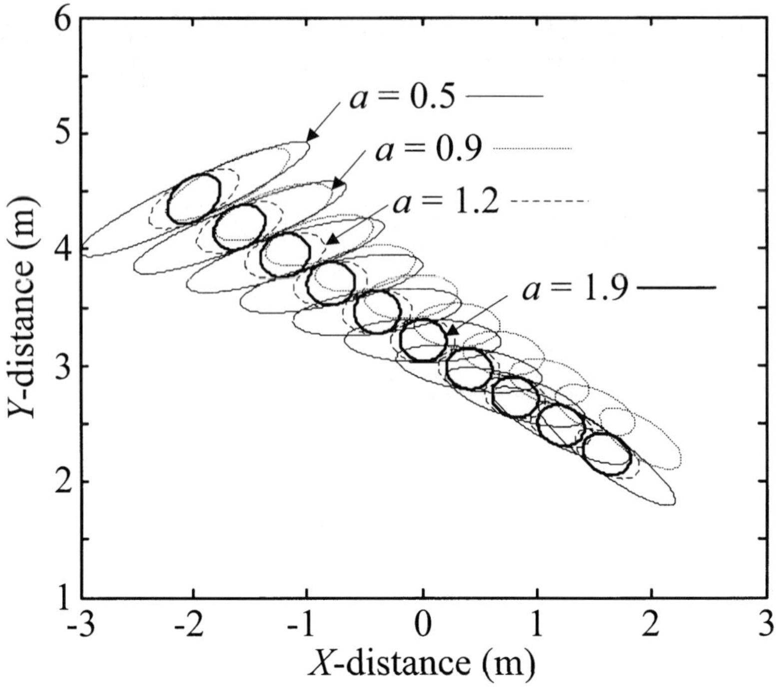

Figure 5. Manipulability ellipsoid changes along the trajectory w.r.t. separation distance a

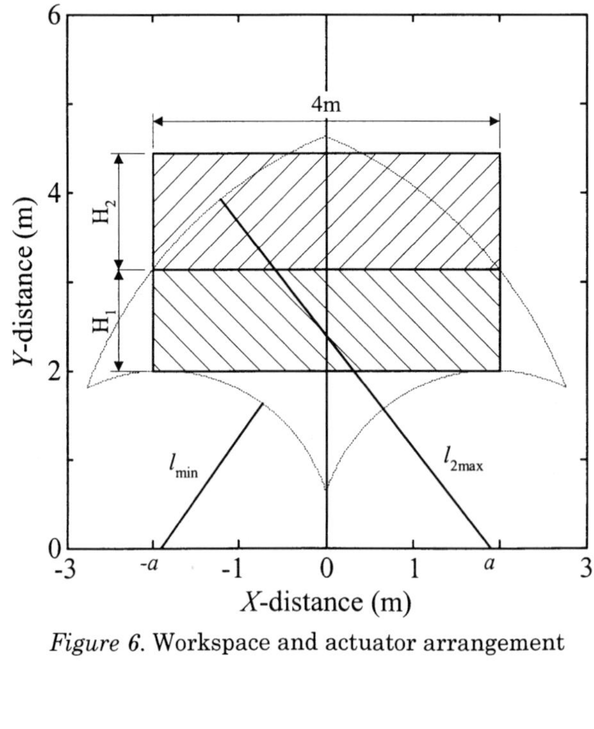

Figure 6. Workspace and actuator arrangement

AN OPTIMAL KINEMATIC DESIGN FOR A HIGHLY DEXTROUS MIS INSTRUMENT

Damien Sallé and Philippe Bidaud
Laboratoire de Robotique de Paris - LRP
University Paris 6, Paris, France
(salle,bidaud,morel)@robot.jussieu.fr

Abstract This paper introduces an original, dextrous, active and modular minimally invasive instrument dedicated to coronary artery bypass grafting surgery. Its design is obtained by a generic evolutionary optimization process using pareto-based multi objective genetic algorithms and including highly realist simulations and experimental models of the surgical gesture. The optimal instrument has 5 intra-body dofs, is actuated by brushless micro motors and is position controlled. Its optimization, mechanical design and performances are discussed.

Keywords: Minimally Invasive Heart Surgery, Optimal Kinematic Design, Multi Objective Genetic Algorithms

1. Introduction

Minimally invasive surgery(MIS) is a rather new technique to perform surgery while limiting the patient's trauma. Trocars (plastic or metallic tubes) are placed on small incisions made in the abdomen or between two ribs, to prevent further damage of the tissues. Then an endoscope and long cylindric instruments are inserted in the body through these trocars. Surgeons are though able to perform intra body procedures with only a few small incisions.

Numerous robotic systems have recently been developed as tools to help surgeons during such MIS procedures. Most of them use four dofs to actuate the trocars and control their orientation and use two more dofs to provide an intra body wrist to increase motion capacities of the instruments. They have brought great improvements and have allowed numerous new minimally invasive abdominal surgeries. However, minimally invasive heart surgery is much more difficult to perform as it is tremendously demanding: very small free space (less than $1dm^3$) to achieve very complex movements. Most of the existing robotic systems are not dedicated to this task and are thus not perfectly adapted to it: for example, only a very few highly skilled surgeons can perform a single coronary artery bypass grafting (CABG) using the Da Vinci™ system.

J. Lenarčič and C. Galletti (eds.), On Advances in Robot Kinematics, 399-408.

The design of instruments dedicated to heart surgery must take into account simultaneously highly demanding geometrical constraints ($\varnothing 10mm$ maximum to be inserted trough a trocar), available technology and environment topology (vital organs in the rib cage). As the robotic system must mimic the surgical gestures, its trajectories are complex and force capacities are critical.

Among existing MIS robotic systems, most of the distal part of the instruments use 2 dof universal joints. Due to their low intra-body mobilities, they are not suited for heart surgery as proved by Da Vinci's limitations. It seems obvious that, for heart surgery, no instrument can afford such a small number of intra-body dofs: they must be highly dextrous and actively controlled redundant mechanisms.

High mobility instruments have been mainly developed for endoscopic applications where low speed and no interaction forces are required (Szewczyk,2000) (Ikuta,1994) or for applications where size is a lower constraint (DAS,1997).

The number of possible kinematics for the distal part of the instrument, its realist control and interactions with the organs make such a redundant instrument almost impossible to design intuitively. The proposed solution is to use a task based optimization procedure using pareto based multi objective genetic algorithms (MOGA) and a realistic simulation of the CABG surgery as evaluation function. MOGA are superior to traditional genetic algorithms as detailed in section 4. They have however never been used yet for controlled robotic mechanism design.

The simulation of the task allows to take into account the thoracic organs, control of the redundant instrument and experimental models of the surgery. According to surgeons, the suture of a graft on the coronary artery, called the anastomosis, is the key gesture of the CABG and an instrument that can achieve the anastomosis can achieve the complete CABG. Intra-body knot tying is an easy technique and any MIS instrument can perform it. The design procedure will though focus on an anastomosis performed on the Intra Ventricular Artery (IVA).

This simulation also allows to continuously evaluate the instruments based on local criteria while they perform the surgery. To cope with the design constraints of this task, 4 contradictory objectives have been set: Capability to perform the gesture (maximized), Dexterity of the instrument (maximized), Joint Torques (minimized) and Distance to organs (maximized).

The optimal solution of this optimization procedure is a highly dexterous, active and redundant modular instrument that has the 4 trocar dofs plus 5 intra-body dofs, is actuated by brushless micro motors and is position controlled.

This paper details key issues in the design of such an instrument: modelling of the surgical gesture, modular mechanical design concept and optimization procedure including description of the realistic simulation used as evaluation. It finally proposes the optimal instrument for this task, details its mechanical design, discusses its performance and introduces the prototype being built.

2. Modelling of CABG Gesture

The Coronary Artery Bypass Grafting is a surgery of the arteries irrigating the heart muscle (the coronary arteries). Their pathology (the atheroma) reduces their cross-section, leading to a stop of the blood flow and heart attack.

The CABG surgery derives the blood from a healthy artery to the coronary artery downstream the lesion, so that the heart muscle remains irrigated. This is achieved by harvesting the mammary or the iliac artery and by suturing this graft on the coronary artery with a "T" connection (figure 1). This is called the anastomosis and is the most difficult and important part of the surgery.

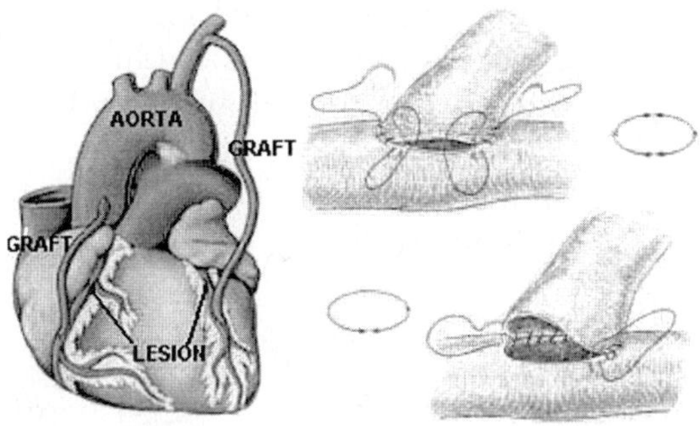

Figure 1. CABG Principle

A surgical gesture recording platform has been developed and models of the CABG anastomosis have been experimentally identified. From the experiments performed by surgeons, the main features of the anastomosis can be extracted: incision on the artery of 7 to 10mm, elliptic motion around this incision (see fig 1); 20 perforation points -1mm stepped-produced by a small translation combined with a rotation of about 120° of the instrument, in a plane normal to the ellipsoid; perforation force of the artery walls of about 0.3N; forces on the thread to tie each point of the anastomosis of about 1.0 N.

These features of the CABG procedure are mapped on the virtual environment used to design the modular instrument.

3. Proposed Mechanical Modular Design

The instrument must have numerous actively controlled intra-cavity degrees of freedom to increase the dexterity of the surgeon. To fulfill size, speed and power constraints, 3mm brushless micro motors have been selected. Due to their size, motion range and fitting capacities in the instrument, magnetic resolvers (Sentron 2SA10) have been chosen as angular sensors.

The module on the left part of Figure 2 shows the proposed mechanical assembly principle. It combines a micro-motor and a worm and gear transmission. A bipolar magnet is fixed on the gear, creating a rotating magnetic field sensed by a chip including 2 perpendicular hall sensors rigidly mounted on the fixed part. This module has a minimal length of 24mm, its range is $\pm 90^o$ and it generates a torque of 6mNm.

Combining 2 of these assemblies leads to 2 dofs modules, with parallel axis or placed at 90°, as shown on the central modules of figure 2. Their minimum length is 36mm, range is $\pm 90^o$ and available torque is 6mNm. The module on the right of figure 2 shows the proposed solution for 2 dofs modules including gross-rotation. It uses the same micro motor with an internal gear. The magnet is fixed on the base part and the sensor rotates with the mobile part of this module. Its length is also 36mm, gross-rotation range is $> 270^o$ and available torque on this axis is 8mNm.
A gripper using a compression spring and SMA wires has also been designed but will not be discussed here.

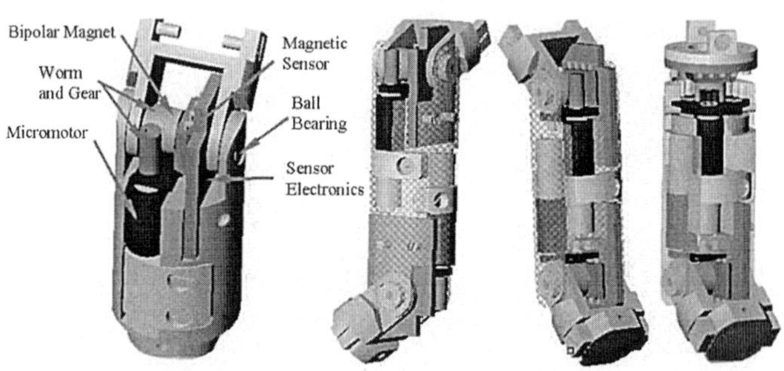

Figure 2. Mechanical modular design

This parametric modular design allows numerous combinations for the MIS instrument's kinematics. Due to its internal motions, finding the best assembly of these modules for such a redundant surgical instrument is not intuitive. An optimization procedure has though been developed to find the best modular instrument dedicated to the CABG surgery.

The instruments are considered as a serial chain of 1 to 7 of the modules discussed above.

4. Task Based optimization of the instrument

The problem to be solved by this optimization procedure is to find non intuitive instruments that are able to perform the surgeon's gesture, considering environment, mechanical and control constraints. It uses a highly realist simulation to evaluate the possible instruments designs, based on the four contradictory objectives specified in section 1 and detailed in section 4.2.

Genetic Algorithms have been widely used in robotics design optimization (Chocron,1997) (Leger,1999) as they are very well adapted for optimization over vast, non continuous search space. However multiple objectives are usually summed up in a single metric, leading to the synthesis of "optimal" solutions usually considered a posteriori "non optimal" by designers. Multi-Objective Genetic Algorithms (MOGA) overcome this problem as they evaluate the solutions with respect to every objective individually and output the optimal pareto front of the problem (Deb,2002). They are superior to the usual algorithms on 2 points: the optimization is based on a non domination criteria (so every solution in the pareto front is optimal with respect to a given combination of the objectives), and they let the designer choose the optimal design according to a posteriori criteria.

Their use in mechanical design has nevertheless been limited to the optimization of structural design parameters such as inlets, airfoils, gearboxes or robot gripper parts and have not yet been used for controlled mechanism design. It is coded in C++ using the NSGAII MOGA algorithm proposed by Deb (Deb,2002), the Evolving Objects library (EO) and is using parallel evaluations to limit the computation time.

4.1 Optimization settings

To represent the modular structure of the instrument, the genotype is defined as a vector of modules, each being a pair $< integer, float >$, standing for respectively the type of module and its length. Genetic operators have been adapted to this genotype. They consist in traditional one point crossover, mutation modifying both a module's type and length and mutations adding or removing a module from the instrument.

Two constraints are set for the considered optimization: the trajectory must be completed and the joints torque must verify the available mechanical capacities.

4.2 Evaluation

It emphasizes realism as it includes 3D models of the thorax and its organs (figure 3), the modelled trajectory and interaction forces obtained in section 2, the candidate instrument and its control during the trajectory following task.

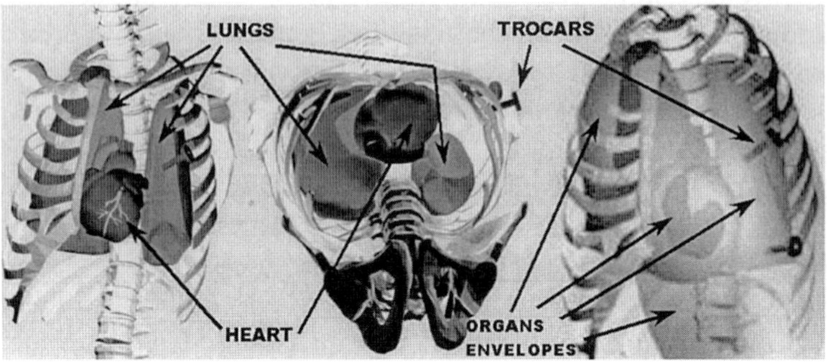

Figure 3. Thorax 3D models and envelopes for distance calculation

Control of the redundant instrument. To follow the suturing trajectory, at least 6 dof are necessary in the instrument. An extra-body instrument holder controls the 4 dofs needed to actuate the trocars and, depending on the its kinematics, the distal part of the instrument has from 1 to 14 additional dofs. Most of the evaluated instruments will though be redundant. This redundancy must be taken into account and even more should be used to add extra functions to the robotic instrument, for example organs avoidance, manipulability maximization, motion range limits avoidance or endoscope view occlusion avoidance.

The high level control of the instrument implemented is based on a trajectory following algorithm using the inverse kinematic differential model:

The jacobian matrix J of the mechanism is calculated at every simulation step. The kinematic differential model of the instrument is then:

$$dX = J\, dq$$

Due to the kinematic redundancy, J is rectangular. Its pseudo inverse J^+ is computed using Greville's algorithm to get the minimal joint displacement solution. Physical inconstancy is avoided using scaling matrices as proposed in (Doty,1993).

The redundancy is used for organs and joint range limits avoidance and manipulability maximization. The control scheme thought becomes

$$dq = J^+\, dX + (I - J^+ J)\nabla V$$

where the second member of this equation is the projection on the kernel of the application of a cost function ∇V to be minimized. This projection allows inner motion of the mechanism while the end effector keeps following the trajectory. ∇V is a gradient function of the joint's position, the mechanism manipulability and the shortest distance between the different modules of the instrument and the organs.

In order to calculate this distance while maintaining acceptable computation time, the organs are approximated by simple envelopes: sphere for the heart, cylinder for left lung and ellipsoid for ribs as shown in figure 3. The distances are calculated between the origin of each module's frame and the organs envelopes. Distances and manipulability are calculated as:

$$F_{dist} = \sum_{modules} dist_{heart}^2 + dist_{lung}^2 + dist_{ribs}^2 \ ; \ Manip = \sqrt{det(JJ^T)}$$

The gradient is then calculated by the following algorithm, where *Range*, $\triangle Fdist$ and $\triangle Manip$ stand respectively for the joint range, distance to organs and manipulability gradients.

Current Position :

$\quad Calculate\ distance \rightarrow F_{dist_{current}}$

$\quad Calculate\ manipulability \rightarrow Manip_{current}$

$For\ i = dof1\ to\ n$

$\quad Rotate\ \triangle q$

$$Calculate\ Range[i] = 2\frac{q[i] - q_{mean}[i]}{(q_{max}[i] - q_{min}[i])^2}$$

$\quad Calculate\ \triangle Fdist[i] = F_{dist_{current}} - F_{dist\triangle q}$

$\quad Calculate\ \triangle Manip[i] = Manip_{current} - Manip_{\triangle q}$

$\quad Rotate\ -\triangle q$

$\quad \nabla V = aRange + b\triangle Fdist + c\triangle Manip$

end

Objective fitness calculation. Each instrument candidate is evaluated with respect to four local criteria objectives, calculated at every simulation step, ie at each of the n points of the trajectory.

- Objective 1: Capability to perform the gesture:
  Position and orientation error is calculated (*error*). The objective score is defined as:
  $if\ error < 0.1,\ score = 5.0;$
  $if\ error > 5.0,\ score = 0;$
  $$else\ score = 5 - \frac{5(error - 0.1)}{5 - 0.1}$$

$$score1 = \sum_{n} score$$

This is a linear decreasing function between 2 error bounds, to account for the variability in the trajectories

- Objective 2: Dexterity of the instrument:
  $score2 = \sum_{n}(\sqrt{det(JJ^T)})$
- Ojective 3: Maximum Joint Torques:
  $score3 = Max(\tau(t) = J(t)^T F(t))$
  where $F(t) = F_{Interaction}(t) + F_{Gravity}(t)$
- Ojective 4: Minimal Distance to organs:
  The shortest distance between the instrument and the organs is calculated at each step ($dist$).
  $score4 = Min(dist(t))$

4.3 Optimization Results

This optimization procedures outputs pareto fronts of the population's scores. To visualize these 4 dimensions fronts, they must be projected on the score planes.

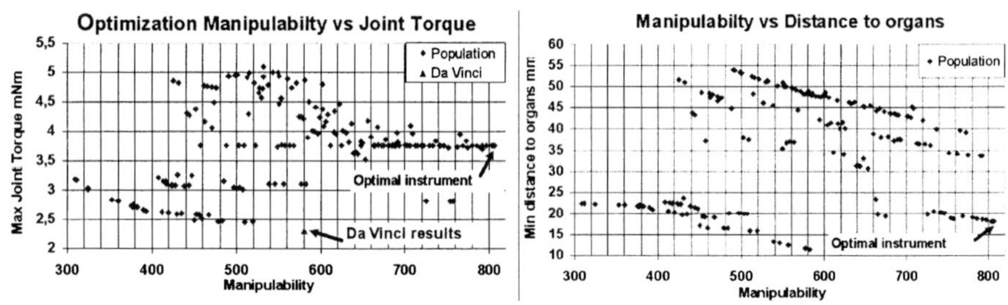

Figure 4. Optimized population: Manipulability vs Torque and Distance to organs

Figure 4 shows a population of 200 instruments, projected on the Manipulability vs Torque plane and on the Manipulability vs Distance to organs plane after 76 generations. It shows that the optimal instrument has a higher dexterity (increase of 28%) than the Da Vinci system but has a higher maximum joint torque, due to the weight and length of the additional modules. It also shows that the module closest to the organs keeps a safety distance of at least 1.8mm. Trajectory performance is not shown as all instruments validated the constraint.

This optimization has been run 30 times with modifications on the crossover and mutation rates and two positions for the trocar.

5. Proposed Instrument

The proposed instrument, resulting from the optimization procedure, is shown in figure 5 as CAD models. It is made of 4 modules (3 for motion and 1 for the gripper) and can be described, using modified DH conventions, by table 1.

Trocar Joint i	α_i	d_i	r_i	θ_i
T1	0	0	0	θ_{T1}
T2	$\pi/2$	0	0	θ_{T2}
T3	$\pi/2$	0	0	θ_{T3}
T4	0	0	λ_{T4} + L	0

Instrument Joint i	α_i	d_i	r_i	θ_i
I1	$-\pi/2$	0	0	θ_{I1}
I2	$\pi/2$	15	0	θ_{I2}
I3	$-\pi/2$	0	36	θ_{I3}
I4	$-\pi/2$	0	0	θ_{I4}
I5	$-\pi/2$	0	36	θ_{I5}

Table 1. Modified DH Table of the optimal instrument

It is designed to be held by a robotic manipulator placed in the operating room. 9 degrees of freedom are though available to control this instrument.

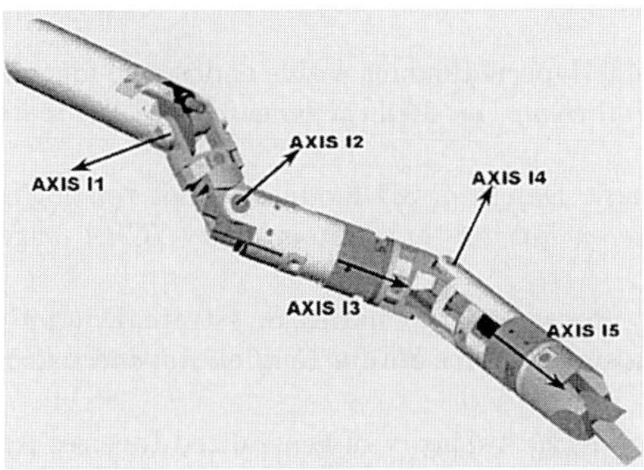

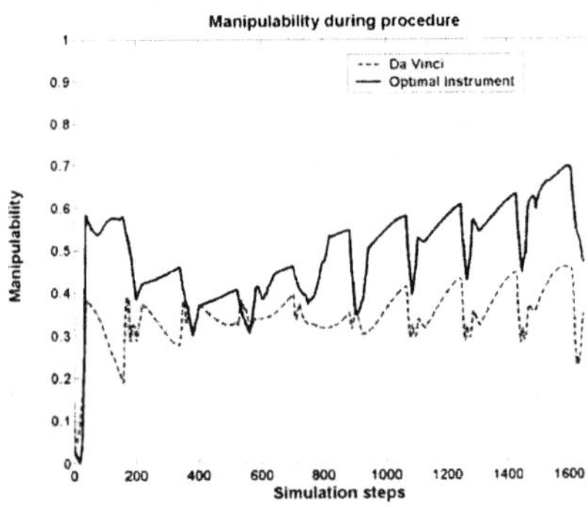

Figure 5. Optimal Instrument and its manupulability during the CABG procedure

Figure 5 shows the compared manipulability between Da Vinci and the optimal instrument during the simulated CABG surgery. For both, it is very low at the beginning of the trajectory as the instrument must be inserted with all modules aligned, and thus close to singularity. Afterwards the manipulability shows periodic variations due to the periodic

suturing motion during the surgery.
The manipulability of the optimal instrument is clearly higher than da Vinci's during all the procedure. Surgeons should though have higher dexterity and should be able to better perform minimally invasive CABG surgeries using the proposed instrument.

6. Conclusion

An original, dextrous, active and modular MIS instrument dedicated to CABG has been proposed. Its design results from a task based optimization method using pareto based multi-objective genetic algorithm. The instruments are evaluated over 4 objectives: gesture completion capacities, manipulability, joint torques and distance to organs.

This redundant modular instrument shows good kinematic properties and fulfills all the CABG procedure constraints. Its higher dexterity, compared to Da Vinci system, should increase the capacities of surgeons to perform minimally invasive CABG surgery and eventually allow double or triple graftings. A prototype of this instrument is being built and experimental evaluations of its performance and benefits will be conducted.

References

Szewczyk, J., de Sars, V. , Bidaud, Ph. (2000), An active tubular polyarticulated microsystem, *Proc International Symposium on Experimental Robotics - ISER 2000*, pages 179-188

Ikuta, K., Nokata, M.,Aritomi, S. (1994), Hyperredundant active endoscope for minimum invasive surgery., *Proc First Int. Symp. on Medical Robotics and Computer Assisted Surgery - MRCAS*

Chocron, O., Bidaud, Ph. (1997), Genetic Design of 3D modular robot topologies, *Proc. IEEE International Conference on Robotics and Automation - ICRA*, pages 223-228

Das, H., Ohm, T.,Boswell, C., Steele, R., Rodriguez,G., Charles, S., Istrate, D. (1997), Dexterity-enhanced Telerobotic Microsurgery, *Proc 8th Int Conf on Advanced Robotics - ICAR*

Doty, K., Melchiorri, C., Bonivento, C. (1993), A Theory of generalized Inverses Applied to Robotics, *International Journal of Robotic research*, volume 12-1, pages 1-19

Leger, C. (1999), Automated Synthesis and Optimization of Robot Configurations: an Evolutionary Approach, *PhD Thesis - Canegie Mellon University*

Deb,K. (2002), Unveiling Innovative Design Principles By Means of Multiple Conflicting Objectives, *Technical Report 2002007 - Kanpur Genetic Algorithms lab - Indian Institute of Technology Kanpur*

ACHIEVABLE CARTESIAN STIFFNESS WITH PASSIVELY COMPLIANT, UNCOUPLED JOINTS

G. Schreiber, A. Albu-Schäffer, M. Fischer, F. Schoeppe and G. Hirzinger
German Aerospace Center - DLR
Institute for Robotics and Mechatronics
82230 Wessling, Germany
[Guenter.Schreiber, Alin.Albu-Schaeffer, Max.Fischer, Florian.Schoeppe]@dlr.de

Abstract In the field of service robotics, whole arm contact with an unstructured environment or human beings becomes a major issue. Therefore soft robots, which means robots with passively (or mechanically) compliant joints, become more and more important. In this work we analyze what Cartesian stiffness at the tool center point one can achieve with a passively compliant, redundant robot with variable joint stiffness. We restrict this work to the special case of uncoupled joint stiffnesses only, as coupling of joint stiffnesses seems to be mechanically difficult to realize.

Keywords: passively compliant device, variable joint stiffness control, redundancy

1. Introduction

In the field of service robotics, compliance and reactivity to unforeseen contacts in a priori unknown environments with the whole arm, not only with the tool center point behind a force torque sensor (FTS), are a major issue. Today there are basically three ways to establish compliance: (1) Use a FTS, possibly in combination with a passively compliant device close to it; (2) build a stiff robot with joint torque sensors and add (actively) compliant behavior through control at joint level; or (3) construct a *soft robot* with mechanically (or passively) compliant joints with variable joint stiffness.

In situations where control and precision in the position domain are less important and the focus is on compliance, soft robots (3) become more and more interesting, as they:

- cause less impact on their environment, especially on human beings, in the case of contact,
- are less prone to structural damage in case of contact, especially in non-nominal situations (power failure; controller failure; sensor failure etc.). This holds especially true for small and lightweight structures as e.g. artificial hands or humanoid robots,

J. Lenarčič and C. Galletti (eds.), On Advances in Robot Kinematics, 409-418.

- yield better compliant performance, as they inherently and instantaneously show a compliant behavior, whereas (1) and (2) can only react compliantly with a time delay through the loop *sensor data processing*, *control* and *actuation*.

There are already a few interesting examples of robots realizing passively compliant behavior (Sugano, 2002; H.Kobayashi, Hyodo & Ogane, 1998; Koganezawa & Ban, 2002). Concerning safety aspects of passively compliant robots, (Zinn, Khatib, Roth & Salisbury, 2002; Bicchi, Bavaro & Piccigallo, 2003) made some important contributions. In this work we would like to give a theoretical analysis of the problem of realizing a desired Cartesian stiffness at the tool center point (TCP) of a passively compliant robot.

2. Problem Statement

With a soft robot, we would like to realize a *desired Cartesian stiffness* at the tool tip. This desired Cartesian stiffness should be realized in a manner, that a force input along specified principal stiffness axes should result in a position displacement as linearly as possibly along these axes.

To realize this behavior, we can use the *mechanically variable joint stiffnesses* and - in case of a redundant robot - the *null space* of the robot (Schöppe, 2003).

Additionally it needs to be stated that when talking about passively compliant robots, we imply that the robot's joints are decoupled in the sense of compliance. Of course, one could establish coupling of joint stiffnesses mechanically (alike the human limbs), but this is beyond the scope of this paper. This leads to the first constraint in this work:

Diagonal Joint Stiffness Matrix: *Joint stiffness matrices* ($\boldsymbol{K}_J$) *have no coupling terms, they are always of diagonal shape in this work.*

For a general robotic application it is most natural to specify the desired stiffness behaviour of the robot in Cartesian coordinates. The desired values would result from a task description with respect to the TCP. The user may specify a desired stiffness matrix $\boldsymbol{K}_C = \frac{\partial \boldsymbol{f}}{\partial \boldsymbol{x}} \in \Re^{m \times m}$ as the relation between the Cartesian force $\boldsymbol{f}$ and the Cartesian position $\boldsymbol{x}$. Here, m is the number of Cartesian degrees of freedom (DoF). On the other hand, the robot will be able to adjust the stiffness $\boldsymbol{K}_J = \frac{\partial \boldsymbol{\tau}}{\partial \boldsymbol{q}} \in \Re^{n \times n}$ of its n joints, where $\boldsymbol{\tau}$ is the joint torque and $\boldsymbol{q}$ is the joint position. Hence the Cartesian stiffness has to be transformed to the joint space using a mapping $\mathcal{T}$: $\boldsymbol{K}_J = \mathcal{T}(\boldsymbol{K}_C)$, an approach known as stiffness control. This transformation can be computed as follows:

$$\boldsymbol{K}_J = \frac{\partial \boldsymbol{\tau}}{\partial \boldsymbol{q}} = \frac{\partial(\boldsymbol{J}(\boldsymbol{q})^T \boldsymbol{K}_C \Delta\, \boldsymbol{x})}{\partial \boldsymbol{q}} = \boldsymbol{J}(\boldsymbol{q})^T \boldsymbol{K}_C \boldsymbol{J}(\boldsymbol{q}) + \frac{\partial \boldsymbol{J}(\boldsymbol{q})^T}{\partial \boldsymbol{q}} \boldsymbol{K}_C\, \Delta\, \boldsymbol{x} \tag{1}$$

$\boldsymbol{J}$ denotes the manipulator Jacobian $\boldsymbol{J}(\boldsymbol{q}) = \frac{\partial f(\boldsymbol{q})}{\partial \boldsymbol{q}}$, where $f(\boldsymbol{q})$ is the forward kinematics mapping. $\Delta\, \boldsymbol{x} = \boldsymbol{x}_d - \boldsymbol{x}$ is the Cartesian displacement between the desired and the actual position. In the early work on stiffness control (Salisbury, 1980), only the first term of (1) is mentioned. The importance of the second term, which reflects the position dependent change of the Jacobian is pointed out in (Hogan, 1990; Chen & Kao, 2000). Notice that in eq. (1) the mapping $\mathcal{T}$ and the matrix $\boldsymbol{K}_J$ have only local meaning and will change with the robot configuration.

As mentioned in the introduction the robot has the ability to adjust mechanically the stiffness in each joint, meaning that a diagonal joint stiffness is available: $\boldsymbol{K}_J = \text{diag}\{k_{Ji}\} \in \Re^{m \times m}$. In general, $\boldsymbol{K}_J$ in (1) is a full matrix, meaning that only an approximation of the desired Cartesian stiffness can be achieved.

Problem statement: *Given a desired Cartesian robot configuration $\boldsymbol{x}_d \in \Re^m$ and a desired Cartesian stiffness matrix $\boldsymbol{K}_C \in \Re^{m \times m}$ for that configuration, find the optimal joint position $\boldsymbol{q}_d$ and the diagonal joint stiffness matrix $\boldsymbol{K}_J \in \Re^{n \times n}$ which provides the best approximation of $\boldsymbol{K}_C$ under the constraint $\boldsymbol{x}_d = f(\boldsymbol{q}_d)$.*

The implementation of the second term in (1) using the passive joint compliance seems difficult, since it needs displacement $\Delta \boldsymbol{x}$ and this in turn would require the fast and continuous change of the joint stiffness during motion, in contrast to the main design idea. Therefore, only the (quite usual) approximation

$$\boldsymbol{K}_J \;=\; \boldsymbol{J}(\boldsymbol{q})^T \boldsymbol{K}_C \boldsymbol{J}(\boldsymbol{q}) \tag{2}$$

will be used for computation of the desired passive joint compliance.

The inverse problem of computing the resulting Cartesian stiffness matrix $\boldsymbol{K}_C = \mathcal{T}^{-1}(\boldsymbol{K}_J)$ for a given $\boldsymbol{K}_J$ can be solved using the pseudo inverse $\boldsymbol{J}^+$ of the Jacobian, defined as $\boldsymbol{J}\boldsymbol{J}^+ = \boldsymbol{I}$:

$$\boldsymbol{K}_C \;=\; \boldsymbol{J}(\boldsymbol{q})^{+T} \boldsymbol{K}_J \boldsymbol{J}(\boldsymbol{q})^+ \tag{3}$$

In order to avoid the usage of the pseudo inverse in (3) for the sake of computational efficiency and the problems around selection of a proper metric, one can compute $\boldsymbol{K}_C$ by its inverse, the Cartesian compliance

$$\boldsymbol{C}_C = \boldsymbol{K}_C^{-1} = \frac{\partial \boldsymbol{x}}{\partial \boldsymbol{f}} = \frac{\partial \boldsymbol{x}}{\partial \boldsymbol{q}}\frac{\partial \boldsymbol{q}}{\partial \boldsymbol{\tau}}\frac{\partial \boldsymbol{\tau}}{\partial \boldsymbol{f}} = \boldsymbol{J}(\boldsymbol{q})\boldsymbol{C}_J\boldsymbol{J}(\boldsymbol{q})^T \tag{4}$$

whereas $\boldsymbol{C}_J = \boldsymbol{K}_J^{-1}$ is the compliance with respect to joint space. Of course, $\boldsymbol{K}_{C,J}$ have to be nonsingular for their inverses $\boldsymbol{C}_{C,J}$ to exist.

3. Analysis of Robots with Compliant Uncoupled Joints

3.1 Analysis of a 2 DoF planar finger

In order to get a first feeling for our problem, the simplest, 2DoF case of a planar finger will be analyzed in this section. Suppose, the robot is fixed at a given configuration. We want to determine the limitations imposed on the Cartesian stiffness by choosing a diagonal joint stiffness $\boldsymbol{K}_J = \mathrm{diag}\{k_{J1}, k_{J2}\}$. Eq. (3) becomes in this case $\boldsymbol{K}_C = \boldsymbol{J}^{+^T} \begin{pmatrix} k_{J1} & 0 \\ 0 & k_{J2} \end{pmatrix} \boldsymbol{J}^{+} = \alpha \cdot \boldsymbol{J}^{+^T} \begin{pmatrix} 1 & 0 \\ 0 & \beta \end{pmatrix} \boldsymbol{J}^{+}$,where the notations $\alpha = k_{J1}$ and $\beta = \frac{k_{J2}}{k_{J1}}$ have been used. This can be rewritten in the following form:

$$
\begin{aligned}
\boldsymbol{K}_C &= \alpha \boldsymbol{P}_{J^+}(\beta)^T \begin{pmatrix} \lambda_1(\beta) & 0 \\ 0 & \lambda_2(\beta) \end{pmatrix} \boldsymbol{P}_{J^+}(\beta) \\
&= \boldsymbol{P}_{J^+}(\beta)^T \begin{pmatrix} \alpha\lambda_1(\beta) & 0 \\ 0 & \alpha\lambda_2(\beta) \end{pmatrix} \boldsymbol{P}_{J^+}(\beta), \qquad (5)
\end{aligned}
$$

where $\lambda_1(\beta)$, $\lambda_2(\beta)$ are the eigenvalues which determine the stiffness along the principal stiffness axes. These axes are given by the matrix $\boldsymbol{P}_{J^+}(\beta)$ which contains the eigenvectors of $\boldsymbol{K}_C$. As a direct consequence of (5) it turns out that *the direction of the principal axes and the ratio of the stiffness values along these axes can not be chosen independently from each other.* Both the eigenvalues $\lambda_1(\beta)$, $\lambda_2(\beta)$ and the matrix $\boldsymbol{P}_{J^+}(\beta)$ are parameterized by the same scalar β. α is linearly scaling both eigenvalues, keeping their ratio and the principal directions constant.

Figure 1 displays an example for this result. The Cartesian stiffness matrices are described by three parameters: the eigenvalues λ_1 and λ_2 and the angle θ between the coordinate system of the TCP and the coordinate system defined by the principal stiffness axes. The implementable stiffness matrices $\boldsymbol{K}_C$ form a two dimensional manifold in this three dimensional space, parameterized by k_{J1} and k_{J2}. The intersection of this manifold with any horizontal plane is a straight line, due to the previously mentioned property. Fig. 2 gives a two dimensional representation of the results. It displays the Cartesian stiffness values which can be obtained with a diagonal upper bounded joint stiffness matrix as a function of the rotation angle θ between the TCP and the principal axes. One can see that there are regions for θ which cannot be chosen.

These results reveal a severe limitation of the method, since independent stiffness values along the different axes are crucial for the use of stiffness control in most applications. It turns out that special attention has to be paid to the design of the manipulator and to the choice of the optimal configuration and stiffness transformation, in order to obtain good results in practice.

These results obtained for this simple planar case give the motivation for a detailed analysis and a solution for the general 6 DoF case.

3.2 Analysis of Cartesian Stiffness in 3D

Orthogonal Principal Stiffness Axes. $\boldsymbol{K}_C$ has eigenvectors which are orthogonal to each other, since it is symmetric and positive definite: (Proof sketch: $\boldsymbol{K}_J$ is positive definite and diagonal, therefore it can be written as $\boldsymbol{K}_J = \boldsymbol{A}^T\boldsymbol{A}$ and $\boldsymbol{K}_C = \boldsymbol{J}^T\boldsymbol{K}_J\boldsymbol{J} = \boldsymbol{J}^T\boldsymbol{A}^T\boldsymbol{A}\boldsymbol{J} = \boldsymbol{B}^T\boldsymbol{B}$).

Structure of Cartesian Stiffness Matrices. An arbitrary Cartesian stiffness matrix can be decomposed in four blocks: The translational and rotational stiffness part $\boldsymbol{K}_T$ and $\boldsymbol{K}_R$, resp. and two coupling matrices, $\boldsymbol{K}_{TR}$ and $\boldsymbol{K}_{RT}$. $\boldsymbol{K}_T$ and $\boldsymbol{K}_R$ can be written as diagonal stiffness matrices $\lambda_\mathbf{T}$ and $\lambda_\mathbf{R}$ in a rotated coordinate system with the rotations $\boldsymbol{P}_T$ and $\boldsymbol{P}_R$ in the translational and rotational stiffness dimensions, resp. (see sec. 3.2).

$$\boldsymbol{K}_C = \begin{pmatrix} \boldsymbol{K}_T & \boldsymbol{K}_{RT} \\ \boldsymbol{K}_{TR} & \boldsymbol{K}_R \end{pmatrix} \begin{pmatrix} \boldsymbol{P}_T^T\lambda_\mathbf{T}\boldsymbol{P}_T & \boldsymbol{K}_{RT} \\ \boldsymbol{K}_{TR} & \boldsymbol{P}_R^T\lambda_\mathbf{R}\boldsymbol{P}_R \end{pmatrix}$$

$$= \begin{pmatrix} \boldsymbol{P}_T^T & 0 \\ 0 & \boldsymbol{P}_R^T \end{pmatrix} \begin{pmatrix} \lambda_\mathbf{T} & \overline{\boldsymbol{K}}_{RT} \\ \overline{\boldsymbol{K}}_{TR} & \lambda_\mathbf{R} \end{pmatrix} \begin{pmatrix} \boldsymbol{P}_T & 0 \\ 0 & \boldsymbol{P}_R \end{pmatrix} \tag{6}$$

To specify a desired Cartesian stiffness $\boldsymbol{K}_{CD}$ it is sufficient for the most (practical) applications to specify (1) translational and rotational stiffnesses in the same coordinate system $\boldsymbol{P}$ and to (2) omit coupling terms between translational and rotational stiffnesses (although one might think of some "screwing" applications where these couplings could make sense).

$$\boldsymbol{K}_{CD} = \begin{pmatrix} \boldsymbol{P}^T\lambda_\mathbf{T}\boldsymbol{P} & 0 \\ 0 & \boldsymbol{P}^T\lambda_\mathbf{R}\boldsymbol{P} \end{pmatrix} \tag{7}$$

Approximation and Linearity Limitations. There are two main sources of errors for the Cartesian stiffness we identified so far:

- Due to the diagonal form of the $\boldsymbol{K}_J$, the desired Cartesian stiffness can not be realized exactly.

- Because of the locality of the Jacobian the desired behaviour is only local: A larger deflection due to a given force will yield a significant position error if the Jacobian and as a result the joint stiffness is not updated.

The first problem will be tackled in the next section, the second problem is beyond the scope of this paper. A possible solution for compensating this error by means of a slower control loop is presented in (Albu-Schäffer & Hirzinger, 2002).

4. The TCP Stiffness Optimization Problem

As stated above, there is a very high probability that a desired Cartesian stiffness cannot be achieved, mostly due to the lack of DoFs and - if we use a robot with at least 14 DoFs - due to the non-surjective mapping from joint stiffnesses to Cartesian stiffnesses. Therefore we can only find a good approximation instead of a perfect solution by formulating the problem as an optimization problem:

$$\text{Minimize } \mathbf{norm}\,(\boldsymbol{K}_{\mathbf{CD}}, \boldsymbol{K}_{\mathbf{C}}(\boldsymbol{K}_{\mathbf{J}}, \boldsymbol{q})) \text{ subject to } \mathbf{f}(\boldsymbol{q}) = \boldsymbol{x}_{\mathbf{D}} \tag{8}$$

This formulation raises three questions: (1) What function can we use as a suitable **norm**, (2) how hard is the optimization problem and (3) what tools can we use?

Ad (1): We use two different measures to quantify the difference between the desired and the resulting Cartesian stiffness: First, we tried to identify a set of intuitive error qualities and came up with an error vector

$$\Delta = \left(\Delta_T^{Rot}, \Delta_R^{Rot}, \Delta_T^{Rel}, \Delta_R^{Rel}, \Delta_K\right)^T \tag{9}$$

where Δ_T^{Rot} and Δ_R^{Rot} mean the angle axis rotations between the main axes coordinate systems of the desired and resulting translational and rotational stiffnesses, respectively, $\Delta_T^{Rel} = \frac{1}{3}\sum_{i=1}^{3} \frac{|\lambda_{Ti} - \tilde{\lambda}_{Ti}|}{\frac{1}{2}(\lambda_{Ti} + \tilde{\lambda}_{Ti})}$ and $\Delta_R^{Rel} = \frac{1}{3}\sum_{i=1}^{3} \frac{|\lambda_{Ri} - \tilde{\lambda}_{Ri}|}{\frac{1}{2}(\lambda_{Ri} + \tilde{\lambda}_{Ri})}$ are a measure for the relative errors of the stiffnesses (i.e. eigenvalues) λ of the translational and rotational dimensions and Δ_K is a measure for the coupling between translational and rotational stiffnesses. For Δ_K we simply use the largest absolute value of the elements of $\boldsymbol{K}_{RT}$ and $\boldsymbol{K}_{TR}$ (6).

We used this error vector as is as an intuitive, human readable error representation. Instead of making a norm out of this error vector (which basically means a weighted sum and implies the selection of a proper set of weighting factors), we used a different norm for the optimization algorithm. This also eases the optimization task dramatically:

Weighted Frobenius- or Schur Norm. Let $\boldsymbol{A} = (a_{ij})$ be a $n \times n$ matrix. Let $\mathbf{G} = (g_{ij})$ (*weight matrix*) be a $n \times n$ matrix of positive scalars. The **weighted Frobenius- or Schur norm** of A is defined by $\|\boldsymbol{A}\|_F^G := \left\{\sum_{i,j}^n g_{ij}|a_{ij}|^2\right\}^{\frac{1}{2}}$. Transforming $\mathbf{G}$ to diagonal form (7) we can choose weights for the coupling terms ($\boldsymbol{K}_{RT}$ and $\boldsymbol{K}_{TR}$) (6), the errors for the translational and rotational stiffness coordinate system orientation (off-diagonal elements of $\boldsymbol{K}_T$ and $\boldsymbol{K}_R$, resp.) and the translational and orientational stiffness values (diagonal elements of $\boldsymbol{K}_T$ and $\boldsymbol{K}_R$, resp.).

An ad-hoc solution would be to use a nonlinear optimization algorithm and optimize joint stiffnesses and the nullspace joint position simultaneously, but looking at the problem more in detail, it turns out that a separate optimization of joint stiffnesses and nullspace joint position makes sense.

4.1 Joint Stiffness Optimization

Let $\boldsymbol{k}_J$ be the vector of joint stiffnesses of the diagonal matrix $\boldsymbol{K}_J$. By choosing the Frobenius norm as **norm** in the general optimization problem statement (8), we can assign a unique optimal joint stiffness to a given robot pose for a desired Cartesian stiffness:

$$\boldsymbol{k}_J^{Best}(\boldsymbol{q}) = \left\{\boldsymbol{k}_J,\ \|\boldsymbol{K}_C(\boldsymbol{q},\boldsymbol{k}_J) - \boldsymbol{K}_{CD}\|_F^G \rightarrow Min!\right\}$$

By searching the extrema of $\boldsymbol{k}_J(\boldsymbol{q})$ (i.e. differentiating the Frobenius norm w.r.t. the joint stiffnesses)

$$\frac{\partial}{\partial \boldsymbol{k}_{Ji}} \|\boldsymbol{K}_C(q_1,\ldots,q_n,k_{J1},\ldots,k_{Jn}) - \boldsymbol{K}_{CD}\|_F^G = 0$$

we get a linear equation system with the unknowns k_{Ji}, $i \in \{1,\ldots,n\}$ (since the k_{Ji} enter linearly in $\boldsymbol{K}_C$) of the structure $\boldsymbol{A} \cdot \boldsymbol{k}_J = \boldsymbol{b}$. The matrix $\boldsymbol{A}$ and the right hand side $\boldsymbol{b}$ depend on the robot pose $\boldsymbol{q}$, the desired Cartesian stiffness $\boldsymbol{K}_J$ and the weight matrix $\mathbf{G}$. So we can calculate $\boldsymbol{k}_J^{Best}$ as

$$\boldsymbol{k}_J^{Best}(\boldsymbol{q}) = \boldsymbol{A}^{-1} \cdot \boldsymbol{b} \tag{10}$$

This extremely efficient joint stiffness optimization step can now be used as a sub-procedure in the following, unfortunately much more complicated optimization step for the nullspace of the robot.

4.2 Nullspace Optimization

At a given Cartesian position within the workspace of a redundant robot, there are an infinite number of joint configurations reaching the respective TCP position. The resulting space is the well-known nullspace. So the question arises, whether there are optimal configurations meeting our desired Cartesian stiffness. Figure 3 shows a plot of the norm $B(\boldsymbol{q}) = \|\boldsymbol{K}_C^{Best}(\boldsymbol{q}) - \boldsymbol{K}_{CD}\|_F^G$ optimized by (10), which was generated by sampling the nullspace of a seven DoF robot.

As one can see, the optimization function contains local minima. The dotted line in figure 3 points out the reachable local minima from the position within the nullspace. This is done by a gradient search along the the Gradient of B.

Due to the existence of the local minima, well-known local optimization methods, like the Moore-Penrose PseudoInverse, will lead to very suboptimal results. Therefore the optimization should be performed in less realtime critical planning layer, where other nonlinear optimization methods can be utilized.

5. Simulation Results

Typical results of the method described in sec. 4 for finding the best approximation for $\boldsymbol{K}_C$ are shown in fig. 4 and fig. 5. A fixed Cartesian desired configuration and the same $\boldsymbol{K}_C$ are commanded for a 7 DoF arm (fig. 4) and for the same arm mounted on a three DoF platform, leading to a 10 DoF system (fig. 5). The following weighting matrix was used for the Frobenius norm:

$$\boldsymbol{G} = \begin{pmatrix} 1 & 500 & 500 & 10 & 10 & 10 \\ 500 & 1 & 500 & 10 & 10 & 10 \\ 500 & 500 & 1 & 10 & 10 & 10 \\ 10 & 10 & 10 & 1 & 160 & 160 \\ 10 & 10 & 10 & 160 & 1 & 160 \\ 10 & 10 & 10 & 160 & 160 & 1 \end{pmatrix} \tag{11}$$

It can be seen that different values have been chosen for the different submatrices in (6). This is related to the different scaling of rotations and translations but is also used to weight the various errors from (9) corresponding to the requirements of a specific application. By increasing the off-diagonal weighting of $\boldsymbol{K}_T$ and $\boldsymbol{K}_R$ with respect to the diagonal elements, a better alignment of the principal axes is enforced, in detriment to an exact stiffness value along these axes.

The simulations showed that with the 7 DoF arm, the resulting approximation is in general very poor. In contrast, with the same arm mounted on a mobile platform, an acceptable accuracy can be achieved, although this system also does not have sufficient DoFs for an exact solution. The errors for the analyzed example are summarized in table 1.

	Δ_T^{Rot}	Δ_R^{Rot}	Δ_T^{Rel}	Δ_R^{Rel}
7 DoF	19°	2°	67.76 %	66.09 %
10 DoF	5°	10°	0.29 %	2.59 %

Table 1. Errors of the Cartesian stiffness approximation.

6. Conclusions

The following conclusions can be drawn from this study:
1.) There are severe limitations concerning the implementation of Cartesian stiffness with an usual (6 or 7 DoF) arm with adjustable diagonal joint compliance. 2.) A mobile system with an arm placed on a vehicle or a humanoid torso has a considerably improved performance. 3.) The optimization problem can be divided into a fast (linear) optimization in the space of the joint stiffness values and a slow (nonlinear) optimization in the null-space of the manipulator. This second part can be computed only offline, in a planning phase for the working posture.

Simple applications like "wiping a table" have very different requirements concerning the stiffness accuracy than challenging assembly tasks (e.g., peg in hole). While the paper provides a framework to weight and optimize the various error components, an absolute evaluation of the resulting quality can be done only in the context of the specific task.

7. Acknowledgements

The research work reported here was made possible by the Bavarian ministry of technology within the framework of the Bavarian network of competence in mechatronics (www.bayern-mechatronik.de).

References

Albu-Schäffer, A., & Hirzinger, G.(2002).Cartesian impedance control techniques for torque controlled light-weight robots. *ICRA*, 657-663.

Bicchi, A., Bavaro, G. T. M. & Piccigallo, M.(2003).Variable stiffness actuators for fast and safe motion control. *11th Int. Symp. of Robotics Research (ISRR).*

Chen, S., & Kao, I.(2000).Simulation of conservative congruence transformation conservative properties in the joint and cartesian spaces. *ICRA*, 1283-1288.

H.Kobayashi, Hyodo, K. & Ogane, D.(1998).On tendon driven robotic mechanisms with redundant tendons. *Int. J. of Robotics Research, 17*(5), 561-571.

Hogan, N.(1990).Mechanical impedance of single- and multi- articular systems. In J. Winters & S. Woo (Eds.), *Multiple Muscle Systems: Biomechanics and Muscle Organization* (p. 149-163).Springer-Verlag, New York.

Koganezawa, K., & Ban, S.(2002).Stiffness control of antagonistically driven redundant d.o.f. manipulator. *IROS*, 2280-2285.

Salisbury, J. K.(1980).Active stiffness control of a manipulator in cartesian coordinates. *19th IEEE Conference on Decision and Control*, 83-88.

Schöppe, F.(2003).*Mathematische Modellierung und numerische Simulation von Robotern mit passiv nachgiebigen Gelenken.*Unpublished master's thesis, Tech. Univ. Munich.

Sugano, S.(2002).Human-robot symbiosis. *Workshop on Human-Robot Interaction, ICRA.*

Zinn, M., Khatib, O., Roth, B. & Salisbury, J.(2002).A new actuation approach for human friendly robot design. *Int. Symp. on Experimental Robotics, Ischia.*

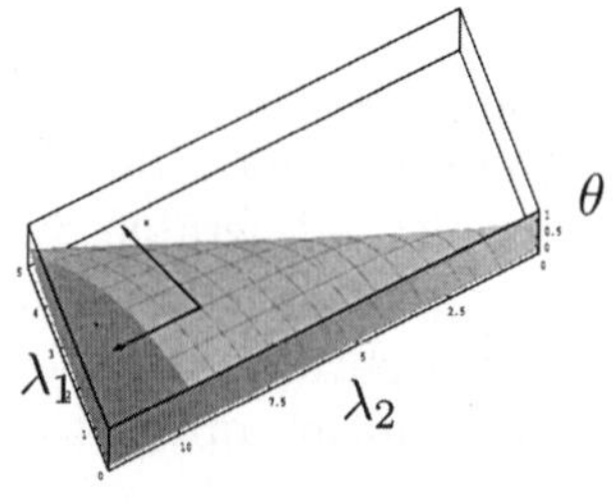

Figure 1. Achievable stiffness with a 2 DoF robot

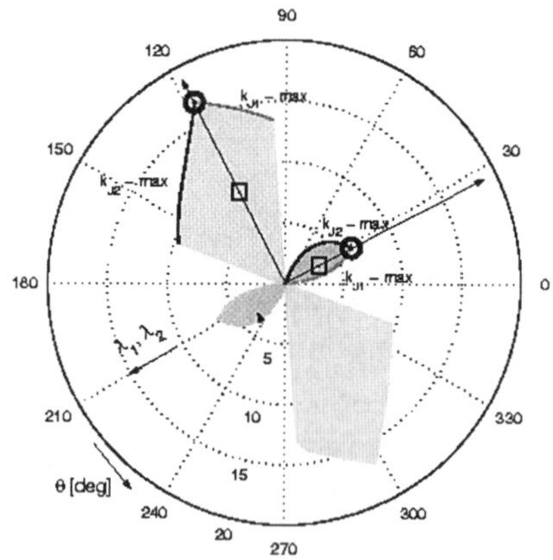

Figure 2. Achievable Cartesian stiffness for $k_{J1}, k_{J2} \in (0, k_{\max}]$. Darker leafs: values for λ_1, lighter leafs: values for λ_2. The principal axes of stiffness are drawn for $\theta = 27.6713°$, $\beta = 1$, what results in $\lambda_2/\lambda_1 = 2.6283$. The values for $\alpha = 1$ are marked with a circle, while the values for $\alpha = 0.5$ are indicated by a rectangle.

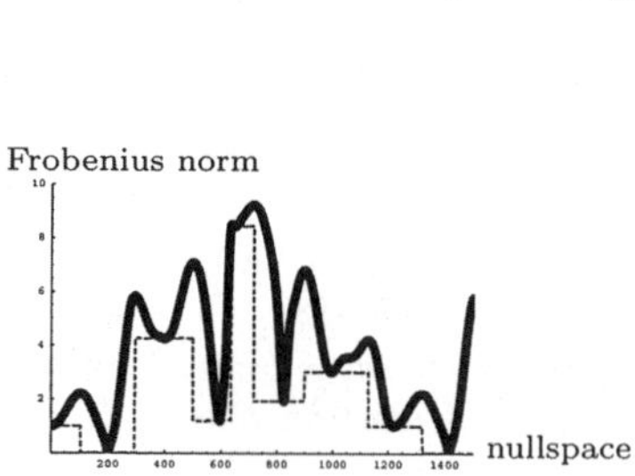

Figure 3. Achievable stiffness quality during a walk through the nullspace of a 7 DoF robot

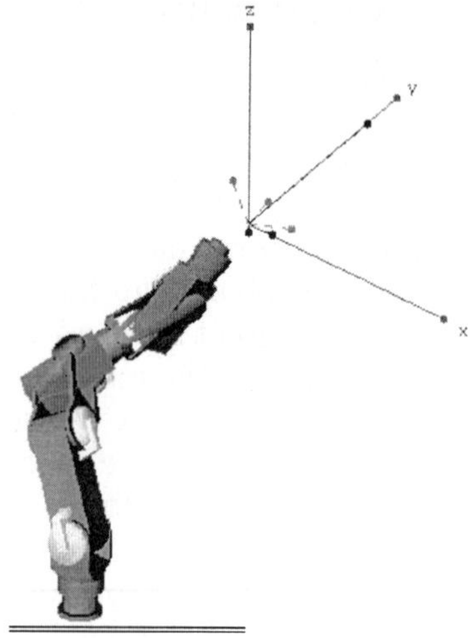

Figure 4. Desired and resulting Cartesian stiffness for a 7 DoF robot. Continuous line: desired stiffness (identical frames and values were chosen for the translational and rotational part); dashed light frame: resulting Cartesian stiffness, translational part; dashed dark frame: resulting Cartesian stiffness, rotational part.

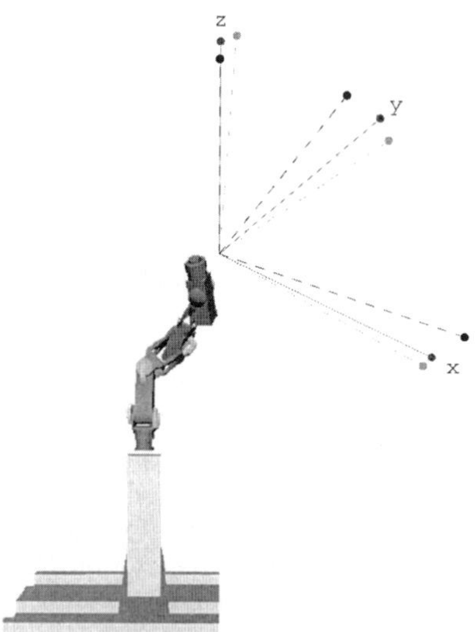

Figure 5. Desired and resulting Cartesian stiffness for a 10 DoF robot. Continuous line: desired stiffness (identical frames and values were chosen for the translational and rotational part); dashed light frame: resulting Cartesian stiffness, translational part; dashed dark frame: resulting Cartesian stiffness, rotational part.

Motion Synthesis and Mobility

MOVEMENT PRIMITIVES AND PRINCIPAL COMPONENT ANALYSIS

Frank C. Park and Kyoosang Jo
School of Mechanical and Aerospace Engineering
Seoul National University
fcp@plaza.snu.ac.kr

Abstract Robot trajectory generation based on the optimization of physical criteria is by now a well-established means of motion programming, but the computational burden still remains a formidable challenge for real-time implementation. Motivated by results from the biological motor control literature, we propose a method for constructing movement primitives based on principal component analysis (PCA) of a set of reference motions. The movement primitives provide an efficient and task-specific set of basis functions for the representation of general motions; by using the basis functions constructed from a set of reference motions associated with a given task, sub-optimal motions that closely resemble the reference motions can be generated rapidly. Experimental results obtained from motion capture data of human arm motions are offered to support the merits of our PCA-based approach.

Keywords: Movement primitive, trajectory generation, motion optimization, principal component analysis.

1. Introduction

The difficulty in generating movements for robots and other complex articulated systems can, we believe, ultimately be traced to the high dimensionality of the problem. The kinematic structure usually has redundant degrees of freedom, a desired task can usually be achieved by infinitely many end-effector trajectories, and both the admissible joint and task configuration spaces exhibit some nonlinearity and irregularity as a result of physical and mechanical constraints.

Yet in spite of these difficulties, humans and animals are able to generate movements instantaneously and with seemingly little effort. To endow robots with similar movement generation capabilities, therefore, it is natural to look to animal motor control for clues. While a widely accepted comprehensive theory of how animals generate movements has yet to emerge, two phenomena seem to be consistently observed in the biological movement literature: (i) the way in which the degrees of freedom appear to be grouped for planning and control purposes, and (ii)

J. Lenarčič and C. Galletti (eds.), On Advances in Robot Kinematics, 421-430.

evidence suggesting the existence of movement primitives, and how they are concatenated to form more complex motions.

There is also strong evidence that some type of optimization is taking place, usually with respect to some physical criterion. In many cases the motions appear to become smoother and richer with practice, as more degrees of freedom are being utilized to enhance performance; this is readily evident, for example, when watching someone learning how to swing a golf club. As movement learning occurs we also see a transition from closed-loop to open-loop control. In the golf swing example, while initially there is a heavy reliance on visual feedback (to ensure proper posture, etc.), with practice one relies less and less on such external sensory feedback as the body "learns" the swing.

Taking these lessons and applying them to robots, we are faced with the following practical issues: (i) how do we select the movement primitives that constitute the vocabulary of motions? (ii) how do we optimize the movements in real-time? This paper offers a framework in which to simultaneously address these two issues. First, movement primitives are defined according to the task, and stored not as a single trajectory or curve, but as a finite family of basis functions. Given a specific task description, the objective is to find the optimal linear combination of basis functions to generate the required movement.

The distinguishing feature of our approach is that the basis functions are determined from principal component analysis (PCA). This has two advantages: because the optimized motions are from a restricted class, they are similar to the benchmark motions used to create the basis functions. This can be useful, for example, when we desire to create human-like (and not necessarily physically optimal) motions rapidly for robots and animated characters. Secondly, by restricting the number of basis functions associated with any given task, the optimal motions can be obtained efficiently, possibly even in real-time.

The relevant literature on biological motor control and movement primitives is extensive (see, *e.g.*, the seminal work of Bernstein, 1967), as are recent approaches to robot movement generation and control inspired by the biological paradigm, *e.g.*, Muss-Ivaldi, 1997, Schaal et al., 2003, Fod et al., 2002. The interpretation of movement primitives as a set of basis functions specific to a given task, combined with the application of principal component analysis to generate these primitives, has, to our knowledge, not been addresssed in the literature.

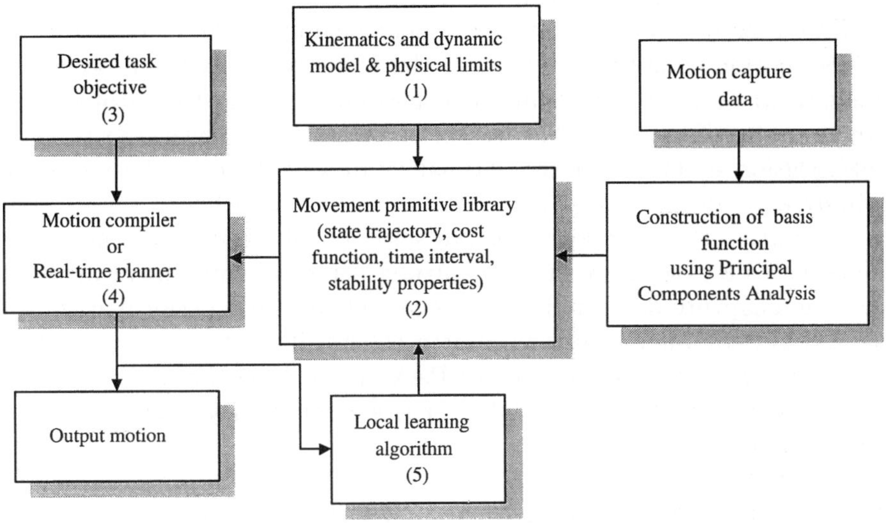

Figure 1. Movement Generation Framework

2. Movement Generation Framework

Our proposed high-level movement coordination and learning framework, in particular how we intend to characterize the problem and organize the data, is shown in Fig. 1. The kinematic and dynamic features of the robot, and all the data needed to define a reasonably good dynamic model, are first defined (Block 1). With the desired class of goal behaviors in mind, we then create a rich set of movement primitives (Block 2) that will form the basis for the motion compiler. Each movement primitive consists of a finite set of basis functions, and will ultimately be paired with an associated control law and its stability properties. A desired task goal (Block 3) is then input to the planner (Block 4).

The planner chooses the motion sequence that best achieves the desired task. This is done at two levels. At a lower level, given the set of basis functions corresponding to a given movement primitive, local optimization is performed to determine the optimal set of weights; the objective function to be used in the optimization is given by the planner. At a higher level, in the event that the task calls for a more complex sequence of movements, the planner first selects and concatenates a set of movement primitives. The concatenation may be done in real-time if optimal behavior is not required. If repeated (on-line) trials are possible, or if planning may be conducted off-line, local learning is used (Block 5) to improve the resulting motion. The learning block refines the entire movement sequence by scaling and warping the selected movement primitives in order to best achieve the desired goal. The refined movement itself may be be used as a new movement primitive if desired.

The motion compiler will convert the high level goal to an objective function to be used in a nonlinear constrained parameter optimization program; in effect, it is the motion compiler that performs the local optimization to determine the optimal set of weights for each movement primitive.

The focus of this paper is on the selection of basis functions corresponding to a movement primitive, using principal component analysis of motion capture data, and on the local optimization performed within the motion compiler. By relying on a set of differential geometric formulations for multibody dynamics Park et al., 1995, Kim et al., 1999, we are able to determine analytic, or exact, gradients of the objective function and the constraints.

3. Principal Component Analysis

Principal Component Analysis (PCA) is a statistical methodology based on singular value decompositions for identifying patterns in, and determining compressed approximate representations of, a set of given data (see, *e.g.*, Jackson, 1993). For our particular movement primitive application, we focus exclusively on three-dimensional arm motions. We first collect motion capture data for a wide array of arm motions corresponding to some task (*e.g.*, swinging a tennis racket, lifting a heavy weight); This usually entails attaching markers to reference points on the arm, and solving the inverse kinematics equations to obtain a set of joint angle trajectories.

To illustrate the general PCA procedure, consider a p-dimensional vector $\vec{X} = (X_1, X_2, \cdots, X_p)^T$, and suppose we have vector time series data for this vector $\vec{X}$. Let the sample mean and covariance matrix associated with this vector time series be respectively given by μ and Σ, respectively. Represent the eigenvalue-normalized eigenvector pairs of the covariance matrix by

$$(\lambda_1, \vec{e}_1), (\lambda_2, \vec{e}_2), \cdots, (\lambda_p, \vec{e}_p)$$

$$\text{where} \quad \begin{array}{l} \vec{e}_i^T \vec{e}_i = 1, \;\; \vec{e}_i^T \vec{e}_k = 0, \;\; i \neq k \\ \lambda_1 \geq \lambda_2 \geq \cdots \geq \lambda_p > 0 \end{array} \tag{1}$$

In the event that Σ is positive definite, the eigenvectors serve as a set of basis vectors; the relative magnitude of the eigenvalues implies how dominant the corresponding eigenvectors are in representing the time series data. These eigenvectors represent the principal components asssociated with the given data, and have the following properties;

- i^{th} principal component:

$$Y_i = \vec{e}_i^T \vec{X} = e_{1i}X_1 + e_{2i}X_2 + \cdots + e_{pi}X_p, \quad i = 1, 2, \cdots, p \tag{2}$$

- variance and covariance of ith principal component:

$$\begin{aligned} Var(Y_i) &= \lambda_i, \quad i = 1, 2, \cdots, p \\ Cov(Y_i, Y_k) &= 0, \quad i \neq k \end{aligned} \tag{3}$$

- variance equality relationship:

$$\sum_{i=1}^{p} Var(X_i) = \sum_{i=1}^{p} Var(Y_i) = \lambda_1 + \lambda_2 + \cdots + \lambda_p \tag{4}$$

- normalized variance equality relationship:

$$\frac{Var(Y_k)}{\sum_{i=1}^{p} Var(X_i)} = \frac{\lambda_k}{\lambda_1 + \lambda_2 + \cdots + \lambda_p}, \quad k = 1, 2, \cdots, p \tag{5}$$

By selecting the principal components with the highest weights, we can use these comoponents as basis functions with which the construct more general trajectories. Of course, this set of basis functions does not constitute a basis in the strict linear algebraic sense, since it can only represent a finite dimensional subset of all possible trajectories. However, the hope is that by capturing the most essential features of a given set of captured motions through its principal components, one can come up with reasonable approximations to this class of motions with great efficiency and speed.

4. Case Study

In this section we explore an example of generating arm raising motions based on PCA of human arm motions obtained from motion capture data. We model the kinematic structure of the human arm using 3 links (upper arm, lower arm, chest) and 2 joints (shoulder, elbow). The shoulder joint is modeled as a three dof spherical joint (obtained by serially arranging three revolute joints in orthogonal fashion), while the elbow joint is a single dof revolute joint. The inertial parameters that we use for the human arm model are shown shown in Table 1.

We first obtain motion capture data for various typical arm motions, such as lifting, pointing, pulling, pushing, rotating, etc.. We then calculate the joint angle trajectories for each motion from the inverse kinematics equations. For our experiment we obtain a total of 52 joint angle

Table 1. The mass and the moment of inertia of the human arm model

Link	*Mass(kg)*	*Moment of inertia(kgm^2)*
Chest	29.27	(0.73, 0.63, 0.32)
Upper arm	2.97	(0.025, 0.025, 0.005)
Lower arm	1.21	(0.005, 0.0054, 0.0012)

trajectory samples. Following the procedure outlined in the previous section, we then obtain the principal components corresponding to our motion data. We discover that the first two principal components explain more than 90% of the data; Table 2 shows the weight of each variance component according to each joint, while Figure 2 shows the graph of each joint's principal components, represented as a joint trajectory in time. For our simulation example we determine the minimum-effort

Table 2. Variance weight according to each joint

Joint	*Shoulder1(%)*	*Shoulder2(%)*	*Shoulder3(%)*	*Elbow(%)*
PC1	92.49	82.37	85.63	74.36
PC2	6.39	12.65	11.90	23.50
PC3	0.82	3.69	1.90	1.65
PC4	0.13	0.81	0.29	0.28
Sum	99.83	99.52	99.72	99.79

control motions for specified beginning and end points in the joint space Martin and Bobrow, 1995, Kim et al., 1999. The trajectories are represented as a linear combination of the basis functions obtained via PCA, approximated by fourth-order polynomials. The cost function is defined as:

$$J = \frac{1}{2} \int_0^{t_f} \|\tau(t)\|^2 dt \tag{6}$$

where τ denotes the torque vector and t_f the final time. The dynamics computations were performed using the geometric formulation described in Park et al., 1995, Kim et al., 1999; The optimizations were performed in Matlab (ver. 6.5) on a 1.8 GHz Pentium 4, making use of the analytic gradient information provided by our geometric dynamics formulation.

To benchmark the gains in computational efficiency afforded by the PCA approach, we also perform the motion optimization using B-splines with 6 control points (as opposed to our basis trajectories obtained from PCA). Our simulation results indicate that, with respect to time cost,

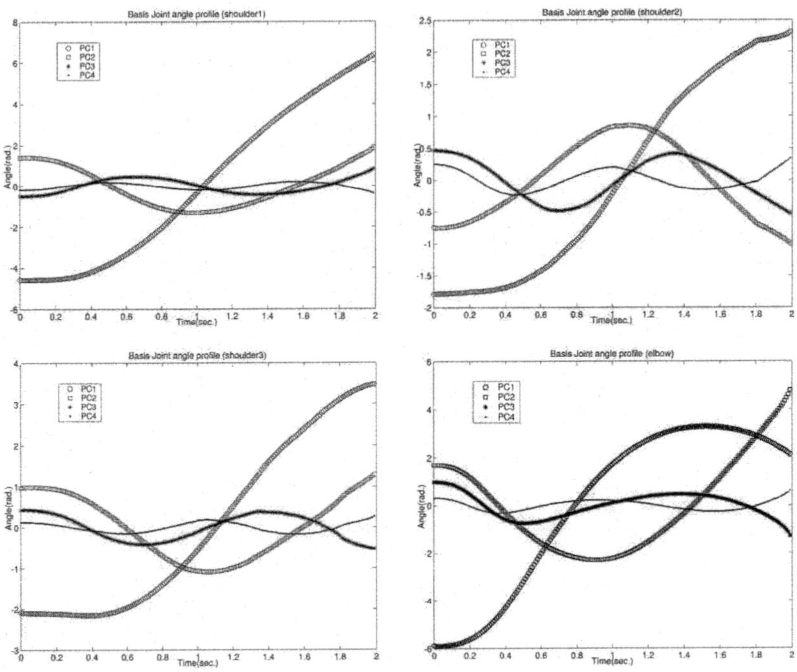

Figure 2. Graph of Each joint' PC

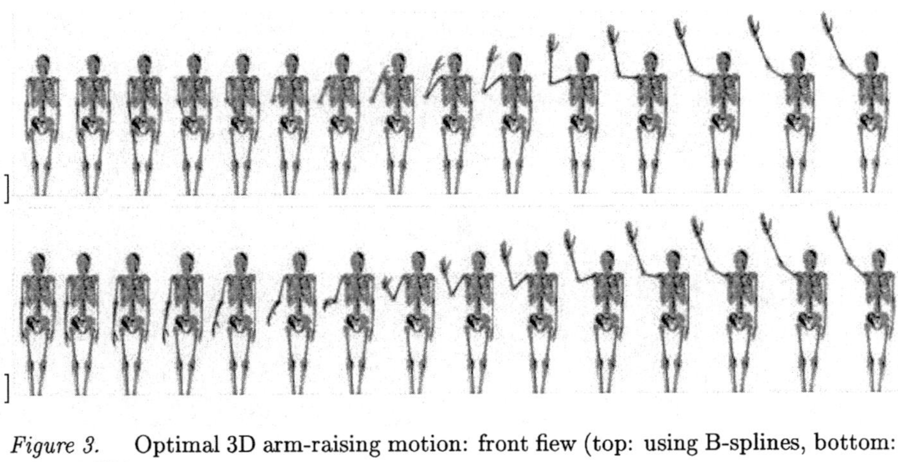

Figure 3. Optimal 3D arm-raising motion: front fiew (top: using B-splines, bottom: using PCA)

the PCA-based approach is more efficient than the B-spline approach (Table 3). Not surprisingly, the B-spline approach leads to smaller values of the cost function, indicating that the PCA based approach leads to suboptimal solutions.

This is not necessarily a disadvantage, and in human animation and computer graphics applications, can in fact be an important advantage. Figures 3 and 4 offer a side-by-side comparison of the optimal motions obtained using PCA and B-splines. In the B-spline case, the arm is first rotated away from the body (front view), and a retraction of the arm behind the body can be seen (side view). This is quite unnatural compared to human arm motions. On the other hand the motion obtained using PCA can be seen to be resemble much more closely to human arm motions; more natural motions can be generated with much less effort, an important benefit from the perspective of realistic human motion animation.

Table 3. Arm-raising motion: performance comparison

Type	*B-spline*	*PCA*
objective function value	5.19594	9.58805
optimization time(sec.)	148.9750	96.5090

5. Conclusions

In this paper we have proposed a means of constructing movement primitives based on principal component analysis (PCA) of a set of reference motions. Inspired by the biological movement control literature, which suggests that human and animal motions are organized into a vocabulary of primitive movements, with mechanisms available for local

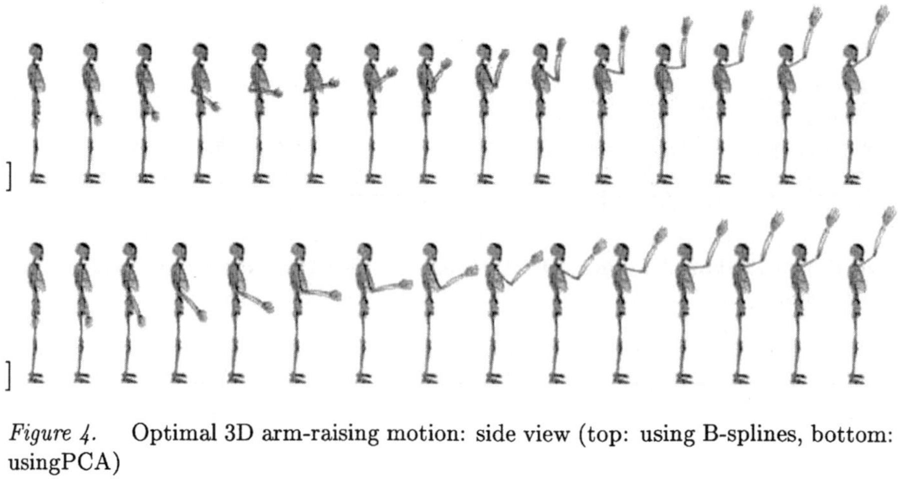

Figure 4. Optimal 3D arm-raising motion: side view (top: using B-splines, bottom: usingPCA)

learning and optimization, we adopt this framework for robot trajectory generation.

The movement primitives provide an efficient and task-specific set of basis functions for the representation of general motions; by using the basis functions constructed from a set of reference motions associated with a given task, sub-optimal motions that closely resemble the reference motions can be generated rapidly. Experimental results were performed using motion capture data of human arm motions, and compared to straightforward minimum torque optimization using standard B-splines. The results suggest that not only can the computational burden of movement generation be significantly reduced, but that much more natural-looking arm motions can be generated.

Our approach can in some sense be viewed as exploiting the classic computational tradeoff between memory and computing resources. By storing the movement primitives as basis functions, we significantly reduce the computational load involved in the optimization, while at the same time limiting the class of generated motions to be close to the set of reference trajectories used in the basis function selection. In future work we hope to examine a richer class of tasks, and extend the framework to more complex kinematic structures.

Acknowledgements

This research was supported by the National Research Laboratory for CAD/CAM at Seoul National University.

References

Bernstein, N.A. (1967), *The Coordination and Regulation of Movements.* London: Pergamon Press.

Fod, A., Mataric, M., Jenkins, O.C. (2002), "Automated derivation of primitives for movement classification," *Autonomous Robots*, vol. 12, no. 1, pp. 39–54.

Jackson, J.E. (1993), *A User's Guide to Principal Components.* New York: John Wiley and Sons.

Kim, J., Baek, J., Park, F.C. (1999), "Newton-type algorithms for robot motion optimization," *Proc. 1999 IEEE Int. Conf. Intelligent Robots and Systems (IROS)*, vol. 3, pp. 1842–1847.

Martin, B.J., Bobrow, J.E. (1995), Determination of minimum-effort motions for general open chains, *Pro. IEEE Int. Conf. Robotics and Automation*, pp. 1160–1165.

Murray, R.M., Li, Z., and Sastry, S.S. (1993), *A Mathematical Introduction to Robotic Manipulation.* Boca-Raton: CRC Press.

Mussa-Ivaldi, F.A. (1997), "Nonlinear force fields: a distributed system of control primitives for representing and learning movements," *Proc. IEEE Int. Symp. Computational Intelligence in Robotics and Automation*, pp. 84–90.

Park, F.C., Bobrow, J.E., Ploen, S.R. (1995), A Lie group formulation of robot dynamics, *Int. J. Robotics Research*, vol. 14, no. 6, pp. 609–618.

Schaal, S., Peters, J., Nakanishi, J., Ijspeert, A. (2003), Control, planning, learning, and imitation with dynamic movement primitives, in Workshop on Bilateral Paradigms on Humans and Humanoids, *IEEE International Conference on Intelligent Robots and Systems (IROS).*

PARALLEL MECHANISMS with PSEUDO-PLANAR MOTION GENERATORS

Jacques M. Hervé
Ecole Centrale Paris
92295 Chatenay-Malabry (France)
jherve@ecp.fr

Abstract The pseudo-planar displacements are defined and their essential properties are disclosed. Several parallel arrangements of pseudo-planar motion generators are proposed and some examples of new parallel manipulators, which can be very simply controlled, are presented.

Keywords: pseudo-planar motion, parallel manipulator, Lie group, kinematics.

1. Introduction

Parallel manipulators (PM) that produce motion with less than six degrees of freedom (dof) are attracting an increasing number of researchers about the type synthesis of mechanisms. Despite the well-known theorem of Mozzi-Chasles, which stipulates that any rigid body motion is a screw motion, most authors ignore the implementation of screw pairs. Nowadays, high quality screw bearings with low friction are readily available. Hence it is worth considering mechanisms, which natively require screw pairs. A special 3-dof motion type, which demands one or several screw pairs for its mechanical generation can be called pseudo-planar motion and is denoted *Y*-motion. A set of pseudo-planar displacements with a given plane direction is a 3-dimensional Lie subgroup of the 6-dimensional Lie group of displacements. As early explained by Hervé, 1978 and Angeles, 1982, the algebraic properties of the displacement set, which is a Lie group, play a key role in the understanding of mobility. Fanghella and Galletti (1994) addressed this approach in an algorithmic manner. In order to allow truly geometrical reasoning, an intrinsic formulation of the displacement transformations has to be employed, Hervé, 1982. In the same way, the Lie algebra of the twist velocity fields has to be presented through the use of intrinsic geometrical entities like points, vectors and linear operators acting on vectors, Hervé, 1994, 1999. Then the algebraic method becomes also a tool for the type synthesis of mechanisms and especially of sub-6-dof PM,

J. Lenarčič and C. Galletti (eds.), On Advances in Robot Kinematics, 431-440.

Hervé, 2000. After recalling some definitions and essential properties of the Y-motion, the article emphasizes examples of PM that benefit from Y-motion generators.

2. Pseudo-planar motion

Among the Lie subgroups of the displacement Lie group $\{D\}$, a 3-dimensional set, which includes any translation parallel to a given plane and helical motions of given pitch about any axis perpendicular to the plane can play an important role in the design of sub-6-dof PM. Such a displacement type can be called a pseudo-planar displacement. As a matter of fact many properties are those of a planar motion and in a special case the pseudo-planar motion becomes a planar motion when the pitch of the helical motions is zero. A Lie group of pseudo-planar displacements is denoted $\{Y(\mathbf{w}, \mathrm{p})\}$ where curly brackets indicate a set, Y stands for pseudo-planar, $\mathbf{w}$ is a given unit vector perpendicular to the plane Pl and p is the given pitch of the helical displacements. (O, $\mathbf{u}$, $\mathbf{v}$, $\mathbf{w}$) being an orthonormal frame of reference and M being a generic point of the 3-dimensional Euclidean affine space, a displacement $Y(\mathbf{w}, \mathrm{p}; \mathrm{a}, \mathrm{b}, \theta)$ belonging to $\{Y(\mathbf{w}, \mathrm{p})\}$ transforms M into the point M' according to the formula

$$\mathrm{M} \longrightarrow \mathrm{M}' = \mathrm{O} + \mathrm{a}\,\mathbf{u} + \mathrm{b}\,\mathbf{v} + (\theta\,\mathrm{p}/2\pi)\,\mathbf{w} + \exp(\theta\,\mathbf{wx})\,(\mathbf{OM}) \quad (1)$$

($\mathbf{OM}$) denotes the vector obtained from the ordered couple of points O and M; $\mathbf{wx}$ is the skew-symmetric linear operator of the vector product by $\mathbf{w}$. a, b and θ are three canonical parameters of the 3-dimensional Lie group. It is worth remarking that planar displacements are obtained for $\mathrm{p} = 0$; in other terms $\{Y(\mathbf{w}, 0)\} = \{G(\mathbf{w})\} = \{G(\mathrm{Pl})\}$, Pl being a plane perpendicular to $\mathbf{w}$. One can write also

$$(\mathbf{OM'}) = \mathrm{a}\,\mathbf{u} + \mathrm{b}\,\mathbf{v} + (\theta\,\mathrm{p}\,/2\pi)\,\mathbf{w} + \exp(\theta\,\mathbf{wx})\,(\mathbf{OM}) \quad (2)$$

$\exp(\theta\,\mathbf{wx})\,(\mathbf{OM})$ is obtained by the action on ($\mathbf{OM}$) of the exponential series of the linear operator $\theta\,\mathbf{wx}$

$$\exp(\theta\,\mathbf{wx})\,(\mathbf{OM}) = (\mathbf{OM}) + \theta\,\mathbf{wx}\,(\mathbf{OM}) + (1/2!)\,\theta^2\,\mathbf{wx}\,[\mathbf{wx}\,(\mathbf{OM})] + \ldots$$

$$+ (1/n!)\,\theta^n\,(\mathbf{wx})^n\,(\mathbf{OM}) + \ldots \quad (3)$$

$$= (\mathbf{OM}) + \sin\theta\,\mathbf{wx}\,(\mathbf{OM}) + (1 - \cos\theta)\,\mathbf{wx}\,[\mathbf{wx}\,(\mathbf{OM})] \quad (4)$$

by using $(\mathbf{wx})^3 = -\,\mathbf{wx}$ that can be demonstrated employing the formula of the double vector product.

The Lie subgroup $\{T(\mathrm{Pl})\}$ of translations that are parallel to the plane Pl , which is perpendicular to $\mathbf{w}$, corresponds to θ identical to 0. If a and b remains equal to 0, the transformations are those of the Lie subgroup $\{H(\mathrm{O}, \mathbf{w}, \mathrm{p})\}$ of helical displacements of pitch p around the axis (O, $\mathbf{w}$), (O, $\mathbf{w}$) characterizing the axis. Any set of helical displacements of pitch p around any axis parallel to (O, $\mathbf{w}$) is included into $\{Y(\mathbf{w}, \mathrm{p})\}$. For example, A being a point not belonging to (O, $\mathbf{w}$), an helical transform of pitch p around (A, $\mathbf{w}$) with the angle φ is determined by

$$(\mathbf{AM'}) = (\varphi\,\mathrm{p}\,/2\pi)\,\mathbf{w} + \exp(\varphi\,\mathbf{wx})\,(\mathbf{AM}) \quad (5)$$

$$(\mathbf{AO}) + (\mathbf{OM'}) = (\varphi\, p / 2\pi)\, \mathbf{w} + \exp(\varphi\, \mathbf{wx})\, [(\mathbf{AO}) + (\mathbf{OM})\,] \qquad (6)$$

The operator $\exp(\varphi\, \mathbf{wx})$ is linear:

$$(\mathbf{OM'}) = (\mathbf{OA}) + \exp(\varphi\, \mathbf{wx})\, (\mathbf{AO}) + (\varphi\, p/2\pi)\, \mathbf{w} + \exp(\varphi\, \mathbf{wx})\, (\mathbf{OM}) \qquad (7)$$

$$(\mathbf{OM'}) = \sin\varphi\, \mathbf{wx}\, (\mathbf{AO}) + (1 - \cos\varphi)\ \mathbf{wx}\, [\mathbf{wx}\, (\mathbf{AO})] + (\varphi\, p/2\pi)\, \mathbf{w} + \exp(\varphi\, \mathbf{wx})\, (\mathbf{OM}) \qquad (8)$$

The vector $\mathbf{wx}$ $(\mathbf{AO})$ is perpendicular to $\mathbf{w}$ and therefore is parallel to the plane Pl. $\mathbf{wx}$ [$\mathbf{wx}$ $(\mathbf{AO})$] is also perpendicular to $\mathbf{w}$ and thus $\sin\varphi\, \mathbf{wx}\, (\mathbf{AO}) + (1 - \cos\varphi)\, \mathbf{wx}[\mathbf{wx}\, (\mathbf{AO})]$ is a vector parallel to Pl. In other terms the tranformation has the form given by (2) and therefore belongs to the group $\{Y(\mathbf{w}, p)\}$.

The Lie subgroups of $\{Y(\mathbf{w}, p)\}$ are consequently $\{H(A,\mathbf{w}, p)\}$, $\forall$ A, $\{T(\mathbf{s})\}$ linear translations parallel to $\mathbf{s}$, $\forall\, \mathbf{s}$ // Pl or $\forall\, \mathbf{s} \perp \mathbf{w}$, and $\{T(\mathrm{Pl})\}$.

Lower kinematic pairs produce motion having a globally invariant surface and any invariant in any group of transformations determines a subgroup. Consequently the lower kinematic pairs generate subgroups of $\{D\}$. A screw pair H of axis (A, $\mathbf{w}$) generates $\{H(A,\mathbf{w}, p)\}$, a prismatic pair P parallel to $\mathbf{s}$ generates $\{T(\mathbf{s})\}$ and there is no lower pair that produces directly $\{T(\mathrm{Pl})\}$.

The set of allowed displacements of a rigid body with respect to another body of the same kinematic chain is called kinematic bond between the bodies. The serial setting of two kinematic pairs generates a kinematic bond between the distal bodies, which bond is the product of the pair bonds. For example the sequence HH of two screw pairs with axes (A, $\mathbf{i}$) and (B, $\mathbf{j}$) and with pitches p and q produces the set product $\{H(A,\mathbf{i}, p)\}.\{H(B,\mathbf{j}, q)\}$. By definition, in any group the product of sets is the set of products. Generally the product of two subgroups is not a subgroup. For example generally $\{H(A,\mathbf{i}, p)\} \cap \{H(B,\mathbf{j}, q)\} = \{E\}$, E being the identical transform, and $\{H(A,\mathbf{i}, p)\}.\{H(B,\mathbf{j}, q)\}$ is a 2-dimensional manifold included in $\{D\}$. If two factors of the product are included in a Lie subgroup then the product is also included in the Lie subgroup because of the product closure in the subgroup. Consequently $\{H(A,\mathbf{w}, p)\}$ and $\{H(B,\mathbf{w}, p)\}$ being included in the subgroup $\{Y(\mathbf{w}, p)\}$, the product $\{H(A,\mathbf{w}, p)\}.\{H(B,\mathbf{w}, p)\}$ is included in $\{Y(\mathbf{w}, p)\}$. Furthermore,we have $\{H(A,\mathbf{w}, p)\}.\{H(B,\mathbf{w}, p)\}.\{H(C,\mathbf{w}, p)\} \subseteq \{Y(\mathbf{w}, p)\}$. If we have $\mathbf{wx}(\mathbf{AB}) \neq \mathbf{0}$, $\mathbf{wx}(\mathbf{BC}) \neq \mathbf{0}$, $\mathbf{wx}(\mathbf{CA}) \neq \mathbf{0}$, then $\{H(A,\mathbf{w}, p)\}.\{H(B,\mathbf{w}, p)\}.\{H(C,\mathbf{w}, p)\}$ is a 3-dimensional manifold included with the broad meaning in the 3-dimensional Lie group $\{Y(\mathbf{w}, p)\}$. Dealing only with motion types and ignoring possible different amplitudes of the parameter variations, the following equality between two finite neighborhoods of the identical displacement E is proven:

$$\{H(A,\mathbf{w}, p)\}.\{H(B,\mathbf{w}, p)\}.\{H(C,\mathbf{w}, p)\} = \{Y(\mathbf{w}, p)\} \qquad (9)$$

The practical consequence is that a serial setting <u>*HHH*</u> of three equal helical pairs with non-coaxial parallel axes generates a subgroup of pseudo-planar displacements. By the same way the other generators of

Y-motion are obtained, namely *HHP*, *HPH*, *PPH*, *PHP* and their kinematic inversions *PHH*, *HPP* provided *P* pairs are not parallel (fig. 1).

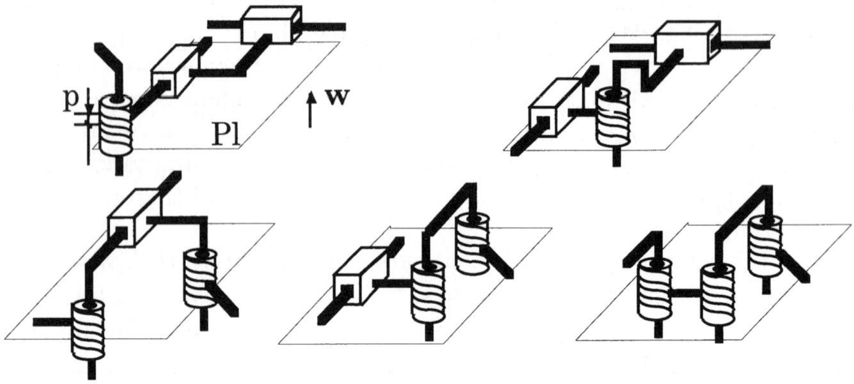

Figure 1. Generators of pseudo-planar displacements

3. Parallel setting of pseudo-planar motion generators

The parallel setting of two generators of two kinematic bonds between a fixed base and a moving platform produces a displacement set (or kinematic bond) between the base and the platform, which is the intersection of the two parallel bonds. It is worth recalling the theorem: the intersection of two subgroups is always a subgroup.

3.1 Equal pseudo-planar motions

A given Lie subgroup $\{Y(\mathbf{w}, \mathrm{p})\}$ has multiple generators that are kinematic equivalencies. Two parallel distinct limbs, each of which generates the same subgroup $\{Y(\mathbf{w}, \mathrm{p})\}$ yields the kinematic bond $\{Y(\mathbf{w}, \mathrm{p})\} \cap \{Y(\mathbf{w}, \mathrm{p})\} = \{Y(\mathbf{w}, \mathrm{p})\}$. The parallel arrangement of three limbs producing the same *Y* bond is a pseudo-planar PM (fig.2).

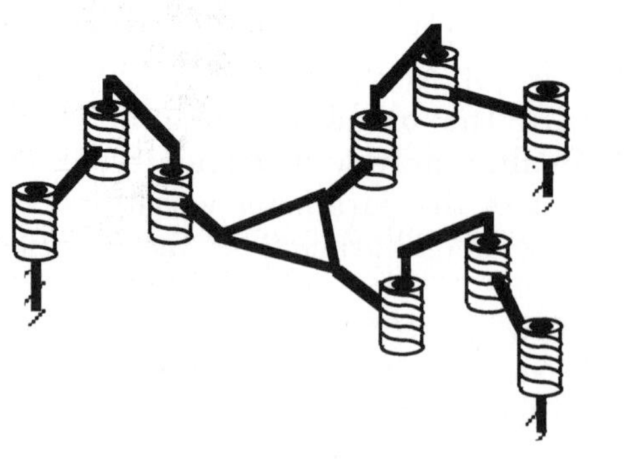

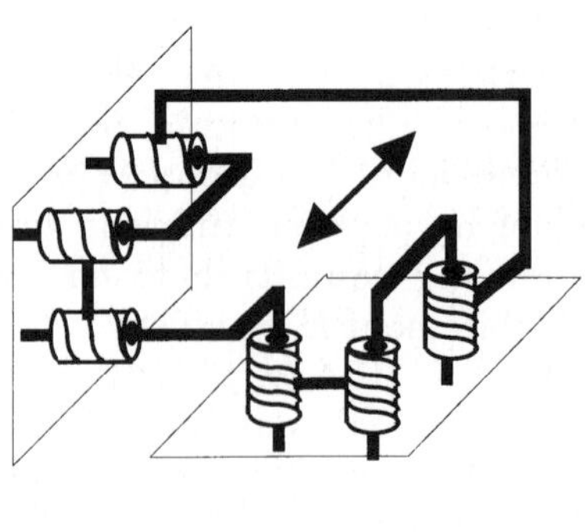

Figure 2. Pseudo-planar parallel mechanism *Figure 3.* 1-dof translation linkage

3.2 Non-parallel pseudo-planar motions

If **u** ≠ **v**, each of two pseudo-planar subgroups {*Y*(**u**, p)} and {*Y*(**v**, q)} contains always a common 1-dimensional subgroup of translation {*T*(**w**)}, if the unit vector **w** is perpendicular to both **u** and to **v** and therefore one can readily establish {*Y*(**u**, p)} ∩ {*Y*(**v**, q)} = {*T*(**w**)}. In other terms, two parallel limbs generating {*Y*(**u**, p)} and {*Y*(**v**, q)} produce {*T*(**w**)},between the two bodies that are connected by the two limbs. For example the linkage *HHH* - *HHH*, figure 3, is the generalization of the Sarrus *RRR* - *RRR* mechanism.

3.3 Parallel pseudo-planar motions

If p ≠ q , {*Y*(**w**, p)} ∩ {*Y*(**w**, q)} = {*T*(Pl)}, Pl being the plane (actually any plane parallel to Pl) that is perpendicular to **w**. Thus 2-dof translational PMs can be designed. Limbs with two prismatic pairs, namely *PPH* , *HPP* and *PHP* , have only a trivial interest because in this case the device is a 2-*PP* PM with idle *H* pairs.

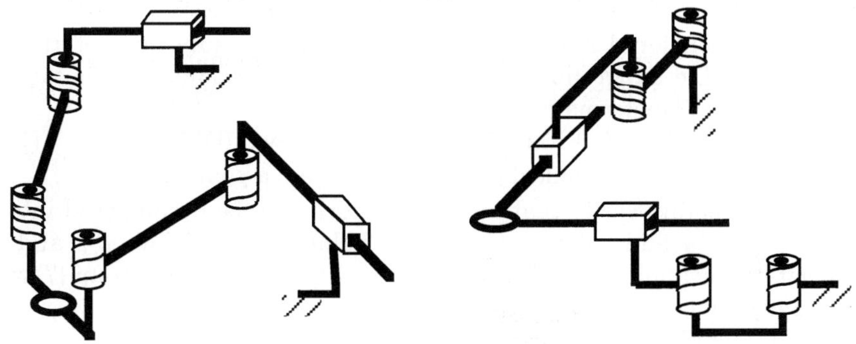

Figure 4. 2-dof translation PMs of types 2-*PHH* and 2-*HHP*

Two limbs of structural type *PHH* yield the mechanism of figure 4(left), which has the properties of the planar "Orthoglide" studied by Majou et al., (2002). Figure 4(right) shows a kinematic inversion of the previous one. Two other PMs can be constructed (figure 5). The reversible working of the screw pairs is required, else a jamming effect can lock these mechanical systems.

Each of the two limbs may include an actuated kinematic pair, which may be either a prismatic pair or a screw pair. Because of the unavailability of direct screwing motors, auxiliary mechanisms are needed for the possible screw actuation. It is worth remarking that two helical pairs fixed on the moving platform can be chosen coaxial thus constituting a kind of compasses.

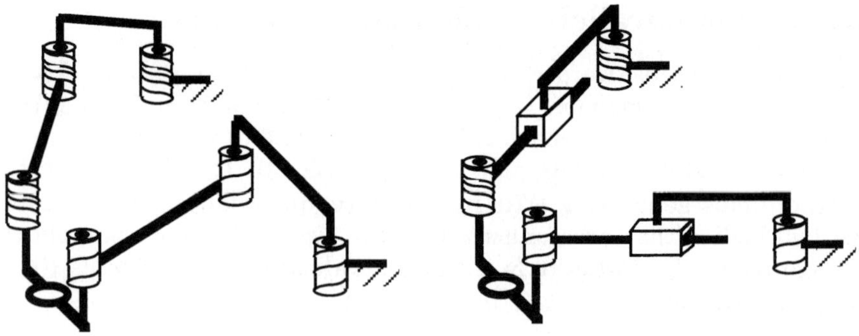

Figure 5. 2-dof translation PMs of types 2-<u>*HHH*</u> and 2-<u>*HPH*</u>

4. Schoenflies motion

4.1 Special serial generation of *X*-motion

The 3-dimensional subgroup $\{Y(\mathbf{w}, \mathrm{p})\}$ is a Lie subgroup of the 4-dimensional Lie subgroup $\{X(\mathbf{w})\}$ of Schoenflies motions, which can be defined by the transformations having the form

$$\mathrm{M} \longrightarrow \mathrm{M}' = \mathrm{O} + \mathrm{a}\,\mathbf{u} + \mathrm{b}\,\mathbf{v} + \mathrm{c}\,\mathbf{w} + \exp(\theta\,\mathbf{w}\mathbf{x})\,(\mathbf{OM}) \tag{10}$$

a, b, c and θ are four canonical parameters of such a point transform.

One can readily verify that any *X*-transform can be decomposed into a product of a pseudo-planar transform belonging to $\{Y(\mathbf{w}, \mathrm{p})\}$ and a translation parallel to **w** belonging to the 1-dimensional subgroup $\{T(\mathbf{w})\}$. As a matter of fact we can write

$$\mathrm{M} \longrightarrow \mathrm{M}_1 = \mathrm{O} + \mathrm{a}\,\mathbf{u} + \mathrm{b}\,\mathbf{v} + (\theta\,\mathrm{p}/2\pi)\,\mathbf{w} + \exp(\theta\,\mathbf{w}\mathbf{x})\,(\mathbf{OM}) \tag{11}$$

$$\mathrm{M}_1 \longrightarrow \mathrm{M}' = \mathrm{M}_1 + \mathrm{c}_1\,\mathbf{w} \tag{12}$$

$$\Rightarrow \mathrm{M} \longrightarrow \mathrm{M}' = \mathrm{O} + \mathrm{a}\,\mathbf{u} + \mathrm{b}\,\mathbf{v} + [\mathrm{c}_1 + (\theta\,\mathrm{p}/2\pi)]\,\mathbf{w} + \exp(\theta\,\mathbf{w}\mathbf{x})\,(\mathbf{OM}) \tag{13}$$

Moreover this product is commutative and we can write

$$\{X(\mathbf{w})\} = \{Y(\mathbf{w}, \mathrm{p})\}.\{T(\mathbf{w})\} = \{T(\mathbf{w})\}.\{Y(\mathbf{w}, \mathrm{p})\} \quad \forall \mathrm{p} \tag{14}$$

and in the special case p = 0

$$\{X(\mathbf{w})\} = \{G(\mathbf{w})\}.\{T(\mathbf{w})\} = \{T(\mathbf{w})\}.\{G(\mathbf{w})\} \tag{15}$$

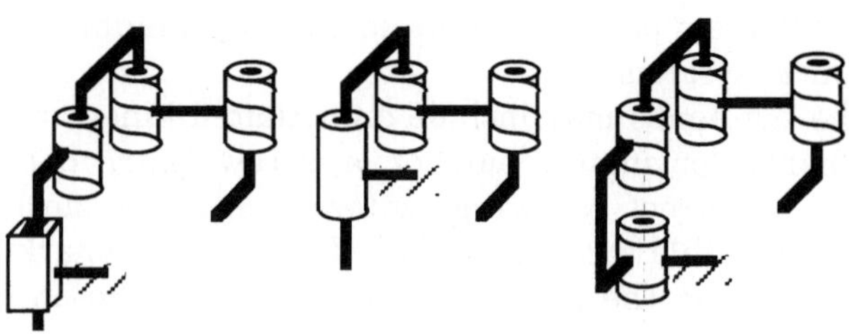

Figure 6. Generation of $\{X(\mathbf{w})\} = \{T(\mathbf{w})\}.\{Y(\mathbf{w}, \mathrm{p})\}$

In the general theory of algebraic groups, this product is called a direct product and $\{T(\mathbf{w})\}$ and $\{Y(\mathbf{w}, \mathrm{p})\}$ are invariant (or normal) subgroups of $\{X(\mathbf{w})\}$.

An example of a kinematic chain *[PH] HH* that generates $\{X(\mathbf{w})\} = \{T(\mathbf{w})\}.\{Y(\mathbf{w}, \mathrm{p})\} = \{T(\mathbf{w})\}.\{H(\mathrm{A},\mathbf{w}, \mathrm{p})\}.\{H(\mathrm{B},\mathbf{w}, \mathrm{p})\}.\{H(\mathrm{C},\mathbf{w}, \mathrm{p})\}$ is shown in figure 6 (left). Using the product closure in the Lie subgroup $\{C(\mathrm{A}, \mathbf{w})\}$ of cylindrical displacements around the axis (A, **w**), we can write (16)

$$\{T(\mathbf{w})\}.\{H(\mathrm{A},\mathbf{w},\mathrm{p})\}.\{H(\mathrm{B},\mathbf{w},\mathrm{p})\}.\{H(\mathrm{C},\mathbf{w},\mathrm{p})\} = \{C(\mathrm{A},\mathbf{w})\}.\{H(\mathrm{B},\mathbf{w},\mathrm{p})\}.\{H(\mathrm{C},\mathbf{w},\mathrm{p})\}$$

That formula is illustrated by the figure 6(center), and we have also

$$\{C(\mathrm{A},\mathbf{w})\}.\{H(\mathrm{B},\mathbf{w},\mathrm{p})\}.\{H(\mathrm{C},\mathbf{w},\mathrm{p})\} = \{R(\mathrm{A},\mathbf{w})\}.\{H(\mathrm{A},\mathbf{w},\mathrm{p})\}.\{H(\mathrm{B},\mathbf{w},\mathrm{p})\}.\{H(\mathrm{C},\mathbf{w},\mathrm{p})\} \quad (17)$$

which equation is materialized by the chain *[R H] HH* of figure 6(right). Squared brackets indicating the equivalency with a cylindrical pair *C*, we can implement all the generators of cylindrical motion, namely *[PR]*, *[RP]*, *[PH]*, *[HP]*,*[RH]*, *[HR]* and *[HH]* where the *R* or *H* axes are coaxial with the equivalent *C* and the *P* direction is parallel to the *C* axis.

All the mechanical generators of a displacement subgroup $\{X(\mathbf{w})\}$ were enumerated by Hervé and Sparacino, 1991, and employed in the design of a 3-dof microfinger, Arai et al.,1996. In the sequel, we will focus on special parallel arrangements of the previous special generators that yield PMs with a simple control.

4.2 A smart parallel generator of *X*-motion: the X-box

A fully PM that produces Schoenflies motion is obtained by setting in parallel four distinct generators of a given subgroup $\{X(\mathbf{w})\}$, **w** being a given unit vector. By a special choice of these generators, the actuation can be made very simple.

In the example of figure 7, which we named a X-box, two limbs have the structural type *[PR] RR* and *[PH] HH*, and two limbs have the type *[R P] RR*. The *H* pairs have the same pitch. The four pairs *H* or *R* that are attached to the moving platform are coaxial; in other words the platform is a moving pin. The fixed *P* pairs and the fixed revolute pairs *R* are actuated by servomotors. The *P* pairs are parallel to the *R* or *H* parallel axes.

The dof of the closed loop with two fixed *P*, namely *[PR]RR*-[*PH]HH* is equal to 4. If the two *P* pairs are locked, the closed loop becomes a 2-dof chain of the type shown in figure 5, which produces planar 2-dof translation. Hence if the two P pairs move with two equal amplitudes, the moving platform undergoes 3-dof translation and therefore the 4-dof of the X-motion implies a difference of the translation amplitudes in the two fixed *P* pairs. Furthermore obviously the rotation of the moving platform doesn't depend on its translation and the rotation axis is the common axis of the coaxial *R* and *H*. Furthermore the angle of rotation is given by the ratio 2π d /p, d being the difference between the two inputted translations and p $\neq$ 0 being the *H* pitch.

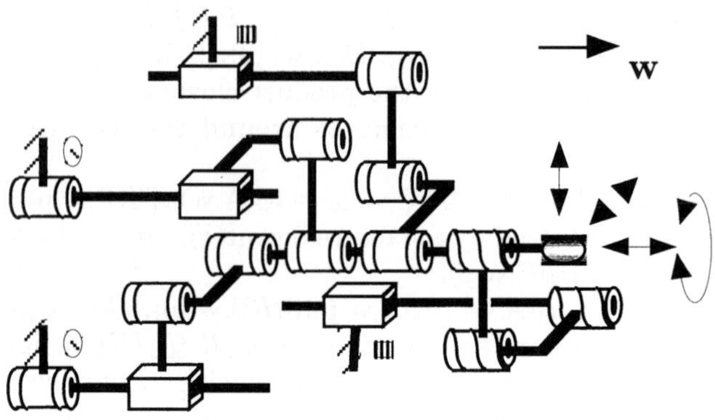

Figure 7. Generation of {*X*(**w**)} by a special parallel manipulator, the X-box

The closed loop *[RP]RR – RR[PR]* have also 4 dof that are reduced to 2 dof when locking the fixed *R* pairs. The set of allowed platform displacements in the remaining *PRR-RRP* loop is evidently the product of linear translations parallel to **w** by the rotations around the common axis of the two R pairs that are attached to the moving platform. Furthermore the projection of the 4-dof loop onto a plane perpendicular to **w** is a planar five-bar linkage and therefore the projection of the moving pin is a point, which undergoes the 2-dof planar motion of a five-bar joint.

Hence in the whole 4-limbed PM, the planar translation perpendicular to **w** is controlled only by the actuation of the two fixed *R* pairs. The translation parallel to **w** is given by the actuation of the *P* pairs in the limbs *[PR] RR* and *[PH] HH*. The rotation about the platform pin is only the result of the translation difference d in the two fixed *P* pairs.

4.3 New Cartesian parallel generators of translation

It is straightforward to prove that the intersection of two sets {*X*(**u**)} and {*X*(**v**)} is the set {*T*} of spatial translations provided the unit vectors **u** and **v** are not equal. Consequently the parallel arrangement between a base and a moving platform of two *X*-motion generators produces 3-dof translation. A third parallel limb, which generates {*X*(**w**)} is compatible with the first two limbs and thus translational fully PMs can be synthesized, Hervé and Sparacino, 1991.

A very simple Cartesian control of the translational motion is obtained by the choice of the direct products that are mentionned in section 4.1, for example {*X*(**u**)} = {*T*(**u**)}.{*Y*(**u**, p)} $\forall$p, {*X*(**v**)} = {*T*(**v**)}.{*Y*(**v**, p')} $\forall$p', and {*X*(**w**)} = {*T*(**w**)}.{*Y*(**w** , p")} $\forall$p" , provided (**u**, **v**, **w**) is an orthonormal vector base. A typical example of the associated parallel manipulator is denoted 3-*[PH] HH* and this mechanism is sketched in figure 8. The actuation is obtained by three fixed linear servomotors, which make up an actuated Cartesian frame of reference (O, **u**, **v**, **w**). If two fixed *P* pairs are locked, the two limbs with locked *P* pairs constitutes a mechanism of the type shown in figure 3 and therefore the moving platform can undergo 1-dof linear translation. Furthermore because of the orthogonality of the fixed linear actuators, this linear translation is parallel to the unlocked prismatic pair. Obviously the translation of the moving platform is equal to the translation in the unlocked prismatic pair.

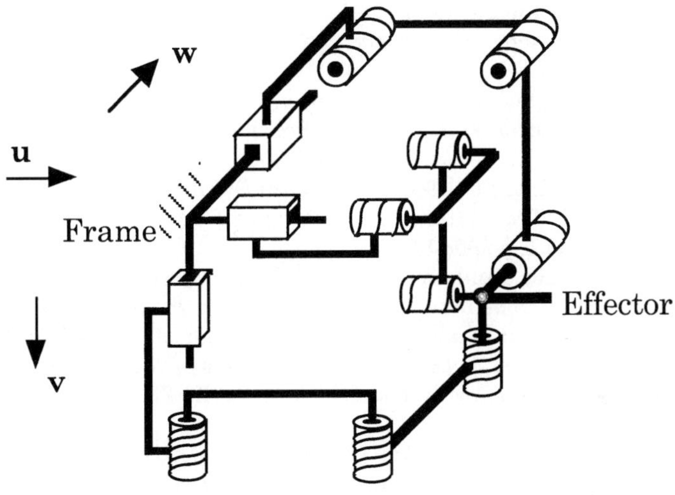

Figure 8. Generation of {*T*} by intersection {*X*(**u**)}∩{*X*(**w**)}∩{*X*(**w**)}

That way, we obtain PMs that produce 3-dof translation such as Cartesian co-ordinates directly control the position of the end-effector. These manipulators are generalizations of those proposed by Kong and Gosselin 2002a, 2002b, and by Kim and Tsai, 2002, 2003.

5. Conclusion

The generation of pseudo-planar motion natively implies the use of helical pairs, which can move freely with no jamming effect. Low friction and large pitches are required. But the present-day technology can afford high quality screw bearings. Hence mechanisms with screw joints may become effective in machine-tools of the future. The current article has introduced new architectures of PMs, which may be effective in time to come.

References

Hervé, J.M. (1978), Analyse structurelle des mécanismes par groupe des déplacements, *Mechanism and Machine Theory*, vol.13, pp.437-450.

Angeles, J. (1982), *Spatial Kinematic Chains*, Springer-Verlag, New-York USA, 1982.

Hervé, J.M. (1982), Intrinsic Formulation of Problems of Geometry and Kinematics of Mechanisms, *Mechanism and Machine Theory*, no.3, vol.17, pp. 179-184.

Hervé, J.M. (1994), The Mathematical Group Structure of the Set of Displacements, *Mechanism and Machine Theory*, no. 1, vol.29, pp. 73-81.

Fanghella, P., and Galletti, C. (1994), Mobility Analysis of Single-Loop Kinematic Chains: an Algorithmic Approach Based on Displacement Groups, *Mechanism and Machine Theory*, vol. 29, pp. 1187-1204.

Hervé, J.M. (1999), The Lie group of rigid body displacements, a fundamental tool for mechanism design, *Mechanism and Machine Theory*, vol.34, pp.719-730.

Hervé, J.M. (2000), Design of New Mechanisms via the Displacement Subgroups, in *Geometrical Foundations of Robotics*, Ed. J.M. Selig, World Scientific, Singapore, pp.39-59.

Hervé, J.M., and Sparacino, Fl. (1991), Structural Synthesis of Parallel Robots Generating Spatial Translation, *Proceedings of IEEE 5th Int. Conf. on Advanced Robotics,* vol.1, pp.808-813.

Arai, T., Hervé, J.M., and Tanikawa, T. (1996), Development of 3 DOF Micro Finger, *Proceedings IROS 96*, Nov. 5-8-1996, Osaka, pp.981-987.

Majou, F., Wenger, P., and Chablat, D. (2002) Design of Two-DOF Parallel Mechanisms for Machining Applications, *Proceedings of the 8th International Symposium on Advances in Robot Kinematics,* 24-28 June, Caldes de Malavella, Spain, pp. 319-328.

Kong, X.-W., and Gosselin, Cl. M. (2002a), Type Synthesis of Linear Translational Parallel Manipulators, *Proceedings of 8th International Symposium on Advances in Robot Kinematics*, 24-28 June, Caldes de Malavella, Spain, pp. 453-462.

Kong, X.-W., and Gosselin, Cl. M. (2002b), Kinematics and Singularity Analysis of a Novel Type of 3-$\underline{C}$RR 3-DOF Translational Parallel Manipulator, *The International Journal of Robotics Research*, vol. 21, no. 9, pp. 791-798.

Kim, H. S., and Tsai, L. W. (2002), Evaluation of a Cartesian Parallel Manipulator, *Proceedings of 8th International Symposium on Advances in Robot Kinematics*, 24-28 June, Caldes de Malavella, Spain, pp. 21-28.

Kim, H. S., and Tsai, L. W. (2003), Design Optimization of a Cartesian Parallel Manipulator, *Journal of Mechanical Design*, vol. 125, no. 1, pp.43-51.

MULTI-AGENT MANIPULATOR CONTROL AND MOVING OBSTACLE AVOIDANCE

George I. Birbilis
Department of Mechanical Engineering & Aeronautics
University of Patras
birbilis@kagi.com

Nikos A. Aspragathos
Department of Mechanical Engineering & Aeronautics
University of Patras
asprag@mech.upatras.gr

Abstract A novel approach to planar serial manipulator motion and reactive moving obstacle avoidance, based on a multi-agent architecture, is presented. A conceptual model of the manipulator is considered, that mimics the motion of a chain of potentially expandable rods, interconnected at their endpoints using pins. Rods and pins are represented by respective software agents. A one-way "Master – Slave" relationship is suggested, with the event of an autonomous motion of an agent propagating to its two neighboring ones in the manipulator chain and progressively further on towards the two endpoints of the chain. A constraint preservation mechanism enforces the respecting of the pin angle and rod length bounds at each propagation step. Thus, the whole chain behaves as if a moving part of the chain is pushing or pulling the two subparts of the chain it connects. To cater for fixed base manipulators, and support replanning in case some slave part of the chain can't adapt to its master's motion cause its trapped in some obstacles or malfunctioning, the notion of a "Master – Vetoable slave" relationship is introduced, where a slave part can object (veto) to the motion of its master part.

Keywords: Redundant manipulators, Reactive motion, Moving obstacle avoidance, Rope-like motion, Multi-agent systems, Event propagation, Constraint satisfaction, Master - Vetoable slave relationship, Push-Pull-Rotate behavior

1. Introduction

In this paper we consider the problem of planning the motion of a *planar serial manipulator*, the workplace of which may contain unknown stationary or moving obstacles. Many approaches to the problem of manipulator motion planning in a *time varying environment* have been proposed, most of which are thoroughly examined in Hwang and Ahuja, 1992. The case of unknown moving obstacles existing in the workplace of

J. Lenarčič and C. Galletti (eds.), On Advances in Robot Kinematics, 441-448.

the manipulator greatly favors local planning methods, since global re-planning would be too expensive, especially in the case of many obstacles or when having a highly redundant manipulator, as pointed out by Chen and Hwang, 1992, and Challou et al., 1998.

The more joints a manipulator has, the higher its capability is, so an approach was sought that easily scales up to highly redundant manipulators, while still supporting those with less DOF. A top-down centralized approach would have increasing complexity and cost as the number of joints and links rises, so a bottom-up, modular approach was favored, modeled as a *multi-agent system*, using the multi-agent system design concepts presented by Liu, 2001.

Almost all the approaches based on a multi-agent platform to address the manipulator's motion planning problem are doing motion planning on each subpart of it and combine the solutions of those sub-problems into a solution for the global problem. In most of the cases, composition is implemented in real-time, via interaction (cooperation or contention) of the respective agents that control the various parts of the manipulator.

Overgaard et al., 1996, proposed a multi-agent system comprised of both joint and link agents, to control a 25-DOF snakelike robot in an environment with obstacles which are modeled using an artificial potential field, pioneered by Khatib, 1986. Bohner and Lüppen, 1997, applied a similar concept on a 7-DOF robot, considering only the robot joints as agents. They are doing both sensor-data integration and motion planning on a per-agent basis, thus decomposing the problem and reducing its complexity to the sum of its subproblems. A major weakness of the artificial potential field approach though is the *trapping in local minima* problem. Current efforts to escape local minima in similar multi-agent approaches are focusing on fuzzy logic, as in Zavlangas and Tzafestas, 2000, and neuro-fuzzy systems, as in Althoefer et al., 2001.

In the proposed multi-agent system, each software agent controls a specific part of the joint-link chain, and agents interact with each other towards the adaptation of manipulator configuration to external events and changing situations in real-time. Our aim is to present a novel approach that reduces the motion planning of a manipulator to motion planning of a single part of it, for instance its tool tip, having the rest of the manipulator's parts react and adapt or object (veto) to the moving part's motion.

2. The kinematic model

We're considering the case of a planar manipulator which can have both rotational and translational joints, in any serial combination. The manipulator works on a 2D plane that may contain obstacles, either stationary ones or moving with some speed comparable to the manipulator's joint motors speed. The case of moving obstacles includes the case of other manipulators also working in the same environment, with which no communication channel or common protocol exists for coordination. The base of the manipulator may be fixed or able to move around either freely or on some predefined path and also possibly rotate around an axis perpendicular to the manipulator plane.

No a-priori knowledge of obstacles is assumed and incoming collisions can be detected using either touch sensors covering the whole body of the manipulator, as in Bohner and Lüppen, 1997, or range sensors that detect the closest obstacle at each side of the manipulator's links, as in Zavlangas and Tzafestas, 2000.

3. The conceptual model

We try to control the manipulator's motion in a way that mimics the motion of a chain of potentially expandable *rods*, interconnected at their endpoints using *pins*, forming a chain. Potentially expandable means that some rods could be resizable (expandable and shrinkable), that is have a minimum and maximum length instead of a fixed one. Any part of the chain that initiates a motion "pushes" or "pulls" the other parts of the chain so that the constraints defining the chain structure are kept.

The chain's two endpoints are pins, with the base joint used to cater for a moveable and rotateable manipulator base. This paper focuses on planar manipulators, where a rotational joint and its outgoing link maps directly to the first pin of a fixed-length rod, whereas a translational joint and both its incoming and outgoing links are represented by a pin and an expandable rod. Each pin agent has a position property and each rod agent has a length property, such that the length bounds of a rod are respected by the positions of the pin agents placed at the rod's two endpoints. The values of joint parameters can be calculated geometrically from the pin positions at any time, using the mapping shown at Fig. 1.

Given a high number of rods and respective pins, this model should behave like a rope dragged from any part of it, slipping through the free areas among a set of vertically elevated areas on a 2D plane.

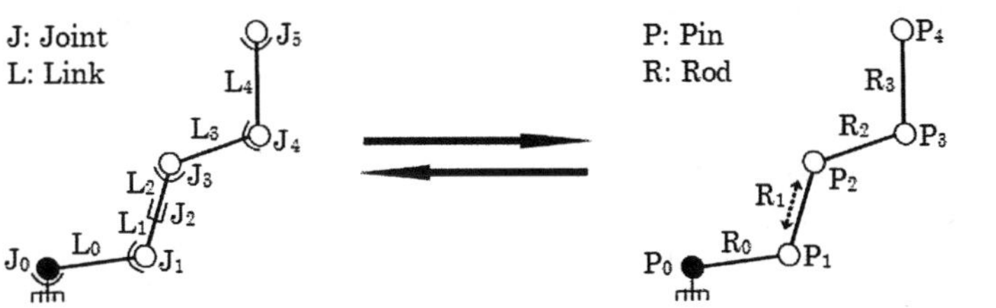

Figure 1. Kinematic to Conceptual (and vice versa) model mapping

3.1 Mapping kinematic to conceptual constraints

Kinematic model's constraints are mapped to respective constraints at the conceptual model. One such constraint is the minimum and maximum joint angle for rotational joints. We are considering an anti-clockwise increment and a signed number for the angle, with the zero position being such that if all such joints were set to have a zero angle, the manipulator would form a straight line, similar to the concept of a virtually straight arm in Bohner and Lüppen, 1997.

Another equally important constraint can be the minimum and maximum value of a translational joint's variable. That constraint is encapsulated in a constraint governing the bounds of the "length" property of the conceptual model's rod entity, since in our conceptual model there are only moveable pins that connect expandable/shrinkable rods.

Finally, kinematic model constraints can also arise from the fact that the manipulator might not be a moveable one, or could have some irregular geometry. The former case is obviously translated to a constraint of the 2D position of the tail pin of the conceptual chain, whereas the later one can be covered by considering rods that connect to each other via a pin at a constrained angle (e.g. a fixed 90 degrees angle).

3.2 Master - Slave relationship

We consider a *Master-Slave relationship* between two entities, a "master" and a "slave", to be such that the slave entity listens for changes of the master entity's state and reacts to them by changing its own state. The listening of changes and reacting to them is also called the *observer pattern* in software engineering.

3.3 Change event propagation

A slave entity of a certain Master-Slave relationship can still take part as a master in other such relationships, having its own slave entities. So, given a set of a finite number of entities, one can define an open chain of Master-Slave relationships, where the first entity is the master of the second entity, the second entity is the master of the third entity and so on, till the entity before the last one, which is master of the last entity in the chain.

In such a case, *state change events* at the head of this Master-Slave relationship chain propagate towards the tail of it, since each slave tries to react and adapt to its master's state change event, potentially changing its own state and thus causing its own slave to react and try itself to change its own state and so on, as shown at the left side of Fig. 2.

In the proposed model, any part of the chain and not just the manipulator tool tip could in fact initiate the chain motion, receiving its own motion commands from a planner module or a human operator via a higher application layer. That is, the chain is treated as two sub-chains, both having as head the part that initiated the motion, and for their tails the head and the tail of the original chain respectively, as shown at the right side of Fig. 2.

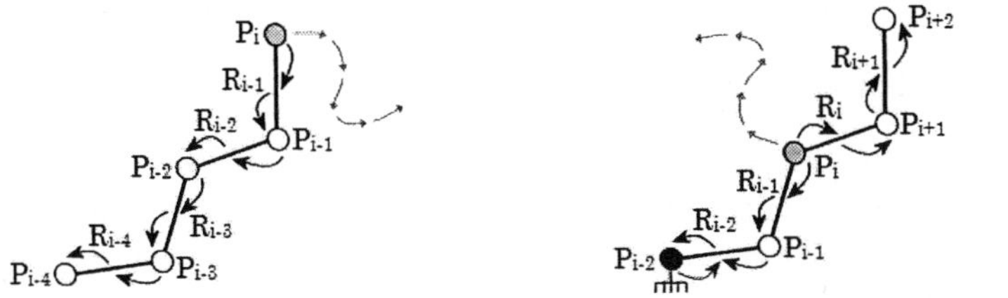

Figure 2. Event propagation to neighbors, and veto back-propagation

3.4 Reaction to master's state change events

Each slave entity follows a predefined *rule set* that defines how it reacts to the changes of its master entity's state. That rule set defines *reactions* that try to preserve some *constraints* imposed on the slave object and its relation to the master one, either by design (internal, hardware constraints) or decided on the fly at runtime (environmental constraints, obstacles or failure of subparts).

The proposed reaction rules are exposing two basic model constraint preservation behaviors, named *Push* and *Pull*, with the former one caused when the direction of a master pin's motion forms an acute angle with the rod that connects it to a slave pin, and the later one caused when that same angle is obtuse.

When a master pin is "pushing" a slave pin, an immediate neighbor of it on the chain, the effect is that the slave pin moves on the guiding line defined by the new position of the master and the old position of the slave respectively. The slave will be pushed away on the guiding line only if its distance from the new position of the master is less than the minimum length of the (potentially shrinkable) rod that connects it to the master pin.

A similar situation exists when a moving master pin is "pulling" a slave one. The slave pin is pulled on that same guiding line as before, but only if its distance from the master exceeds the maximum length of the (potentially expandable) rod that connects it to the master pin. The pushing and pulling actions of a master pin on its slave pin neighbors are obtained via the respective slave rods that connect the slave pins to their master. This is clearly illustrated at Fig. 3.

Figure 3. Push and Pull behaviors

3.5 Reaction to sensed obstacles

During the reaction of a slave pin to the motion of a master pin towards a new position in the free space, a potential obstacle collision for the rod that connects the master and slave pins may be blocking the realization of the new placement of the slave pin that is obtained using the push or pull behaviors described at section 3.4. In such a case, the rod rotates around the master pin as if the motion of the master pin was causing a rotation (at the rod's obstacle collision point that is closest to the master) of the guiding line on which the link would have lied according to the respective Push or Pull behavior if the obstacle wasn't there, as shown at Fig. 4. We name these environmental constraint preservation behaviors *Push-Rotate* and *Pull-Rotate* respectively.

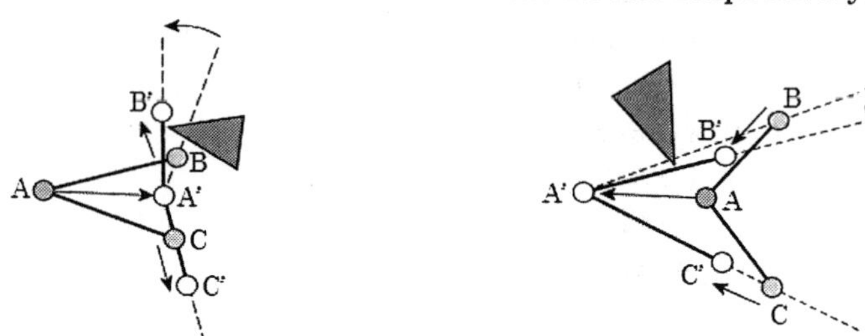

Figure 4. Push-Rotate and Pull-Rotate behaviors

3.6 Master - Vetoable slave relationship

We define a *Master – Vetoable slave relationship* between two entities, a "master" and a "slave", to be such that the slave entity listens for changes of the master entity's state and reacts to them by changing its own state, as at the Master-Slave relationship defined at section 3.2, but where the slave entity can object to the state changing of the master (*veto*). This pattern is usually called a *VetoableChangeListener* in Object Oriented Programming.

3.7 Veto back-propagation

To cater for fixed base manipulators, and support replanning in case some slave part of the chain can't adapt to its master's motion because it's trapped in some obstacles or malfunctioning, we make use of the Master - Vetoable slave relationship. A slave part of the chain forced to react and adapt to an initial motion of a master neighboring part of the chain, can object (veto) to the motion of that master part if itself cannot move to adapt to the master's motion while still preserving the chain model and environmental constraints. At the right side of Fig. 2, a manipulator with a fixed base is shown, where the pin agent at the manipulator's base always objects to any relocation of it.

4. The software model

In Liu et al., 2002, an agent is defined as an entity that is able to live and act in an environment, is able to sense its local environment, is driven by certain objectives and has some reactive behaviors. Also, a multi-agent system is defined as a system that has an environment, that is a space where the agents live, a set of reactive rules, governing the interaction between the agents and their environment (the laws of the agent universe) and of course a set of agents.

These definitions map directly to the entities defined in the conceptual model, as presented at section 3, making the modeling of the multi-agent system a straightforward process. The event and veto propagation can be implemented as remote events and exceptions over a data channel at an embedded platform like Java or .NET, or using plain hardware interrupts at lower level implementations.

5. Current development and future plans

A testbed application for defining 2D polygonal moving obstacles has been developed using the Visual Basic.NET programming language. The proposed multi-agent architecture is currently being integrated, together with a virtual planar manipulator, so that qualitative and quantitative observations are made, resulting in possible adjustments to the model and environmental constraint preservation behaviors presented above.

We plan to extend the proposed architecture into 3D space and apply it to real-word manipulators. We expect that given the modular design used, extension into 3D space will be smooth, with changes focusing mostly on the local environment sensing mechanism and the constraint preservation rules, having the rest of the multi-agent system and event propagation design kept intact.

6. Conclusions

The most important advantage of the proposed architecture is that it reduces the problem of path planning for a serial manipulator to that of path planning for only a master part of it (usually the tool-tip). The rest of the manipulator parts adjust in real-time to the motion of the master, while at the same time reacting to any sensed obstacles. Its best application should be at highly redundant manipulators, where it can scale up efficiently, since an agent only needs to sense obstacles locally and interact with its immediately neighboring agents in the manipulator chain. The support for a moving base also adds to the system's flexibility.

However, the proposed architecture doesn't deal directly with the situation where a veto from a slave agent back-propagates as far as the original agent that had tried to perform a motion. In that case, the mover should replan its motion if it's autonomous, or propagate the veto further to a planner module or human operator that had commanded it to move at the first place, since the rest of the manipulator chain can't follow it.

References

Althoefer, K., Krekelberg, B., Husmeier, D., and Seneviratne, L. (2001), Reinforcement learning in a rule-based navigator for robotic manipulators, *Neurocomputing*, no. 37, pp.51-70.

Bohner, P., and Lüppen, R. (1997), Reactive Multi-Agent Based Control of Redundant Manipulators, *Proceedings of the 1997 IEEE Int. Conf. on Robotics and Automation*, Albuquerque, New Mexico.

Challou, D., Bolcy, D., Gini, M., Kumar, V., Olson, C. (1998), Parallel Search Algorithms for Robot Motion Planning, *Practical Motion Planning in Robotics: Current Approaches and Future Directions*, eds. Kamal Gupta and Angel P. del Pobil.

Chen, P., and Hwang, Y. (1992), Sandros, a motion planner with performance proportional to task difficulty, *Proceedings of the 1992 IEEE Int. Conf. on Robotics and Automation*.

Hwang, Y., and Ahuja, N. (1992), Gross Motion Planning – A Survey, ACM *Computing Surveys*, no. 3, vol. 24, pp.219-291.

Khatib, O. (1986), Real-time obstacle avoidance for manipulators and mobile robots, International Journal of Robotic Research, no. 5, pp.90-98.

Liu, J. (2001), Autonomous Agents and Multi-Agent Systems: Explorations in Learning, Self-Organization, and Adaptive Computation, *World Scientific*, Singapore.

Liu, J., Jing, H., and Tang Y. (2002), Multi-agent oriented constraint satisfaction, *Artificial Intelligence*, no. 136, pp.101-144.

Overgaard, L., Petersen, H., and Perram, J. (1996), Reactive Motion Planning: A Multi-agent Approach, *Applied Artificial Intelligence*, no. 10, pp.35-51.

Zavlangas, P., and Tzafestas, S. (2000), Industrial Robot Navigation and Obstacle Avoidance Employing Fuzzy Logic, *Journal of Intelligent and Robotic Systems*, no. 27, pp.85-97.

KINEMATICS OF A PARALLEL MECHANISM FOR THE GENERATION OF SPHERICAL MOTIONS

Massimo Callegari, Paolo Marzetti, Babila Olivieri
Department of Mechanics, Polytechnic University of Marche
Via Brecce Bianche, 60131 Ancona, Italy
m.callegari@univpm.it

Abstract The paper presents the concept of a novel spherical mechanism based on the 3-CPU topology and its main geometrical parameters are evidenced. Both IPA and DPA have been solved in closed-form and a mobility analysis revealed the possible occurrence of constraint singularities allowing for translational motions. Jacobian analysis has been developed as well, with the discussion of all the singular configurations for the rotational motions. The results of the analysis suggest the possible utilisation of this mechanical architecture for the development of spherical wrists for automated machines or for parallel pointing systems.

Keywords: Spherical Parallel Machine (SPM), kinematics, singularities, wrists, 3-CPU

1. Introduction

The complex kinematics of 6-dof parallel machines pushes for the design of new minor-mobility mechanisms that are able to perform elementary motions, like pure translations, pure rotations or even planar displacements. The industrial relevance of the matter is evident: in fact some important applications may well be accomplished by these kinds of reduced mobility, while in other cases hybrid machines may be designed (e.g. a conventional "serial" wrist on top of a parallel shoulder) or mini-maxi architectures can be experimented; alternatively, a full-mobility task may be decomposed into elemental sub-tasks, to be performed by separate minor mobility machines, like done already in conventional machining operations (Callegari and Suardi, 2004). In this contest, much research has been developed in recent years for the synthesis of Translating Parallel Machines (TPMs) but comparably few results are presently available for the design of parallel machines for pure rotations, or spherical wrists. Notwithstanding the useful enumerations by Tsai (2000) and Merlet (2000), the most important mechanism of this type is the *agile eye* by Gosselin and Angeles (1989), upon which many prototype machines have been designed; among the first studies on Spherical Parallel Machines (SPMs) the work of Asada and Granito (1985), Lee and

J. Lenarčič and C. Galletti (eds.), On Advances in Robot Kinematics, 449-458.

Chang (1992), Innocenti and Parenti-Castelli (1993) and Alizade *et al.* (1994) may be cited. In more recent years, interesting studies have been developed on the subject by Di Gregorio (2000 and 2001), Ji and Wu (2001), Karouia and Hervè (2000 and 2002).

The Authors, following the approach outlined by Karouia and Hervè (2000), have devised the concept of a novel SPM (Olivieri, 2003) characterised by the 3-CPU topology, see Fig. 1. As a matter of fact, the proposed architecture is implicitly suggested in the referenced paper by Karouia and Hervè (2000), where they describe a spherical 3-UPU wrist: in fact if legs' kinematic chains are reduced to their equivalent simple forms (i.e. only R and P pairs are considered) in both cases the three middle joints in each leg (here PPR, there RPR) form an equivalent planar pair and first and last revolute pairs meet at a common point. On the other hand, the 3-CPU topology can also be found in the long listing of SPMs made by Kong and Gosselin (2002), since leg's topology $\dot{R}\overline{PPR}\dot{R}$, which they list as number 70, is equivalent to the present CPU topology for the above mentioned reasons.

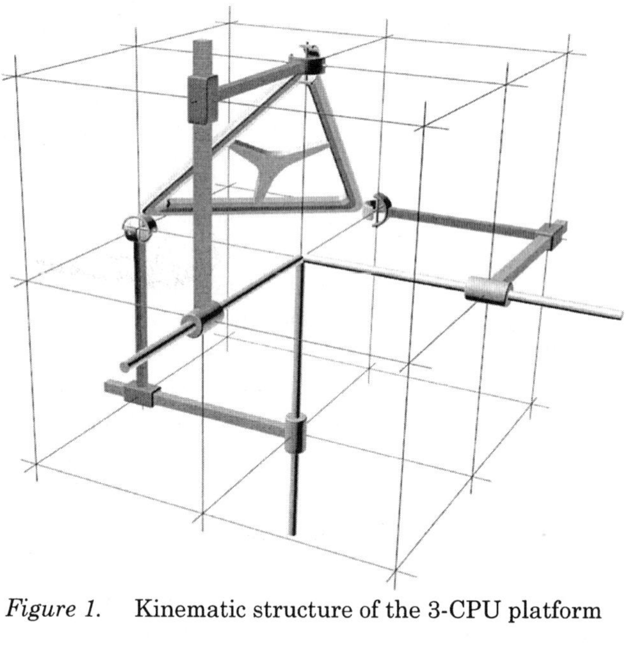

Figure 1. Kinematic structure of the 3-CPU platform

2. Geometry of the spherical wrist

Fig. 1 shows a sketch of the spherical mechanism; three identical serial chains connect the rotating platform to the fixed base: each chain is composed by two links, connected to the frame by a cylindrical pair (C) and to the mobile base by a universal pair (U) and linked together by a prismatic joint (P); for this reason, this architecture is called 3-CPU.

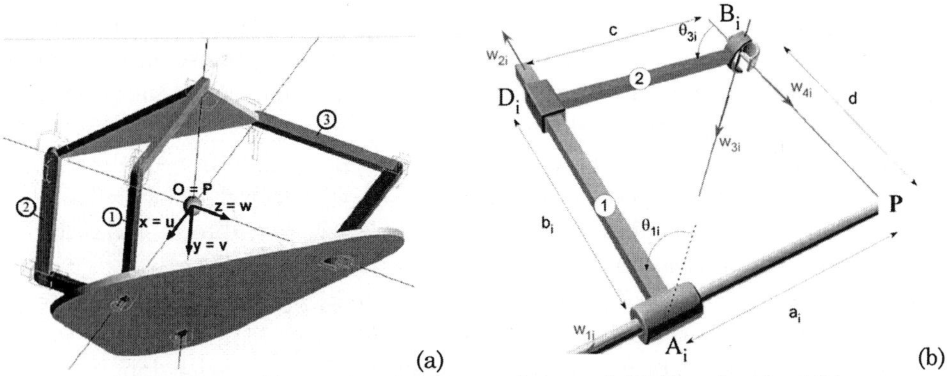

Figure 2. Placement of reference frames, initially coincident (a) and geometry of a single limb (b)

Making reference to Fig. 2b, the axes of cylindrical joints A_i, *i=1,2,3* intersect at point *O* (centre of the motion) and are aligned to the axes **x**, **y**, **z** respectively of a *(fixed)* cartesian frame located in *O*. The first member of each link (*1*) is perpendicular to A_i and has a variable length b_i due to the presence of the prismatic joint D_i: the second link (*2*) of the leg is set parallel the cylindrical pair. The universal joint B_i is composed by two revolute pairs with orthogonal axes: one is perpendicular to leg's plane while the other intersects at a common point *P* with the corresponding joints of the other limbs; such directions, for the legs *i=1,2,3* orderly, are aligned to the axes **v**, **w**, **u** respectively of a (*mobile*) cartesian frame, located in *P* and solid with the rotating platform (Fig. 2a). For a successful functioning of the mechanism, such *manufacturing conditions* must be accompanied by a proper *mounting condition*: assembly should be operated in such a way that the two frames *O*(**x**,**y**,**z**) and *P*(**u**,**v**,**w**) come to coincide. Finally, it is assumed an initial configuration such that the linear displacements a_i of the cylindrical joints are equal to the constant length *c* (that is the same for all the legs): in this case also the linear displacements b_i of the prismatic joints are equal to the constant length *d*. It is also evident that, for practical design considerations, the actuation of machines based on the 3-CPU concept will be probably provided by driving the linear displacements of the cylindrical pairs coupling the limbs with the frame: therefore in the following kinematic analysis it will be made reference to this case (i.e. joint variables a_i, *i=1,2,3* will be considered the actuated parameters).

3. Mobility analysis

It can be easily seen that the 3-CPU mechanism has 3 dof's, since it is composed by 3 serial chains of connectivity 5 that link the platform to the frame. If the manufacturing and assembly conditions previously specified are satisfied, the mobile platform is characterised by motions of pure rotation, as shown in the following paragraphs. The mentioned conditions can be geometrically expressed by:

i. $\hat{w}_{1i}$ and $\hat{w}_{4i}$ incident in P;

ii. $\hat{w}_{3i}$ normal to the plane $<\hat{w}_{1i}, \hat{w}_{4i}>$, i.e. $\hat{w}_{3i} \cdot \hat{w}_{4i} = 0$ and $\hat{w}_{3i} \cdot \hat{w}_{1i} = 0$;

iii. $\hat{w}_{2i}$ lying on the plane $<\hat{w}_{1i}, \hat{w}_{4i}>$, i.e. $\hat{w}_{3i} \cdot \hat{w}_{2i} = 0$; due to condition (ii) must also hold: $\hat{w}_{3i} = \hat{w}_{1i} \times \hat{w}_{2i}$;

iv. $\hat{w}_{2i}$ not parallel to $\hat{w}_{1i}$ and therefore: $\hat{w}_{1i} \times \hat{w}_{2i} \neq \hat{0}$ (for simplicity, the condition $\hat{w}_{1i} \cdot \hat{w}_{2i} = 0$ has been posed).

Making reference to Fig. 2b, if the point P is considered belonging to the ith leg, its velocity can be written in three different ways as follows:

$$\dot{P} = \dot{P}_{2i} + \dot{P}_{ri} \quad \text{for } i=1,2,3 \tag{1}$$

where $\dot{P}_{2i}$ is the velocity of point P if considered fixed to link *2*:

$$\dot{P}_{2i} = \dot{B}_i + \omega_{2i} \times (P - B_i) = \dot{B}_i + \omega_{2i} \times d\, \hat{w}_{4i} \tag{2}$$

and $\dot{P}_{ri}$ is the velocity of point P relative to a frame fixed to link *2* and with origin in B_i:

$$\dot{P}_{ri} = \dot{\theta}_{3i}\hat{w}_{3i} \times (P - B_i) = \dot{\theta}_{3i}\hat{w}_{3i} \times d\hat{w}_{4i} \tag{3}$$

In Eq. (2), ω_{2i} is the angular velocity of link 2:

$$\omega_{2i} = \dot{\theta}_{1i}\hat{w}_{1i} \tag{4}$$

In the same way, with obvious meaning of the symbols, the vector $\dot{B}_i$ can be expressed as:

$$\dot{B} = \dot{B}_{1i} + \dot{B}_{ri} \quad \text{for } i=1,2,3 \tag{5}$$

where:

$$\dot{B}_{1i} = \dot{a}_i\hat{w}_{1i} + \omega_{1i} \times (B_i - A_i) = \dot{a}_i\hat{w}_{1i} + \dot{\theta}_{1i}\hat{w}_{1i} \times (a_i\hat{w}_{1i} - d\hat{w}_{4i}) = \dot{a}_i\hat{w}_{1i} - \dot{\theta}_{1i}\hat{w}_{1i} \times d\hat{w}_{4i} \tag{6}$$

$$\dot{B}_{ri} = \dot{b}_i\hat{w}_2 \tag{7}$$

If Eqs. (2-7) are substituted back in (1), it is found:

$$\dot{P} = \dot{b}_i\hat{w}_{2i} + \dot{a}_i\hat{w}_{1i} + \dot{\theta}_{3i}\hat{w}_{3i} \times d\hat{w}_{4i} \quad \text{for } i=1,2,3 \tag{8}$$

By dot-multiplying Eq. (8) by $\hat{w}_{3i}$ and by taking into account the conditions (i)-(iv), it is finally obtained:

$$\hat{w}_{3i} \cdot \dot{P} = 0 \tag{9}$$

that can be differentiated to yield:

$$\hat{w}_{3i} \cdot \ddot{P} + \dot{\hat{w}}_{3i} \cdot \dot{P} = 0 \tag{10}$$

Eqs. (9-10), written for the 3 legs, build up a system of 6 linear algebraic equations in 6 unknowns, the scalar components of $\dot{P}$ and $\ddot{P}$. Such a system can be written in matrix form as follows:

$$M\begin{bmatrix}\dot{P}\\ \ddot{P}\end{bmatrix}=0 \tag{11}$$

where the *6x6* matrix *M* can be partitioned as:

$$M=\begin{bmatrix}H & O\\ \dot{H} & H\end{bmatrix} \tag{12}$$

with:

$$H=\begin{bmatrix}\hat{w}_{31}^T\\ \hat{w}_{32}^T\\ \hat{w}_{33}^T\end{bmatrix}=\begin{bmatrix}w_{31i} & w_{31j} & w_{31k}\\ w_{32i} & w_{32j} & w_{32k}\\ w_{33i} & w_{33j} & w_{33k}\end{bmatrix} \qquad O=\begin{bmatrix}0 & 0 & 0\\ 0 & 0 & 0\\ 0 & 0 & 0\end{bmatrix} \tag{13-14}$$

If the matrix *M* is non singular, the system (11) only admits the trivial null solution:

$$\dot{P}=\ddot{P}=0 \tag{15-16}$$

which means that the point *P* does not move in space, i.e. the moving platform only rotates around *P*. The singular configurations, on the other hand, can be identified by posing:

$$\det(M)=[\det(H)]^2=0 \tag{17}$$

that leads to:

$$\det(H)=\hat{w}_{31}\cdot\hat{w}_{32}\times\hat{w}_{33}=0 \tag{18}$$

Eq. (18) is satisfied only when the three unit vectors $\hat{w}_{31}$, $\hat{w}_{32}$, $\hat{w}_{33}$ are linearly dependent; therefore the platform incurs in a *constraint singularity* if and only if:

- all the planes containing the three legs are perpendicular to the base;
- such planes are coincident with the base plane (that is unfeasible);
- at least two out of the three aforementioned planes admit parallel normal unit vectors.

This justifies the choice previously operated of having the legs laid on mutual orthogonal planes, since this configuration is the most far from singularities.

4. Orientation kinematics

Orientation kinematics is based on the definition of the relative rotation between fixed frame *O*(**x**,**y**,**z**) and the mobile frame *P*(**u**,**v**,**w**), where is always *P*≡*O*, Fig. 2a; to this aim the following set of Cardan angles is used:

$$ {}^{O}_{P}R(\alpha,\beta,\gamma)=R_x(\alpha)\cdot R_y(\beta)\cdot R_z(\gamma)=\begin{bmatrix} c\beta c\gamma & -c\beta s\gamma & s\beta \\ s\alpha s\beta s\gamma+c\alpha s\gamma & -s\alpha s\beta s\gamma+c\alpha c\gamma & -s\alpha c\beta \\ -c\alpha s\beta c\gamma+s\alpha s\gamma & c\alpha s\beta s\gamma+s\alpha c\gamma & c\alpha c\beta \end{bmatrix} \quad (19) $$

Moreover, a local frame $O_i(\mathbf{x}_i,\mathbf{y}_i,\mathbf{z}_i)$, $i=1,2,3$ is defined in O for each leg: the $\mathbf{x}_i$ axis is aligned with cylindrical joint's axis and the $\mathbf{y}_i$ axis is chosen parallel to limb's first link, when it is laid in the initial configuration.

One loop-closure equation can be written for each leg as follows:

$$ (A_i-P)+(D_i-A_i)+(B_i-D_i)+(P-B_i)=0 \quad \text{for } i=1,2,3 \quad (20) $$

Eq. (20) can be easily expressed in the local frame $O_i(\mathbf{x}_i,\mathbf{y}_i,\mathbf{z}_i)$:

$$ \begin{bmatrix} a_i \\ 0 \\ 0 \end{bmatrix}+\begin{bmatrix} 0 \\ -b_i\cdot c\theta_{1i} \\ -b_i\cdot s\theta_{1i} \end{bmatrix}+\begin{bmatrix} -c \\ 0 \\ 0 \end{bmatrix}+{}^{i}(P-B_i)=0 \quad (21) $$

The last term in (21) is actually evaluated in the global frame $O(\mathbf{x},\mathbf{y},\mathbf{z})$, then it is transported to limb's frame $O_i(\mathbf{x}_i,\mathbf{y}_i,\mathbf{z}_i)$:

$$ {}^{i}(P-B_i)={}^{i}_{P}R\cdot{}^{P}(P-B_i)={}^{i}_{O}R\cdot{}^{O}_{P}R\cdot{}^{P}(P-B_i) \quad (22) $$

where the introduced terms assume the following values:

$$ {}^{1}_{O}R=\begin{bmatrix} 1 & 0 & 0 \\ 0 & 1 & 0 \\ 0 & 0 & 1 \end{bmatrix} \qquad {}^{2}_{O}R=\begin{bmatrix} 0 & 1 & 0 \\ 0 & 0 & 1 \\ 1 & 0 & 0 \end{bmatrix} \qquad {}^{1}_{O}R=\begin{bmatrix} 0 & 0 & 1 \\ 1 & 0 & 0 \\ 0 & 1 & 0 \end{bmatrix} \quad (23\text{-}25) $$

$$ {}^{P}(P-B_1)=d\cdot[0\ 1\ 0]^T \quad {}^{P}(P-B_2)=d\cdot[0\ 0\ 1]^T \quad {}^{P}(P-B_3)=d\cdot[1\ 0\ 0]^T \quad (26\text{-}28) $$

In inverse kinematics the values α, β, γ of Cardan angles (or equivalently the elements r_{ij} of the rotation matrix ${}^{O}_{P}R$) are know and the joint variables a_i must be found; loop closure equations (22) for $i=1,2,3$ represent three decoupled systems of non linear algebraic equations in the unknowns a_i, θ_{1i} and b_i, that can be solved to find the single solution:

$$ \begin{cases} a_1=c-d\cdot r_{12} \\ \theta_{11}=\operatorname{atan2}(r_{32},r_{22}) \\ b_1=\dfrac{d\cdot r_{22}}{c\theta_{11}} \end{cases} \quad \begin{cases} a_2=c-d\cdot r_{23} \\ \theta_{12}=\operatorname{atan2}(r_{13},r_{33}) \\ b_2=\dfrac{d\cdot r_{33}}{c\theta_{12}} \end{cases} \quad \begin{cases} a_3=c-d\cdot r_{31} \\ \theta_{13}=\operatorname{atan2}(r_{21},r_{11}) \\ b_3=\dfrac{d\cdot r_{11}}{c\theta_{13}} \end{cases} \quad (29\text{-}31) $$

The direct kinematic problem, on the other hand, assumes the knowledge of joint variables a_i, $i=1,2,3$ and aims at finding the corresponding attitudes of the platform in the space (i.e. the expression of the Cardan angles α, β, γ or the rotation matrix ${}^{O}_{P}R$). The analysis is performed by means of simple trigonometric manipulations; by substituting in (29-31) the expression of r_{ij} given in (19), it is obtained:

$$\begin{cases} c\beta s\gamma = \dfrac{c-a_1}{d} = k_1 \\ s\alpha c\beta = \dfrac{c-a_2}{d} = k_2 \\ c\alpha s\beta c\gamma - s\alpha s\gamma = \dfrac{c-a_3}{d} = k_3 \end{cases} \tag{32}$$

where the k_i, $i=1,2,3$ are known values. The 3 equations in (32) can be solved to find up to 4 admissible values for $s\gamma$:

$$\left(2\frac{k_3 k_2}{k_1} - k_2^2 + 1 + \frac{k_2^2}{k_1^2}\right)s^4\gamma + \left(k_3^2 - k_2^2 - k_1^2 - 1\right)s^2\gamma + k_1^2 = 0 \tag{33}$$

For each angle γ that solves Eq. (33), 2 different values can be found for angles β and α:

$$c\beta = \frac{k_1}{s\gamma} \qquad\qquad s\alpha = \frac{k_1}{k_2}s\gamma \tag{34-35}$$

therefore system (32) admits up to 32 different solutions: direct kinematics of the mechanism, however, is characterised by a maximum number of 16 different configurations, since angle β can be restricted in the range $[-\pi/2, \pi/2]$ without any loss of information.

5. Differential kinematics

By direct differentiation of the first 3 equations in (29-31), the expression of the analytic Jacobian J_A is directly derived:

$$\begin{bmatrix} \dot{a}_1 \\ \dot{a}_2 \\ \dot{a}_3 \end{bmatrix} = d \cdot \begin{bmatrix} 0 & -s\beta s\gamma & c\beta c\gamma \\ c\alpha c\beta & -s\alpha s\beta & 0 \\ -s\alpha s\beta c\gamma - c\alpha s\gamma & c\alpha c\beta c\gamma & -c\alpha s\beta s\gamma - s\alpha c\gamma \end{bmatrix} \cdot \begin{bmatrix} \dot{\alpha} \\ \dot{\beta} \\ \dot{\gamma} \end{bmatrix} = J_A \begin{bmatrix} \dot{\alpha} \\ \dot{\beta} \\ \dot{\gamma} \end{bmatrix} \tag{36}$$

The geometric Jacobian J_G can be worked out by expressing the relation between the derivatives of Cardan angles and the components of angular velocity ω:

$$\begin{bmatrix} \omega_x \\ \omega_y \\ \omega_z \end{bmatrix} = \begin{bmatrix} 1 & 0 & s\beta \\ 0 & c\alpha & -s\alpha c\beta \\ 0 & s\alpha & c\alpha c\beta \end{bmatrix} \cdot \begin{bmatrix} \dot{\alpha} \\ \dot{\beta} \\ \dot{\gamma} \end{bmatrix} = T_\varphi \begin{bmatrix} \dot{\alpha} \\ \dot{\beta} \\ \dot{\gamma} \end{bmatrix} \tag{37}$$

$$\begin{bmatrix} \dot{a}_1 \\ \dot{a}_2 \\ \dot{a}_3 \end{bmatrix} = d \cdot \begin{bmatrix} 0 & -c\alpha s\beta s\gamma - s\alpha c\gamma & c\alpha c\gamma - s\alpha s\beta s\gamma \\ c\alpha c\beta & 0 & -s\beta \\ -s\alpha s\beta c\gamma - c\alpha c\beta s\gamma & c\beta c\gamma & 0 \end{bmatrix} \begin{bmatrix} \omega_x \\ \omega_y \\ \omega_z \end{bmatrix} = J_G \begin{bmatrix} \omega_x \\ \omega_y \\ \omega_z \end{bmatrix} \tag{38}$$

The singular configurations can be found by letting the determinant of J_G vanish:

$$|J_G|/d = -2s\alpha c\alpha s\beta s\gamma c\gamma - c^2\alpha c^2\beta c^2\gamma - c^2\alpha - c^2\gamma + c^2\beta c^2\gamma + c^2\alpha c^2\beta + 2c^2\alpha c^2\gamma \tag{39}$$

The zeros of Eq. (39) all lie on closed surfaces in the 3-dimensional space α, β, γ; their intersections with the coordinate planes are straight lines (see also Fig. 4), as given by:

$$\begin{cases} \alpha = 0 \rightarrow \beta \pm \gamma = \pm \pi/2 \\ \beta = 0 \rightarrow \alpha = \pm \pi/2,\ \gamma = \pm \pi/2 \\ \gamma = 0 \rightarrow \alpha \pm \beta = \pm \pi/2 \end{cases} \tag{40}$$

The analysis of singular configurations has been performed by means of numerical simulations; Fig. 3 shows the value of the determinant of the geometric Jacobian matrix (39), normalised in the range [-1, +1] after division by the constant d: the black regions are characterised by determinant values within the range [-0,05, +0,05]. All the singularity maps are plot against the β and γ angles, α being a parameter of the representation; the configuration of the mechanism for $\beta=\gamma=0$ is represented aside. The analysis of the graphs shows the presence of singularity-free regions inside the workspace, Fig. 4, where the planning of a motion could be performed: for the mechanism under test, that has not been optimised, a sphere with a radius of about 50° can be inscribed.

6. Conclusions

The concept of a novel spherical wrist has been presented: it consists of an in-parallel actuated mobile platform with a 3-CPU architecture, characterised by an orthogonal setting of the three cylindrical joints' axes. After having chosen the linear displacements of the frame cylindrical pairs as the most convenient actuated variables, direct and inverse orientation kinematics has been studied and the related solutions have been worked out in closed form. The geometrical Jacobian matrix has been derived and the occurrence of rotation singularities has been identified through computer simulation: it results that the three dimensional orientation space is separated by singularity surfaces, but relevant dextrous volumes can be identified close to the home configuration. In the same manner, also the occurrence of constraint singularities has been studied, evidencing all the configurations that allow the mechanism to translate.

It is noted that, differently from many SPMs that have been proposed so far, the 3-CPU architecture is non-overconstrained, which should allow the mechanism to work even when precise geometrical conditions are not accurately met: therefore the Authors are presently developing the concept to arrive at the design of a physical prototype. To this aim, since the Jacobian matrix does not depend on geometrical parameters, as shown in (38), the dynamic model is being presently developed to allow an optimisation based on the dynamic manipulability concept.

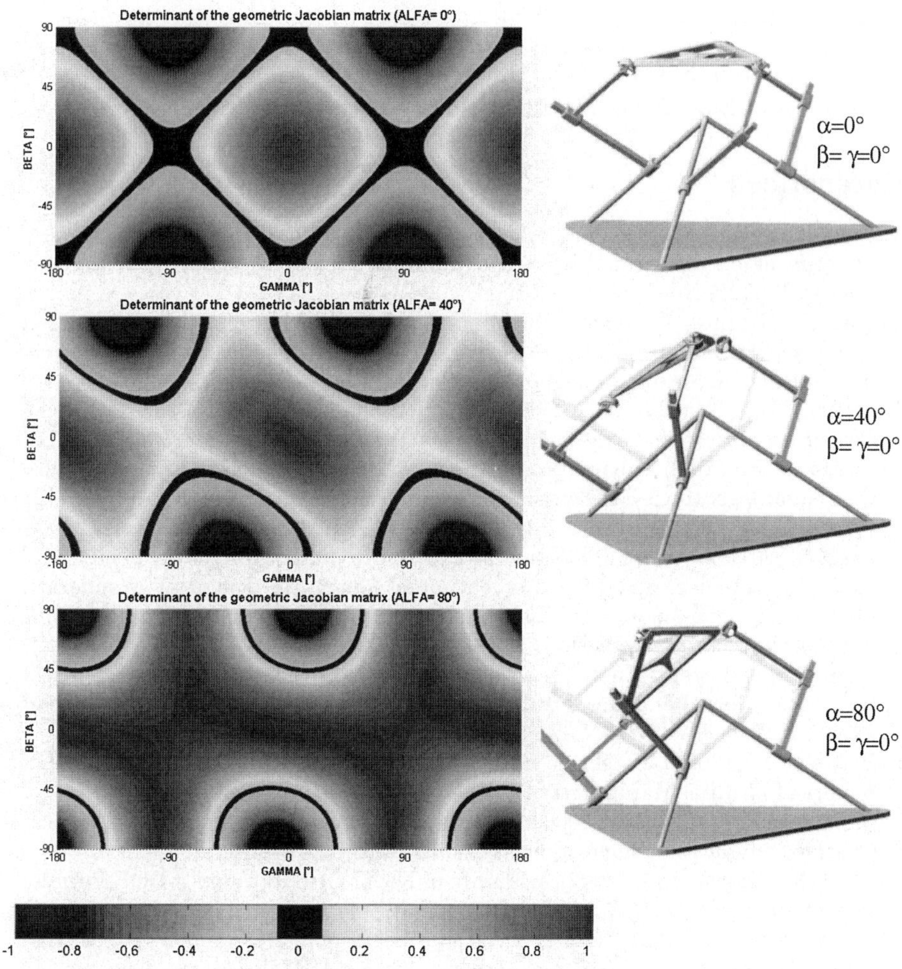

Figure 3. Singularity maps and mechanism configurations

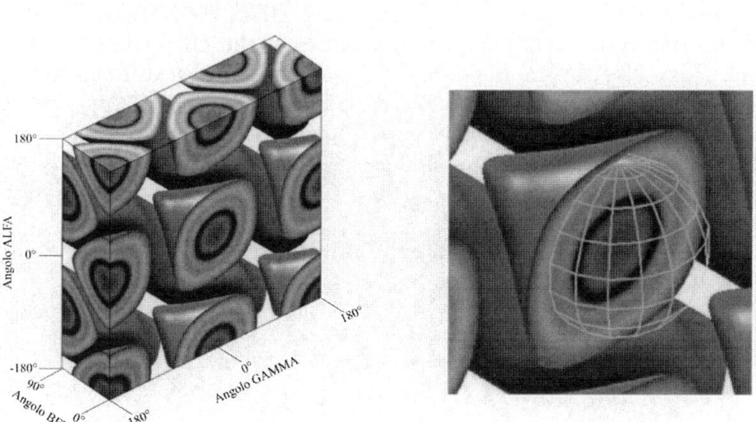

Figure 4. Singularity-free regions inside workspace

Acknowledgment

The research has been partially supported the Italian Ministry of Research and Education (MIUR) within the MiniPAR Project.

References

Alizade, R.I., Tagiyev, N.R., Duffy, J. (1994) A forward and reverse displacement analysis of an in-parallel spherical manipulator, *Mechanism and Machine Theory*, Vol. 29, No. 1, pp.125-137.

Asada, H., Granito, C. (1985) Kinematic and static characterization of wrist joints and their optimal design, *Proc. IEEE Conf. Robotics and Automation*, St. Louis, USA, March 25-28, pp.244-250.

Callegari, M., Suardi, A. (2004) Functionally-Oriented PKMs for Robot Cooperation, *Proc. Intelligent Autonomous Systems*, Amsterdam, March 10-13.

Di Gregorio, R. (2000) Kinematics of a new spherical parallel manipulator with thee equal legs: the 3-URC wrist, *J. Robotic Systems*, pp.213-219

Di Gregorio, R. (2001) Statics and singularity loci of the 3-UPU wrist, *Proc. IEEE/ASME Advanced Intelligent Mechatronics*, Como, Italy, July 8-11.

Gosselin, C. Angeles, J. (1989) The optimum kinematic design of a spherical three-degree-of-freedom parallel manipulator, *J. Mechanisms, Transmissions and Automation in Design*, Vol. 111. n°2. pp.202-207.

Innocenti, C., Parenti-Castelli V. (1993) Echelon form solution of direct kinematics for the general fully-parallel spherical wrist, *Mechanism and Machine Theory*, Vol. 28, No. 4, pp.553-561.

Ji, P., Wu, H. (2001) Algebraic Solution to Forward Kinematics of a 3-DOF Spherical Parallel Manipulator, *J. Robotic Systems*, pp.251-257.

Karouia, M., Hervè, J.M. (2000) A three-dof tripod for generating spherical rotation, *Advances in Robot Kinematics*, Kluwer, pp.395-402.

Karouia, M., Hervè, J.M. (2002) A Family of Novel Orientational 3-DOF Parallel Robots, *Proc. RoManSy 14*, Udine, Italy, July 1-4.

Kong, X., Gosselin, C.M., (2002) Type synthesis of 3-DOF spherical parallel manipulators based on screw theory, *Proc. ASME Design Engineering Technical Conferences and Computers and Information in Engineering Conference*, Montreal, Canada, Sept. 29-Oct. 2, DETC2002/MECH- 21152.

Lee, J.J., Chang, S.-L. (1992) On the kinematics of the UPS wrist for real time control, *Proc. 22nd ASME Biennial Mechanisms Conference: Robotics, Spatial Mechanisms and Mechanical Systems*, Scottsdale, Sept. 13-16, pp.305-312.

Merlet, J.-P. (2000) *Parallel Robots*, Kluwer Academic Publ.

Olivieri, B. (2003) *Study of a novel parallel kinematics spherical robot for cooperative applications*, Tesi di laurea (in Italian), Università Politecnica delle Marche, Ancona, Italy.

Tsai, L.-W. (2000) *Mechanism Design: Enumeration of Kinematic Structures According to Function*, CRC Press.

CONTROL OF A MANEUVERING MOBILE ROBOT BY TRANSVERSE FUNCTIONS

Guillaume Artus, Pascal Morin, Claude Samson
INRIA - U.R. Sophia-Antipolis
2004, route des Lucioles
06 902 Sophia Antipolis Cedex, France
firstname.lastname@sophia.inria.fr

Abstract Based on the transverse function (t.f.) approach recently developed by two of the authors for the control of general nonlinear driftless systems, a control strategy for tracking an omnidirectional target with a unicycle-like robot is proposed. An original feature of the approach is the capacity to comply with a target which moves freely in the plane and performs motions which are not feasible by the nonholonomic robot. With respect to a previous publication on the subject, the proposed control solution involves two extensions in order to i) limit the control magnitude and the number of maneuvers when the initial tracking errors are large, and ii) adapt automatically the tracking precision depending on whether or not the target's motion corresponds to a feasible trajectory for the nonholonomic robot.

Keywords: Practical stabilization, target tracking, nonholonomic robot, transverse functions

1. Introduction

As in (Artus et al., 2003) we consider the problem of tracking a vehicle (called *target* from now on) with a nonholonomic unicycle-like robot. Unlike the robot, the target is not subjected to non-holonomic constraints. It can move freely in the plane and thus perform motions which may not be feasible for the robot. Such a situation occurs, for example, when the control problem consists in tracking a (virtual) frame —representing the target— located behind, and rigidly linked to, the body of a wheeled vehicle —a car, for instance. This would correspond to a classical car-platooning problem except that we would like to extend the operating domain of the control to the case when the leading vehicle moves backward and makes maneuvers. It is then not difficult to verify that such maneuvers usually yield trajectories of the target frame which are not feasible for a nonholonomic vehicle.

To our knowledge, the problem of tracking non-feasible trajectories has seldom been addressed in the control literature devoted to nonholo-

J. Lenarčič and C. Galletti (eds.), On Advances in Robot Kinematics, 459-468.

nomic systems. This is much related to the fact that feedback control studies traditionally focus on asymptotic stabilization (see e.g. (Samson, 1995; de Luca et al., 1998)), whereas a non-feasible trajectory cannot, by definition, be asymptotically stabilized. In (Morin and Samson, 2003), a general framework, based on the use of so-called transverse functions (t.f.), has been proposed for the design of feedback laws yielding practical stabilization of controllable driftless systems subjected to additive perturbations. This approach has been applied in (Artus et al., 2003) to solve the target tracking problem here considered. In the present paper, the control methodology is further extended along two directions. The first issue is related to the determination of a vector field term which, in the control expression, ensures the stability of the controlled system and the property of practical stabilization. The extension here proposed aims at monitoring the transient behavior associated with large initial tracking errors in order to reduce the control effort and the number of robot maneuvers. The method relies on the constrained minimization of the control norm. The second issue involves the introduction of new transverse functions and additional design parameters which are used to automatically adjust the tracking precision depending on the nature of the target's motion. The underlying motivation is that small traking errors are expected when the target's trajectory is feasible, whereas larger tracking errors are required to avoid fast oscillatory maneuvers otherwise.

The paper is organized as follows. Models for control design are introduced in Section 2. Control design issues are addressed in Section 3. Simulation results are reported in Section 4. Due to space limitations, the proofs are omitted.

2. Modeling

Let us consider the three frames represented on Fig. 1: $\mathcal{F}_0$ is a fixed frame, $\mathcal{F}$ is a frame attached to the unicycle, and $\mathcal{F}_r$ is a frame attached to the target. Let (x_m, y_m) denote the coordinates of $\overrightarrow{\mathrm{OP}}$ in $\mathcal{F}_0$, and α_m denote the angle between $\vec{\imath}_0$ and $\vec{\imath}$ (see Fig. 1). The control inputs of the robot are the longitudinal velocity u_1 along the vector $\vec{\imath}$ of $\mathcal{F}$ and the angular velocity $u_2 = \dot{\alpha}_m$. With these notations, the well known kinematic equations of the unicycle are

$$\left\{\begin{array}{lcl} \dot{x}_m & = & u_1 \cos\alpha_m \\ \dot{y}_m & = & u_1 \sin\alpha_m \\ \dot{\alpha}_m & = & u_2 \end{array}\right. \tag{1}$$

For stabilization purposes, we want to describe the kinematics of the robot with respect to the target. Let (x_r, y_r) denote the coordinates

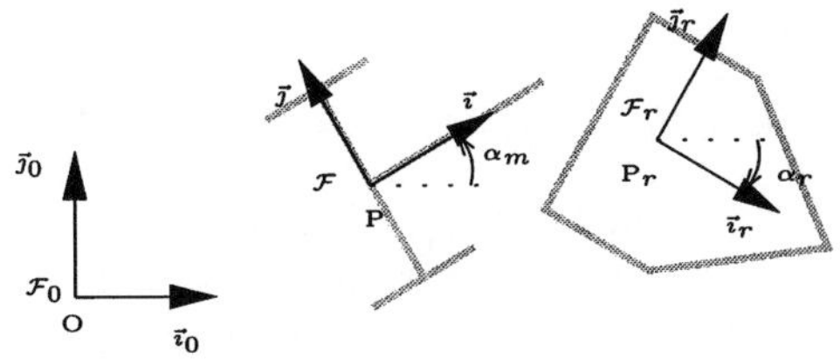

Figure 1. Configuration frames

of $\overrightarrow{\mathrm{OP}}_r$ in $\mathcal{F}_r$, and α_r denote the angle between $\vec{\imath}_0$ and $\vec{\imath}_r$. The two components of the velocity of P_r, expressed in the basis of the target frame, are denoted as a and b, and the target angular velocity is denoted as c, i.e.

$$\left\{ \begin{array}{rcl} \frac{d\overrightarrow{\mathrm{OP}}_r}{dt} & = & a\vec{\imath}_r + b\vec{\jmath}_r \\ \dot{\alpha}_r & = & c \end{array} \right. \tag{2}$$

The target velocity vector is $u_t = (a, b, c)^T$. With x and y denoting the coordinates of the position error vector between the robot and the target, expressed in the target frame $\mathcal{F}_r$, i.e. $\overrightarrow{\mathrm{P}_r\mathrm{P}} = x\vec{\imath}_r + y\vec{\jmath}_r$, and $\alpha = \alpha_m - \alpha_r$, one infers from (1) and (2)

$$\left\{ \begin{array}{rcl} \dot{x} & = & u_1 \cos\alpha + cy - a \\ \dot{y} & = & u_1 \sin\alpha - cx - b \\ \dot{\alpha} & = & u_2 - c \end{array} \right. \tag{3}$$

System (3) represents the error system associated with the target tracking problem. It can also be written as

$$\dot{g} = u_1 b_1(g) + u_2 b_2 + b_0(g, u_t) \tag{4}$$

with $g = (x, y, \alpha)^T \in G := \mathbb{R}^2 \times \mathbb{T} \approx SE(2)$, $\mathbb{T} := \mathbb{R}/2\pi\mathbb{Z}$, and

$$b_0(g, u_t) = (cy - a, -cx - b, -c)^T \tag{5}$$

$$b_1(g) = (\cos\alpha, \sin\alpha, 0)^T,\ b_2 = (0, 0, 1)^T \tag{6}$$

Note that $b_0 = 0$ iff the target is motionless.

3. Control design

The proposed control strategy is based on the *transverse function* approach (Morin and Samson, 2003) here applied to a unicycle-like robot, as in (Artus et al., 2003). Prior to presenting new developments of interest for our application, let us first review a few basic features of the approach (see the above mentioned references for more details).

3.1 Recalls

Let us temporarily focus on System (4) without the drift term b_0, i.e.

$$\dot{g} = u_1 b_1(g) + u_2 b_2 \tag{7}$$

A smooth function $f : \mathbb{T}^2 \longrightarrow G$ is called a *transverse function* for System (7) if, for any $(\theta, \beta) \in \mathbb{T}^2$, the matrix

$$H\left(f(\theta,\beta)\right) \triangleq \begin{pmatrix} b_1\left(f(\theta,\beta)\right) & b_2 & -\frac{\partial f}{\partial \theta}(\theta,\beta) \end{pmatrix} \tag{8}$$

is invertible. The dependence on the second variable β is not required in the original formulation of the approach, but it is here introduced for the sake of generalization. From now on, the variables θ and β should be interpreted as additional state components which are controlled via their time derivatives $\dot{\theta}$ and $\dot{\beta}$.

Recall that a Lie group G is a differentiable manifold equipped with a smooth group operation. The natural Lie group involved in the present application is $G = \mathbb{R}^2 \times \mathbb{T} \approx SE(2)$ endowed with the group operation $(g, g') \longmapsto gg'$ defined by

$$gg' = \begin{pmatrix} p + R(\alpha)p' \\ \alpha + \alpha' \end{pmatrix} \tag{9}$$

with $g = (p, \alpha)^T$, $g' = (p', \alpha')^T$, p and $p' \in \mathbb{R}^2$, and $R(\alpha)$ the rotation matrix of angle α. We denote by $e = (0,0,0)^T$ the unit element of this group. Note that we could equivalently define G as the group of homogeneous matrices in the plane, and the group operation as the product of these matrices.

It is well known, and straightforward to verify, that System (7) defines a left-invariant control system on G, i.e., for any solution $g(.)$ of (7) and any $g_0 \in G$, $g_0 g(.)$ is also a solution to (7).

Let us finally define the variable z as the group product of g and $f^{-1}(\theta,\beta)$, with $f^{-1}(\theta,\beta)$ denoting the inverse of $f(\theta,\beta)$ on the group, i.e.

$$z := g f^{-1}(\theta,\beta) \tag{10}$$

The core of the control approach is the asymptotic stabilization of z to e by using the control variables u_1, u_2, and $\dot{\theta}$. The following proposition shows that this is a simple task, when f is a transverse function.

Proposition 1 *Let $f := (f_1, f_2, f_3)^T$ denote a transverse function, and $\bar{u}$ denote the "extended" control vector defined by $\bar{u} := (u_1, u_2, \dot{\theta})^T$. Then,*

i) along any solution $g(.)$ of (4), and any smooth time functions $\theta(.)$ and $\beta(.)$,

$$\dot{z} = \bar{H}\bar{u} - w \tag{11}$$

with $\bar{H} = AH(f)$, $w = A\left(\frac{\partial f}{\partial \beta}\dot{\beta} - Bb_0(g, u_t)\right)$, and

$$A = \begin{pmatrix} R(\alpha - f_3) & R(\alpha - f_3)\begin{pmatrix} f_2 \\ -f_1 \end{pmatrix} \\ 0 & 1 \end{pmatrix}, \quad B = \begin{pmatrix} R(f_3 - \alpha) & 0 \\ 0 & 1 \end{pmatrix}$$

ii) given an arbitrary vector valued function $v \in C^0(G \times \mathbb{T}^2 \times \mathbb{R}; \mathbb{R}^3)$, the application of the following dynamic feedback law

$$\bar{u} = \bar{H}^{-1}(w + v) \tag{12}$$

applied to System (11) yields $\dot{z} = v$.

The part ii) of the above proposition tells us that any function v which ensures the asymptotic stability of the origin of the system $\dot{z} = v$ yields the convergence of the solutions of the controlled system (4,12) to $f(\mathbb{T}^2)$, *whatever* the "perturbation" term b_0 in (4) —and thus *whatever* the target's velocity. In other words the vector error g tends to $f(\mathbb{T}^2)$ when the control defined by (12) is used, provided that v asymptotically stabilizes the origin of the decoupled linear system $\dot{z} = v$. Moreover, the rate of convergence of z to zero is the same as the rate of convergence of g to $f(\mathbb{T}^2)$. To obtain exponential convergence, a simple possibility obviously consists in setting

$$v(z, \theta, \beta, t) = -Kz \tag{13}$$

with $(-K)$ denoting a Hurwitz stable matrix. In (Artus et al., 2003), this possibility is further simplified by choosing K diagonal and positive definite. Now, since $\mathbb{T}^2$ is compact and f is smooth, the set $f(\mathbb{T}^2)$ is bounded. In particular, by choosing f such that this set is contained in a small neighborhood of zero, then g is ultimately close to zero.

3.2 Transient dynamics adjustment

As explained above, the feedback control law used in (Artus et al., 2003) is

$$\bar{u}_{zl} = \bar{H}^{-1}(w - Kz) \tag{14}$$

with K a positive definite matrix, and the index *zl* reminding us that this control yields a linear system in the z variable. When no further requirement is made on the controlled system, choosing the control term v as in (13) makes sense, since it combines simplicity and fast (exponential) convergence. However, there are many other possible choices.

Also, nothing proves that, during the transient phase when $|z(t)|$ converges to zero from an initially large value $z(0)$, the robot's trajectory in the cartesian space resulting from this choice has interesting properties. As a matter of fact, simulations show that this choice tends to produce an unnecessary large number of maneuvers. In order to improve on this aspect, we propose that v be instead chosen so as to minimize the norm of a function of the control $\bar{u}$, under the constraint of having $|z|$ still converge exponentially to zero. More precisely, we consider the following optimization problem

$$\begin{cases} \min_{\bar{u}} J := \frac{1}{2} \left\| \bar{u} - \dot{\beta} H^{-1} \frac{\partial f}{\partial \beta} \right\|_{\delta}^{2} \\ z^T(\bar{H}\bar{u} - w) = -z^T K z \end{cases} \tag{15}$$

with $\|.\|_\delta$ defined by $\|x\|_\delta^2 = \delta_1(x_1^2 + x_2^2) + \delta_2 x_3^2$, $\delta_1, \delta_2 > 0$, and K a positive definite matrix. Note that, by (11), the constraint in (15) is equivalent to $z^T \dot{z} = -z^T K z$. The solution to this problem is easily derived by applying the standard method of Lagrange multipliers.

Lemma 1 *The solution to the optimization problem* (15) *is*

$$\begin{pmatrix} u \\ \dot{\theta} \end{pmatrix}^{\star} = \bar{u}^{\star} = \begin{pmatrix} -\frac{1}{\delta_1 \eta}(z^T K z + z^T \tilde{w}) \bar{H}_1^T z \\ -\frac{1}{\delta_2 \eta}(z^T K z + z^T \tilde{w}) \bar{H}_2^T z \end{pmatrix} + \dot{\beta} H^{-1} \frac{\partial f}{\partial \beta} \tag{16}$$

with

$$\tilde{w} := ABb_0(g, u_t) \quad , \quad \eta = z^T \left(\delta_1^{-1} \bar{H}_1 \bar{H}_1^T + \delta_2^{-1} \bar{H}_2 \bar{H}_2^T \right) z \tag{17}$$

and $\bar{H}_1 \in \mathbb{R}^{3\times 2}, \bar{H}_2 \in \mathbb{R}^{3\times 1}$ *the matrices defined by* $\bar{H} = (\bar{H}_1, \bar{H}_2)$.

When $\tilde{w} = 0$, the feedback law (16) can be defined at $z = 0$ by continuity. Otherwise, this feedback is not defined at $z = 0$. To circumvent this difficulty, and retain the qualities of the above optimal solution when $|z|$ is large, we propose to combine the feedback laws (14) and (16) as specified in the following proposition.

Proposition 2 *Let*

$$\bar{u} = \left(1 - \frac{\eta}{\eta + \psi}\right) \bar{u}_{zl} + \frac{\eta}{\eta + \psi} \bar{u}^{\star} \tag{18}$$

with $\bar{u}_{zl}$, $\bar{u}^\star$, *and* η *defined by* (14), (16), *and* (17) *respectively, and* ψ *a positive constant number used for regularization purposes. Then, the feedback control law* (18) *is well defined everywhere, and the origin* $z = 0$ *of the closed-loop system* (11)–(18) *is exponentially stable (because* $z^T \dot{z} = -z^T K z$ *along any trajectory of this system).*

3.3 Monitoring of the tracking precision

The choice of the transverse function f is obviously of central importance. In (Artus et al., 2003) the following functions, depending on a single variable θ, and two positive parameters $(\varepsilon_1, \varepsilon_2)$, were used:

$$f(\theta) = (\varepsilon_1 \sin\theta, \frac{\varepsilon_1\varepsilon_2}{4} \sin 2\theta, \varepsilon_2 \cos\theta)^T \tag{19}$$

with $\varepsilon_1, \varepsilon_2 > 0$. Note that by setting $\bar{f}(\theta, \beta) = f(\theta + \beta)$ the above function is formally transformed into a transverse function which depends on two variables, as specified at the beginning of Section 3.1. Therefore, in order to use the function (19) in the control laws derived earlier, one only has to set $\dot{\beta} = 0$ in the control expressions. Note also that the "size" of f, which determines the ultimate bound of the tracking errors, is directly related to ε_1 and ε_2.

The next proposition points out another set of transverse functions which depend on the variable β in a more useful way, and also on a third parameter γ.

Lemma 2 *For any $\varepsilon_1, \varepsilon_2 > 0$, and any $\gamma \in [0, 1]$, $f : \mathbb{T}^2 \longrightarrow G$ defined by*

$$f = (\bar{f}_1, \bar{f}_2, \arctan \bar{f}_3)^T \tag{20}$$

with

$$\begin{pmatrix} \bar{f}_1 \\ \bar{f}_2 \\ \bar{f}_3 \end{pmatrix} = \begin{pmatrix} \varepsilon_1 (\sin(\theta+\beta) - \gamma \sin\beta) \\ \frac{\varepsilon_1\varepsilon_2}{2} ((\sin(\theta+\beta) - \gamma\sin\beta)(\cos(\theta+\beta) - \gamma\cos\beta) - \gamma\sin\theta) \\ \varepsilon_2 (\cos(\theta+\beta) - \gamma\cos\beta) \end{pmatrix} \tag{21}$$

is a transverse function.

The variable β and the parameter γ represent extra degrees of freedom which can be used to "monitor" the size of $f(\theta, \beta)$ and, in doing so, achieve complementary control objectives. This possibility is illustrated by the following proposition.

Proposition 3 *Assume that the dynamic feedback law $\bar{u}_{zl}$ given by (14), or $\bar{u}$ given by (18), is applied to the control system (4), with a transverse function f defined by (20,21). Assume also that the target velocity vector $u_t = (a, b, c)^T$ is bounded, then:*

1) For $\gamma = 1$ and $\dot{\beta} = k\tan(\theta/2)$ $(k > 0)$ with $\theta(0) = 0$, $g = 0$ is an asymptotically stable equilibrium point if $b_0 = 0$ (i.e. if the target does not move).

2) For $\gamma \in (0, 1)$, and

$$\dot{\beta} = \frac{k}{1+\gamma^2 - 2\gamma\cos\theta} \tan\frac{\theta}{2} + \frac{2}{\varepsilon_1\varepsilon_2} \frac{\bar{f}_3(c\bar{f}_2 - a) + c\bar{f}_1}{1+\gamma^2 - 2\gamma\cos\theta} \tag{22}$$

if the component b of the target velocity vector in (2) is equal to zero, then $\theta(t)$ converges to zero exponentially. In this case, the convergence of $g(t)$ to $f(\mathbb{T}^2)$ implies that the norm of the tracking error g is ultimately bounded by $(1-\gamma)\|(\varepsilon_1, \varepsilon_2, \frac{\varepsilon_1\varepsilon_2}{4})\|$.

3) In both cases, the convergence of $g(t)$ to $f(\mathbb{T}^2)$ ensures that the norm of the tracking error g is ultimately bounded by $2\|(\varepsilon_1, \varepsilon_2, \varepsilon_1\varepsilon_2)\|$.

The first part of the proposition indicates that asymptotic stabilization of the target frame can be achieved when the target is motionless, whereas the second part points out the controls's capacity to automatically augment the tracking precision when the target's trajectory is feasible for the robot. Recall from (2) that it is feasible when $b = 0$. By choosing γ close to one, small tracking errors are obtained in this case. When the trajectory is not feasible ($b \neq 0$), it is intuitively preferable to reduce the tracking precision so as to lower the frequency of the maneuvers and, subsequently, the overall control effort. In all cases, as pointed out in the third part of the proposition, the tracking errors are ultimately bounded by a number whose size is adjustable via the choice of the parameters ε_1 and ε_2.

4. Simulation results

Figure 2 shows trajectories of the robot's point P when the target is motionless and the initial tracking errors are large. The control applied to the robot is given by (14) in the case of Figure 2(a), and (18) in the case of Figure 2(b). The matrix K used in both cases is equal to $0.5\mathrm{I}_3$, with I_3 the identity matrix. The parameters of the control (18) are as follows: $\delta_1 = 1$, $\delta_2 = 10$, $\psi = 0.01$. In both cases the transverse function (20,21) is used with ε_1, $\varepsilon_2 = 0.3$, and $\gamma = 0.9$. The time derivative of β is given by (22) with $k = 0.2$. One can observe that, as anticipated, the control (18) produces much fewer maneuvers than the control (14). Once z has converged to zero, using one control or the other is indifferent, because they yield the same zero dynamics.

Figure 3 illustrates what happens in the long range (i.e. on the zero dynamics $z = 0$), and show how the monitoring mechanism described in Section 3.3, associated with the second variable β in the transverse function, modifies the tracking precision depending on the target's motion. The t.f. (19), which depends on a single variable, is used for the simulation results of Figure 3(a), with no monitoring of the tracking precision in this case. The t.f. function defined by (20,21), and the adaptation mechanism as specified by (22), are used for the simulation results of Figure 3(b). All parameters are the same as before. In these simulations the target starts moving along a trajectory which is feasible

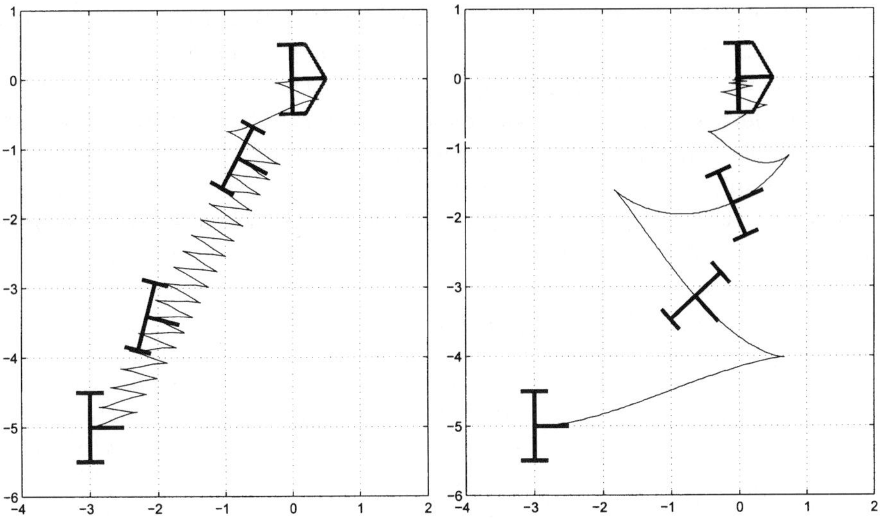

(a) Control law (12) (b) Control law (18)

Figure 2. Cartesian motion: robot and target

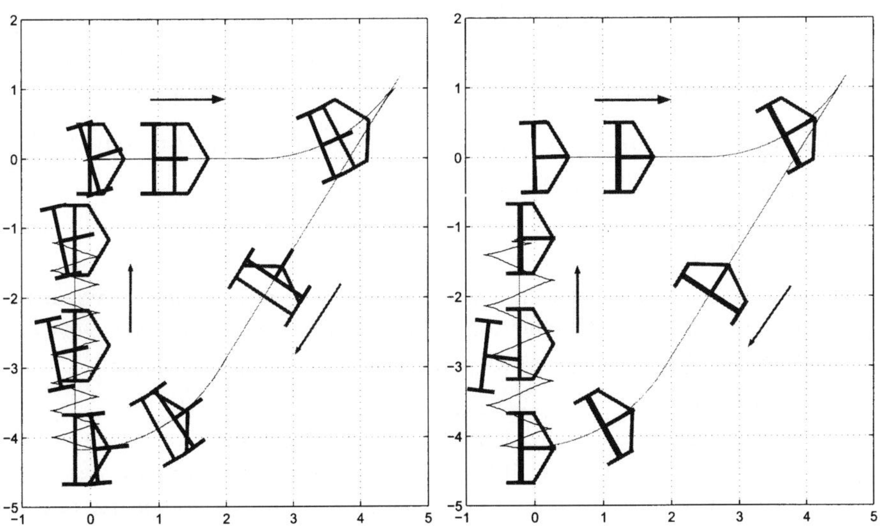

(a) Control law using the transverse function (19) (b) Control law using the transverse function (20,21)

Figure 3. Cartesian motion: robot and target

for the unicycle (i.e. $b = 0$), with a positive longitudinal velocity at first (forward motion, $a > 0$), then with a negative one (backward motion,

$a < 0$). One can observe that the tracking errors are much smaller in the second figure, due to the action of the adaptation mechanism. During the last part of the simulations, the target performs a lateral motion ($b > 0$, $a = c = 0$) which is not feasible for the robot. One can then observe the maneuvers that the robot has to perform to track the target. Whereas the tracking precision has not changed in the case of the control simulated in the first figure, it has automatically been lowered in the case of the control simulated in the second figure so as to reduce the frequency of the robot's maneuvers.

5. Concluding remarks

We have presented extensions of the t.f. control approach applied to the problem of tracking an omnidirectional target frame with a nonholonomic unicycle-like mobile robot, with simulations illustrating their practical usefulness. The proposed control solutions have also been tested on our experimental benchmark —a description of which can be found in (Tsakiris et al., 1997). Experimenting on a physical system involves complementary issues (control discretization, control saturation, state reconstruction, target velocity estimation, etc.) which have been addressed in (Artus et al., 2003). The extension of the proposed control approach to car-like mobile robots will be the subject of future articles.

6. Acknowledgments

This work is supported by the European Commission through the IST research program CyberCars.

References

Artus, G., Morin, P., and Samson, C. (2003). Tracking of an omnidirectional target with a nonholonomic mobile robot. In *The 11th International Conference on Advanced Robotics*, volume 3, pages 1468–1473.

de Luca, A., Oriolo, G., and Samson, C. (1998). Feedback control of a nonholonomic car-like robot. In Laumond, J.-P., editor, *Robot motion planning and control*, volume 229 of *LNCIS*. Springer Verlag.

Morin, P. and Samson, C. (2003). Practical stabilization of driftless systems on lie groups: The transverse function approach. *IEEE Trans. on Automatic Control*, 48(9):1496–1508.

Samson, C. (1995). Control of chained systems. Application to path following and time-varying point-stabilization. *IEEE Trans. on Automatic Control*, 40:64–77.

Tsakiris, D., Kapellos, K., Samson, C., Rives, P., and Borrelly, J. (1997). Experiments in real-time vision-based point stabilization of a nonholonomic mobile manipulator. In *5th International Symposium on Experimental Robotics*.

ACCELERATION-DRIVEN KINEMATICS OF MOBILE MANIPULATORS: AN ENDOGENOUS CONFIGURATION SPACE APPROACH

Krzysztof Tchoń
Institute of Engineering Cybernetics
Wrocław University of Technology
ul. Janiszewskiego 11/17
50–372 Wrocław, Poland
tchon@ict.pwr.wroc.pl

Janusz Jakubiak
Institute of Engineering Cybernetics
Wrocław University of Technology
ul. Janiszewskiego 11/17
50–372 Wrocław, Poland
jjakubia@ict.pwr.wroc.pl

Abstract Using the endogenous configuration space approach we study the inverse kinematic problem for mobile manipulators represented by an affine (not driftless) control system with outputs. A particular case of acceleration-driven mobile platform is addressed, and solved by means of a Jacobian pseudoinverse inverse kinematics algorithm. Performance of this algorithm has been illustrated with computer simulations involving the acceleration-driven kinematic car platform with RTR manipulator aboard.

Keywords: Mobile manipulator, affine representation, inverse kinematics

1. Introduction

We shall define a mobile manipulator as a robotic system composed of a nonholonomic mobile platform and a stationary manipulator mounted on the platform. In the last decade there has been a considerable interest in mobile manipulators, motivated both by theoretical as well as by applicational reasons. The origin of interest in mobile manipulators can be traced back to space robotics Papadopoulos and Dubowsky, 1991. Problems of motion coordination, dynamic control, and performance evaluation of mobile ma-

J. Lenarčič and C. Galletti (eds.), On Advances in Robot Kinematics, 469-476.

nipulators with nonholonomic platform have been addressed in Yamamoto and Yun, 1994; Yamamoto and Yun, 1996; Seraji, 1998; Yamamoto and Yun, 1999; Gardner and Velinsky, 2000; Bayle et al., 2001. The motion and trajectory planning problems of mobile manipulators with nonholonomic platform have been dealt with in Pin et al., 1996; Furuno et al., 2003 for single mobile manipulators, and in Desai and Kumar, 1997; Sugar et al., 2001 for a group of mobile manipulators. Mobile manipulators equipped with holonomic platform have been treated in Khatib et al., 1996; Khatib et al., 1999.

Recently, in the papers Tchoń and Muszyński, 2000; Tchoń et al., 2001 we have elaborated a uniform theory of mobile manipulators, called the endogenous configuration space approach. Within this approach the requirement of nonslipping motion of the platform has been described by Pfaffian phase constraints that define a driftless control theoretic representation of the mobile manipulator kinematics. The control signal of the platform and the joint position of the aboard manipulator compose the endogenous configuration of the mobile manipulator. Special attention has been paid to performance assessment of regular and singular Jacobian inverse kinematics algorithms Tchoń and Jakubiak, 2003, their repeatability Tchoń, 2002, and to dexterity measures of mobile manipulators Tchoń and Zadarnowska, 2003.

Differently to what we have done until now, in this paper we shall apply the endogenous configuration space approach to mobile manipulators whose task accomplishment conditions impose affine, not Pfaffian, phase constraints on the motion of the platform. This includes the case of a mobile manipulator moving on a rotating table. However, in this paper we focus on the inverse kinematic problem associated with the platform kinematics driven by accelerations instead of velocities. In this case we solve the problem using the basic Jacobian pseudoinverse algorithm. The performance of this algorithm is illustrated with computer simulations involving the kinematic car platform carrying a 3 d.o.f. RTR manipulator. The simulations have shown that the endogenous configuration space may offer efficient tools for solving the inverse kinematic problem for mobile manipulators with affine control system representations.

The composition of this paper is the following. Section 2 introduces the affine control system representation of mobile manipulators, and recalls basic concepts of the endogenous configuration space approach. Section 3 introduces the affine representation of the mobile manipulator kinematics driven by accelerations. Section 4 presents results of computer simulations. The paper is concluded with section 5.

2. Basic concepts

Given a mobile manipulator consisting of a nonholonomic mobile platform and a holonomic stationary manipulator mounted on the platform, let $q \in R^n$

denote platform coordinates, and $x \in R^p$ be joint positions of the aboard manipulator. Denote by $y \in R^r$ the position and orientation of the end effector with respect to an inertial coordinate frame. Suppose that associated with the mobile manipulator is the following control system representation

$$\begin{cases} \dot{q} = f(q) + G(q)u = f(q) + \sum_{i=1}^{m} g_i(q)u_i, \\ y = k(q,x) = (k_1(q,x), \ldots, k_r(q,x)). \end{cases} \tag{1}$$

The system (1) will be regarded as an analytic affine control system with controls $u(t) \in R^m$, $x \in R^p$, and the output map $k(q,x)$.

Following the endogenous configuration space approach, we define a pair $(u(\cdot),x) \in L^2_m[0,T] \times R^p \cong \mathcal{X}$ as the endogenous configuration of the mobile manipulator, and endow the endogenous configuration space $\mathcal{X}$ with inner product

$$\langle (u_1(\cdot),x_1),(u_2(\cdot),x_2)\rangle = \int_0^T u_1^T(t)u_2(t)dt + x_1^T x_2, \tag{2}$$

and the corresponding norm. The space $\mathcal{X}$ consists of control functions $u(\cdot)$ of the platform and constant control functions $x(\cdot) = x$ of the aboard manipulator. By definition, the endogenous configuration space is a Hilbert space.

The kinematics $K_{q_0,T} : \mathcal{X} \longrightarrow R^r$ of the mobile manipulator determine the end effector position and orientation at time T achieved at an endogenous configuration $(u(\cdot),x)$, provided that the initial platform state has been q_0, i.e.

$$K_{q_0,T}(u(\cdot),x) = y(T) = k(\varphi_{q_0,T}(u(\cdot)),x), \tag{3}$$

where $\varphi_{q_0,t}(u(\cdot))$ denotes the flow of the system (1) driven by $u(\cdot)$. We shall assume that this flow exists for every $t \in [0,T]$. Having introduced the kinematics (3), we define the analytic Jacobian of the mobile manipulator at $(u(\cdot),x) \in \mathcal{X}$ as the derivative

$$J_{q_0,T}(u(\cdot),x)(v(\cdot),w) = \left.\frac{d}{d\alpha}\right|_{\alpha=0} K_{q_0,T}(u(\cdot)+\alpha v(\cdot), x+\alpha w) =$$

$$C(T,x)\int_0^T \Phi(T,s)B_1(s)v(s)ds + D(T,x)w, \tag{4}$$

where $(v(\cdot),w) \in \mathcal{X}$, and $\Phi(t,s)$ denotes the transition matrix of the linear approximation to (1)

$$\dot{\xi} = A(t)\xi + B(t)v(t), \quad \eta = C(t,x)\xi + D(t,x)w \tag{5}$$

along a control-trajectory triple $(u(t),x,q(t) = \varphi_{q_0,t}(u(\cdot)))$, satisfying the customary differential equation $\frac{\partial}{\partial t}\Phi(t,s) = A(t)\Phi(t,s),\ \Phi(s,s) = I_n$. The matrices appearing in (5) are defined below:

$$\begin{aligned} A(t) &= \tfrac{\partial (f(q(t))+G(q(t))u(t))}{\partial q}, & B(t) &= G(q(t)), \\ C(t,x) &= \tfrac{\partial k(q(t),x)}{\partial q}, & D(t,x) &= \tfrac{\partial k(q(t),x)}{\partial x}. \end{aligned} \tag{6}$$

It is easily seen that the Jacobian (4) computes the output value at T of the linear approximation (5) initialized at $\xi_0 = 0$.

Given an endogenous configuration $(u(\cdot),x) \in \mathcal{X}$, the analytic Jacobian maps the endogenous configuration space into the taskspace. As long as this map is surjective, the configuration is referred to as regular, otherwise it is a singular configuration. The regularity of a configuration $(u(\cdot),x)$ means that the dexterity matrix

$$\mathcal{D}_{q_0,T}(u(\cdot),x) = D(T,x)D^T(T,x) + C(T,x)\int_0^T \Phi(T,s)B(s)B^T(s)\Phi^T(T,s)ds\, C^T(T,x) \quad (7)$$

has full rank r. At regular endogenous configurations there exists the Jacobian pseudoinverse $J^{\#}_{q_0,T}(u(\cdot),x) : R^r \longrightarrow \mathcal{X}$, defined as

$$(J^{\#}_{q_0,T}(u(\cdot),x)\eta)(t) = \begin{bmatrix} B^T(t)\Phi^T(T,t)C^T(T,x) \\ D^T(T,x) \end{bmatrix} \mathcal{D}^{-1}_{q_0,T}(u(\cdot),x)\eta. \quad (8)$$

The inverse kinematic problem for a mobile manipulator with kinematics (3) is formulated in the standard way: **given a desirable end effector position and orientation $y_d \in R^r$, find an endogenous configuration $(\mathbf{u}(\cdot),\mathbf{x}) \in \mathcal{X}$ such that $\mathbf{K}_{\mathbf{q_0},\mathbf{T}}(\mathbf{u}(\cdot),\mathbf{x}) = \mathbf{y_d}$.** The Jacobian inverse kinematics algorithms can be obtained directly from inverse kinematics algorithms proposed in Tchoń and Jakubiak, 2003. Below we shall derive the basic Jacobian pseudoinverse algorithm.

Suppose that the inverse kinematic problem has a solution. Choose a smooth curve $(u_\theta(\cdot),x(\theta))$, $\theta \in R$, in the endogenous configuration space, passing through a certain configuration $(u_0(\cdot),x_0)$. Compute the taskspace error

$$e(\theta) = K_{q_0,T}(u_\theta(\cdot),x(\theta)) - y_d \quad (9)$$

along this curve, and make it decrease exponentially with a decay rate $\gamma > 0$, so that $\frac{d}{d\theta}e(\theta) = -\gamma e(\theta)$. After suitable mathematical manipulations we shall get a differential equation involving the curve $(u_\theta(\cdot),x(\theta))$, that at regular configurations can be solved by means of the Jacobian pseudoinverse (8). Eventually, this procedure results in the basic Jacobian pseudoinverse algorithm providing a solution $(u_d(\cdot),x_d)$ to the inverse task problem as the limit at $\theta \to +\infty$ of the trajectory of the dynamic system

$$\frac{d}{d\theta}\begin{pmatrix} u_\theta(t) \\ x(\theta) \end{pmatrix} = -\gamma(J^{\#}_{q_0,T}(u_\theta(\cdot),x(\theta))e(\theta))(t) \quad (10)$$

whose right hand side is defined by formulas (8) and (9) above.

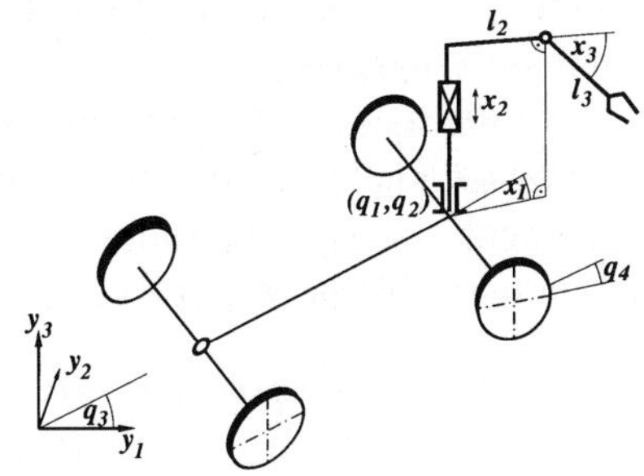

Figure 1. RTR manipulator atop of kinematic car

3. Mobile manipulator driven by accelerations

When dealing with the kinematics of a mobile manipulator equipped with a nonholonomic platform, the phase constraints are usually Pfaffian, i.e. they take the form $A(q)\dot{q} = 0$, and lead to a driftless control system representation of the kinematics

$$\dot{q} = G(q)u = \sum_{i=1}^{m} g_i(q)u_i, \quad y = k(q,x). \tag{11}$$

By design, the control actions $u(t)$ in (11) have the meaning of certain platform velocities. Suppose that we have extended this system in such a way that the controls get the interpretation of accelerations. To this aim we make a substitution $u = \eta$, define new control $\dot{\eta} = u$, and obtain the representation

$$\begin{pmatrix} \dot{q} \\ \dot{\eta} \end{pmatrix} = \begin{pmatrix} G(q)\eta \\ 0 \end{pmatrix} + \begin{bmatrix} 0 \\ I_m \end{bmatrix} u, \quad y = k(q,x) \tag{12}$$

that is affine, of the form (1). An important and useful consequence of extending (11) toward (12) is a desingularization of the zero velocity control, that allows one to use regular Jacobian inverse kinematics algorithm in order to solve the inverse kinematic problem with zero initial or final velocity. Another advantage of the extension (12) shows up when the task of a mobile manipulator is decomposed into a sequence of subtasks, and we need to avoid velocity jumps at switching points between subsequent subtasks. It might be expected that such a decomposition will considerably alleviate the effect of local convergence of Jacobian inverse kinematics algorithms for mobile manipulators. As an illustration, in the following section we shall apply the Jacobian pseudoinverse algorithm to a specific mobile manipulator steered by accelerations of the platform and joint positions of the aboard manipulator.

4. Computer simulations

By using the basic Jacobian pseudoinverse algorithm we intend to solve the inverse kinematic problem for the mobile manipulator consisting of a kinematic car type platform equipped with RTR manipulator, portrayed in figure 1. This object has been examined extensively in Tchoń and Jakubiak, 2003.

The control variables have the sense of forward and directing accelerations of front wheels. In order to set final platform velocities, we have added to the output map new components y_4, y_5. The following affine control system representation of the mobile manipulator results from this extension

$$\begin{cases} \dot{q}_1 = q_5 \cos q_3 \cos q_4, \quad \dot{q}_2 = q_5 \sin q_3 \cos q_4, \quad \dot{q}_3 = q_5 \sin q_4, \\ \qquad \dot{q}_4 = q_6, \qquad\qquad \dot{q}_5 = u_1, \qquad\qquad \dot{q}_6 = u_2, \\ y = (y_1, y_2, y_3, y_4, y_5) = (q_1 + (l_2 + l_3 \cos x_3) \cos(q_3 + x_1), \\ \qquad q_2 + (l_2 + l_3 \cos x_3) \sin(q_3 + x_1), x_2 + l_3 \sin x_3, q_5, q_6). \end{cases} \tag{13}$$

For link lengths $l_2 = 0.6$, $l_3 = 0.4$, control horizon $T = 1$, fixed initial platform coordinates $q_{03} = q_{05} = q_{06} = 0$, $q_{04} = \pi/6$, and initial joint positions $x_{01} = 0$, $x_{02} = 2$, $x_{03} = \pi/4$ of the aboard manipulator, as well as for prescribed desirable value of the output map $y_d = (0, 0, 1, 0, 0)$, we have solved the inverse kinematic problem of reaching y_d from 8 points $(q_{01}, q_{02}) = A, B, \ldots, H$ distributed uniformly on a circle of radius 10 centered at the origin of the inertial frame, such that $A = (10, 0)$. Control functions (u_1, u_2) in (13) are taken in the form of truncated Fourier series

$$u_i(t) = \lambda_{i0} + \sum_{k=1}^{\frac{s_i-1}{2}} \lambda_{i2k-1} \sin k\omega t + \lambda_{i2k} \cos k\omega t, \tag{14}$$

where $\omega = 2\pi/T$, $i = 1, 2$, containing either $(s_1, s_2) = (5, 3)$ or $(s_1, s_2) = (7, 3)$ terms. Initial values of nonzero control coefficients in (14) are $\lambda_{10} = 0.01$, $\lambda_{20} = -0.01$. Selected results of computer simulations are displayed in figures 2-4 below. The platform apparently starts and stops still.

5. Conclusions

The main result of this paper lies in extending the endogenous configuration space approach to mobile manipulators whose control system representation of kinematics takes the form of an affine control system with outputs. Special attention has been paid to the inverse kinematic problem for acceleration-driven mobile manipulators, that has been solved by means of the basic Jacobian pseudoinverse algorithm. Computer simulations involving the kinematic car platform equipped with RTR manipulator have demonstrated that the Jacobian pseudoinverse inverse kinematics algorithm works satisfactorily when the mobile manipulator is acceleration-driven. Further applications of the endogenous configuration space approach will be directed toward practical tasks of achieving the motion with prescribed velocities and smooth velocity changes at waypoints, and of eliminating turning points at platform trajectories.

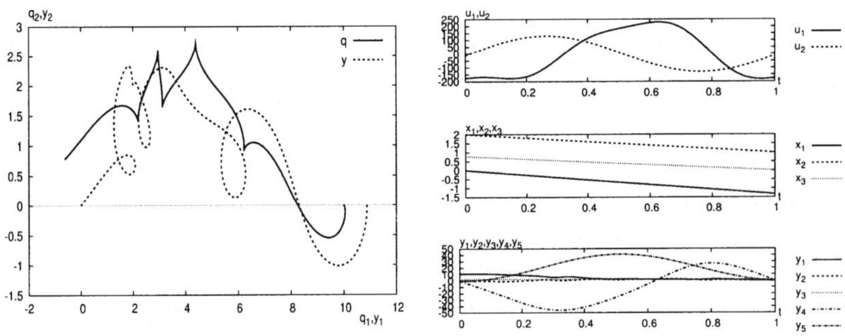

Figure 2. Point A: $(s_1, s_2) = (7,3)$, control energy $E = \int_0^1 (u_1^2(t) + u_2^2(t))dt = 33216.5$, squared configuration norm $||(u(\cdot),x)||^2 = E + x^T x = 33219.3$, no. of iterations 170, target accuracy 8.91663e-07.

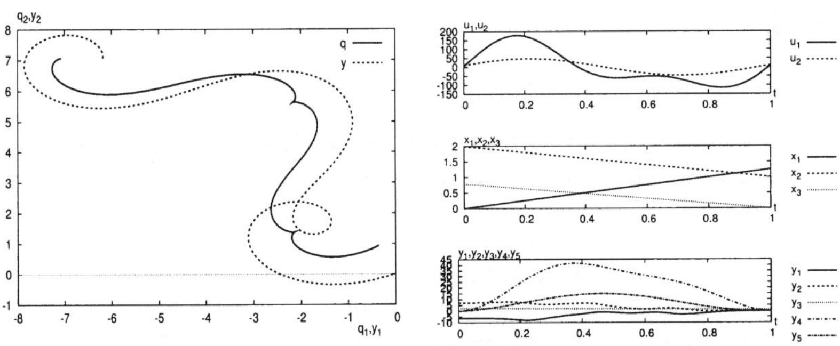

Figure 3. Point D: $(s_1, s_2) = (5,3)$, control energy $E = \int_0^1 (u_1^2(t) + u_2^2(t))dt = 10122.0$, squared configuration norm $||(u(\cdot),x)||^2 = E + x^T x = 10124.6$, no. of iterations 88, target accuracy 6.42785e-07.

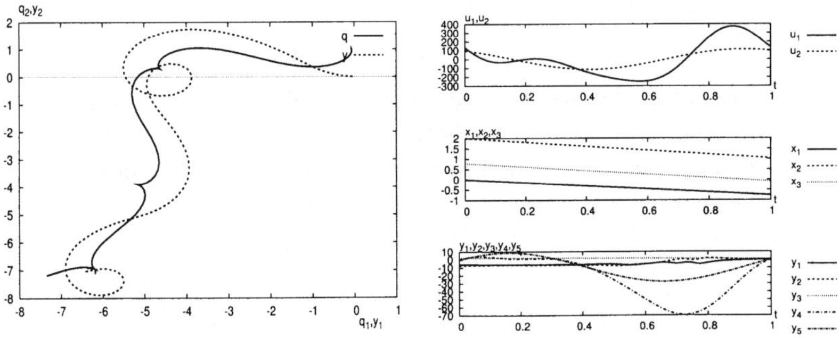

Figure 4. Point F: $(s_1, s_2) = (7,3)$, control energy $E = \int_0^1 (u_1^2(t) + u_2^2(t))dt = 20726.5$, squared configuration norm$||(u(\cdot),x)||^2 = E + x^T x = 20728$, no. of iterations 64, target accuracy 8.10538e-07.

6. Acknowledgements

This research has been supported by the Polish State Committee of Scientific Research under a project realized in the years 2002-2004.

References

Bayle, B., Fourquet, J.-Y., and Renaud, M. (2001), Manipulability analysis for mobile manipulators, *Proc. 2001 IEEE Int. Conf. Robot. Automat.*, Seoul, Korea.

Desai, J. P., and Kumar, V. (1997), Nonholonomic motion planning for multiple mobile manipulators, *Proc. 1997 IEEE Int. Conf. Robot. Automat.*, Albuquerque, New Mexico, USA.

Furuno, S., Yamamoto, M., and Mohri, A. (2003), "Trajectory planning of mobile manipulator with stability considerations, *Proc. 2003 IEEE Int. Conf. Robot. Automat.*, Taipei, Taiwan.

Gardner, J. F., and Velinsky, S. A. (2000), Kinematics of mobile manipulators and impications for design, *J. Robotic Systems*, vol. 17, no. 6, pp. 309-320.

Khatib, O., Yokoi, K., Chang, K., Ruspini, D., Holberg, R., and Casal, A. (1996), Coordination and decentralized cooperation of mobile manipulators, *J. Robotic Systems*, vol. 13, no. 11, pp. 755-764.

Khatib, O., K. Yokoi, K., Brock, O., Chang, K., and Casal, A. (1999), Robots in human environments: Basic autonomous capabilities, *Int. J. Robotics Res.*, vol. 18, no. 7, pp. 684-696.

Papadopoulos, E., and Dubowsky, S. (1991), Coordinated manipulator/spacecraft motion control for space robotics, *Proc. 1991 IEEE Int. Conf. Robot. Automat.*, Sacramento, CA, USA.

Pin, F. G., Morgansen, K. A., Tulloch, F. A., Hacker, Ch. J., and Gower, K. B. (1996), Motion planning for mobile manipulators with a non-holonomic constraint using the FSP method, *J. Robotic Systems*, vol. 13, no. 11, pp. 723-736.

Sugar, T., Desai, J. P., Kumar, V., and Ostrowski, P. (2001), Coordination of multiple mobile manipulators, *Proc. 2001 IEEE Int. Conf. Robot. Automat.*, Seoul, Korea.

Seraji, H. (1998), A unified approach to motion control of mobile manipulators, *Int. J. Robotics Res.*, vol. 17, no. 2, pp. 107-118.

Tchoń, K. (2002), Repeatability of inverse kinematics algorithms for mobile manipulators, *IEEE Trans. Autom. Control*, vol. 47, no. 8, pp. 1376-1380.

Tchoń, K., and Muszyński, R. (2000), Instantaneous kinematics and dexterity of mobile manipulators, *Proc. 2000 IEEE Int. Conf. Robot. Automat.*, San Francisco, CA, USA.

Tchoń, K., Jakubiak, J., and R. Muszyński, R. (2001), Kinematics of mobile manipulators: a control theoretic perspective, *Archives of Control Sciences*, vol. 11, no. 3-4, pp. 79-105.

Tchoń, K., and Jakubiak, J. (2003), Endogenous configuration space approach to mobile manipulators: a derivation and performance assessment of Jacobian inverse kinematics algorithms, *Int. J. Control*, vol. 76, no. 14, pp. 1387-1419.

Tchoń, K., and Zadarnowska, K. (2003), Kinematic dexterity of mobile manipulators: an endogenous configuration space approach, *Robotica*, vol. 21, pp. 521-530.

Yamamoto, Y., and Yun, X. (1994), Coordinating locomotion and manipulation of a mobile manipulator, *IEEE Trans. Autom. Control*, vol. 39, no. 6, pp. 1326-1332.

Yamamoto, Y., and Yun, X. (1996), Effect of the dynamic interaction on coordinated control of mobile manipulators, *IEEE Trans. Robot. Automat.*, vol. 12 , no. 5, pp. 816-824.

Yamamoto, Y., and Yun, X. (1999), Unified analysis on mobility and manipulability of mobile manipulators, *Proc. 1999 IEEE Int. Conf. Robot. Automat.*, Detroit, Michigan, USA.

DYNAMIC DEXTERITY AND ISOTROPY OF MOBILE ROBOTS: AN ENDOGENOUS CONFIGURATION SPACE APPROACH

Katarzyna Zadarnowska
Institute of Engineering Cybernetics
Wrocław University of Technology
ul. Janiszewskiego 11/17
50-372 Wrocław, Poland
kz@ict.pwr.wroc.pl

Abstract By extension of kinematic dexterity and isotropy concepts for mobile manipulators, defined within the endogenous configuration space approach, we introduce corresponding dynamic performance measures for nonholonomic mobile robots. As an illustration dynamically dextrous and isotropic configurations have been computed for the mobile robot MK.

Keywords: Mobile robot, kinematics, dynamics, dynamic dexterity, dynamic isotropy

1. Introduction

The performance evaluation of stationary manipulators in terms of dexterity and isotropy belongs to a well established domain of robotics. Looking through the robotic literature concerned with stationary manipulators, we find kinematic as well as dynamic measure of manipulability proposed by Yoshikawa, 1985a and Yoshikawa, 1985b. This measure has been exploited further in Doel and Pai, 1998. Kinematic isotropy has been introduced for the first time by Salisbury and Craig, 1982. A dynamic concept of isotropy has been used by Ma and Angeles, 1990 in optimal design of stationary manipulators. Both kinematic manipulability and kinematic isotropy have been studied in Klein and Blaho, 1987, whereas dynamic isotropy was dealt with in Matone and Roth, 1997 and used to optimize the manipulator performance.

An attempt at extending the concept of kinematic and dynamic manipulability from stationary to mobile manipulators was undertaken by Yamamoto and Yun, 1999. Kinematic performance measures of wheeled mobile robots have been treated in Bayle et al., 2003. An application of kinematic isotropy of mobile manipulators characterized by its manipulability ellipsoid can be found in Gardner and Velinsky, 2000. All performance measures mentioned above refer to the classical concept of exogenous configuration space. Below we shall take a different perspective, called the endogenous configuration space approach, and presented in detail in Tchoń and Jakubiak, 2003. An endoge-

J. Lenarčič and C. Galletti (eds.), On Advances in Robot Kinematics, 477-484.

nous configuration of a mobile manipulator comprises the control function of the platform and the joint position of the aboard manipulator. A new view of kinematic performance evaluation of mobile manipulators, based on this approach, was developed in Tchoń and Zadarnowska, 2003 and Zadarnowska and Tchoń, 2003, culminating with introducing kinematic performance measures. These measures refer to the local output controllability of the control system representation of the kinematics of a mobile manipulator. In particular they establish how much a given endogenous configuration differs from singular, and assess applicability of Jacobian inverse kinematics algorithms.

In this paper we make a step toward extending the endogenous configuration space approach to mobile robots whose control system representation includes not only kinematics but also dynamics of the platform. Relying on corresponding kinematic concepts we define dynamic equivalents of dexterity, dexterity ellipsoid, and isotropy of mobile robots. Dynamic performance measures have a similar interpretation to the kinematic ones: they measure the output controllablity of the control system representation of both kinematics and dynamics of the mobile platform, increasing away from singular configurations.

The paper is composed as follows. Section 2 collects basic ingredients of the endogenous configuration space approach. In section 3 we define dynamic dexterity, dexterity ellipsoids, and the condition number. Section 4 presents the mathematical model of MK mobile robot. Section 5 gives a sample of computations for MK mobile robot. The paper is concluded with section 6.

2. Basic concepts

Consider a mobile platform with generalized coordinates $q \in \mathbb{R}^n$, subject to l nonholonomic Pfaffian constraints $A(q)\dot{q} = 0$. The control system representation of the platform kinematics is defined by a driftless control system

$$\dot{q} = G(q)\eta = \sum_{i=1}^{m} g_i(q)\eta_i, \tag{1}$$

where $A(q)G(q) = 0$, $\eta \in \mathbb{R}^m$, $m = n - l$. Using the Lagrange-d'Alembert principle, we can write down the dynamics of the mobile robot in the form

$$Q(q)\ddot{q} + C(q,\dot{q})\dot{q} + D(q) = A^T(q)\lambda + B(q)u \tag{2}$$

with $\lambda \in \mathbb{R}^l$ denoting a vector of Lagrange multipliers, $Q(q)$, $C(q,\dot{q})$, $D(q)$ being usual components of the unconstrained dynamics, and $B(q)$ determining the control matrix. After elimination of the multipliers and addition of an output function $y = k(q) \in \mathbb{R}^r$ describing task coordinates of the mobile robot, we obtain an affine control system representation of the mobile platform

$$\begin{pmatrix} \dot{q} \\ \dot{\eta} \\ y \end{pmatrix} = \begin{pmatrix} G(q)\eta \\ P(q,\eta) \\ k(q) \end{pmatrix} + \begin{bmatrix} 0 \\ R(q) \\ 0 \end{bmatrix} u, \tag{3}$$

where $P(q,\eta) = (G^T(q)Q(q)G(q))^{-1}G^T(q)(-Q(q)\dot{G}(q)\eta - C(q,G(q)\eta)\eta - D(q))$ and $R(q) = (G^T(q)Q(q)G(q))^{-1}G^T(q)B(q)$. The control functions $u(\cdot)$ in (3), have the sense of forces or torques, and therefore may be called dynamic endogenous configurations of the mobile platform. These configurations belong to the Hilbert space $\mathcal{X} = L^2_m[0,T]$ referred to as the dynamic endogenous configuration space. Let $q(t) = \varphi_{q_0,\eta_0,t}(u(\cdot))$ and $\eta(t) = \psi_{q_0,\eta_0,t}(u(\cdot))$ denote trajectories of the control system driven by $u(\cdot) \in \mathcal{X}$. By analogy to the instantaneous kinematics we define a dynamic task map of the mobile platform

$$K_{q_0,\eta_0,T} : \mathcal{X} \to \mathbb{R}^r, \quad K_{q_0,\eta_0,T}(u(\cdot)) = k(\varphi_{q_0,\eta_0,T}(u(\cdot))). \tag{4}$$

The dynamic task Jacobian can be defined as the derivative

$$J_{q_0,\eta_0,T}(u(\cdot))v(\cdot) = \frac{d}{d\alpha}|_{\alpha=0} K_{q_0,\eta_0,T}(u(\cdot) + \alpha v(\cdot)). \tag{5}$$

For a given configuration $u(\cdot) \in \mathcal{X}$ and corresponding trajectories $(q(t),\eta(t))$, the variational system associated with (3) is a linear approximation

$$\dot{\xi} = A(t)\xi + B(t)v, \quad \eta = C(t)\xi \tag{6}$$

to (3) along the triple $(u(t),q(t),\eta(t))$, initialized at $\xi_0 = 0$, where

$$A(t) = \begin{bmatrix} \frac{\partial(G(q(t))\eta(t))}{\partial q} & G(q(t)) \\ \frac{\partial(P(q(t),\eta(t))+R(q(t))u(t))}{\partial q} & \frac{\partial P(q(t),\eta(t))}{\partial \eta} \end{bmatrix},$$

$$B(t) = \begin{bmatrix} 0 \\ B_2(t) \end{bmatrix} = \begin{bmatrix} 0 \\ R(q(t)) \end{bmatrix}, \; C(t) = \begin{bmatrix} C_1(t), & C_2(t) \end{bmatrix} = \begin{bmatrix} \frac{\partial k(q(t))}{\partial q}, & 0 \end{bmatrix}.$$

Taking into account (3) and (6) we can express the task Jacobian in the form

$$J_{q_0,\eta_0,T}(u(\cdot))v(\cdot) = C_1(T)\int_0^T \Phi_{12}(T,s)B_2(s)v(s)ds, \tag{7}$$

where $\Phi(t,s) = \begin{bmatrix} \Phi_{11}(t,s) & \Phi_{12}(t,s) \\ \Phi_{21}(t,s) & \Phi_{22}(t,s) \end{bmatrix}$ denotes the fundamental matrix of (6), satisfying the differential equation $\frac{\partial \Phi(t,s)}{\partial t} = A(t)\Phi(t,s), \quad \Phi(s,s) = I_{n+m}$. Eventually, we define the dynamic dexterity matrix of the mobile platform at the endogenous configuration $u(\cdot) \in \mathcal{X}$

$$\mathcal{D}_{q_0,\eta_0,T}(u(\cdot)) = C_1(T)\int_0^T \Phi_{12}(T,s)B_2(s)B_2^T(s)\Phi_{12}^T(T,s)dsC_1^T(T). \tag{8}$$

A configuration $u(\cdot) \in \mathscr{X}$ will be called regular, if $\det \mathcal{D}_{q_0,\eta_0,T}(u(\cdot)) \neq 0$, otherwise $u(\cdot) \in \mathscr{X}$ is singular.

3. Performance measures of mobile robot

At a regular endogenous configuration $u(\cdot) \in \mathscr{X}$ of the mobile robot, let us define the unit sphere of velocities

$$S_{q_0,\eta_0,T}(u(\cdot)) = \{v(\cdot) \in \mathscr{X} \mid \| v(\cdot) \|^2 = 1\}, \tag{9}$$

where $||\cdot||$ is the norm in $\mathcal{X}$, and compute its image by the task Jacobian (7)

$$E_{q_0,\eta_0,T}(u(\cdot)) = J_{q_0,\eta_0,T}(u(\cdot))S_{q_0,\eta_0,T}(u(\cdot)) = \{\zeta \in \mathbb{R}^r \mid \zeta^T \mathcal{D}^{-1}_{q_0,\eta_0,T}(u(\cdot))\zeta = 1\}.$$

This object will be called the dynamic dexterity ellipsoid of the mobile robot. Minimal and maximal half axes of this ellipsoid are equal, respectively, to the square root of the minimal $\underline{\lambda}^{1/2}_{\mathcal{D}_{q_0,\eta_0,T}u(\cdot)}$, and the maximal $\overline{\lambda}^{1/2}_{\mathcal{D}_{q_0,\eta_0,T}u(\cdot)}$ eigenvalues of the dynamic dexterity matrix. The square root of the determinant

$$\mathsf{d}_{q_0,\eta_0,T}(u(\cdot)) = \sqrt{\det \mathcal{D}_{q_0,\eta_0,T}(u(\cdot))} \tag{10}$$

we shall call the dynamic dexterity of the endogenous configuration $u(\cdot) \in \mathscr{X}$. In the case when all the eigenvalues of dexterity matrix $\mathcal{D}_{q_0,\eta_0,T}(u(\cdot))$ are equal, i.e. $\underline{\lambda}_{\mathcal{D}_{q_0\eta_0,T}(u(\cdot))} = \overline{\lambda}_{\mathcal{D}_{q_0,\eta_0,T}(u(\cdot))}$, $u(\cdot)$ becomes an isotropic configuration of the mobile robot. Otherwise, we speak of an anisotropic configuration.

In order to quantatively deal with the isotropy and the anisotropy, we introduce a dynamic condition number of the endogenous configuration

$$cond_{q_0,\eta_0,T}(u(\cdot)) = \frac{\sqrt{\overline{\lambda}_{\mathcal{D}_{q_0,\eta_0,T}(u(\cdot))}}}{\sqrt{\underline{\lambda}_{\mathcal{D}_{q_0,\eta_0,T}(u(\cdot))}}}. \tag{11}$$

At the isotropic configuration $cond_{q_0,\eta_0,T}(u(\cdot)) = 1$, while approaching a singular configuration implies that $cond_{q_0,,\eta_0,T}(u(\cdot)) \to +\infty$.

To simplify the computation of isotropic configurations, we transform the problem to the following optimization problem. Let

$$\det(\lambda I_r - \mathcal{D}_{q_0,\eta_0,T}(u(\cdot))) = \lambda^r + a_{r-1}(u(\cdot))\lambda^{r-1} + a_{r-2}(u(\cdot))\lambda^{r-2} + \ldots + a_0(u(\cdot))$$

denote the characteristic polynomial of $\mathcal{D}_{q_0,\eta_0,T}(u(\cdot))$. Its coefficients are symmetric polynomials in the eigenvalues, of which the first two are

$$a_{r-1} = -(\lambda_1 + \lambda_2 + \ldots + \lambda_r), \;\; a_{r-2} = \lambda_1\lambda_2 + \ldots + \lambda_1\lambda_r + \cdots + \lambda_{r-1}\lambda_r \tag{12}$$

Due to (12) finding an isotropic endogenous configuration amounts to solving the optimization problem $\min_{u(\cdot)\in\mathscr{X}} F_r(u(\cdot))$, with the objective function

$$F_r(u(\cdot)) = (\lambda_1-\lambda_2)^2+\cdots+(\lambda_1-\lambda_r)^2+(\lambda_2-\lambda_3)^2+\cdots+(\lambda_2-\lambda_r)^2+\cdots+ (\lambda_{r-1}-\lambda_r)^2 = (r-1)a_{r-1}^2(u(\cdot)) - 2ra_{r-2}(u(\cdot)).$$

4. MK mobile robot

In order to illustrate the concepts proposed in the previous two sections, we have chosen MK mobile robot, recently designed in our Laboratory. The robot has been shown schematically in Fig. 1. MK mobile robot is driven with respect to its swinging cylindric body. The coordinate vector $q = (x, y, \varphi_1, \varphi_2, \alpha)$ describes position of the middle point on the robot axle in the inertial frame, the revolution angles of the wheels, and the swing angle of the robot body. The meaning of geometric parameters of the robot follows directly from the figure. The mathematical model of kinematics and dynamics of the platform becomes

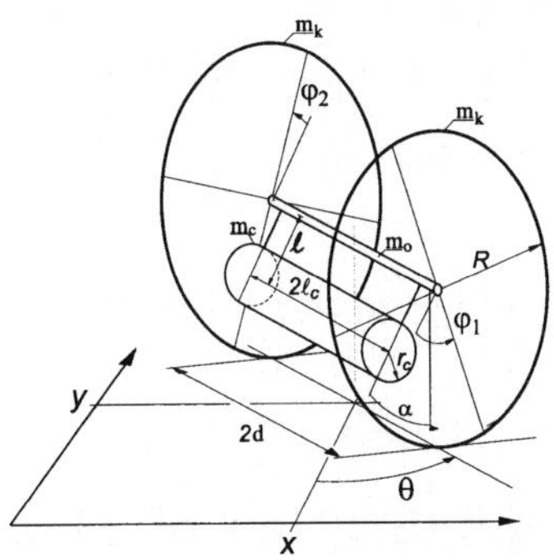

Figure 1. MK mobile robot

$$\begin{cases} \dot{x} = \eta_{123}\cos\theta,\ \dot{y} = \eta_{123}\sin\theta,\ \dot{\varphi}_1 = \frac{2}{R}\eta_1,\ \dot{\varphi}_2 = \frac{2}{R}\eta_2,\ \dot{\alpha} = \frac{1}{R}\eta_3 \\ M(\alpha)\dot{\eta} + N(\alpha,\eta) = Bu \\ y = (x+(\ell+r_c)\sin\alpha\cos\theta, y+(\ell+r_c)\sin\alpha\sin\theta, R-(\ell+r_c)\cos\alpha), \end{cases} \tag{13}$$

where $\eta_{123} = \eta_1+\eta_2+\eta_3$ and $\theta = \frac{R}{2d}(\varphi_2-\varphi_1)$. The inertia matrix $M(\alpha)$, vector of gravity, Coriolis, centripetal and frictional forces $N(\alpha,\eta)$, and control matrix B are the following Zadarnowska and Tchoń, 2001

$$M(\alpha) = \begin{bmatrix} \left(m_c + \frac{4}{3}m_o + 8m_k\right)d^2 + m_k R^2 + m_c\left(\ell^2\sin^2\alpha + \frac{1}{3}\ell_c^2 + \frac{1}{2}r_c^2\right) \\ \left(m_c + \frac{2}{3}m_o\right)d^2 - m_k R^2 - m_c\left(\ell^2\sin^2\alpha + \frac{1}{3}\ell_c^2 + \frac{1}{2}r_c^2\right) \\ \left(m_c + m_o + 4m_k + \frac{m_c\ell}{R}\cos\alpha\right)d^2 \end{bmatrix.$$

$$\begin{matrix} \left(m_c + \frac{2}{3}m_o\right)d^2 - m_k R^2 - m_c\left(\ell^2\sin^2\alpha + \frac{1}{3}\ell_c^2 + \frac{1}{2}r_c^2\right) \\ \left(m_c + \frac{4}{3}m_o + 8m_k\right)d^2 + m_k R^2 + m_c\left(\ell^2\sin^2\alpha + \frac{1}{3}\ell_c^2 + \frac{1}{2}r_c^2\right) \\ \left(m_c + m_o + 4m_k + \frac{m_c\ell}{R}\cos\alpha\right)d^2 \end{matrix} \quad (14)$$

$$\begin{bmatrix}. \left(m_o + m_c + 4m_k + \frac{m_c\ell}{R}\cos\alpha\right)d^2 \\ \left(m_c + m_o + 4m_k + \frac{m_c\ell}{R}\cos\alpha\right)d^2 \\ \left(m_c + m_o + 4m_k + \frac{2m_c\ell}{R}\cos\alpha + \frac{m_c}{R^2}\left(\ell^2 + \frac{1}{2}r_c^2\right)\right)d^2 \end{bmatrix},$$

$$N(\alpha,\eta) = \begin{pmatrix} -m_c\ell\sin\alpha(2\eta_2+\eta_3)(\eta_2-\eta_1) - \\ \frac{2m_c\ell^2}{R}\sin\alpha\cos\alpha\eta_3(\eta_2-\eta_1) - \frac{m_c d^2\ell\sin\alpha}{R^2}\eta_3^2 + \frac{4d^2k_1}{R^2}\eta_1, \\ m_c\ell\sin\alpha(2\eta_1+\eta_3)(\eta_2-\eta_1) + \\ \frac{2m_c\ell^2}{R}\sin\alpha\cos\alpha\eta_3(\eta_2-\eta_1) - \frac{m_c d^2\ell\sin\alpha}{R^2}\eta_3^2 + \frac{4d^2k_2}{R^2}\eta_2, \\ -\frac{m_c\ell\sin\alpha}{R}(\eta_2-\eta_1)^2(R+\ell\cos\alpha) - \frac{m_c\ell d^2}{R^2}\sin\alpha\eta_3^2 + \\ \frac{m_c g\ell d^2}{R}\sin\alpha \end{pmatrix}, \quad B = \frac{2d^2}{R}\begin{bmatrix} I_2 \\ 0 \end{bmatrix}. \quad (15)$$

5. Optimal configuration of MK robot

In the control system representation (13) we have chosen the controls $u_1(t)$, $u_2(t)$ in the form of a truncated Fourier series $u_i(t) = \lambda_{i0} + \sum_{k=1}^{\frac{s_i-1}{2}} \lambda_{i2k-1}\sin k\omega t + \lambda_{i2k}\cos k\omega t$ where $\omega = \frac{2\pi}{T}$, and $(s_1,s_2) = (3,3)$ (first order harmonics). Specific calculations have been accomplished for geometric parameters of MK robot: $R = 0.254[m]$, $d = 0.225[m]$, $\ell = 0.12[m]$, $r_c = 0.035[m]$, $l_c = 0.165[m]$, $m_o = 0.5[kg]$, $m_k = 2[kg]$, $m_c = 7[kg]$, fixed initial posture $q_0 = 0$ of the platform, and initial velocities $\eta_0 = 0$. Configurations with maximal dexterity have been computed under constraints $|\lambda_{i0}| \le 0.5$, $|\lambda_{11}|, |\lambda_{12}|, |\lambda_{21}|, |\lambda_{22}| \le 0.25$. Trajectories of MK platform corresponding to maximal dexterity along with the associated dynamic dexterity ellipsoids are shown in figures Fig. 2. In 3-dimensional taskspace (y_1,y_2,y_3), isotropic configurations exist only for small values of the control horizon T (an increase of the time horizon improves the platform mobility only in (y_1,y_2) directions, whereas the y_3-axis mobility remains the same). Therefore, in order to determine isotropic configurations for larger values of T, we assume 2-dimensional taskspace (y_1,y_2), and speak of subisotropic configurations of mobile robot. Figure 3 illustrates exemplary subisotropic configurations of MK robot. It turns out that at dextrous as well as isotropic configurations the mobile robot may move along one of four curves.

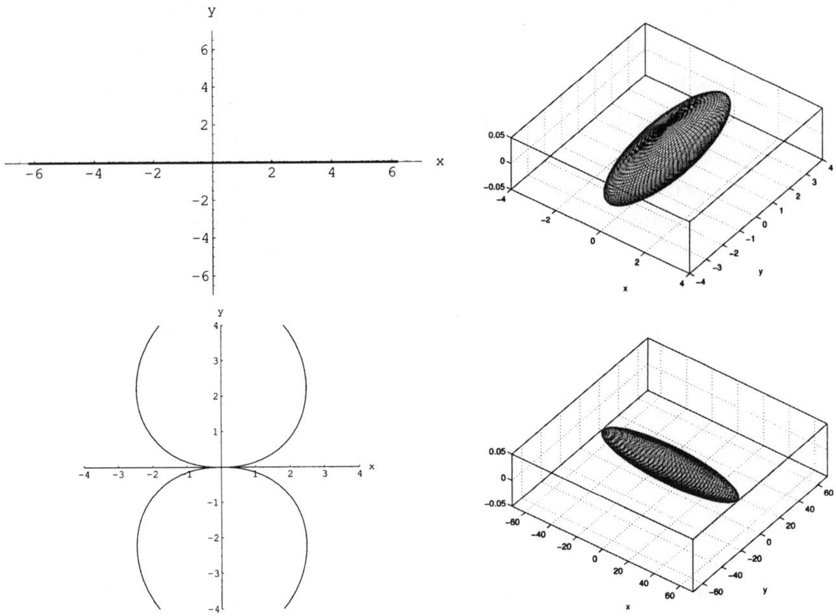

a)

b)

Figure 2. Trajectory corresponding to the dextrous configuration of MK robot along with dynamical dexterity ellipsoid a) $T = 3$; b) $T = 7$.

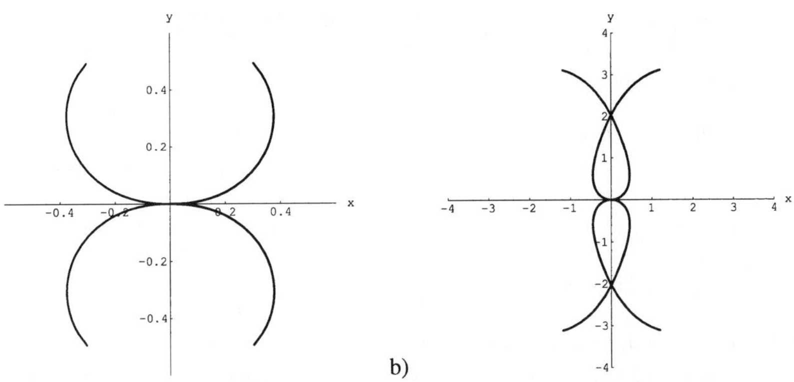

a) b)

Figure 3. Trajectories corresponding to isotropic configurations of MK robot: a) $T = 3$; b) $T = 7$.

6. Conclusions

By means of the endogenous configuration space approach, we have proposed dynamic performance measures of mobile robots. Special attention has been paid to determining isotropic and dextrous configurations. From simulation results we conclude that after crossing the time horizon $T = 3$, dextrous trajectories change from straight lines into curves. In the case of dextrous configurations their anisotropy increases along with the time horizon. Also, after a suitable restriction of the taskspace, the mobile robot achieves isotropic configurations very easily. Furthermore, we have found out that for large values of the time horizon and less restrictive bounds on control functions (e.g. $| u | \leq 3$), the mobile robot behaves chaotically; this kind of behaviour of mobile manipulators has been observed in Zadarnowska and Tchoń, 2003.

7. Acknowledgements

This research has been supported by the Polish State Committee of Scientific Research under a project realized in the year 2002-2004.

References

Bayle, B., Fourquet, J.-Y., and Renaud, M. (2003), Manipulability of wheeled mobile manipulators: application to motion generation, *Int. J. Robotics Res.*, vol. 22, no. 7-8, pp. 565-581.

van den Doel, K., and Pai, D. K. (1998), Performance measures for robot manipulators: A unified approach, *Int. J. Robotics Res.*, vol. 15, no. 1, pp. 92-111.

Gardner, J. F., and Velinsky, S. A. (2000), Kinematics of mobile manipulators and implications for design, *J. Robotic Syst.*, vol. 17, no. 6, pp.. 309-320.

Klein, Ch. A., and Blaho, B. E. (1987), Dexterity measures for the design and control of kinematically redundant manipulators, *Int. J. Robotics Res.*, vol. 6, no. 2, pp. 72-82.

Ma, O., and Angeles, J. (1990), The concept of dynamic isotropy and its applications to inverse kinematics and trajectory planning, *Proc. 1990 IEEE Int. Conf. Robot. Automat*, San Francisco, CA, pp. 10-15.

Matone, R., and Roth, B. (1997), Designing manipulators for both kinematic and dynamic isotropy, *Proc. 11th CISM-IFToMM Symp. RoManSy*, Udine, Italy, pp. 10-15.

Salisbury, J. K., and Craig, J. J. (1982), Articulated hands: force control and kinematic issue, *Int. J. Robotics Res.*, vol. 1, no. 1, pp. 4-17.

Tchoń, K., and Zadarnowska, K. (2003), Kinematic dexterity of mobile manipulators: an endogenous configuration space approach, *Robotica*, vol. 21, pp. 521-530.

Tchoń, K., and Jakubiak, J. (2003), Endogenous configuration space approach to mobile manipulators: A derivation and performance assessment of Jacobian inverse kinematics algorithms, *Int. J. Control*, vol. 76, no. 14, pp. 1387-1419.

Yamamoto, Y., and Yun, X. (1999), Unified analysis on mobility and manipulability of mobile manipulators, *Proc. 1999 IEEE Int. Conf. Robot. Automat.*, Detroit, MI, pp. 1200-1206.

Yoshikawa (1985), Manipulability of robotic mechanisms, *Int. J. Robotics Res.*, vol. 4, no. 2, pp. 3-9.

Yoshikawa (1985), Dynamic manipulability of robotic mechanisms, *J. Robotic Syst.*, vol. 1, no. 1, pp. 113-124.

Zadarnowska, K., and Tchoń, K. (2002), Dynamic dexterity of mobile robots, *Proc. 14th Nat. Automat. Conf.*, Zielona Góra, Poland, pp. 669-674 (in Polish).

Zadarnowska, K., and Tchoń, K. (2003), Isotropic configurations of mobile manipulators, *Proc. 7th IFAC Int. Symp. on Robot Control Syroco 2003*, Wrocław, Poland (in print).

Author Index